Biological Physics
Energy, Information, Life

Philip Nelson
University of Pennsylvania
with the assistance of Marko Radosavljević and Sarina Bromberg

W. H. Freeman and Company
New York

Publisher and Acquisitions Editor: Susan Finnemore Brennan
Marketing Manager: Mark Santee
Project Manager and Text Designer: Leslie Galen, Integre Technical Publishing Company, Inc.
Project Editor: Jane O'Neill
Cover Designer: Blake Logan
Illustrations: Sarina Bromberg, Larry Gonick, Felice Macera
Illustration Coordinator: Bill Page
Photo Editor: Patricia Marx
Production Coordinator: Susan Wein
Composition: Integre Technical Publishing Company, Inc.
Printing and Binding: RR Donnelley & Sons Company

Front cover: Purkinje neuron from rat brain, visualized by two-photon laser scanning microscopy. The scale bar represents $15\,\mu$m. The neuron shown is alive and surrounded by a dense network of other neurons; a fluorescent dye has been injected into the cell from the micropipette at lower left, to reveal only the one cell of interest. The dendritic tree of this neuron (*top*) receives over 100 000 synaptic inputs. Dendritic spines are visible as tiny bumps on the dendrites. A single axon (*lower left*) sends output signals on to other neurons. [Digital image kindly supplied by K. Svoboda; see also Svoboda et al., 1996.]

Title page: DNA from a bacterium that has been lysed (burst) by osmotic shock. The bacterial genome that once occupied a small region in the center of the figure now extends in a series of loops from the core structure. Top to bottom, about $10\,\mu$m. [Electron micrograph by Ruth Kavenoff.]

ISBN 0-7167-4372-8 (EAN 9780716743729)

Library of Congress Control Number: 2003105929

Third printing

W. H. Freeman and Company
41 Madison Avenue
New York, NY 10010
Houndmills, Basingstoke RG21 6XS, England

www.whfreeman.com

For Dr. Shirley M. Davidson, 1930–2002

Not chaos-like together crush'd and bruis'd,
But, as the world, harmoniously confus'd:
Where order in variety we see,
And where, though all things differ, all agree.
— Alexander Pope, 1713

Contents

Part II   Diffusion, Dissipation, Drive

Chapter 6 | Entropy, Temperature, and Free Energy   195

Chapter 7	Entropic Forces at Work	245

Chapter 8	Chemical Forces and Self-Assembly	294

Part III   Molecules, Machines, Mechanisms

<u>Chapter 9</u> | ## Cooperative Transitions in Macromolecules   **341**

Chapter 10 | Enzymes and Molecular Machines   401

Chapter 11 | Machines in Membranes   469

| Chapter 12 | Nerve Impulses | 505 |

| | Epilogue | 557 |

| Appendix A | Global List of Symbols and Units | 559 |

To the Student

This is a book for life science students who are willing to use calculus. This is also a book for physical science and engineering students who are willing to think about cells. I believe that in the future every student in both groups will need to know the essential core of the others' knowledge.

In the past few years, I have attended many conferences and seminars. Increasingly, I have found myself surrounded not only by physicists, biologists, chemists, and engineers, but also by physicians, mathematicians, and entrepreneurs. These people come together to learn from one another, and the traditional academic distinctions between their fields are becoming increasingly irrelevant to this exciting work. I want to share some of their excitement with you.

I began to wonder how this diverse group managed to overcome the Tower-of-Babel syndrome. Slowly I began to realize that, even though each discipline carries its immense load of experimental and theoretical detail, still the headwaters of these rivers are manageable, and come from a common spring, a handful of simple, general ideas. Armed with these few ideas, I found that one can understand an enormous amount of front line research. This book explores these first common ideas, ruthlessly suppressing the more specialized ones for later.

I also realized that my own undergraduate education had postponed the introduction of many of the basic ideas to the last year of my degree (or even later) and that many programs still have this character: We meticulously build a sophisticated mathematical edifice before introducing many of the Big Ideas. My colleagues and I became convinced that this approach did not serve the needs of our students. Many of our undergraduate students start research in their very first year and need the big picture early. Many others create interdisciplinary programs for themselves and may never even get to our specialized, advanced courses. In this book, I hope to make the big picture accessible to any student who has taken first-year physics and calculus (plus a smattering of high school chemistry and biology), and who is willing to stretch. When you're done, you should be in a position to read current work in *Science* and *Nature*. You won't get every detail, of course. But you will get the sweep.

When we began to offer this course, we were surprised to find that many of our graduate students wanted to take it, too. In part this reflected their own compartmentalized education: The physics students wanted to read the biology part and see it integrated with their other knowledge; the biology students wanted the reverse. To our amazement, we found that the course became popular with students at all levels from sophomore to third-year graduate, with the latter digging more deeply into the details. Accordingly, many sections in this book have "Track–2" addenda addressing this more mathematically experienced group.

Physical science versus life science At the dawn of the twentieth century, it was already clear that, chemically speaking, you and I are not much different from cans of soup. And yet we can do many complex and even fun things we do not usually see cans of soup doing. At that time, people had very few correct ideas about how living organisms create order from food, do work, and even compute things—just a lot of inappropriate metaphors drawn from the technology of the day.

By mid-century, it began to be clear that the answers to many of these questions would be found in the study of very big molecules. Now, as we begin the twenty-first century, ironically, the situation is inverted: The problem is now that we have *way too much information* about those molecules! We are drowning in information; we need an armature, a framework, on which to organize all those zillions of facts.

Some life scientists dismiss physics as 'reductionist', tending to strip away all the details that make frogs different from, say, neutron stars. Others believe that right now some unifying framework is essential to see the big picture. I think that the *tension* between the developmental/historical/complex sciences and the universal/ahistorical/reductionist ones has been enormously fruitful and that the future belongs to those who can switch fluidly between both kinds of brains.

Setting aside philosophy, it's a fact that the past decade or two has seen a revolution in physical techniques to get inside the nanoworld of cells, tweak them in physical ways, and measure quantitatively the results. At last, a lot of physical ideas lying behind the cartoons found in cell biology books are getting the precise tests needed to confirm or reject them. At the same time, even some mechanisms not necessarily used by Nature have proved to be of immense technological value.

Why all the math?

> *I said it in Hebrew, I said it in Dutch,*
> *I said it in German and Greek;*
> *But I wholly forgot (and it vexes me much)*
> *That English is what you speak!*
> — Lewis Carroll, *The Hunting of the Snark*

Life science students may wonder whether all the mathematical formulas in this book are really needed. This book's premise is that the way to be sure that a theory is correct is to make quantitative predictions from a simplified model, then test those predictions experimentally. The following chapters supply many of the tools to do this. Ultimately, I want you to be able to walk into a room with an unfamiliar problem, pull out the right tool, and solve the problem. I realize this is not easy, at first.

Actually, it's true that physicists sometimes overdo the mathematical analysis. In contrast, the point of view in this book is that beautiful formulas are usually a means, not an end, in our attempts to understand Nature. Usually only the simplest tools, like dimensional analysis, suffice to see what's going on. Only when you've been a very good scientist, do you get the reward of carrying out some really elaborate mathematical calculation and seeing your predictions come to life in an experiment.

Your other physics and math courses will give you the background you'll need for that.

Features of this book I have tried to adhere to some principles while writing the book. Most of these are boring and technical, but there are four that are worth pointing out here:

1. When possible, *relate the ideas to everyday phenomena.*
2. *Say what's going on.* Instead of just giving a list of steps, I have tried to explain *why* we are taking these steps, and how we might have guessed that a step would prove fruitful. This exploratory (or discovery-style) approach involves more words than you may be used to in physics texts. The goal is to help you make the difficult transition to *choosing your own steps.*
3. *No black boxes.* The dreaded phrase "it can be shown" hardly ever appears in Track–1. Almost all mathematical results mentioned are actually derived here, or taken to the point where you can get them yourself as homework problems. When I could not obtain a result in a discussion at this level, I usually omitted it altogether.
4. *No fake data.* When you see an object that looks like a graph, almost always it really is a graph. That is, the points are somebody's actual laboratory data, usually with a citation. The curves are some actual mathematical function, usually derived in the text (or in a homework problem). Graphlike *sketches* are clearly labeled as such. In fact, every figure carries a pedantic little tag giving its logical status, so you can tell which are actual data, which are reconstructions, and which are an artist's sketches.

Real data are generally not as pretty as fake data. You need the real thing in order to develop your critical skills. For one thing, some simple theories *don't work* as well as you might believe just from listening to lectures. On the other hand, some unimpressive-looking fits of theory to experiment actually do support strong conclusions; you need practice looking for the relevant features.

Many chapters contain a section titled "Excursion." These sections lie outside the main story line. Some are short articles by leading experimentalists about experiments they did. Others are historical or cultural essays. There are also two appendices. Please take a moment now to check them. They include a list of all the symbols used in the text to represent physical quantities, definitions of all the units, and numerical values for many physical quantities, some of them useful in working the problems.

Why the history? This is not a history book, and yet you will find many ancient results discussed. (Many people take "ancient" to mean "before Internet," but in this book I use the more classical definition "before television.") The old stuff is not there just to give the patina of scholarship. Rather, a recurring theme of the book is the way in which physical measurements have often disclosed the existence and nature of molecular devices in cells long before traditional biochemical assays nailed down their precise identities. The historical passages document case studies where this has happened; in some cases, the gap has been measured in decades!

Even today, with our immensely sophisticated armamentum of molecular biology, the traditional knock-out-the-gene-and-see-what-kind-of-mouse-you-get experimental strategy can be much slower and more difficult to perform and interpret than a more direct, reach-in-and-grab-it approach. In fact, the menu of ingenious new tools for applying *physical stresses* to functioning cells or their constituents (all the way down to the single-molecule level) and *quantitatively measuring* their responses has grown rapidly in the last decade, giving unprecedented opportunities for indirectly deducing what must be happening at the molecular level. Scientists who can integrate the lessons of both the biochemical and biophysical approaches will be the first ones to see the whole picture. Knowing how it has worked in the past prepares you for your turn.

Learning this subject If your previous background in physical science is a first-year undergraduate course in physics or chemistry, this book will have a very different feel from the texts you've read so far. This subject is rapidly evolving; my presentation won't have that authoritative, stone-tablets feeling of a fixed, established subject, nor should it. Instead, I offer you the excitement of a field in flux, a field where you personally can make new contributions without first hacking through a jungle of existing formalism for a decade.

If your previous background is in life sciences, you may be accustomed to a writing style in which facts are delivered to you. But in this book, many of the assertions, and most of the formulas, are supposed to follow from the previous ones, in ways you can and must check. In fact, you will notice the words *we, us, our, let's* throughout the text. Usually in scientific writing, these words are just pompous ways of saying *I, me, my,* and *watch me*; but in this book, they refer to a team consisting of you and me. You need to figure out which statements are new information and which are deductions, and work out the latter ones. Sometimes, I have flagged especially important logical steps as "Your Turn" questions. Most of these are short enough that you can do them on the spot before proceeding. It is essential to work these out yourself in order to get the skill you need in constructing new physical arguments.

Each time the text introduces a formula, take a moment to look at it and think about its reasonableness. If it says $x = yz/w$, does it make sense that increasing w should decrease x? How do the units work out? At first, I'll walk you through these steps; but from then on, you need to do them automatically. When you find me using an unfamiliar mathematical idea, please talk to your instructor as soon as possible instead of just bleeping over it. Another helpful resource is the book by Shankar (Shankar, 1995).[1]

Beyond the questions in the text, you will find problems at the ends of the chapters. They are not as straightforward as they were in first-year physics; often you will need some common sense, some seat-of-the-pants qualitative judgment, even some advice from your instructor to get off to the right start. *Most* students are uncomfortable with this approach at first—it's not just you!—but in the end this skill is going to be one of the most valuable ones you'll ever learn, no matter what you do later in life.

---

[1] See the Bibliography at the back of this book.

It's a high-technology world out there, and it will be your oyster when you develop the agility to solve open-ended, quantitative problems.

The problems also get harder as you go on in the text, so do the early ones even if they seem easy.

T_2 Some sections and problems are flagged with this symbol. These are For Mature Audiences Only. Of course, I say it that way to make you want to read them, whether or not your instructor assigns them. These Track–2 sections take the mathematical development a bit further. They forge links to what you are learning/will learn in other physics courses. They also advertise some of the cited research literature. The main (Track–1) text does not rely on these sections; it is self-contained. Even Track–2 readers should skip the Track–2 sections on the first reading.

Many students find this course to be a stiff challenge. The physics students have to digest a lot of biological terminology; the biology students have to brush up on their math. It's not easy, but it's worth the effort: Interdisciplinary subjects like this one are among the most exciting and fertile. I've noticed that the happiest students are the ones who team up to work together with another student from a different background and do the problems together, teaching each other things. Give it a try.

To the Instructor

A few years ago, my department asked their undergraduate students what they needed but were not getting from us. One of the answers was, "a course on biological physics." Our students could not help noticing all the exciting articles in *The New York Times*, all the cover articles in *Physics Today*, and so on; they wanted a piece of the action. This book emerged from their request.

Around the same time, many of my friends at other universities were beginning to work in this field and were keenly interested in teaching a course, but they felt uncomfortable with the existing texts. Some were brilliant but decades old; none seemed to cover the beautiful new results in molecular motors, self-assembly, and single-molecule manipulation and imaging that were revolutionizing the field. My friends and I were also daunted by the vastness of the literature and our own limited penetration of the field; we needed a synthesis. This book is my attempt to answer that need.

The book also serves to introduce much of the conceptual material underlying the young fields of nanotechnology and soft materials. It's not surprising—the molecular and supramolecular machines in each of our cells are the inspiration for much of nanotechnology, and the polymers and membranes from which they are constructed are the inspiration for much of soft-materials science.

This text was intended for use with a wildly diverse audience. It is based on a course I have taught to a single class containing students majoring in physics, biology, biochemistry, biophysics, materials science, and chemical, mechanical, and bioengineering. I hope the book will prove useful as a main or adjunct text for courses in any science or engineering department. My students also vary widely in experience, from sophomores to third-year graduate students. You may not want to try such a broad group, but it works at Penn. To reach them all, the course is divided into two sections; the graduate section has harder and more mathematically sophisticated problems and exams. The structure of the book reflects this division, with numerous Track–2 sections and problems covering the more advanced material. These sections are placed at the ends of the chapters and are introduced with a special symbol: $\boxed{T_2}$. The Track–2 sections are largely independent of one another, so you can assign them à la carte. I recommend that *all* students skip them on the first reading.

The only prerequisites for the core, Track–1, material are first-year calculus and calculus-based physics, and a distant memory of high school chemistry and biology. The concepts of calculus are used freely, but very little of the technique; only the very simplest differential equations need to be solved. More important, the student needs to possess or acquire a fluency in throwing numbers around, making estimates, keeping track of units, and carrying out short derivations. The Track–2 material and problems should be appropriate for senior physics majors and first-year graduate students.

For a one-semester class of less experienced students, you will probably want to skip one or both of Chapters 9 and 10 (or possibly 11 and 12). For more experienced students, you can instead skim the opening chapters quickly, then spend extra time on the advanced chapters.

When teaching this course, I also assign supplementary readings from one of the standard cell biology texts. Cell biology inevitably contains a lot of nomenclature and iconography; both students and instructor must make an investment in learning these. The payoff is clear and immediate: Not only does this investment allow one to communicate with professionals doing exciting work in many fields, it is also crucial for seeing what physical problems are relevant to biomedical research.

I have made a special effort to keep the terminology and notation unified, a difficult task when spanning several disciplines. Appendix A summarizes all the notation in one place. Appendix B contains many useful numerical values, more than are used in the text. (You may find these data useful in making new homework and exam problems.)

More details about how to get from this book to a full course can be found in the *Instructor's Guide*, available from W. H. Freeman and Company. The *Guide* also contains solutions to all the problems and "Your Turn" questions, suggested class demonstrations, and the computer code used to generate many of the graphs found in the text. You can use this code to create computer-based problems, do class demos, and so on. Errata to this book will appear at

http://www.whfreeman.com/biologicalphysics

Why doesn't my favorite topic appear?

> *A garden is finished when there is nothing left to remove.*
> — Zen aphorism

It's probably one of my favorite topics, too. But the text reflects the relentless pursuit of a few maxims:

- Keep it a course, not an encyclopedia. The book corresponds to what I actually manage to cover (that is, what the students actually manage to learn) in a typical 42-hour semester, plus about 20% more to allow flexibility.
- Keep a unified story line.
- Maintain a balance between recent results and the important classical topics. Choose those topics that *open the most doors* into physics, biology, chemistry, and engineering.
- Make practically no mention of quantum theory, which our students encounter only after this course. Fortunately, a huge body of important biological physics

(including the whole field of soft biomaterials) makes no use of the deep quantum ideas.

• Restrict the discussion to concrete problems where the physical vision leads to falsifiable, quantitative predictions and where laboratory data are available. Every chapter presents some real experimental data.

• But choose problems that illuminate, and are illuminated by, the big ideas. Students want that—that's why they study science.

There are certainly other topics meeting all these criteria but not covered in this book. I look forward to your suggestions as to which ones to add to the next edition.

Underlying the preceding points is a determination to present physical ideas as beautiful and important in their own right. Respect for these foundational ideas has kept me from relegating them to the currently fashionable utilitarian status of a mere toolbag to help out with other disciplines. A few apparently dilatory topics, which pursue the physics beyond the point (currently) needed to explain biological phenomena, reflect this conviction.

Standard disclaimers This is a textbook, not a monograph. I am aware that many subtle subjects are presented in this book with important details burnished off. No attempt has been made to sort out historical priority, except in those sections titled "history." The experiments described here were chosen simply because they fit some pedagogical imperative and seemed to have particularly direct interpretations. The citation of original works is haphazard, except for my own work, which is systematically not cited. No claim is made that anything in this book is original, although at times I just couldn't stop myself.

Is this stuff really physics? Should it be taught in a physics department? If you've come this far, probably you have made up your mind already. But I'll bet you have colleagues who ask this question. The text attempts to show, not only that many of the founders of molecular biology had physics background, but conversely that historically the study of life has fed crucial insights back into physics. It's true at the pedagogical level as well. Many students find the ideas of statistical physics to be most vivid in the life science context. In fact, some students take my course *after* courses in statistical physics or physical chemistry; they tell me that it puts the pieces together for them in a new and helpful way.

More important, I have found a group of students who are interested in studying physics but feel turned away when their physics departments offer no connections to the excitement in the life sciences. It's time to give them what they need.

At the same time, your life sciences colleagues may ask, "Do our students need this much physics?" The answer is, maybe not in the past, but certainly in the future. Your colleagues may enjoy two recent eloquent articles on this subject (Alberts, 1998; Hopfield, 2002), and the comprehensive NRC report (National Research Council, 2003). This book tries to show that there is a quantitative, physical sciences approach to problems, and it's versatile. It's not the only toolbox in the well-educated scientist's mind, but it's one of the powerful ones. We need to teach it to everyone, not just to physical science majors. I believe that the recent insularity of physics is only a temporary aberration; both sides can only stand to prosper by renewing their once-tight linkage.

Last I had the great good fortune to see statistical physics for the first time through the beautiful lectures of Sam Treiman (1925–1999). Treiman was a great scientist and one of the spiritual leaders of a great department. From time to time, I still go back to my notes from that course. And there he is, just as before.

Acknowledgments

I think I wrote the first draft—it's hard to remember—but the book you now hold was shaped by many people, including many students. Like the cells in your body, nearly every sentence of this book has turned over at least once, thanks in part to the help of these readers. I am deeply grateful to all of them.

The book grew out of Chapters 7 and 9, which themselves grew out of a set of lectures I gave at the Institut d'Etudes Scientifiques in Cargèse, on the island of Corsica. I thank Bertrand Fourcade for inviting me to give these lectures.

Of course, there is a certain distance between a few lectures and a book; Rama-murti Shankar and Joseph Dan pitilessly explained to me why I had to go this whole distance. Throughout the journey, I benefited from Nily Dan's incisive suggestions, which covered every issue of strategy and indeed the book's whole purpose. The road would also have been much darker without the constant support and insight of my friends Gino Segrè and Scott Weinstein.

In an intense collaboration lasting several months, Sarina Bromberg made extensive improvements to both the science and the presentation, including the permutation of entire chapters and sections. Dr. Bromberg's background in bio-chemistry saved me from many missteps, small and large; she also created several of the book's most complex graphics.

The book owes a very direct debt to the authors who contributed the Excursions in Chapters 6 and 11, to Marko Radosavljević, who wrote many of the problem solutions, and to many other colleagues who contributed, or helped me find, striking graphics: Howard Berg, Paul Biancaniello, Scott Brady, David Deamer, David Derosier, Tony Dinsmore, Dennis Discher, Ken Downing, Deborah Kuchnir Fygenson, David Goodsell, Julian Heath, John Heuser, Nobutaka Hirokawa, A. James Hudspeth, Sir Andrew Huxley, Miloslav Kalab, Trevor Lamb, Jan Liphardt, Berenike Maier, Elisha Moses, Steve Nielsen, Iwan Schaap, Christoph Schmidt, Cornelis Storm, Karel Svoboda, and Jun Zhang.

Among the many readers, I'd especially like to mention major contributions from David Busch, Michael Farries, Rodrigo Guerra, Rob Phillips, and Tom Pologruto, each of whom generously committed their time and zeal to making this a better book. Beyond helping me to catch errors, these readers also made many suggestions about the book's pedagogy.

Many other colleagues answered my endless questions, supplied their experimental data, explained their research to me, and gave me helpful suggestions or criticism, including Ralph Amado, Charles Asbury, Howard Berg, Steven Block, Robijn Bruinsma, Vincent Croquette, David Deamer, Dennis Discher, David Fung, Raymond Goldstein, A. James Hudspeth, Wolfgang Junge, Randall Kamien,

David Keller, Matthew Lang, Tom Lubensky, Marcelo Magnasco, John Marko, Simon Mochrie, Alan Perelson, Charles Peskin, Dan Rothman, Jeffery Saven, Mark Schnitzer, Udo Seifert, Cornelis Storm, Kim Sharp, Edwin Taylor, Koen Visscher, Donald Voet, Michelle Wang, Eric Weeks, and John Weeks.

Still other colleagues reviewed one or more chapters and pounced on falsehoods, typos, scams, sloppiness, obfuscation, and missed opportunities, including Clay Armstrong, John Broadhurst, Russell Composto, M. Fevzi Daldal, Isard Dunietz, Bret Flanders, Jeff Gelles, Mark Goulian, Thomas Gruhn, David Hackney, Steve Hagen, Donald Jacobs, Ponzy Lu, Kristina Lynch, John Marko, Eugene Mele, Tom Moody, John Nagle, Lee Peachey, Scott Poethig, Tom Powers, Steven Quake, M. Thomas Record, Jr., Sam Safran, Brian Salzberg, Mark Schnitzer, Paul Selvin, Peter Sterling, Steven Vogel, Roy Wood, Michael Wortis, and Sally Zigmond.

This book has been extensively tested on live organisms. I am grateful to those colleagues who taught courses using draft versions of the book and, of course, to their long-suffering students as well: Anjum Ansari (University of Illinois at Chicago), Thomas Duke (University of Cambridge), Michael Fisher (University of Maryland), Bret Flanders (Oklahoma State), Erwin Frey (Hahn–Meitner Institut, Freie Universität Berlin), Steve Hagen (University of Florida), Gus Hart (Northern Arizona University), John Hegseth (University of New Orleans), Jané Kondev (Brandeis University), Serge Lemay (Delft University of Technology), Bob Martinez (University of Texas at Austin), Carl Michal (University of British Columbia), John Nagle (Carnegie Mellon University), David Nelson (Harvard University), Rob Phillips (California Institute of Technology), Tom Powers (Brown University), Daniel Reich (Johns Hopkins University), Michael Schick (University of Washington), Ulrich Schwarz (Max-Planck-Institut für Kolloid- und Grenzflächenforschung), Harvey Shepard (University of New Hampshire), Holger Stark (Universität Konstanz), Koen Visscher (University of Arizona), Z. Jane Wang (Cornell University), Shimon Weiss (University of California, Berkeley), Chris Wiggins (New York University), Charles Wolgemuth (University of Connecticut Health Center), and Jun Zhang (Courant Institute). I'd especially like to thank my own students at Penn, particularly Cristian Dobre, Thomas Pologruto, and Kathleen Vernovsky, and my class assistants David Busch, Alper Corlu, Corey O'Hern, and Marko Radosavljević for their enthusiasm and spirit of collaboration.

This book could not have been written without the support of several institutions. My greatest debt is to the University of Pennsylvania. Besides being a nonstop circus of terrific science, Penn has for sixteen years supported my ideas about teaching and research, in very tangible ways. Certainly the completion of this book has depended on having colleagues and two department Chairs who supported a vision that involved my dropping many other balls over the last three years. The crucial earliest stage of the project also benefited from the warm hospitality of the Weizmann Institute of Science; I am grateful to Elisha Moses for creating this opportunity. Finally, some of the most arduous revision was done in the infinite calm of the Aspen Center for Physics, and the infinite hubbub of the Philadelphia Museum of Art.

Partial financial support for this work was provided by the National Science Foundation's Division of Undergraduate Education (Course, Curriculum, and Laboratory Improvement Program) and by the Division of Materials Research. I'd particularly like to thank G. Bruce Taggart and Herbert Levitan at NSF for their initiative in supporting me at two tricky junctures in my ever-unpredictable career.

A large team of dedicated professionals have helped make the book a reality, right from its very beginnings. At W. H. Freeman and Company, I'm particularly grateful to Susan Finnemore Brennan, for her constant advice and support. (The reader will thank Susan for her gentle but persistent reminders that a shorter book is a better book.) I'd also like to thank Ellen Cash, Kathleen Civetta, Julia DeRosa, Brian Donnellan, Blake Logan, Patricia Marx, Philip McCaffrey, Eileen McGinnis, Jane O'Neill, Bill Page, Nancy Walker, Susan Wein, and Tobi Zausner at Freeman for their expert help. Finally, Jodi Simpson's contribution just can't be reduced to the words *copy editor.* Words like *mastermind* or *nerve center* spring to mind when I recall her unifying influence on practically every sentence.

At Penn, it's been a great pleasure working with Felice Macera as he created many of the drawings. Annette Day and Jean O'Boyle relentlessly tracked down countless references and copyright permissions. Steven Nelson offered his expertise in color reproduction. Last, I've been fortunate to have the help of Leslie Galen, who created the book's graceful design. Donald DeLand, Leslie Galen, and the rest of the Integre Technical Publishing team then carried out the intricate typography on a crushing schedule, despite all my attempts to derail them with endless revisions.

Of course, my biggest debt is to the people who invented all these beautiful stories and to those friends and strangers who first told them to me, through their books, articles, and talks; in their classrooms; while watching meteor showers or riding on the subway; and so on. Surely most of these people don't even realize how happy they have made me. Thank you.

Philip Nelson
Philadelphia, March 2003

PART I

Mysteries, Metaphors, Models

Transduction of free energy. [Drawing by Eric Sloane, from Eric Sloane, *Diary of an early American boy* (Funk and Wagnalls, New York, 1962).]

CHAPTER 1

What the Ancients Knew

*Although there is no direct connection between beer and the
First Law of thermodynamics, the influence of Joule's
professional expertise in brewing technology on his scientific
work is clearly discernible.*

— Hans Christian von Baeyer,
Warmth disperses and time passes

The modest goal of this book is to take you from the mid-nineteenth century, where first-year physics courses often end, to the science headlines you read this morning. It's a long road. To get to our destination on time, we'll need to focus tightly on just a few core issues involving the interplay between energy, information, and life.

We will eventually erect a framework, based on only a few principles, in which to begin addressing these issues. It's not enough simply to enunciate a handful of key ideas, of course. If it were, then this book could have been published on a single wallet card. The pleasure, the depth, the craft of our subject lie in the *details* of how living organisms work out the solutions to their challenges within the framework of physical law. The aim of the book is to show you a few of these details.

Each chapter of this book opens with a biological question, and a terse slogan encapsulating a physical idea relevant to the question. Think about these as you read the chapter.
Biological question: How can living organisms be so highly ordered?
Physical idea: The *flow* of energy can leave behind increased order.

1.1 HEAT

Living organisms eat, grow, reproduce, and compute. They do these things in ways that appear totally different from man-made machines. One key difference involves the role of temperature. For example, if you chill your vacuum cleaner, or even your television, to a degree above freezing, these appliances continue to work fine. But try this with a grasshopper, or even a bacterium, and you find that life processes practically stop. (After all, that's why you own a freezer in the first place.) Understanding the interplay of heat and work will become a central obsession of this book. This chapter will develop some plausible but preliminary ideas about this interplay; Part II of the book will sharpen these ideas into precise, quantitative tools.

3

1.1.1 Heat is a form of energy

When a rock of mass m falls freely, its altitude z and velocity v change together in just such a way as to ensure that the quantity $E = mgz + \frac{1}{2}mv^2$ stays constant, where g is the acceleration of gravity at Earth's surface.

Example: Show this.
Solution: We need to show that the time derivative $\frac{dE}{dt}$ equals 0. Taking v to be the velocity in the upward direction $\hat{z}$, we have $v = \frac{dz}{dt}$. Applying the chain rule from calculus then gives $\frac{dE}{dt} = mv(g + \frac{dv}{dt})$. But the acceleration, $\frac{dv}{dt}$, is always equal to $-g$ in free fall. Hence, $\frac{dE}{dt} = 0$ throughout the motion: The energy is a constant.

Gottfried Leibnitz obtained this result in 1693. We call the first term of E (that is, mgz) the **potential energy** of the rock, and the second term ($\frac{1}{2}mv^2$) its **kinetic energy**. We'll call their sum the **mechanical energy** of the rock. We express the constancy of E by saying that "energy is conserved."

Now suppose our rock lands in some mud at $z = 0$. The instant before it lands, its kinetic energy is nonzero, so E is nonzero, too. An instant later, the rock is at rest in the mud and its total mechanical energy is zero. Apparently, mechanical energy is *not* conserved in the presence of mud! Every first-year physics student learns why: A mysterious "frictional" effect in the mud drained off the mechanical energy of the rock. The genius of Isaac Newton lay in part in his realizing that the laws of motion were best studied in the context of the motions of cannonballs and planets, where complications like frictional effects are tiny: Here the conservation of energy, so apparently false on Earth, is most clearly seen. It took another two centuries before others would arrive at a precise statement of the more subtle idea that

> *Friction converts mechanical energy into thermal form. When thermal energy is properly accounted for, the energy accounts balance.* (1.1)

That is, the actual conserved quantity is not the mechanical energy, but the *total* energy, the sum of the mechanical energy plus heat.

But what *is* friction? What *is* heat? On a practical level, if energy is conserved, if it cannot be created or destroyed, then why must we be careful not to "waste" it? Indeed, what could "waste" mean? We'll need to look a bit more deeply before we really understand Idea 1.1.[1]

Idea 1.1 says that friction is not a process of energy *loss* but rather of energy *conversion*, just as the fall of a rock converts potential to kinetic energy. You may have seen an illustration of energy conversion in a grammar school exercise exploring the pathways that could take energy from the Sun and convert it to useful work, for example, a trip up a hill (Figure 1.1).

A point your schoolteacher may not have mentioned is that, in principle, all the energy conversions in Figure 1.1 are two-way: Light from the Sun can generate electricity in a solar cell, that energy can be partially converted back to light with a

---

[1]Throughout this book, the references Equation *n.m*, Idea *n.m*, and Reaction *n.m* all refer to a single sequence of numbered items. Thus Equation 1.2 comes after Idea 1.1; there is no Idea 1.2.

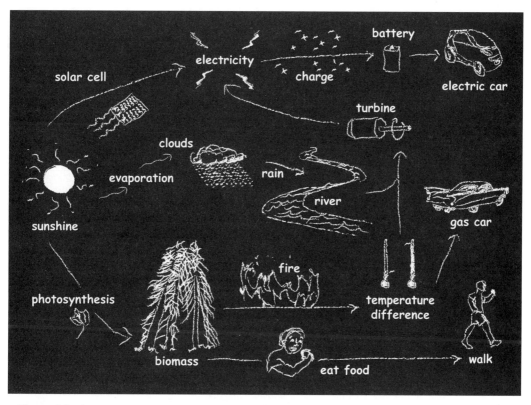

Figure 1.1: (Diagram.) Various ways to get up a hill. Each arrow represents an energy-conversion process.

light bulb, and so on. The key word here is *partially*. We never get *all* the original energy back in this way: Some is lost as heat, in both the solar cell and the light bulb. The word *lost* doesn't imply that energy disappears, but rather that *some of it makes a one-way conversion to heat.*

The same idea holds for the falling rock. We could let it down on a pulley, taking some of its gravitational potential energy to run a lawnmower. But if we just let it plop into the mud, its mechanical energy is lost. Nobody has ever seen a rock sitting in warm mud suddenly fly up into space, leaving cold mud behind, even though such a process is perfectly compatible with the conservation of energy!

So, even though energy is strictly conserved, *something* has been wasted when we let the rock plop. To make a scientific theory of this something, we'd like to find an independent, measurable quantity describing the "quality" or "usefulness" of energy; then we could assert that sunlight, or the potential energy of a rock, has high quality, whereas thermal energy (heat) has poor quality. We could also try to argue that the quality of energy always degrades in any transaction, and thus explain why the conversions indicated by arrows in Figure 1.1 are so much easier than those moving against the arrows. Before doing these things, though, it's worthwhile to recall how the ancients arrived at Idea 1.1.

1.1.2 Just a little history

Physicists like a tidy world with as few irreducible concepts as possible. If mechanical energy can be converted to thermal energy, and (partially) reconverted back again, and the sum of these forms of energy is always constant, then it's attractive to suppose that in some sense these two forms of energy are really the same thing. But we can't build scientific theories on æsthetic, culturally dependent judgments—Nature cares little for our prejudices, and other eras have had different prejudices. Instead, we must anchor Idea 1.1 on some firmer ground.

An example may help to underscore this point. We remember Benjamin Franklin as the great scientist who developed a theory of electricity as an invisible *fluid*. Franklin proposed that a positively charged body had "too much" of this fluid[2] and a negative body "too little." When such bodies were placed in contact, the fluid flowed from one to the other, much like joining a cylinder of compressed air to a balloon and opening the valve. What's less well remembered is that Franklin, and most of his contemporaries, had a similar vision of *heat*. In this view, heat also was an invisible fluid. Hot bodies had "too much," cold bodies "too little." When one placed such bodies in contact, the fluid flowed until the fluid was under the same "pressure" in each—or in other words, until both were at the same temperature.

The fluid theory of heat made some superficial sense. A large body would need more heat fluid to increase its temperature by one degree than would a small body, just as a large balloon needs more air than does a small one to increase its internal pressure to, say, 1.1 times atmospheric pressure. Nevertheless, today we believe that *Franklin's theory of electricity was exactly correct, but the fluid theory of heat was dead wrong.* How did this change in attitudes come about?

Franklin's contemporary Benjamin Thompson was also intrigued by the problem of heat. After leaving the American colonies in a hurry in 1775 (he was a spy for the British), Thompson eventually became a major general in the court of the Duke of Bavaria. In the course of his duties, Thompson arranged for the manufacture of weapons. A curious phenomenon in the boring (drilling) of cannon barrels aroused his curiosity. Drilling takes a lot of work, at that time supplied by horses. It also generates a lot of frictional heat. If heat were a fluid, one might expect that rubbing would transfer some of it from one body to another, just as brushing your cat leaves cat and brush with opposite electrical charges. But the drill bit doesn't grow cold while the cannon barrel becomes hot! *Both* become hot.

Moreover, the fluid theory of heat seems to imply that eventually the cannon barrel would become depleted of heat fluid and that no more heat could be generated by additional friction. This is not what Thompson observed. One barrel could generate enough heat to boil a surrounding bath of water. The bath could be replaced by cool water, which would also eventually boil, ad infinitum. A fresh cannon barrel proved neither better nor worse at heating water than one that had already boiled many liters. Thompson also weighed the metal chips cut out of the barrel and found

---

[2]Franklin's convention for the sign of charge was unfortunate. Today we know that the main carriers of charge—electrons—each carry a *negative* quantity of charge in his convention. Thus, it's more accurate to say that a positively charged body has too few electrons, and a negatively charged body too many.

their mass plus that of the barrel to be equal to the original mass of the barrel: No material substance had been lost.

What Thompson noticed instead was that *heat production from friction ceases the moment we stop doing mechanical work on the system.* This was a suggestive observation. But later work, presented independently in 1847 by James Joule and Hermann von Helmholtz, went much further. Joule and Helmholtz upgraded Thompson's qualitative observation to a *quantitative* law: *The heat produced by friction is a constant times the mechanical work done against that friction,* or

$$\text{(heat produced)} = \text{(mechanical energy input)} \times (0.24 \, \text{cal/J}). \qquad (1.2)$$

Let's pause to sort out the shorthand in this formula. We measure heat in **calories**: One calorie is roughly the amount of heat needed to warm a gram of water by one degree Celsius.[3] The mechanical energy input, or **work** done, is the force applied (in Thompson's case, by the horse), times the distance (walked by the horse); we measure it in joules just as in first-year physics. Multiplying work by the constant $0.24 \, \text{cal/J}$ creates a quantity with units of calories. The formula asserts that this quantity is the amount of heat created.

Equation 1.2 sharpens Idea 1.1 into a quantitative assertion. It also succinctly predicts the outcomes of several different kinds of experiments: It says that the horse will boil twice as many liters of water if it walks twice as far, or walks equally far while exerting twice the force, and so on. It thus contains vastly more information than the precise but limited statement that heat output stops when work input stops. Scientists like hypotheses that make such a sweeping web of interlocking predictions, because the success of such a hypothesis is hard to brush aside as a mere fluke. We say that such hypotheses are highly **falsifiable**, because any one of the many predictions of Equation 1.2, if disproved experimentally, would kill the whole thing. The fluid theory of heat made no comparably broad, correct predictions. Indeed, as we have seen, it does make some wrong qualitative predictions. This sort of reasoning ultimately led to the demise of the fluid theory, despite the strenuous efforts of its powerful adherents to save it.

Suppose that we use a very dull drill bit, so that in one revolution we make little progress in drilling; that is, the cannon barrel (and the drill itself) are not changed very much. Equation 1.2 says that the net work done on the system equals the net heat given off. More generally,

> *Suppose that a system undergoes a process that leaves it in its original state (that is, a* **cyclic process***). Then the net of the mechanical work done on the system, and by the system, equals the net of the heat it gives off and takes in, once we convert the work into calories using Equation 1.2.* (1.3)

[3]The modern definition of the calorie acknowledges the mechanical equivalent of heat: One calorie is now *defined* as the quantity of thermal energy created by converting exactly 4.184 J of mechanical work. (The "Calorie" appearing on nutritional statements is actually one thousand of the physical scientist's calories, or one kilocalorie.)

It doesn't matter whether the mechanical work was done by a horse, or by a coiled spring, or even by a flywheel that was initially spinning.

What about processes that *do* change the system under study? In this case, we'll need to amend Idea 1.3 to account for the energy that was stored in (or released from) the system. For example, the heat released when a match burns represents energy initially stored in chemical form. A tremendous amount of nineteenth-century research by Joule and Helmholtz (among many others) convinced scientists that when every form of energy is properly included, the accounts balance for *all* the arrows in Figure 1.1, and for every other thermal/mechanical/chemical process. This generalized form of Idea 1.3 is now called the **First Law** of thermodynamics.

1.1.3 Preview: The concept of free energy

This subsection is just a preview of ideas to be made precise later. Don't worry if these ideas don't seem firm yet. The goal is to build up some intuition, some expectations, about the interplay of order and energy. Chapters 3–5 will give many concrete examples of this interplay, to get us ready for the abstract formulation in Chapter 6.

The quantitative connection between heat and work lent strong support to an old idea (Newton had discussed it in the seventeenth century) that heat *really is* nothing but a particular form of mechanical energy, namely, the kinetic energy of the individual molecules constituting a body. In this view, a hot body has a lot of energy stored in an (imperceptible) jiggling of its (invisible) molecules. Certainly we'll have to work hard to justify claims about the imperceptible and the invisible. But before doing this, we must deal with a more direct problem.

Equation 1.2 is sometimes called the "mechanical equivalent of heat." The discussion in Section 1.1.1 makes it clear, however, that this phrase is a slight misnomer: Heat is *not* fully equivalent to mechanical work, because one cannot be fully converted to the other. Chapter 3 will explore the view that slowly emerged in the late nineteenth century, which is that thermal energy is the portion of the total energy attributable to *random* molecular motion (all molecules jiggling in random directions) and so is distinct from the *organized* kinetic energy of a falling rock (all molecules have the same average velocity).

Thus, the random character of thermal motion must be the key to its low quality. In other words, we are proposing that *the distinction between high- and low-quality energy is a matter of organization.* Everyone knows that an orderly system tends to degrade into a disorganized, random mess. Sorting it back out again always seems to take work, both in the colloquial sense (sorting a big pile of coins into pennies, nickels, and so on is a lot of work) and in the strict sense. For example, an air conditioner *consumes* electrical energy to suppress random molecular motion in the air of your room; hence, it heats the outside world more than it cools your room.

The idea in the preceding paragraph may be interesting, but it hardly qualifies as a testable physical hypothesis. We need a quantitative measure of the *useful* energy of a system, the part of the total that can actually be harnessed to do useful work. A major goal of Chapter 6 will be to find such a measure, which we will call free energy and denote by the symbol F. But we can already see what to expect. The idea we are considering is that F is less than the total energy E by an amount related to the randomness, or disorder, of the system. More precisely, Chapter 6 will show how to characterize this disorder by using a quantity called entropy and denoted by the letter S. The free energy will turn out to be given by the simple formula

$$F = E - TS, \tag{1.4}$$

where T is the temperature of the system. We can now state the proposal that F measures the "useful" energy of a system a bit more clearly:

> *A system held at a fixed temperature T can spontaneously drive a process if the net effect of the process is to reduce the system's free energy F. Thus, if the system's free energy is already at a minimum, no spontaneous change will occur.* (1.5)

According to Equation 1.4, a decrease in free energy can come about *either* by lowering the energy E (rocks tend to fall) *or* by increasing the entropy S (disorder tends to increase).

We can also use Equation 1.4 to clarify our idea of the "quality" of energy: A system's free energy is always less than its mechanical energy. If the disorder is small, though, so that TS is much smaller than E, then $F \approx E$; we then say that the system's energy content is of "high quality." (More precisely still, we should discuss *changes* of energy and entropy; see Section 6.5.4.)

Again, Equation 1.4 and Idea 1.5 are provisional—we haven't even defined the quantity S yet. Nevertheless, they should at least seem reasonable. In particular, it makes sense that the second term on the right side of Equation 1.4 should be multiplied by T, because hotter systems have more thermal motion and so should be even more strongly influenced by the tendency to maximize disorder than cold ones. Chapters 6 and 7 will make these ideas precise. Chapter 8 will extend the idea of free energy to include chemical forms of energy; these are also of high quality.

1.2 HOW LIFE GENERATES ORDER

1.2.1 The puzzle of biological order

The ideas of the previous section have a certain intuitive appeal. When we put a drop of ink in a glass of water, the ink eventually mixes, a process we will study in great detail in Chapter 4. We never see an ink–water mixture spontaneously unmix. Chapter 6 will make this intuition precise, formulating a principle called the Second Law of thermodynamics. Roughly speaking, it says that in an isolated system molecular disorder never decreases spontaneously.

But now we are in a bit of a bind. We have just concluded that a mixture of hydrogen, carbon, oxygen, nitrogen, phosphorus, and traces of a few other elements, left alone and isolated in a beaker, will never organize spontaneously to make a living organism. After all, even the lowliest bacterium is full of exquisite structure (see Chapter 2), whereas physical systems tend relentlessly toward greater disorder. Yet the Earth is teeming with life, even though long ago it was barren. How indeed does any organism manage to remain alive, let alone create progeny, and even evolve to more sophisticated organisms? Stated bluntly, our puzzle is, *Must we suppose that living organisms somehow lie outside the jurisdiction of physical law?*

At the end of the nineteenth century, many respected scientists still answered "yes" to this question. Their doctrine was called "vitalism." Today vitalism has gone the way of the fluid theory of heat, as answers to the paradox of *how living things generate order* have emerged. Sketching a few of the details of these answers, along with their precise quantitative tests, is the goal of this book. It will take some time to reach that goal. But we can already propose the outlines of an answer in the language developed so far.

It's encouraging to notice that living creatures obey at least *some* of the same physical laws as inanimate matter, even those involving heat. For example, we can measure the heat given off by a mouse, and add the work it does on its exercise wheel by using the conversion formula (Equation 1.2). Over the course of a few days, the mouse doesn't change. The First Law of thermodynamics, Idea 1.3, then says that the total energy output must be proportional to the food intake of the mouse, and indeed it's roughly true. (The bookkeeping can get a bit tricky—see Problem 1.7.)

Thus, living organisms don't manage to create energy from nothing. Still, when we look around, it seems obvious that life is constantly generating *order* from nothing (that is, from disorder). To escape from vitalism, then, we must reconcile this commonplace observation with the Second Law of thermodynamics.

Such a reconciliation is easier than it at first sounds. After all, a sealed jar full of dense water vapor changes spontaneously into a jar with a puddle of water at the bottom and very little vapor. After this transformation, the inside of the jar is more organized than before: Most of the water molecules are stuck in a very thin layer instead of moving freely throughout the interior of the jar. But nobody would be tempted to believe that an unphysical, occult influence ordered the water molecules!

To see what is happening, we must recall that the Second Law applies only to an *isolated* system. Even though the jar with water vapor is sealed, it gave off *heat* to its surroundings as the water condensed; so it's not isolated. And there is nothing paradoxical about a *subsystem* of the world spontaneously increasing its order. Indeed, Section 1.1.3 proposed that a system (in this case, the contents of the jar) will tend spontaneously to move toward lower free energy F, which is not necessarily the same as moving toward higher disorder. According to our proposed formula for F (Equation 1.4), the subsystem's entropy S can indeed decrease (the water can condense) without raising F, if the internal energy E also decreases by a large enough amount (via heat loss).

The Earth, like our jar, is not an isolated system. Hence, the increasing organization of molecules as life began to develop does not necessarily contradict the Second Law. To make that statement more precise, let us look globally at what flows into and out of Earth. Figure 1.2a depicts the stream of *solar energy* impinging on Earth. Because Earth's temperature is roughly stable over the long term, all of this energy must also *leave* the Earth (along with a bit of geothermal energy generated here). Some of this energy is just reflected into space. The rest leaves when the Earth radiates it away as thermal energy to the rest of the Universe. Thus, Earth constantly accepts energy from the Sun, a very hot body, and exports it as radiation at its own surface

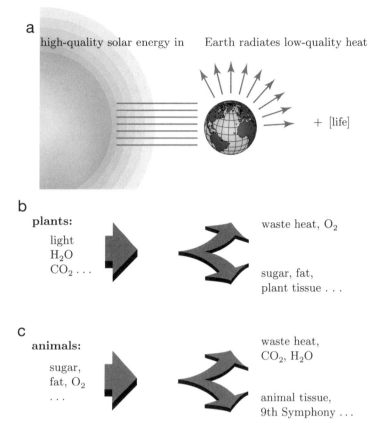

a
high-quality solar energy in   Earth radiates low-quality heat

+ [life]

b
plants:

light
H_2O
$CO_2 \ldots$

waste heat, O_2

sugar, fat,
plant tissue . . .

c
animals:

sugar,
fat, O_2
. . .

waste heat,
CO_2, H_2O

animal tissue,
9th Symphony . . .

Figure 1.2: (Diagram.) (a) Energy budget of Earth's biosphere. Most of the incident high-quality energy is degraded to thermal energy and radiated into space, but some gets captured and used to create the order we see in life. (b) What plants do with energy: High-quality solar energy is partly used to upgrade low-energy molecules to high-energy molecules and the ordered structures they form; the rest is released in thermal form. (c) What animals do with energy: The high-quality energy in food molecules is partly used to do mechanical work and create ordered structures; the rest is released in thermal form.

temperature. On a dead rock like the Moon, this is the whole story. But, as depicted symbolically in Figure 1.2b,c, there is a more interesting possibility.

Suppose that the incoming energy is of higher "quality" than the outgoing energy and hence represents a net *flow of order* into the Earth (Chapter 6 will sharpen this statement). Then we can imagine some enterprising middleman inserting itself in the middle of this process and *skimming off some of the incoming flow of order*, using it to create more and better middlemen! Looking only at the middle layer, it would *seem* as though order were magically increasing. That is,

The flow of energy through a system can leave behind increased order.   (1.6)

This is life's big trick. The middle zone is our biosphere; we are the middlemen.[4] Green plants ingest a high-quality form of energy (sunlight), passing it through their bodies to exit as thermal energy (Figure 1.2b). The plant needs some of this energy just to resist the degrading tendency of thermal disorder to turn its tissues into well-mixed chemical solutions. By processing even more energy through its body than this minimum, the plant can grow and do some "useful work," for example, upgrading some of its input matter from a low-energy form (carbon dioxide and water) to a high-energy form (carbohydrate). *Plants consume order, not energy.*

Closer to home, each of us must constantly process about 100 joules per second (100 W) of high-quality energy through our bodies (for example, by eating the carbohydrate molecules manufactured by plants), even at rest. If we eat more than that, we can generate some excess mechanical (ordered) energy to build our homes and so on. As shown in Figure 1.2c, the input energy again leaves in a low-quality form (heat). *Animals, too, consume order, not energy.*

Again, life doesn't really create order from nowhere. Life *captures* order, ultimately from the Sun. This order then trickles through the biosphere in an intricate set of processes that we will refer to generically as **free energy transductions**. Looking only at the biosphere, it *seems* as though life has created order.

1.2.2 Osmotic flow as a paradigm for free energy transduction

If the trick described in Section 1.2.1 were unique to living organisms, then we might still feel that they sat outside the physical world. But nonliving systems can transduce free energy, too: The drawing on page 1 shows a machine that processes solar energy and performs useful work. Unfortunately, this sort of machine is not a very precise metaphor for the processes driving living cells.

Figure 1.3 sketches another sort of machine, more closely related to what we are looking for. A sealed tank of water has two freely sliding pistons. When one piston moves to the left, so does the other, because the water between them is practically incompressible (and unstretchable). Across the middle of the chamber, we place a membrane permeable to water but not to dissolved sugar molecules. The whole system is kept at room temperature: Any heat that must be added or removed to hold it at this temperature comes from (or goes into) the surrounding room. Initially, a lump of sugar is uncovered on the right side. What happens?

At first, nothing seems to happen at all. But as the sugar dissolves and spreads throughout the right-hand chamber, a mysterious force begins to push the pistons to the right. This is an honest, mechanical force; we could use it to lift a weight, as shown in Figure 1.3a. The process is called **osmotic flow**.

Where did the energy to lift the weight come from? The only possible source of energy is the outside world. Indeed, careful measurements show that the system absorbs *heat* from its surroundings; somehow this thermal energy gets converted to mechanical work. Didn't Section 1.1.3 argue that it is impossible to convert heat

---

[4]A second, largely independent, biosphere exists in hot ocean vents, fueled not by the Sun but by high-energy chemicals escaping from inside the Earth.

a

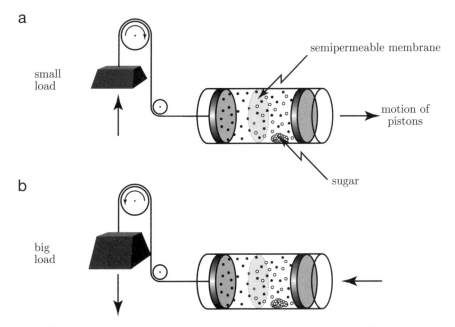

small
load

semipermeable membrane

motion of
pistons

sugar

b

big
load

Figure 1.3: (Schematic.) A machine transducing free energy. A cylinder filled with water is separated into two chambers by a semipermeable membrane. The membrane is anchored to the cylinder. Two pistons slide freely, thus allowing the volumes of the two chambers to change as water molecules (*solid dots*) cross the membrane. The distance between the pistons stays fixed, however, because the water between them is incompressible. Sugar molecules (*open circles*) remain confined to the right-hand chamber. (a) Osmotic flow: As long as the weight is not too heavy, when we release the pistons, water crosses the membrane, thereby forcing both pistons to the right and lifting the weight. The sugar molecules then spread out into the increased volume of water on the right. (b) Reverse osmosis: If we pull hard enough, however, the pistons will move to the *left*, thereby increasing the concentration of the sugar solution in the right-hand chamber and generating heat.

completely back into mechanical work? Yes, but we *are* paying for this transaction; something *is* getting used up. That something is order. Initially, the sugar molecules are partially confined: Each one moves freely, and randomly, throughout the region between the membrane and the right-hand piston. As water flows through the membrane, forcing the pistons to the right, the sugar molecules lose some of their order (or gain some disorder), being no longer confined to just one-half of the total volume of water. When finally the left side has shrunk to zero, the sugar molecules have free run of the entire volume of water between the pistons; their disorder can't increase any more. Our device then stops and will yield no more work, even though there's plenty of thermal energy left in the surrounding world. Osmotic flow sacrifices *molecular order* to organize random thermal motion into gross mechanical motion against a load.

We can rephrase the above argument in the language introduced in Section 1.1.3. Idea 1.5 introduced the idea that the osmotic machine will spontaneously move in the direction that lowers its free energy F. According to Equation 1.4, F can decrease *even if the potential energy of the weight increases,* as long as the entropy increases by a compensating amount. But the previous paragraph argued that, as the pistons move to the right, the disorder (and hence the entropy) increases. So, indeed, Idea 1.5 predicts that the pistons will move to the right, as long as the weight is not too heavy.

Now suppose we pull very hard on the left piston, as in Figure 1.3b. This time, a rightward movement of the piston would increase the potential energy of the weight so much that F *increases,* despite the second term of Equation 1.4. Instead, the pistons will move to the *left,* the region of concentrated solution will shrink and become more concentrated, and the system will *gain* order. This really works—it's a common industrial process called **reverse osmosis** (or ultrafiltration). You could use it to purify water before drinking it.

Reverse osmosis (Figure 1.3b) is just the sort of process we were looking for. An input of high-quality energy (in this case, mechanical work) suffices to upgrade the order of our system. The energy input must go somewhere, according to the First Law (Idea 1.3), and indeed it does: The system gives off heat in the process. *We passed energy through our system, which degraded the energy from mechanical form to thermal form while increasing its own order.* We could even make our machine cyclic. After pulling the pistons all the way to the left, we dump out the contents of each side, move the pistons all the way to the right (lifting the weight), refill the right side with sugar solution, and repeat everything. Then our machine continuously accepts high-quality (mechanical) energy, degrades it into thermal energy, and creates molecular order (by separating the sugar solution into sugar and pure water).

But that's the same trick we ascribed to living organisms, as summarized in Figure 1.2! It's not precisely the same—in Earth's biosphere, the input stream of high-quality energy is sunlight, whereas our reverse-osmosis machine runs on externally supplied mechanical work. Nevertheless, much of this book will be devoted to showing that at a deep level these processes, one from the living and one from the nonliving world, are essentially the same. In particular, Chapters 6, 7, and 10 will pick up this story and parlay our understanding into a view of biomolecular machines. The motors found in living cells differ from our osmotic machine by being *single molecules,* or collections of a few molecules. But we'll argue that these "molecular motors" are again just free energy transducers, essentially like Figure 1.3. *They work better than simple machines because evolution has engineered them to work better, not because of some fundamental exemption from physical law.*

1.2.3 Preview: Disorder as information

The osmotic machine illustrates another key idea, on which Chapter 6 will build, namely, the connection between disorder and information. To introduce this concept, consider again the case of a small load (Figure 1.3a). Suppose that we measure experimentally the maximum work done by the piston, by integrating the maximum force the piston can exert over the distance it travels. Doing this experiment at room temperature yields an empirical observation:

$$\text{(maximum work)} \approx N \times (4.1 \times 10^{-21} \text{ J} \times \gamma). \tag{1.7}$$

Here N is the number of dissolved sugar molecules. (γ is a numerical constant whose value is not important right now; you will find it in Your Turn 7B.)

In fact, Equation 1.7 holds for *any* dilute solution at room temperature, not just sugar dissolved in water, regardless of the details of the size or shape of the container and the number of molecules. Such a universal law must have a deep meaning. To interpret it, we return to Equation 1.4. We get the maximum work when we let the pistons move gradually, always applying the biggest possible load. According to Idea 1.5, the largest load we can apply without stalling the machine is the one for which the free energy F hardly decreases at all. In this case, Equation 1.4 claims that the change in potential energy of the weight (that is, the mechanical work done) just equals the temperature times the change of entropy. Writing ΔS for the entropy change, Equation 1.7 says $T\Delta S \approx N \times (4.1 \times 10^{-21} \text{ J} \times \gamma)$.

We already have the expectation that entropy involves disorder, and indeed, some order does disappear when the pistons move all the way to the right in Figure 1.3a. Initially, each sugar molecule was confined to half the total volume, whereas in the end they are not so confined. Thus, what's lost as the pistons move is a knowledge of which half of the chamber each sugar molecule was in—a binary choice. If there are N sugar molecules in all, we need to specify N binary digits (bits) of information to specify where each one sits in the final state, to the same accuracy that we knew it originally. Combining this remark with the result of the previous paragraph gives

$$\Delta S = \text{constant} \times \text{(number of bits lost)}.$$

Thus, the entropy, which we have been thinking of qualitatively as a measure of disorder, turns out to have a quantitative interpretation. If we find that biomolecular motors also obey some version of Equation 1.7, with the *same numerical constant*, then we will be on firm ground when we assert that they really are free energy transduction devices; and we can make a fair claim to have learned something fundamental about how they work. Chapter 10 will develop this idea.

1.3 EXCURSION: COMMERCIALS, PHILOSOPHY, PRAGMATICS

And oftentimes, to winne us to our harme
The Instruments of Darkness tell us Truths
Winne us with honest trifles, to betray's
In deepest consequence.
— Shakespeare, *Macbeth*

Cell and tissue, shell and bone, leaf and flower, are so many portions of matter, and it is in obedience to the laws of physics that their particles have been moved, moulded, and conformed.
— D'Arcy Thompson, *On growth and form*, 1917

Section 1.2 dove directly into the technical issues that we'll wrestle with throughout this book. But before we begin our exploration in earnest, a very few words are in order about the relation between physical science and biology.

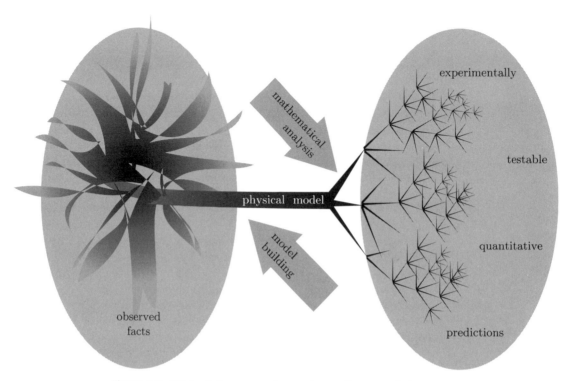

Figure 1.4: (Vision.) One approach to understanding natural phenomena.

The quotes above were chosen to highlight a fruitful tension between the two cultures:

- The physical scientist's impulse is to look for the forest, not the trees, to see that which is universal and simple in any system.
- Traditionally, life scientists have been more likely to emphasize that, in the inherently complex living world, frozen accidents of history often dominate what we see, not universal laws. In such a world, often it's the details that really matter most.

The views are complementary; one needs the agility to use whichever approach is appropriate at any given moment and a willingness to entertain the possibility that the other one is valuable, too.

How can one synthesize these two approaches? Figure 1.4 shows the essential strategy. The first step is to look around at the rich fabric of the phenomena around us. Next, we selectively ignore nearly everything about these phenomena, snipping the fabric down to just a few threads. This process involves (a) selecting a simplified but real model system for detailed study and (b) representing the simple system by an equally simple mathematical model, with as few independent constructs and relations as possible. The steps (a) and (b) are not deductive; words like *mystery* and *insight* apply to this process.

The last step is to (c) deduce from the mathematical model some nonobvious, quantitative, and experimentally testable predictions. If a model makes many such successful predictions, we gain conviction that we have found the few key ingredients in our simplifying steps (a) and (b). Words like *hygiene* and *technique* apply to step (c). Even though this step is deductive, again imagination is needed to find those consequences of the model that are both nontrivial and practical to test. The best, most striking results are those for which the right side of the figure opens up to embrace phenomena that had previously seemed unrelated. We have already foreshadowed an example of such a global linkage of ideas: The physics of osmotic flow is linked to the biology of molecular machines.

In the best case, the results of step (c) give the sense of getting something for nothing: The model generates more structure than was present in its bare statement (the middle part of Figure 1.4), a structure, moreover, that is usually buried in the mass of raw phenomena we began with (left end of Figure 1.4). In addition, we may in the process find that the most satisfactory physical model involves some threads, or postulated physical entities (middle part of the figure), whose *very existence wasn't obvious from the observed phenomena* (left part) but can be substantiated by making and testing quantitative predictions (right part). One famous example of this process is Max Delbrück's deduction of the existence of a hereditary molecule, to be discussed in Chapter 3. We'll see another example in Chapters 11 and 12, namely, the discovery of ion pumps and channels in cells.

Physics students are heavily trained on the right end of the figure, the techniques for working through the consequences of a mathematical model. But this technical expertise is not enough. Uncritically accepting someone's model can easily lead to a large body of both theory and experiment culminating in irrelevant results. Similarly, biology students are heavily trained in the left side, the amassing of many details of a system. For them, the risk is that of becoming an archivist who misses the big picture. To avoid both these fates, one must usually know all the details of a biological system, then transcend them with an *appropriate* simple model.

Is the physicist's insistence on simplicity, concreteness, and quantitative tests on model systems just an immature craving for certainty in an uncertain world? Certainly, at times. But at other times, this approach lets us perceive connections not visible "on the ground" by viewing the world "from above." When we find such universality, we get a sense of having *explained* something. We can also get more pragmatic benefits:

- Often, when we forge such a link, we find that powerful theoretical tools useful to solve one problem have already been created in the context of another. An example is the mathematical solution of the helix-coil transition model discussed in Chapter 9.
- Similarly, we can carry over powerful existing *experimental* techniques as well. For example, the realization that DNA and proteins were molecules led Max Perutz, Linus Pauling, Maurice Wilkins, and others to study the structure of these molecules with X-ray diffraction, a technique invented to find the structure of simple, nonbiological crystals like quartz.

- Finally, perceiving a link between two circles of ideas can lead us to *ask new questions* that later prove to be important. For example, even after James Watson and Francis Crick's discovery that the DNA molecule was a very long sentence written in an alphabet with four letters (see Chapter 3), attention did not focus at once on the importance of finding the dictionary, or code, relating sequences of those letters to the 20-letter alphabet of amino acids that constitute proteins. Thinking about the problem as one in information transfer led George Gamow, a physicist interested in biology, to write an influential paper in 1954 asking this question and suggesting that answering it might not be so difficult as it at first seemed.

It may seem that we need no longer content ourselves with simple models. Can't massive computers now follow the fine details of any process? Yes and no. Many low-level processes can now be followed in molecular detail. Nevertheless, our ability to get a detailed picture of even simple systems is surprisingly limited, in part by the rapid increase of computational complexity when we study large numbers of particles. Surprisingly, though, many physical systems have simple "emergent properties" not visible in the complex dynamics of their individual molecules. The simple equations we'll study seek to encapsulate these properties and often manage to capture the important features of the whole complex system. Examples in this book will include the powerful property of hydrodynamic scale invariance to be explored in Chapter 5, the mean-field behavior of ions in Chapter 7, and the elasticity of macromolecules in Chapter 9. The need to exploit such simplicity and regularity in the collective behavior of many similar actors becomes even more acute when we begin to study even larger systems than the ones discussed in this book.

1.4 HOW TO DO BETTER ON EXAMS (AND DISCOVER NEW PHYSICAL LAWS)

Equation 1.2 and the discussion following it made use of some simple ideas involving units. Students often see units, and the associated ideas of **dimensional analysis**, presented with a brush-your-teeth attitude. This is regrettable. Dimensional analysis is more than just hygiene. It's a *shortcut to insight,* a way to organize and classify numbers and situations, and even to guess new physical laws. Working scientists eventually realize that, when faced with an unfamiliar situation, dimensional analysis is always step one.

1.4.1 Most physical quantities carry dimensions

A physical quantity generally has abstract **dimensions** that tell us *what kind of thing* it represents. Each kind of dimension can be measured by using a variety of different **units**. The choice of units is arbitrary. People once used the size of the king's foot. This book will instead use primarily the Système International d'Unités, or **SI units**. In this system, lengths are measured in meters, masses in kilograms, time in seconds, and electric charge in coulombs. The distinction between dimensions and units becomes clearer when we look at some examples:

1. Length has dimensions of $\mathbb{L}$, by definition. In SI units, we measure it in meters, abbreviated in this book as m.
2. Mass has dimensions of $\mathbb{M}$, by definition. In SI units, we measure it in kilograms, abbreviated as kg.
3. Time has dimensions of $\mathbb{T}$, by definition. In SI units, we measure it in seconds, abbreviated as s.
4. Velocity has dimensions of $\mathbb{L}\mathbb{T}^{-1}$. In SI units, we measure it in m s^{-1} (pronounced "meters per second").
5. Acceleration has dimensions of $\mathbb{L}\mathbb{T}^{-2}$. In SI units, we measure it in m s^{-2}.
6. Force has dimensions of $\mathbb{M}\mathbb{L}\mathbb{T}^{-2}$. In SI units, we measure it in kg m s^{-2}, which we also call **newtons** and abbreviate as N.
7. Energy has dimensions of $\mathbb{M}\mathbb{L}^2\mathbb{T}^{-2}$. In SI units, we measure it in kg m^2 s^{-2}, which we also call **joules** and abbreviate as J.
8. Electric charge has dimensions of $\mathbb{Q}$, by definition. In SI units, we measure it in coulombs, abbreviated in this book as coul to avoid confusion with the symbol C. The *flow rate* of charge, or *electric current*, then must have dimensions of $\mathbb{Q}\mathbb{T}^{-1}$. In SI units, we measure it in coulombs per second, or coul s^{-1}, also called **amperes**, abbreviated as A.
9. We defer a discussion of temperature units to Section 6.3.2.

Notice that in this book *all units are set in a special typeface*, to help you distinguish them from named quantities (such as m for the mass of an object).

We also create related units by attaching prefixes giga ($=10^9$, or billion), mega ($=10^6$, or million), kilo ($=10^3$, or thousand), milli ($=10^{-3}$, or thousandth), micro ($=10^{-6}$, or millionth), nano ($=10^{-9}$, or billionth), pico ($=10^{-12}$). In writing, we abbreviate these prefixes to G, M, k, m, μ, n, and p, respectively. Thus, 1 Gy is a billion years, 1 pN is a trillionth of a newton, and so on. Forces in cells are usually in the pN range.

A few non-SI units, like cm and kcal, are so traditional that we'll occasionally use them as well. You will constantly find these units in the research literature, so you might as well get good at interconverting them now. See Appendix A for a list of all the units in this book; Appendix B presents the hierarchy of length, time, and energy scales of interest to cell biology and pulls together the numerical values of many useful constants.

In any quantitative problem, it is absolutely crucial to keep units in mind at all times. Students sometimes don't take dimensional analysis too seriously because it seems trivial, but it's a very powerful method for catching algebraic errors.

A few physical quantities are **dimensionless** (they are also called "pure numbers"). For example, a geometrical angle is dimensionless; it expresses the circumference of a part of a circle divided by the circle's radius. Nevertheless, we sometimes use dimensionless units to describe such quantities. A dimensionless unit is just an abbreviation for some pure number. Thus the degree of angle, represented by the symbol °, denotes the number $2\pi/360$. From this point of view, the "radian" is nothing but the pure number 1 and may be dropped from formulas; we sometimes retain it just to emphasize that a particular quantity is an angle.

A quantity with dimensions is sometimes called **dimensional**. It's important to understand that the units are an integral part of such a quantity. Thus, when we use a named variable for a physical quantity, the units are part of what the name represents. For example, we don't say, "A force equal to f newtons" but rather, "A force equal to f" where, say, $f = 5\,\mathrm{N}$.

In fact, a dimensional quantity should be thought of as the *product* of a "numerical part" times some units; this viewpoint makes it clear that the numerical part depends on the units chosen. For example, the quantity 1 m is *equal to* the quantity 1000 mm. Similarly, the phrase "ten square micrometers," or "$10\,\mu\mathrm{m}^2$," refers to $10 \times (\mu\mathrm{m})^2 = 10^{-11}\,\mathrm{m}^2$, not $(10\,\mu\mathrm{m})^2 = 10^{-10}\,\mathrm{m}^2$.

To convert from one unit to another, we take any equivalence between units, for example 1 inch $= 2.54$ cm, and reexpress it as

$$\frac{1\,\text{inch}}{2.54\,\text{cm}} = 1.$$

Then, we take any expression and multiply or divide by 1, canceling the undesired units. For example, we can convert the acceleration of gravity to inch s^{-2} by writing

$$g = 9.8\frac{\cancel{\mathrm{m}}}{\mathrm{s}^2} \times \frac{100\,\cancel{\mathrm{cm}}}{\cancel{\mathrm{m}}} \times \frac{1\,\text{inch}}{2.54\,\cancel{\mathrm{cm}}} = 386\,\frac{\text{inch}}{\mathrm{s}^2}.$$

Finally, no dimensional quantity can be called "large" in any absolute sense. Thus, a speed of $1\,\mathrm{cm\,s}^{-1}$ may seem slow to you, but it's impossibly fast to a bacterium. In contrast, dimensionless quantities do have an absolute meaning: When we say that they are "large" or "small," we implicitly mean "compared with 1." Finding relevant dimensionless combinations of parameters is often a key step to classifying the qualitative properties of a system. Section 5.2 will illustrate this idea, defining the "Reynolds number" to classify fluid flows.

1.4.2 Dimensional analysis can help you catch errors and recall definitions

Isn't this a lot of pedantic fuss over something trivial? Not really. Things can get complicated pretty quickly; for example, on an exam. Training yourself to carry all the units explicitly, through *every* calculation, can save you from many errors.

Suppose you need to compute a force. You write down a formula that contains various quantities. To check your work, write down the dimensions of each of the quantities in your answer, cancel whatever cancels, and make sure the result is $\mathbb{MLT}^{-2}$. If it's not, you probably forgot to copy something from one step to the next. It's easy, and it's amazing how many errors you can find in this way. (You can also catch your instructors' errors.)

When you multiply two quantities, the dimensions just pile up: force $(\mathbb{MLT}^{-2})$ times length $(\mathbb{L})$ has dimensions of energy $(\mathbb{ML}^2\mathbb{T}^{-2})$. But you can *never* add or subtract terms with different dimensions in a valid equation, any more than you can add dollars to kilograms. You can add euros to rupees, with the appropriate conversion factor, and similarly meters to miles. Meters and miles are different *units* that both carry the same *dimension*, namely, length ($\mathbb{L}$).

Another useful rule of thumb involving dimensions is that *you can only take the exponential of a dimensionless number*. The same thing holds for other familiar functions, such as sin, cos, and ln. One way to understand this rule is to recall that $\exp x = 1 + x + \frac{1}{2}x^2 + \cdots$. According to the previous paragraph, this sum makes no sense unless x is dimensionless. (Recall also that the sine function's argument is an *angle*, and angles are dimensionless.)

Suppose you run into a new constant in a formula. For example, the force between two point charges q_1 and q_2 in vacuum, separated by distance r, is

$$f = \frac{1}{4\pi\varepsilon_0}\frac{q_1 q_2}{r^2}. \tag{1.8}$$

What are the dimensions of the constant ε_0? Just compare:

$$\mathbb{M}\mathbb{L}\mathbb{T}^{-2} = [\varepsilon_0]^{-1}\mathbb{Q}^2\mathbb{L}^{-2}.$$

In this formula, the notation $[\varepsilon_0]$ means "the dimensions of ε_0"; it's some combination of $\mathbb{L}, \mathbb{M}, \mathbb{T}, \mathbb{Q}$ that we want to find. Remember that numbers like 4π have no dimensions. (After all, π is the ratio of two lengths, the circumference and the diameter of a circle.) So right away, we find $[\varepsilon_0] = \mathbb{Q}^2\mathbb{T}^2\mathbb{L}^{-3}\mathbb{M}^{-1}$, which you can then use to check other formulas containing ε_0.

Finally, dimensional analysis helps you remember things. Suppose you're faced with an obscure SI unit, say, "farad" (abbreviated F). You don't remember its definition. You know it measures capacitance, and you have some formula involving it, say, $E = \frac{1}{2}q^2/C$, where E is the stored electrostatic energy, q is the stored charge, and C is the capacitance. Starting from the dimensions of energy and charge, you find that the *dimensions* of C are $[C] = \mathbb{T}^2\mathbb{Q}^2\mathbb{M}^{-1}\mathbb{L}^{-2}$. Substituting the SI *units* second, coulomb, kilogram, and meter, we find that the natural SI unit for capacitance is $s^2\text{coul}^2\text{kg}^{-1}\text{m}^{-2}$. That's what a farad really is.

Example: Appendix B lists the units of the permittivity of empty space ε_0 as F/m. Check this statement.

Solution: You could use Equation 1.8, but here's another way. The electrostatic potential $V(r)$ a distance r away from a point charge q is

$$V(r) = \frac{q}{4\pi\varepsilon_0 r}. \tag{1.9}$$

The potential energy of another charge q sitting at r equals $qV(r)$. Because we know the dimensions of energy, charge, and distance, we work out $[\varepsilon_0] = \mathbb{T}^2\mathbb{Q}^2\mathbb{M}^{-1}\mathbb{L}^{-3}$, as we already found. Also using what we found earlier for the dimensions of capacitance gives $[\varepsilon_0] = [C]/\mathbb{L}$, so the SI units for ε_0 are the same as those for capacitance per length, or F m^{-1}.

1.4.3 Dimensional analysis can also help you formulate hypotheses

Dimensional analysis has other uses. For example, it can actually help us to *guess new physical laws.*

Chapter 4 will discuss the "viscous friction coefficient" ζ for an object immersed in a fluid. This parameter equals the force applied to the object, divided by its resulting speed; so its dimensions are $\mathbb{M}/\mathbb{T}$. We will also discuss another quantity, the "diffusion constant" D of the same object, which has dimensions $\mathbb{L}^2/\mathbb{T}$. Both ζ and D depend in very complicated ways on the temperature, the shape and size of the object, and the nature of the fluid.

Suppose now that someone tells you that, despite this great complexity, the *product ζD* is very simple: This product depends only on the temperature, not on the nature of the object nor even on the kind of fluid it's in. What could the relation be? You work out the dimensions of the product to be $\mathbb{ML}^2/\mathbb{T}^2$. That's an *energy*. What sort of energy scales are relevant to our problem? It occurs to you that the energy of thermal motion, E_{thermal} (to be discussed in Chapter 3), is relevant to the physics of friction, because friction makes heat. So you could guess that if there is any fundamental relation, it must have the form

$$\zeta D \overset{?}{=} E_{\text{thermal}}. \tag{1.10}$$

You win. You have just guessed a true law of Nature, one that we will derive in Chapter 4. In this case, Albert Einstein got there ahead of you, but maybe next time you'll have priority. As we'll see, Einstein had a specific goal: By measuring both ζ and D experimentally, he realized, one could find E_{thermal}. We'll see how this gave Einstein a way to measure how big atoms are, without ever needing to manipulate them individually. And ... *atoms really are that size!*

What did we really accomplish here? This isn't the end, it's the beginning: We didn't find any *explanation* of frictional drag, nor of diffusion, yet. But we know a *lot* about how that theory should work. It has to give a relation that looks like Equation 1.10. This result helps in figuring out the real theory.

1.4.4 Some notational conventions involving flux and density

To illustrate how units help us disentangle related concepts, consider a family of related quantities that will be used throughout the book. (See Appendix A for a complete list of symbols used in the book.)

- We will often use the symbols N to denote the number of discrete things (a dimensionless integer), V to denote volume (with SI units m^3), and q to denote a quantity of electric charge (with SI unit coul).
- The rates of change of these quantities will generally be written dN/dt (with units s^{-1}), Q (the **volume flow rate**, with units $m^3\,s^{-1}$), and I (the **electric current**, with units $coul\,s^{-1}$), respectively.
- If we have five balls in a room of volume $1000\,m^3$, we say that the **number density** (or **concentration**) of balls in the room is $c = 0.005\,m^{-3}$. Densities of dimensional quantities are traditionally denoted by the symbol ρ; a subscript will indicate what

sort of quantity. Thus, **mass density** is ρ_m (units kg m^{-3}), whereas **charge density** is ρ_q (units coul m^{-3}).

- Similarly, if we have five checkers on a $1\,\text{m}^2$ checkerboard, the **surface number density** σ is $5\,\text{m}^{-2}$. Similarly, the **surface charge density** σ_q has units coul m^{-2}.
- Suppose we pour sugar down a funnel and 40 000 grains fall each second through an opening of area $1\,\text{cm}^2$. We say that the **number flux** (or simply "flux") of sugar grains through the opening is $j = (40\,000\,\text{s}^{-1})/(10^{-2}\,\text{m})^2 = 4 \cdot 10^8\,\text{m}^{-2}\,\text{s}^{-1}$. Similarly, the fluxes of dimensional quantities are again indicated by using subscripts; thus, j_q is the **electric charge flux** (with units $\text{coul m}^{-2}\,\text{s}^{-1}$) and so on.

If you accidentally use number density in a formula requiring mass density, you'll notice that your answer's units are missing a factor of kg; this discrepancy is your signal to go back and find your error.

1.5 OTHER KEY IDEAS FROM PHYSICS AND CHEMISTRY

Our story will rest on a number of other points known to the ancients.

1.5.1 Molecules are small

Ordinary molecules, like water, must be very small—we never perceive any grainy quality to water. But how small, exactly, are they? Once again we turn to Benjamin Franklin.

Around 1773, Franklin's attention turned to, of all things, oil slicks. What intrigued him was the fact that a certain quantity of oil could spread only so far on water. Attempting to spread it farther caused the film to break up into patches. Franklin noticed that a given quantity of olive oil always covered about the same area of water; specifically, he found that a teaspoon of oil ($\approx 5\,\text{cm}^3$) covered half an acre of pond ($\approx 2000\,\text{m}^2$). Franklin reasoned that if the oil were composed of tiny irreducible particles, then it could only spread until these particles formed a single layer, or "monolayer," on the surface of the water. It's easy to go one step further than Franklin did and find the thickness of the layer, and hence the size scale of a single molecule. Dividing the volume of oil by the area of the layer, we find the linear size of one oil molecule to be about 2.5 nm. Remarkably, Franklin's eighteenth-century experiment gives a reasonable estimate of the molecular size scale!

Because molecules are so tiny, we find ourselves discussing inconveniently big numbers when we talk about, say, a gram of water. Conversely, we also find ourselves discussing inconveniently small numbers when we try to express the energy of one molecule in human-size units like joules—see, for example, the constant in Equation 1.7. Chemists have found it easier to define, once and for all, one huge number expressing the smallness of molecules and then relate everything to this one number. That number is **Avogadro's number** N_{mole}, defined as the number of carbon atoms needed to make up twelve grams of (ordinary) carbon. Thus, N_{mole} is also roughly the number of hydrogen atoms in one gram of hydrogen, because a carbon atom has a mass about 12 times that of hydrogen. Similarly, there are roughly N_{mole} oxygen molecules, O_2, in 32 g of oxygen, because each oxygen atom's mass is about 16 times that of a hydrogen atom and each oxygen molecule consists of two of them.

Note that N_{mole} is dimensionless.[5] Any collection of N_{mole} molecules is called a **mole** of that type of molecule. In our formulas, the word *mole* will simply be a synonym for the number N_{mole}, just as the word *million* can be thought of as a synonym for the number 10^6.

Returning to Franklin's estimate, suppose water molecules are similar to oil molecules, roughly tiny cubes 2.5 nm on a side.[6] Let's see what we can deduce from this observation.

Example: Find an estimate for Avogadro's number starting from this size.
Solution: We won't get lost if we carry all the dimensions along throughout the calculation. One cubic meter of water contains

$$\frac{1 \text{ m}^3}{(2.5 \cdot 10^{-9} \text{ m})^3} = 6.4 \cdot 10^{25}$$

molecules. That same cubic meter of water has a mass of a thousand kilograms, because the density of water is 1 g cm^{-3} and

$$1 \text{ m}^3 \times \left(\frac{100 \text{ cm}}{1 \text{ m}}\right)^3 \times \frac{1 \text{ g}}{1 \text{ cm}^3} \times \frac{1 \text{ kg}}{1000 \text{ g}} = 1000 \text{ kg}.$$

We want to know how many molecules of water make up a mole. Because each water molecule consists of one oxygen and two hydrogen atoms, its total mass is about $16 + 1 + 1 = 18$ times that of a single hydrogen atom. So we must ask, if $6.4 \cdot 10^{25}$ molecules have mass 1000 kg, then how many molecules does it take to make 18 g, or 0.018 kg?

$$N_{\text{mole}} = 0.018 \text{ kg} \times \frac{6.4 \cdot 10^{25}}{1000 \text{ kg}} = 0.011 \cdot 10^{23}. \qquad \text{(estimate)}$$

The estimate for Avogadro's number just found is not very accurate (the modern value is $N_{\text{mole}} = 6.0 \cdot 10^{23}$). But it's amazingly good, considering that the data on which it is based were taken nearly a quarter of a millennium ago. Improving on this estimate, and hence nailing down the precise dimensions of atoms, proved surprisingly difficult. Chapter 4 will show how the dogged pursuit of this quarry led Albert Einstein to a key advance in our understanding of the nature of heat.

Your Turn 1A

Using the modern value of Avogadro's number, turn the above calculation around and find the volume occupied by a single water molecule.

---

[5] $\boxed{T_2}$ See Section 1.5.4′ on page 30 for more about notational conventions.
[6] Really they're more like slender *rods*. The cube of the length of such a rod is an overestimate of its volume, so our estimate here is rough.

1.5.2 Molecules are particular spatial arrangements of atoms

There are only about a hundred kinds of atoms. Every atom of a given element is exactly like every other: Atoms have no individual personalities. For example, every atom of (ordinary) hydrogen has the same mass as every other one. The mass of N_{mole} atoms of a particular species is called that atom's **molar mass**.

Similarly, every molecule of a given chemical compound has a fixed, definite composition, a rule we attribute to J. Dalton and J. Gay-Lussac. For example, carbon dioxide always consists of exactly two oxygen atoms and one carbon, in a fixed spatial relationship. Every CO_2 molecule is like every other, for example, equally ready or unwilling to undergo a given chemical change.

There may be more than one allowed arrangement for a given set of atoms, yielding two or more chemically distinct molecules called **isomers**. Some molecules flip back and forth rapidly between their isomeric states: They are "labile." Others do so very rarely: They are rigid. For example, Louis Pasteur discovered in 1857 that two sugars containing the same atoms, but in mirror-image arrangements, are chemically different and essentially never spontaneously interconvert (Figure 1.5). A molecule

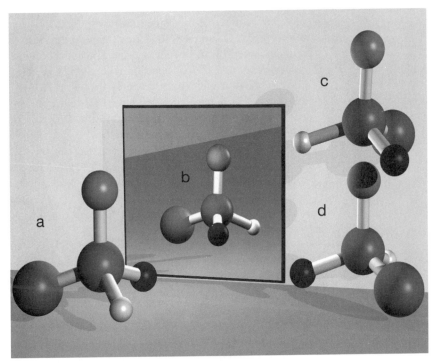

Figure 1.5: (Molecular structure sketches.) (a) The molecule shown is chiral. (b) To show this property, this panel shows the mirror image of (a). (c,d) No rotated version of (a) coincides with its mirror image (b), even though (b) has the same atoms, bonds, and bond angles as (a). However, if the original molecule had had two identical groups (for example, two white groups in place of one white and one black), then the molecule would have been nonchiral: (b) would then coincide with (a).

whose mirror image is an inequivalent stereoisomer is called chiral; such molecules will play a key role in Chapter 9.

T_2 *Section 1.5.2′ on page 30 discusses the division of elements into isotopes.*

1.5.3 Molecules have well-defined internal energies

Section 1.1.2 briefly alluded to the chemical energy stored in a match. Indeed, the atoms making up a molecule carry a definite amount of stored energy, which is said to reside in chemical bonds between the atoms. The chemical bond energy drives toward lower values just as any other form of stored energy does (for example, the potential energy of the weight in Figure 1.3). In fact, the chemical bond energy is just another contribution to the quantity E appearing in the formula for free energy $F = E - TS$ (Equation 1.4). Molecules generally prefer to indulge in heat-liberating (**exothermic**) reactions rather than heat-accepting (**endothermic**) ones, but we can nevertheless get them to adopt higher energy states by adding energy from outside. For example, we can split (or **hydrolyze**) water by passing electric current through it. More precisely, Chapter 8 will show that *chemical reactions proceed in the direction that tends to lower the free energy,* just as in the osmotic machine.

Even an unstable molecule may not spontaneously split up until a large "activation energy" is supplied; thus for example an explosive stores its chemical energy until detonated by an outside agency. The activation energy can be delivered to a molecule mechanically, by collision with a neighbor. But this is not the only possibility. In one of his five historic papers written in 1905, Albert Einstein showed that *light, too, comes in packets of definite energy,* called **photons**. A molecule can absorb such a packet and then hop over its activation energy barrier, perhaps even ending in a higher energy state than its initial state.

The explanations for all the familiar facts in this subsection and the previous one come from a branch of physics called "quantum mechanics." Quantum mechanics also explains the numerical values of the typical atomic sizes and bond energies in terms of a fundamental physical constant, the Planck constant $\hbar$. This book will take all these values simply as experimentally determined facts, sidestepping their quantum origins altogether.

How can there be a "typical" bond energy? Don't some reactions (say, in a stick of dynamite) liberate a lot more energy than others (burning a match)? No, the dynamite just liberates its energy much *faster*; the energy liberated per chemical bond is roughly comparable to that liberated in any other reaction.

Example: One important chemical reaction is the one happening inside the batteries in your channel changer. Estimate the chemical energy released in this reaction.
Solution: Printed on the battery, we find that its terminals differ in potential by $\Delta V = 1.5$ volt. This statement means that the battery imparts an energy of roughly $e\Delta V = 1.6 \cdot 10^{-19}$ coul $\times$ 1.5 volt $= 2.4 \cdot 10^{-19}$ J to each electron passing through it. (The value of the fundamental charge e is listed in Appendix B.) If we suppose that each electron passing across the battery enables the chemical reaction inside to take one step, then the energy just calculated is the change in chemical bond energies (minus any thermal energy given off).

In contrast to chemical reactions, the radioactive decay of plutonium liberates about a million times more energy per atom than the value just found. Historically, this discovery was the first solid clue that something very different from chemistry was going on in radioactive decay.

1.5.4 Low-density gases obey a universal law

The founders of chemistry arrived at the idea that atoms combine in definite proportions by noticing that gases combine in simple, fixed ratios of volume. Eventually it became clear that this observation reflects the fact that the *number* of gas molecules in a box at atmospheric pressure is just proportional to its volume. More precisely, one finds experimentally that the pressure p, volume V, number of molecules N, and temperature T of any gas (at low enough density) are related in a simple way called the **ideal gas law**:

$$pV = Nk_{\mathrm{B}}T. \qquad (1.11)$$

Here the temperature T is understood to be measured relative to a special point called **absolute zero**; other equations in this book, such as Equation 1.4, also use T measured from this point. In contrast, the Celsius scale assigns $0°\mathrm{C}$ to the freezing point of water, which is $273°\mathrm{C}$ above absolute zero. Thus, room temperature T_{r} corresponds to about 295 degrees above absolute zero (Section 6.3.2 will define temperature more carefully). The quantity k_{B} appearing in Equation 1.11 is called the **Boltzmann constant**; it turns out to be about $1.38 \cdot 10^{-23}$ joules per degree. Thus, the numerical value of $k_{\mathrm{B}}T$ at room temperature is $k_{\mathrm{B}}T_{\mathrm{r}} = 4.1 \cdot 10^{-21}$ J. A less cumbersome way of quoting this value, and an easier way to memorize it, is to express it in units relevant to cellular physics (piconewtons and nanometers):

$$k_{\mathrm{B}}T_{\mathrm{r}} \approx 4.1 \; \mathrm{pN \, nm}. \qquad \text{(most important formula in this book)} \qquad (1.12)$$

Take a minute to think about the reasonableness of Equation 1.11: If we pump in more gas (N increases), the pressure goes up. Similarly, if we squeeze the box (V decreases) or heat it up (T increases), p again increases. The detailed form of Equation 1.11 may look unfamiliar, however. Chemistry texts generally write it as $pV = nRT$, where n is the "amount of substance" (number of moles) and RT is about 2500 joules per mole at room temperature. Dividing 2500 J by N_{mole} indeed gives the quantity $k_{\mathrm{B}}T_{\mathrm{r}}$ in Equation 1.12.

The remarkable thing about Equation 1.11 is that it holds *universally*: Any gas, from hydrogen to vaporized steel, obeys it (at low enough density). All gases (and even mixtures of gases) have the *same numerical value* of the constant k_{B} and all agree about the value of absolute zero. In fact, even the osmotic work formula, Equation 1.7, involves this same quantity! Physical scientists sit up and take notice when a law or a constant of Nature proves to be universal (Section 1.3). Accordingly, our first order of business in Part II of this book will be to tease out the deep meaning of Equation 1.11 and its constant k_{B}.

$\boxed{T_2}$ *Section 1.5.4' on page 30 makes more precise this book's use of the word mole and relates it to other books' usage.*

THE BIG PICTURE

Let's return to this chapter's Focus Question. Section 1.2 discussed the idea that the flow of energy, together with its degradation from mechanical to thermal energy, could create order. We saw this principle at work in a humble process (reverse osmosis, Section 1.2.2), then claimed that life, too, exploits this loophole in the Second Law of thermodynamics to create—or rather, capture—order. Our job in the following chapters will be to work out a few of the details. For example, Chapter 5 will describe how tiny organisms, even single bacteria, carry out purposeful motion in search of food, enhancing their survival, despite the randomizing effect of their surroundings. We will need to expand and formalize our ideas in Chapters 6–8. Chapter 8 will then consider the self-assembly of compound molecular structures. Finally, Chapters 10–12 will discuss how two paragons of orderly behavior—namely, the motion of molecular machines and nerve impulses—emerge from the disorderly world of single-molecule dynamics.

Before attempting any of these tasks, however, we should pause to appreciate the sheer immensity of the biological order puzzle. Accordingly, the next chapter will give a tour of some of the extraordinarily ordered structures and processes present even in single cells. Along the way, we will meet many of the devices and interactions to be discussed in later chapters.

KEY FORMULAS

Each chapter of Parts II and III of this book ends with a summary of the key formulas appearing in that chapter. The list below is slightly different; it focuses mainly on formulas from first-year physics that will be used throughout the book. You may want to review these, referring to an introductory physics text.

1. *First-year physics:* Make sure you recall these formulas from first-year physics, and what all their symbols mean. Most of these have not been used yet, but they will appear in the coming chapters.
   momentum = (mass) × (velocity).
   centripetal acceleration in uniform circular motion = $r\omega^2$.
   force = rate of transfer of momentum.
   torque = (moment arm) × (force).
   work = transferred mechanical energy = (force) × (distance) = (torque) × (angle).
   pressure = (force)/(area).
   kinetic energy = $\frac{1}{2}mv^2$.
   force and potential energy of a spring, $f = -kx$, $E = \frac{1}{2}kx^2$.

potential energy in Earth's gravity = (mass) $\times$ g $\times$ (height).

potential energy of a charged object in an electrostatic field = qV.

electric field, $\mathcal{E} = -dV/dx$.

force on a charged body, $f = q\mathcal{E}$.

electrostatic potential created by a single point charge q in an infinite, uniform, insulating medium, $V(\mathbf{r}) = q/(4\pi\varepsilon|\mathbf{r}|)$, where ε is the permittivity of the medium.

electrostatic self-energy of a charged sphere of radius a, $q^2/(8\pi\varepsilon a)$.

Ohm's law, $V = IR$; power loss from a resistor, I^2R.

electrostatic potential drop across a capacitor, $V = q/C$.

electrostatic potential energy stored in a capacitor, $E = \frac{1}{2}q^2/C$.

capacitance of a parallel-plate capacitor of area A and thickness d, $C = A\varepsilon/d$.

2. *Mechanical equivalent of heat:* One joule of mechanical energy, when completely converted to heat, can raise the temperature of $1\,\mathrm{g}$ of water by about $0.24\,^\circ\mathrm{C}$ (Equation 1.2).

3. *Ideal gas:* The pressure, volume, number of molecules, and temperature of a confined ideal gas are related by $pV = Nk_{\mathrm{B}}T$ (Equation 1.11). At room temperature T_{r}, the quantity $k_{\mathrm{B}}T_{\mathrm{r}} \approx 4.1\,\mathrm{pN\,nm}$ (Equation 1.12).

FURTHER READING

Semipopular:

Heat: von Baeyer, 1999; Segrè, 2002.

The Second Law: Atkins, 1994.

Franklin's oil experiment: Tanford, 1989.

Intermediate:

Biophysics, and general physics with biological applications: Benedek & Villars, 2000a; Benedek & Villars, 2000b; Benedek & Villars, 2000c; Hobbie, 1997; Cotterill, 2002; Vogel, 2003.

Technical:

The Biophysical Society's On-Line Textbook: http://www.biophysics.org/btol/

$\boxed{T_2}$ **1.5.2′ Track 2**

There is an important elaboration of the rule that atoms of a given species are all identical. Atoms that behave identically *chemically* may nevertheless subdivide into a few distinct classes of slightly different mass, the "isotopes" of that chemical element. Thus, we specified *ordinary* hydrogen in Section 1.5.2 on page 25 to acknowledge the existence of two other, heavier forms (deuterium and tritium). Despite this complication, however, there are only a handful of different stable isotopes of each element, so the number of distinct species is still small, a few hundred. The key point is that the distinction between them is discrete, not continuous.

$\boxed{T_2}$ **1.5.4′ Track 2**

Physics textbooks generally use molecular quantities, whereas chemistry textbooks generally use the corresponding molar versions. Like most artificial barriers to friendship, this one is easily overcome. The SI gives "amount of substance" its own dimension, with a corresponding fundamental unit called mol. *This book will not use any quantities containing this unit.* Thus, we will not measure amounts by using the quantity n, with units mol, nor will we use the quantities $RT_r = 2470 \, \mathrm{J \, mol^{-1}}$ or $\mathcal{F} = 96\,000 \, \mathrm{coul \, mol^{-1}}$; instead, we will use respectively the number of molecules N, the molecular thermal energy, $k_B T_r$, and the charge on one proton, e. Similarly, we will not use the quantity $N_0 = 6.0 \cdot 10^{23} \, \mathrm{mol^{-1}}$; our N_{mole} is the dimensionless number $6.0 \cdot 10^{23}$. And we don't use the unit dalton, defined as $1 \, \mathrm{g \, mol^{-1}}$; instead, we measure masses in kilograms.

A more serious notational problem arises from the fact that different books use the same symbol μ (the "chemical potential" defined in Chapter 8) to mean two slightly different things: μ can represent the derivative of the free energy either with respect to n (so $[\mu] \sim \mathrm{J \, mol^{-1}}$), or with respect to N (so $[\mu] \sim \mathrm{J}$). *This book always uses the second convention* (see Chapter 8). We choose this convention because we will frequently want to study *single molecules*, not mole-sized batches.[7]

In this book the word *mole* in formulas is just an abbreviation for the number N_{mole}. When convenient, we can express molecular energies as multiples of mole^{-1}; then the numerical part of our quantities just equals the numerical part of the corresponding molar quantities. For example, we can write

$$k_B T_r = 4.1 \cdot 10^{-21} \, \mathrm{J} \times \frac{6.0 \cdot 10^{23}}{\mathrm{mole}} \approx 2500 \, \mathrm{J/mole},$$

whose numerical part agrees with that of RT_r.

---

[7]Similar remarks apply to the standard free energy change ΔG.

PROBLEMS

1.1   Dorm-room dynamics

a. An air conditioner cools down your room, removing thermal energy. Yet it *consumes* electrical energy. Is there a contradiction with the First Law?

b. Could you design a high-tech device that sits in your window, continuously converting the unwanted thermal energy in your room to electricity, which you then sell to the power company? Explain.

1.2   Thompson's experiment

Long ago, people did not use SI units.

a. Benjamin Thompson actually said that his cannon-boring apparatus could bring 25.5 pounds of cold water to the boiling point in 2.5 hours. Supposing that "cold" water is at 20 °C, find the power input into the system by his horses, in watts. [*Hint:* A kilogram of water weighs 2.2 pounds. That is, Earth's gravity pulls it with a force of $1 \text{ kg} \times g = 2.2$ pound.]

b. James Joule actually found that 1 pound of water increases in temperature by one degree Fahrenheit (or 0.56 °C) after he input 770 foot pounds of work. How close was he to the modern value of the mechanical equivalent of heat?

1.3   Metabolism

Metabolism is a generic term for all of the chemical reactions that break down and "burn" food, thereby releasing energy. Here are some data for metabolism and gas exchange in humans.

food	kcal/g	liters O_2/g	liters CO_2/g
carbohydrate	4.1	0.81	0.81
fat	9.3	1.96	1.39
protein	4.0	0.94	0.75
alcohol	7.1	1.46	0.97

The table gives the energy released, the oxygen consumed, and the carbon dioxide released upon metabolizing the given food, per gram of food.

a. Calculate the energy yield per liter of oxygen consumed for each food type and note that it is roughly constant. Thus, we can determine a person's metabolic rate simply by measuring her rate of oxygen consumption. In contrast, the CO_2/O_2 ratios are different for the different food groups; this circumstance allows us to estimate what is actually being used as the energy source, by comparing oxygen intake to carbon dioxide output.

b. An average adult at rest uses about 16 liters of O_2 per hour. The corresponding heat release is called the "basal metabolic rate" (BMR). Find it, in kcal/hour and in kcal/day.

c. What power output does this correspond to in watts?

d. Typically, the CO_2 output rate might be 13.4 liters per hour. What, if anything, can you say about the type of food materials being consumed?

e. During exercise, the metabolic rate increases. Someone performing hard labor for 10 hours a day might need about 3500 kcal of food per day. Suppose the person does mechanical work at a steady rate of 50 W over 10 hours. We can define the body's efficiency as the ratio of mechanical work done to excess energy intake (beyond the BMR calculated in (b)). Find this efficiency.

1.4 Earth's temperature

The Sun emits energy at a rate of about $3.9 \cdot 10^{26}$ W. At Earth, this sunshine gives an incident energy flux I_e of about $1.4 \, \mathrm{kW \, m^{-2}}$. In this problem, you'll investigate whether any other planets in our solar system could support the sort of water-based life we find on Earth.

Consider a planet orbiting at distance d from the Sun (and let d_e be Earth's distance). The Sun's energy flux at distance d is $I = I_e(d_e/d)^2$, because energy flux decreases as the inverse square of distance. Call the planet's radius R, and suppose that it absorbs a fraction α of the incident sunlight, reflecting the rest back into space. The planet intercepts a disk of sunlight of area πR^2, so it absorbs a total power of $\pi R^2 \alpha I$. Earth's radius is about 6400 km.

The Sun has been shining for a long time, but Earth's temperature is roughly stable: The planet is in a steady state. For this to happen, *the absorbed solar energy must get reradiated back to space as fast as it arrives* (see Figure 1.2). Because the rate at which a body radiates heat depends on its temperature, we can find the expected mean temperature of the planet, using the formula

$$\text{radiated heat flux} = \alpha \sigma T^4.$$

In this formula, σ denotes the number $5.7 \cdot 10^{-8} \, \mathrm{W \, m^{-2} \, K^{-4}}$ (the "Stefan–Boltzmann constant"). The formula gives the rate of energy loss per unit area of the radiating body (here, the Earth). You needn't understand the derivation of this formula but make sure you do understand how the units work.

a. Using this formula, work out the average temperature at the Earth's surface and compare your answer to the actual value of 289 K.

b. Using the formula, work out how far from the Sun a planet the size of Earth may be, as a multiple of d_e, and still have a mean temperature greater than freezing.

c. Using the formula, work out how close to the Sun a planet the size of Earth may be, as a multiple of d_e, and still have a mean temperature below boiling.

d. *Optional:* If you know the planets' orbital radii, which ones are then candidates for water-based life, using this rather oversimplified criterion?

1.5 Franklin's estimate

The estimate of Avogadro's number in Section 1.5.1 came out too small partly because we used the molar mass of water, not of oil. We can look up the molar mass and mass density of some sort of oil available in the eighteenth century in the *Handbook of chemistry and physics* (Lide, 2001). The *Handbook* tells us that the principal component of olive oil is oleic acid and gives the molar mass of oleic acid (also known as 9-octadecenoic acid or $CH_3(CH_2)_7CH{=}CH(CH_2)_7COOH$) as 282 g mole^{-1}. We'll see in Chapter 2 that oils and other fats are triglycerides, made up of three fatty acid chains, so we estimate the molar mass of olive oil as a bit more than three times the value for oleic acid. The *Handbook* also gives the density of olive oil as $0.9 \, \text{g cm}^{-3}$.

Make an improved estimate of N_{mole} from these facts and Franklin's original observation.

1.6   *Atomic sizes, again*

In 1858, J. Waterston found a clever way to estimate molecular sizes from macro-scopic properties of a liquid, by comparing its surface tension and heat of vaporiza-tion.

The surface tension of water, Σ, is the work per unit area needed to create more free surface. To define it, imagine breaking a brick in half. The two pieces have two new surfaces. Let Σ be the work needed to create these new surfaces, divided by their total area. The analogous quantity for liquid water is the surface tension.

The **heat of vaporization** of water, Q_{vap}, is the energy per unit volume we must add to liquid water (just below its boiling point) to convert it completely to steam (just above its boiling point). That is, the heat of vaporization is the energy needed to separate every molecule from every other one.

Picture a liquid as a cubic array with N molecules per centimeter in each of three directions. Each molecule has weak attractive forces to its six nearest neighbors. Sup-pose it takes energy ϵ to break one of these bonds. Then the complete vaporization of $1 \, \text{cm}^3$ of liquid requires that we break all the bonds. The corresponding energy cost is $Q_{vap} \times (1 \, \text{cm}^3)$.

Next consider a molecule on the *surface* of the fluid. It has only five bonds—the nearest neighbor on the top is missing (suppose this is a fluid–vacuum interface). Draw a picture to help you visualize this situation. Thus, to create more surface area requires that we break some bonds. The energy needed to do that, divided by the new area created, is Σ.

a. For water, $Q_{vap} = 2.3 \cdot 10^9 \, \text{J m}^{-3}$ and $\Sigma = 0.072 \, \text{J m}^{-2}$. Estimate N.

b. Assuming the molecules are closely packed, estimate the approximate molecule diameter.

c. What estimate for Avogadro's number do you get?

1.7   *Tour de France*

A bicycle rider in the Tour de France eats a lot. If his total daily food intake were burned, it would liberate about 8000 kcal of heat. Over the three or four weeks of the race, his weight change is negligible, less than 1%. Thus, his energy input and output must balance.

Let's first look at the mechanical work done by the racer. A bicycle is incredibly efficient. The energy lost to internal friction, even including the tires, is negligible. The expenditure against air drag is, however, significant, amounting to 10 MJ per day. Each day, the rider races for 6 hours.

a. Compare the 8000 kcal input to the 10 MJ of work done. Something's missing! Could the missing energy be accounted for by the altitude change in a hard day's racing?

Regardless of how you answered (a), next suppose that on one particular day of racing there's no net altitude change, so that we must look elsewhere to see where the missing energy went. We have so far neglected another part of the energy equation: the rider gives off *heat*. Some of this is radiated. Some goes to warm up the air he breathes in. But by far the greatest share goes somewhere else.

The rider *drinks a lot of water*. He doesn't need this water for his metabolism—he is actually creating water when he burns food. Instead, nearly all that liquid water leaves his body as water *vapor*. The thermal energy needed to vaporize water appeared in Problem 1.6.

b. How much water would the rider have to drink for the energy budget to balance? Is this reasonable?

Next let's go back to the 10 MJ of mechanical work done by the rider each day.

c. The wind drag for a situation like this is a backward force of magnitude $f = Bv^2$, were B is some constant. We measure B in a wind-tunnel to be $1.5\,\mathrm{kg\,m^{-1}}$. If we simplify by supposing a day's racing to be at constant speed, what is that speed? Is your answer reasonable?

CHAPTER 2

What's Inside Cells

Architecture is the learned game, correct and magnificent, of forms assembled in the light.
— Le Corbusier, 1887–1965

Chapter 1 exposed an apparent incompatibility between physical law and the living world (the apparently spontaneous generation of order by living things) and proposed the outline of a reconciliation (living things ingest high-quality energy and give off low-quality energy). With this physical backdrop, we're now ready to look a bit more closely into the organization of a living cell, where the same ideas play out over and over. This chapter sketches the *context* for the various phenomena that will concern us in the rest of the book:

- Each device we will study is a physical object; its *spatial* context involves its location in the cell relative to the other objects.
- Each device also participates in some processes; its *logical* context involves its role in these processes relative to other devices.

Certainly this introductory chapter can only scratch the surface of this vast topic.[1] But it is useful to collect some visual images of the main characters in our story, so that you can flip back to them as they appear in later chapters. Figures 2.1–2.4 give an overall sense of the relative sizes of the objects we'll be studying.

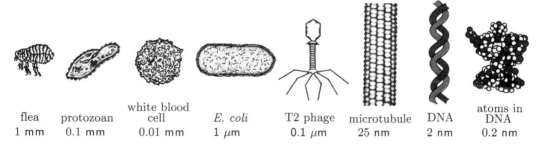

flea	protozoan	white blood cell	*E. coli*	T2 phage	microtubule	DNA	atoms in DNA
1 mm	0.1 mm	0.01 mm	1 μm	0.1 μm	25 nm	2 nm	0.2 nm

Figure 2.1: (Icons.) *Dramatis personæ.* Approximate relative sizes of some of the actors in our story. T2 phage is a virus that infects bacteria, for example, *Escherichia coli.* Much of this book will be occupied with phenomena relevant at length scales from the protozoan down to the DNA helix. [Adapted from Kornberg, 1989.]

---

[1]If you're not familiar with the vocabulary of this chapter, you will probably want to supplement it by reading the opening chapters of any cell biology book; see for example the list at the end of this chapter.

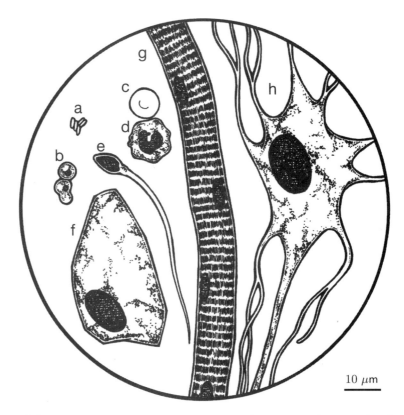

Figure 2.2: (Drawing, based on light microscopy.) Relative sizes. (a) Five *Escherichia coli* bacteria cells (enlarged in Figure 2.3). (b) Two cells of baker's yeast. (c) Human red blood cell. (d) Human white blood cell (lymphocyte). (e) Human sperm cell. (f) Human epidermal (skin) cell. (g) Human striated muscle cell (myofibril). (h) Human neuron (nerve cell). [From Goodsell, 1993.]

This chapter has a very different flavor from the others. For one thing, there will be no formulas. Most of the assertions will appear with no attempt to justify them. Most of the figures have detailed captions, whose meaning may not be clear to you until we study them in detail in a later chapter. Don't worry about this. Right now, your goal should be to finish this chapter knowing a lot of the vocabulary we will use later. You should also come away with a general feeling for the *hierarchy of scales* in a cell and a sense of how the governing principles at each scale emerge from, but have a character different from, those at the next deeper scale.

Finally, the exquisite structures on the following pages practically beg us to ask: How can a cell keep track of everything, when there's nobody in there running the factory? This question has a very long answer, of course. Among the many physical ideas relevant to this question, however, three will dominate this chapter and the rest of the book:

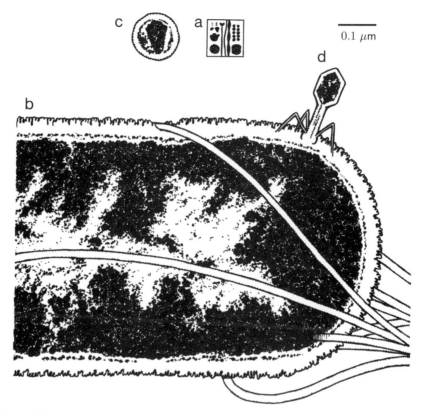

Figure 2.3: (Drawing, based on electron microscopy.) Relative sizes. (a) Several molecules and macromolecules (enlarged in Figure 2.4). (b) A bacterial cell (see Figures 2.1 and 2.2a). Visible structures include flagella (trailing to the right), the nucleoid (white region in center), and the thick, rigid cell wall. The flagella propel the bacterium by a mechanism discussed in Chapter 5; they are, in turn, driven by motors discussed in Chapter 11. (c) Human immunod-eficiency virus. (d) A bacterial virus, or phage. [From Goodsell, 1993.]

Biological question: How do cells organize their myriad ongoing chemical processes and reactants?

Physical ideas: a. Bilayer membranes *self-assemble* from their component molecules; the cell uses them to partition itself into separate compartments. b. Cells use *active transport* to bring synthesized materials to particular destinations. c. Biochemical processes are highly *specific:* Most are mediated by enzymes, which select one par-ticular target molecule and leave the rest alone.

2.1   CELL PHYSIOLOGY

Roadmap Section 2.1 will begin our story by recalling some of the characteristic activities of living cells, then turn to their overall structural features. The physical

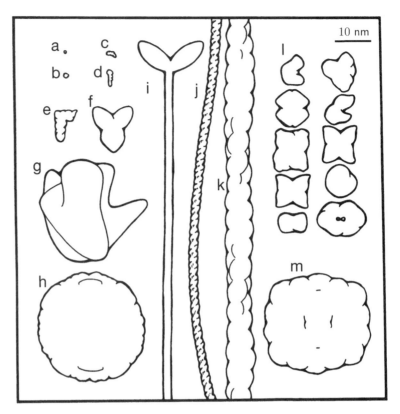

Figure 2.4: (Drawing, based on structural data.) Relative sizes of the objects shown in panel (a) of Figure 2.3. (a) Single carbon atom. (b) Glucose, a simple sugar molecule. (c) ATP, a nucleotide. (d) Chlorophyll molecule. (e) Transfer RNA, or tRNA. (f) An antibody, a protein used by the immune system. (g) The ribosome, a complex of protein and RNA. (h) The virus responsible for polio. (i) Myosin, a molecular machine discussed in Chapter 10. (j) DNA, a nucleic acid. Chapter 9 will discuss the mechanical properties of long molecules like this one. (k) F-actin, a cytoskeletal element. (l) Ten enzymes (protein machines) involved in glycolysis, which is a series of coupled chemical reactions that produce ATP, the energy currency molecule, from glucose. Chapter 11 will discuss ATP production. (m) Pyruvate dehydrogenase, a large enzyme complex also discussed in Chapter 11. [From Goodsell, 1993.]

aspects of cell function and structure are sometimes called cell physiology. Section 2.2 will turn to the ultimate molecular constituents of cells, progressively building from the smallest to the largest. By this point, we will have a beautiful, but static, picture of the cell as a collection of architectural elements. To close the circle of logic, we'll need to understand something about how these elements get constructed and, more generally, how the cell's other *activities* come about. Thus, Section 2.3 will introduce the world of molecular devices. *This third aspect of cells is the primary focus of this book,* although along the way, we will touch on the others, and even occasionally go beyond cells to organisms.

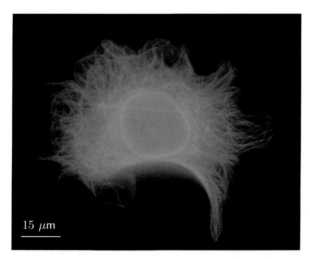

15 μm

Color Figure 1: (Fluorescence micrograph.) Newt lung cell in which the DNA is stained blue and microtubules in the cytoplasm are stained green. This network of rigid cytoskeletal filaments helps maintain the cell's required shape as well as supplying the tracks along which kinesin and other motors walk. Chapter 10 will discuss these motors.

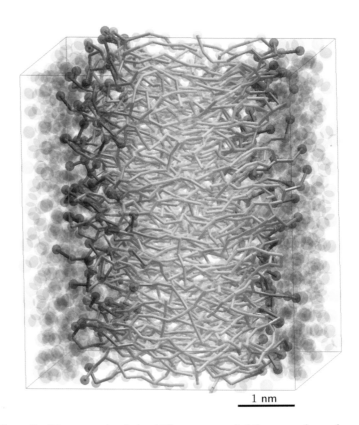

1 nm

Color Figure 2: (Computer simulation.) The structure of a bilayer membrane formed by the self-assembly of phospholipid molecules. Imagine repeating the arrangement of molecules upward and downward on the page, and into and out of the page, to form a double layer. The phospholipid molecules are free to move about in each layer, but they remain oriented with their polar head groups (*red*) facing outward, toward the surrounding water (*blue*), and their nonpolar hydrocarbon tails (*yellow*) pointing inward. Chapter 8 will discuss the self-assembly of structures like these. For computational simplicity the molecules have been simplified: Each yellow segment represents four carbon atoms in the real molecule. [Digital image kindly supplied by S. Nielsen; see Nielsen & Klein, 2002.]

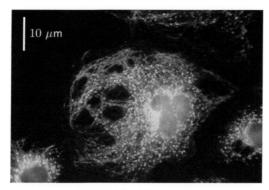

Color Figure 3: (Fluorescence optical micrograph.) Experimental demonstration that kinesin and microtubules are found in the same places within cells. This cell has been doubly labeled with fluorescent antibodies labeling both kinesin (*yellow*) and tubulin (*green*). The kinesin, attached to transport vesicles, is mostly associated with the microtubule network, as seen from the orange color where fluorescence from the two kinds of antibodies overlap. [Digital image kindly supplied by S. T. Brady; see Brady & Pfister, 1991.]

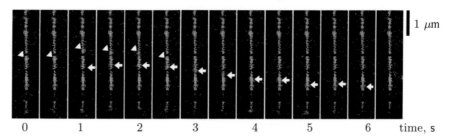

Color Figure 4: (Video photomicrograph frames.) Motility assay of the fluorescently labeled molecular motor C351, a single-headed member of the kinesin family. A solution of C351 with concentration between 1–10 pm was washed over a set of microtubules fixed to a glass slide. The microtubules were also fluorescently labeled; one of them is shown here (*green*). The motors (*red*) attached to the microtubule, moved along it for several seconds, then detached and wandered away. Two individual motors have been chosen for study; their successive locations are marked by *triangles* and *arrows*, respectively. Generally the motors moved strictly in one direction, but backward stepping was also observed (*triangles*), in contrast to ordinary, two-headed kinesin. [From Okada & Hirokawa, 1999.]

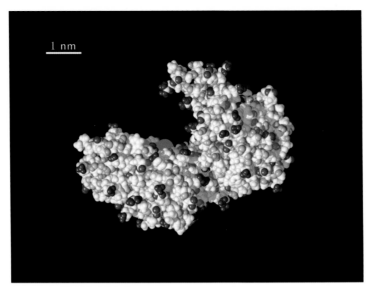

Color Figure 5: (Structure rendered from atomic coordinates.) Phosphoglycerate kinase. This enzyme performs one of the steps in the glycolysis reaction; see Section 10.4. In this figure and Color Figure 6, hydrophobic carbon atoms are white, mildly hydrophilic atoms are pastel (light blue for nitrogen and pink for oxygen), and strongly hydrophilic atoms carrying a full electric charge are brightly colored (blue for nitrogen and red for oxygen). The concept of hydrophobicity and the behavior of electrostatic charges in solution are discussed in Chapter 7. Sulfur and phosphorus atoms are colored yellow. Hydrogen atoms are colored according to the atom to which they are bonded. The enzyme manufactures one ATP molecule (*green object*) with each cycle of its action. [Digital image kindly supplied by D. Goodsell; see Goodsell, 1993.]

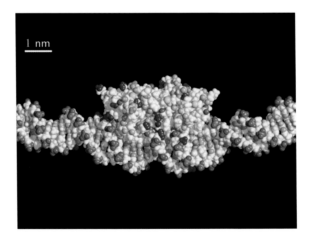

Color Figure 6: (Composite of structures rendered from atomic coordinates.) A DNA-binding protein. The color scheme is the same as Color Figure 5. Repressor proteins like this one bind directly to the DNA double helix, physically blocking the polymerase that makes messenger RNA. They recognize a specific sequence of DNA, generally blocking a region of 10–20 basepairs. The binding does not involve the formation of chemical bonds; instead it uses the weaker interactions discussed in Chapter 7. Repressors form a molecular switch, turning off the synthesis of a given protein until it is needed. [Digital image kindly supplied by D. Goodsell; see Goodsell, 1993.]

Cells are the fundamental functional units of life. Whether alone or integrated into communities (organisms), individual cells perform a common set of activities. Even though a particular cell may not do everything on the following list—there are a couple hundred distinct, specialized cell types in our bodies, for example—still there is enough overlap between all cells to make it clear that all are basically similar.

- Like entire organisms, individual cells take in chemical or solar energy. As discussed in Chapter 1, most of this energy gets discarded as heat, but a fraction turns into useful mechanical activity or the synthesis of other energy-storing molecules, via a set of processes collectively called metabolism. Chapter 11 will examine one aspect of this remarkably efficient free energy transduction process.

- In particular, each cell manufactures more of its own internal structure in order to grow. Much of this structure consists of a versatile class of macromolecules called proteins. Our bodies contain about 100 000 different protein types. We will return many times to the interactions responsible for protein structure and function.

- Most cells can reproduce by mitosis, a process of duplicating their contents and splitting in two. (One cell type instead creates germ cells by meiosis; see Section 3.3.2.)

- All cells must maintain a particular internal composition, sometimes in spite of widely varying external conditions. Cells generally must also maintain a fixed interior volume (see Chapter 7).

- By maintaining concentration differences of electrically charged atoms and molecules (generically called ions), most cells also maintain a resting electrical potential difference between their interiors and the outside world (see Chapter 11). Nerve and muscle cells use this resting potential for their signaling needs (see Chapter 12).

- Many cells move about, for example, by crawling or swimming. Chapter 5 discusses the physics of such motions.

- Cells sense environmental conditions for a variety of purposes:
  1. Sensing the environment can be one step in a feedback loop that regulates the cell's interior composition.
  2. Cells can alter their behavior in response to opportunities (such as a nearby food supply) or hardships (such as drought).
  3. Single cells can even engage in attack, self-defense, and evasive maneuvers upon detecting other cells.
  4. The highly specialized nerve and muscle cells obtain input from neighboring nerve cells by sensing the local concentration of particular small molecules, the **neurotransmitters**, secreted by those neighbors. Chapter 12 will discuss this process.

- Cells can also sense their own *internal* conditions as part of feedback and control loops. For example, an abundant supply of a particular product effectively shuts down further production of that product. One way feedback is implemented is by the physical distortion of a molecular machine when it binds a messenger molecule, a phenomenon called **allosteric control** (see Chapter 9).

- As an extreme form of feedback, a cell can even destroy itself. This mechanism, called apoptosis, is a normal part of the development of higher organisms, for example, removing unneeded neurons in the developing brain.

2.1.1 Internal gross anatomy

Paralleling the large degree of overlap between the *functions* of all cells, we find a correspondingly large overlap between their gross internal *architecture:* Most cells share a common set of quasipermanent structures, many of them visible in optical microscopy. (Electron microscopy reveals finer substructure, sometimes down to a fraction of a nanometer, but its use involves killing the cell.)

Prokaryotes and eukaryotes The simplest and most ancient types of cells are the **prokaryotes**, including the familiar **bacteria** (Figure 2.3b).[2] Bacteria are typically about one micrometer long; their gross anatomy consists mainly of a thick, rigid **cell wall** that surrounds a single interior compartment. The wall may be studded with a variety of structures, such as one or several **flagella**, long appendages used for swimming (Chapter 5). Just inside the wall lies a thin layer called the plasma membrane.

Plants, fungi, and animals are collectively called **eukaryotes**. Baker's yeast, or *Saccharomyces cerevisiæ,* is an example of a simple eukaryotic cell (Figure 2.5). Eukaryotic cells are bigger than prokaryotes, typically $10\,\mu$m or more in diameter. They too are bounded by a plasma membrane, although the cell wall may be either absent (in animal cells) or present (in plants and fungi). Eukaryotes contain various well-defined internal compartments (examples of **organelles**), each bounded by one or more membranes roughly similar to the plasma membrane.[3] In particular, eukaryotic cells are defined by the presence of a **nucleus**. The nucleus contains the genetic material, which condenses into visible **chromosomes** during cell division (Section 3.3.2); the rest of the cell's contents is collectively called the **cytoplasm**. The nucleus loses its definition during division, then re-forms.

Membrane-bounded structures in eukaryotes In addition to a nucleus, eukaryotic cells contain **mitochondria**, sausage-shaped organelles about $1\,\mu$m wide (Figure 2.6). The mitochondria carry out the final stages of the metabolism of food and the conversion of its chemical energy into molecules of ATP, the internal energy currency of the cell (see Chapter 11). Mitochondria divide independently of the surrounding cell; when the cell divides, each daughter cell gets some of the parent's intact mitochondria.

---

[2]Because prokaryotes were originally defined only by the absence of a well-defined nucleus, it took some time to realize that they actually consist of two distinct kingdoms, the bacteria (including the familiar human pathogens) and the archæa (including many of those found in environments with extreme acidity, salt concentration, or high temperature).

[3]One definition of organelle is a discrete structure or subcompartment of a cell specialized to carry out a particular function.

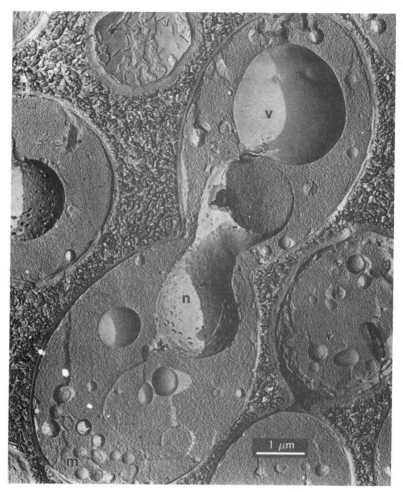

Figure 2.5: (Electron micrograph.) Budding yeast cell, a simple eukaryote. The nucleus (*n*) is in the process of dividing. Pores in the nuclear surface are visible. Also shown is a vacuole (*v*) and several mitochondria (*m, lower left*). The sample was prepared by flash-freezing, cleaving the frozen block, then heating gently in a vacuum chamber to remove outer layers of ice. A replica in a carbon-platinum mixture was then made from the surface thus revealed and finally examined in the electron microscope. [From Dodge, 1968.]

Eukaryotic cells also contain several other classes of organelles:

- The endoplasmic reticulum is a labyrinthine structure attached to the nucleus. It serves as the main factory for the synthesis of the cell's membrane structures, as well as most of the products destined for export outside the cell.

- Products from the endoplasmic reticulum in turn get sent to a set of organelles called the Golgi apparatus for further processing, modification, sorting, and packaging.

a

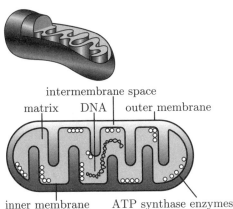

intermembrane space

matrix   DNA   outer membrane

inner membrane   ATP synthase enzymes

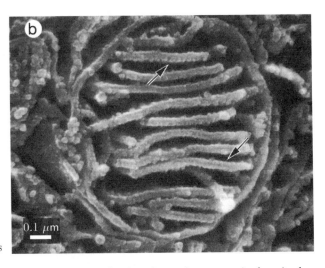

Figure 2.6: (Schematic; scanning electron micrograph.) (a) Locations of various internal structures in the mitochondrion. The ATP synthase particles are molecular machines where ATP production takes place (see Chapter 11). They are studded throughout the mitochondrion's inner membrane, a partition between an interior compartment (the matrix) and an intermembrane space. (b) Interior of a mitochondrion. The sample has been flash-frozen, fractured, and etched to show the convoluted inner membrane (*arrows*). [(a) Adapted from Karp, 2002. (b) From Tanaka, 1980.]

- Green plants contain **chloroplasts**. Like mitochondria, chloroplasts manufacture the internal energy-carrying molecule ATP. Instead of metabolizing food, however, they obtain high-quality energy by capturing sunlight.
- The cells of fungi, such as yeast, as well as those of plants also contain internal storage areas called vacuoles (see Figure 2.5). Like the cell itself, vacuoles also maintain an electrostatic potential difference across their bounding membranes (see Problem 11.3).

The part of the cytoplasm not contained in any membrane-bounded organelle is collectively called the cell's **cytosol**.

In addition, cells create a variety of **vesicles** (small bags). Vesicles can form by endocytosis, a process occurring when a part of the cell's outer membrane engulfs some exterior object or fluid, then pinches off to form an internal compartment. The resulting vesicle then fuses with internal vesicles containing digestive enzymes, which break down its contents. Another class of vesicles are the secretory vesicles, bags containing products destined for delivery outside the cell. A particularly important class of secretory vesicles is the **synaptic vesicles**, which hold neurotransmitters at the ends of nerve cells. When triggered by an arriving electrical impulse, the synaptic vesicles fuse with the outer membrane of the nerve cell (Figure 2.7), release their contents, and thus stimulate the next cell in a neural pathway (see Chapter 12).

Other elements In addition to the membrane-bounded structures listed above, eukaryotes construct various other structures that are visible with the light microscope. For example, during mitosis, the chromosomes condense into individual objects,

100 nm

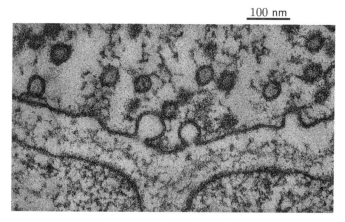

Figure 2.7: (Transmission electron micrograph.) Fusion of synaptic vesicles with the nerve cell membrane (upper *solid line*) at the junction, or synapse, between a neuron (*above*) and a muscle fiber (*below*). A vesicle at the left has arrived but not yet fused; two in the center are in the process of fusion, releasing their contents; one on the right is almost completely incorporated into the cell membrane. Vesicle fusion is the key event in the transmission of nerve impulses from one neuron to the next (see Chapter 12). [Digital image kindly supplied by J. Heuser.]

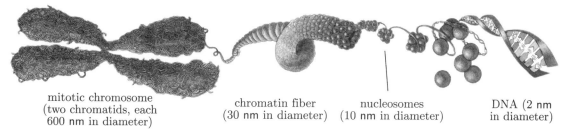

mitotic chromosome
(two chromatids, each
600 nm in diameter)

chromatin fiber
(30 nm in diameter)

nucleosomes
(10 nm in diameter)

DNA (2 nm
in diameter)

Figure 2.8: (Schematic.) One of the 46 chromosomes of a somatic (ordinary, or nongerm) human cell. Just prior to mitosis, every chromosome consists of two copies called chromatids, each consisting of tightly folded fibers called chromatin. Each chromatin fiber consists of a long DNA molecule wrapped around a chain of proteins called histones forming complexes called nucleosome particles. [From Nelson & Cox, 2000.]

each with a characteristic shape and size (Figure 2.8). Another class of structures, the cytoskeletal elements, will appear in Section 2.2.4.

2.1.2 External gross anatomy

Although many cells have simple spherical or brick-shaped forms, still others can have a much richer external anatomy. For example, the fantastically complex,

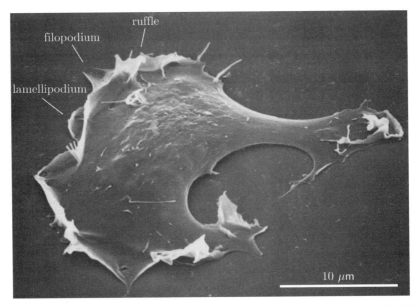

Figure 2.9: (Scanning electron micrograph.) Crawling cell. At the leading edge of this fibro-blast cell (*upper left*), filopodia, lamellipodia, and ruffles project from the cell surface. The cell crawls by extending its leading edge to the left. [Digital image kindly supplied by J. Heath.]

branched form of nerve cells (see the cover of this book) allows them to connect to their neighbors in a correspondingly complex way. Each nerve cell, or **neuron**, has a central cell body (the **soma**) with a branching array of projections (or **processes**). The processes on a neuron are subdivided into many "input lines," the **dendrites**, and one "output line," the **axon**. The entire branched structure has a single interior compartment filled with cytoplasm. Each axon terminates with one or more **axon terminals** (or boutons), containing synaptic vesicles. A narrow gap, or **synapse**, sep-arates the axon terminal from one of the next neuron's dendrites. Chapter 12 will discuss the transmission of information along the axon and from one neuron to the next.

Still other elements of the external anatomy of a cell are transient. For example, consider the cell shown in Figure 2.9. This cell is a fibroblast; its job is to crawl be-tween other cells, laying down a trail of protein that then forms connective tissue. Other crawling cells include the osteoblasts, which lay down mineral material to make bones, and Schwann cells and oligodendroglia, which wrap themselves around nerve axons, creating layers of electrical insulation.

The fibroblast in Figure 2.9 has many protrusions on its leading edge. Some of these protrusions, called filopodia, are fingerlike, about $0.1\,\mu$m in diameter and several micrometers long. Others, the lamellipodia, are sheetlike. Single-celled or-ganisms such as *Amœba* push out thicker protrusions called pseudopodia. All these protrusions form and retract rapidly, for example, searching for other cells with ap-propriate signaling molecules on their surfaces. When such a surface is found, the

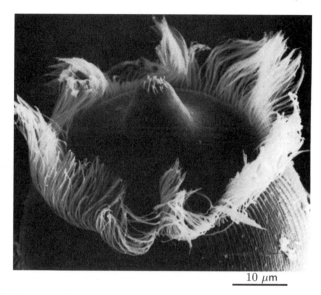

10 μm

Figure 2.10: (Scanning electron micrograph.) The ciliate *Didinium*, a single-cell animal found in still fresh water. *Didinium*'s "mouth" is at the end of a small projection, surrounded by a ring of cilia. Chapter 5 will discuss how cilia drive fluid flow. [From Shih & Kessel, 1982.]

crawling cell adheres to it, pulling the rest of its body along. In this way, cell crawling can lead to the construction of complex multicellular tissues: Each cell searches for a proper neighbor, then sticks to it.

Other specialized cells, such as those lining the human intestine, have hundreds of tiny fingerlike projections, called microvilli, to increase their surface area for fast absorption of food. Other cells have similarly shaped projections (cilia and eukaryotic flagella) that actively beat back and forth (Figure 2.10). For example, the protozoan *Paramecium* has cilia that propel it through fluid; conversely, the stationary cells lining your lungs wash themselves by constantly transporting a layer of mucus upward. Chapter 5 will discuss this process. Figure 2.10 shows yet another use for cilia: These appendages bring food particles to the "mouth" of a single-celled animal.

Another class of small anatomical features includes the fine structure of the dendrite on a neuron. The actual synapse frequently involves not the main body of the dendrite, but a tiny **dendritic spine** projecting from it (fine bumps in the cover illustration of this book).

2.2 THE MOLECULAR PARTS LIST

As promised at the start of this chapter (Roadmap, page 37), we now take a brief tour of the chemical world, from which all the beautiful biological structures shown earlier arise. We will not be particularly concerned with the chemical details of the molecules shown in this section. Nevertheless, a certain minimum of terminology is needed to express the ideas we will study.

2.2.1 Small molecules

Of the hundred or so chemically distinct atoms, our bodies consist mostly of just six: carbon, hydrogen, nitrogen, oxygen, phosphorus, and sulfur. Other atoms (such as sodium and chlorine) are present in smaller amounts. A subtle change in spelling communicates a key property of many of these single-atom chemicals: In water, neutral chlor*ine* atoms (abbreviated Cl) take on an extra electron from their surroundings, becoming chlor*ide* ions (Cl^-). Other neutral atoms *lose* one or more electrons in water, such as sodium atoms (abbreviated Na), which become sodium ions (Na^+).

Of the small molecules in cells, the most important is water, which constitutes 70% of our body mass. Chapter 7 will explore some of the remarkable properties of water. Another important inorganic (that is, containing no carbon) molecule is phosphoric acid (H_3PO_4); in water, this molecule separates into the doubly charged **inorganic phosphate** (HPO_4^{2-}, also called P_i) and two positively charged hydrogen ions (called **protons**). (You'll look more carefully at the dissociation of phosphate in Problem 8.6.)

An important group of organic (containing carbon) molecules have atoms bonded into rings:

- Simple sugars include glucose and ribose (compounds with one ring), and sucrose (cane sugar, with two rings).
- The four bases of DNA (see Section 2.2.3) also have a ring structure. One class (the pyrimidines: cytosine and thymine) has one ring; the other (the purines: guanine and adenine) has two. See Figure 2.11.
- A slightly different set of four bases is used to construct RNA: Thymine is replaced by the similar one-ring molecule uracil.

The ring structures of all these molecules give them a fixed, rigid shape. The bases are *flat* (planar) rings. Joining a base to a simple sugar (ribose or deoxyribose) and a phosphate yields a **nucleotide**. For example, the nucleotide formed from the base adenine, the sugar ribose, and a single phosphate is called adenosine monophosphate, or **AMP**. The corresponding molecules with two or three phosphate groups in a row are called adenosine diphosphate (**ADP**) or adenosine triphosphate (**ATP**), respectively (Figure 2.12). Such molecules are sometimes referred to generically as nucleoside triphosphates, or NTPs.

Nucleoside triphosphates such as ATP carry a lot of stored energy, due in part to the self-repulsion of a large electric charge (equivalent to three protons) held in close proximity by the chemical bonds of the molecule. (Chapter 8 will discuss the idea of stored chemical energy and its utilization.) In fact, cells use ATP as a nearly universal internal energy currency; they maintain high interior concentrations of ATP for use by all their molecular machines as needed.[4]

Two more classes of small molecules are of special interest to us. The first of these, the **fatty acids**, have a simple structure: They consist of a chain of carbon

---

[4]Cells also use guanosine triphosphate (**GTP**) and a handful of other small molecules for similar purposes. Nucleotides also serve as internal signaling molecules in the cell. A modified form of AMP, called cyclic AMP or cAMP, is particularly important in this regard.

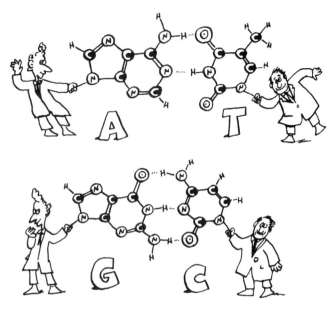

Figure 2.11: (Molecular structure.) J. Watson and F. Crick demonstrate the complementarity of DNA basepairs. The dotted lines denote hydrogen bonds (see Chapter 7). The shapes and chemical structure of the bases allow hydrogen bonds to form optimally only between adenine (A) and thymine (T) and between guanine (G) and cytosine (C); in these pairings, atoms that are able to form hydrogen bonds can be brought close together without distorting the bases' geometries. [Cartoon by Larry Gonick, from Gonick & Wheelis, 1991.]

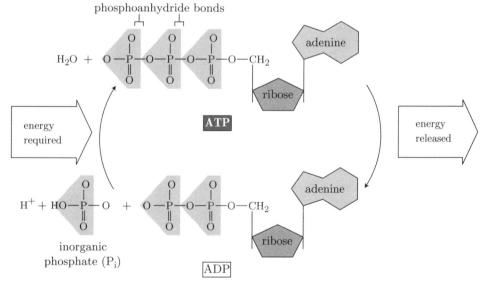

Figure 2.12: (Molecular structure diagrams.) Adenosine triphosphate is hydrolyzed as part of many biochemical processes. An ATP and a water molecule are both split, yielding ADP, inorganic phosphate (P_i), and a proton (H^+). A similar reaction yielding about the same amount of free energy splits ATP into adenosine monophosphate (AMP), a compound with one phosphate group, and pyrophosphate, or PP_i. Chapter 8 will discuss chemical energy storage; Chapter 10 will discuss molecular motors fueled by ATP. [Adapted from Alberts et al., 1997.]

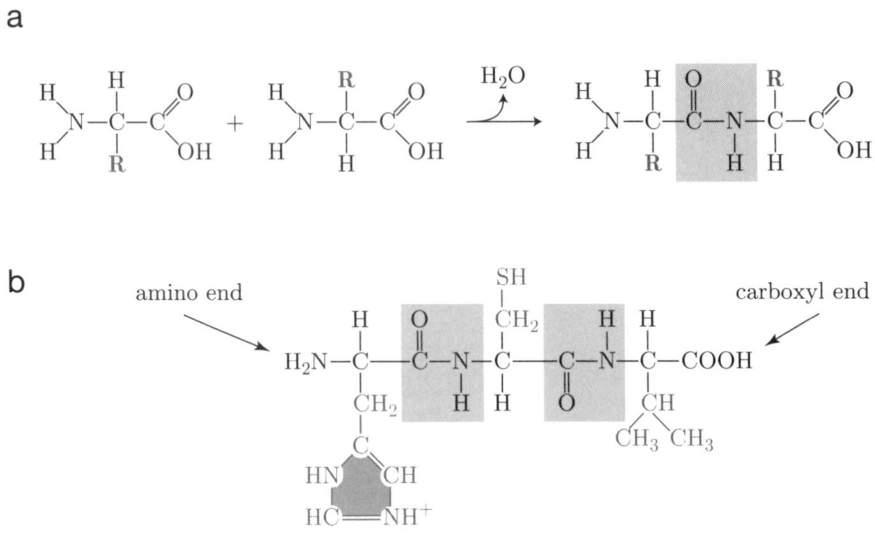

a

b

Figure 2.13: (Molecular structure diagrams.) (a) Formation of a polypeptide from amino acids by the condensation reaction, essentially the reverse of the hydrolysis reaction shown in Figure 2.12. The four atoms in the gray box constitute the peptide bond. (b) A short segment of a polypeptide chain, showing three residues (amino acid monomers) joined by two peptide bonds. The residues consist of a common backbone, with various side groups attached to it. The residues shown are respectively histidine, cysteine, and valine. Chapters 7 and 8 will discuss the interactions between the residues that determine the protein's structure; Chapter 9 will briefly discuss the resulting complex arrangement of protein substates. [Adapted from Alberts et al., 2002 .]

atoms (for example, 15 for palmitic acid, derived from palm oil), with a carboxyl group (–COOH) at the end. Fatty acids are partly important as building blocks of the phospholipids to be discussed in Section 2.2.2. Finally, the **amino acids** are a group of about 20 building blocks from which proteins are constructed (Figure 2.13). As shown in the figure, each amino acid has a common central backbone, with a "plug" at one end (the carboxyl group) and a "socket" at the other (the amino group, –NH_2). Attached to the side of the central carbon atom (called the α-carbon) is a side group (generically denoted by R in Figure 2.13a) determining the identity of the amino acid; for example, alanine is the amino acid with the side group –CH_3. Protein synthesis consists of successively attaching the socket of the next amino acid (or **residue**) to the plug of the previous one by the **condensation reaction** in Figure 2.13a, thereby creating a polymer called a **polypeptide**. The C–N bond formed in this process is called the **peptide bond**. Section 2.2.3 and Chapter 9 will sketch how polypeptides turn into functioning proteins.

2.2.2 Medium-sized molecules

A huge number of medium-sized molecules can be formed from the handful of atoms used by living organisms. Remarkably, only a tiny subset of these are actually used by living organisms. Indeed, the list of possible compounds with mass under 25 000

Table 2.1: Molecular composition of bacterial cells, by weight.

molecular class	percentage of total cell weight
Small molecules	(74% of total cell weight)
ions, other inorganic small molecules	1.2
sugars	1
fatty acids	1
individual amino acids	0.4
individual nucleotides	0.4
water	70
Medium and big molecules	(26% of total cell weight)
protein	15
RNA	6
DNA	1
lipids	2
polysaccharides	2

[From Alberts et al., 1997.]

times that of water probably runs into the billions, and yet fewer than a hundred of these (and their polymers) account for most of the weight of any given cell (see Table 2.1).

Figure 2.14 shows a typical **phospholipid** molecule. Phospholipids are formed by joining one or two fatty acid chains ("tails"), via a glycerol molecule, to a phosphate and thence to a "head group." As described in Section 2.3.1 and Chapter 8, phospholipids self-assemble into thin membranes, including the one surrounding every cell. Phospholipid molecules have long but informative names; for example, dipalmitoyl phosphatidylcholine (or **DPPC**) consists of two ("*di*") *palmit*ic acid chains joined by a *phosphat*e to a *choline* head group. Similarly, most fats consist of three

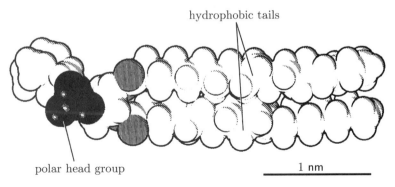

hydrophobic tails

polar head group

1 nm

Figure 2.14: (Structure.) Space-filling model of a phospholipid molecule. Two hydrocarbon tails (*right*) join to a head group (*left*) via phosphate and glycerol groups (*middle*). Molecules like this one self-assemble into bilayer membranes (Color Figure 2 and Figure 2.20), which in turn form the partitions between cell compartments. Chapter 8 will discuss self-assembly. [From Goodsell, 1993.]

fatty acid chains, each joined by a chemical bond to one of the three carbon atoms in a glycerol molecule, to form a triglyceride. The joining is accomplished by a condensation reaction similar to the one shown in Figure 2.13.

2.2.3 Big molecules

Cells create giant molecules as **polymers**, long chains of similar units.

Polynucleotides Just as amino acids can be joined into polypeptide chains, so, too can chains of nucleotides be strung together to form **polynucleotides**. A polynucleotide formed from nucleotides containing ribose is called a ribonucleic acid, or **RNA**; the analogous chain with deoxyribose is called a molecule of deoxyribonucleic acid, or **DNA**. Watson and Crick's insight (Section 3.3.3) was that not only do the flat bases of DNA fit each other precisely, like jigsaw puzzle pieces (Figure 2.11); but they also can nest neatly in a helical stack (Figure 2.15). In this helix, the bases point inward and the sugar and phosphate groups form two backbones on the outside. Cells do not manufacture double-stranded RNA; but a single RNA strand can have short tracts that complement others along the chain, a situation giving rise to a partially folded structure (Figure 2.16).

Each of your cells contains a total of about a meter of DNA, consisting of 46 pieces. Manipulating such long threads, without turning them into a useless tangle, is not easy. Part of the solution is a hierarchical packaging scheme: The DNA is wound onto protein "spools," to form complexes called **nucleosomes**. The nucleosomes in turn wind into higher order structures, and so on up to the level of entire condensed chromosomes (Figure 2.8).[5]

Polypeptides Section 2.2.1 mentioned the formation of polypeptides. The genetic message in DNA encodes only the polypeptide's **primary structure**, or linear sequence of amino acids. After the linear polypeptide chain has been synthesized, it folds into an elaborate three-dimensional structure—a **protein**—such as those seen in Figure 2.4f, i, k, l. The key to understanding this process is to note that individual amino acid residues on a protein may attract or repel each other. Later chapters will discuss how the polypeptide's primary structure thus determines the protein's final, three-dimensional folded structure. (In contrast, the monomer units composing DNA are all negatively charged, so they repel each other uniformly: DNA by itself does not spontaneously fold.)

The lowest level of folding (the **secondary structure**) involves interactions between residues near each other along the polypeptide chain. An example that will interest us in Chapter 9 is the **alpha helix**, shown in Figure 2.17. At the next higher level, the secondary structures (along with other, disordered regions) assemble to give the protein's **tertiary structure**, the overall shape visible in the examples of Figure 2.4. A simple protein consists of a single chain of 30–400 amino acids, folded into a tertiary structure that is dense, roughly spherical, and a few nanometers in diameter (a "globular" protein).

---

[5]Simpler forms of DNA packaging have also been found in prokaryotic cells.

1 nm

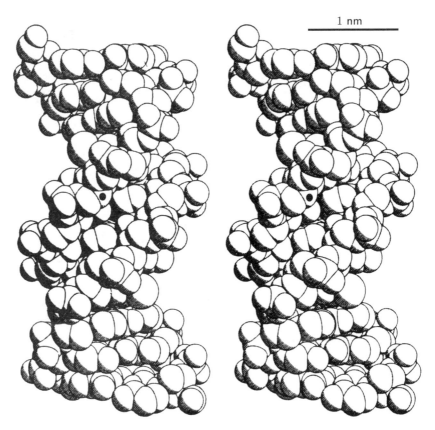

Figure 2.15: (Structure rendered from atomic coordinates.) Stereo image of the DNA double helix. To view this image, begin with your nose a few centimeters from the page (if you're nearsighted, remove your glasses). Imagine staring through the page at a distant object. If necessary, rotate the page a few degrees, so that the two dots near the centers of each panel are aligned horizontally. Wait until the dots fuse. Concentrate on holding the dots fused as you slowly move the page away from your nose. When the page is far enough away for your eyes to focus on it, the three-dimensional image will jump off the page at you. The structure is about 2 nm wide. The portion shown consists of twelve basepairs in a vertical stack. Each basepair is roughly a flat, horizontal plate about 0.34 nm thick. The stack twists through slightly more than one full revolution from top to bottom. [From Dickerson et al., 1982.]

More complex proteins consist of multiple polypeptide chain subunits, usually arranged in a symmetrical array—the **quaternary structure**. A famous example is hemoglobin, the carrier of oxygen in your blood (Chapter 9), which has four subunits. Many membrane channels (see Section 2.3.1) also consist of four subunits.

Polysaccharides **Polysaccharides** form a third class of biopolymers (after nucleic acids and proteins). These are long chains of sugar molecules. Some, like glycogen, are used for long-term energy storage. Others help cells to identify themselves to one

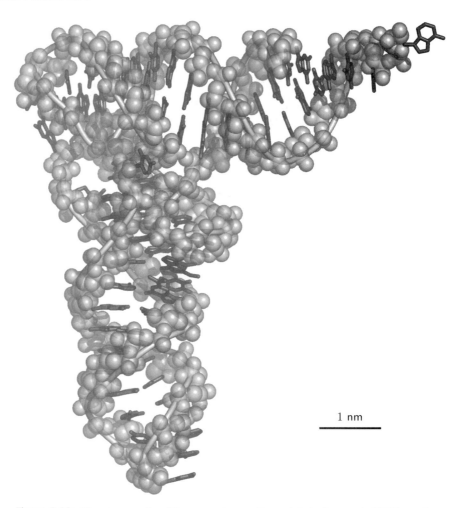

Figure 2.16: (Structure rendered from atomic coordinates.) A single strand of RNA uses base-pairing and other interactions to form a unique three-dimensional structure. The molecule shown is a transfer RNA from yeast; it binds the amino acid phenylalanine, transports it to the ribosome, then releases it (see Figure 2.24). The flat, stacked nucleotides are shown as stick structures mostly on the interior; the sugar-phosphate backbone atoms are instead shown as spheres, to reveal the double helical nature of parts of the folded molecule. Longer strands of RNA can have several pairs of complementary stretches, leading to more complex folded structures than the one shown here. Section 6.7 will discuss how the folding and unfolding of RNA can be controlled by external forces.

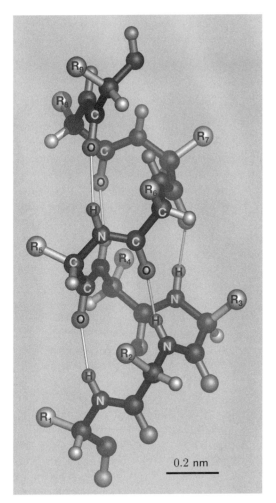

Figure 2.17: (Molecular structure from crystallography data.) A segment of the alpha helix structure. Nine successive residues are shown. Each residue's side group has been replaced by a single ball, labeled $R_1, \ldots, R_9$. Each residue has a hydrogen atom bound to one of the nitrogens on the chain. Each of these hydrogens is attracted to an oxygen located four units farther down the chain, to form a hydrogen bond (*thin lines*). The hydrogen bonds help to stabilize the ordered, helical structure against thermal disruption. Chapter 9 will discuss the formation and loss of ordered structures like this one under changes in environmental conditions. The structure shown is "right-handed" in the following sense: Choose either direction along the helix axis, for example, upward in the figure. Point your right thumb along this direction. Then as you proceed in the direction of your thumb, the peptide backbone rotates around the axis in the same direction as your fingers point (opposite to the direction you'd have gotten using your left hand).

another. When crosslinked by short peptides, polysaccharides can also form a tough two-dimensional mesh, the **peptidoglycan layer** that gives the bacterial cell wall its strength.

2.2.4 Macromolecular assemblies

The previous section mentioned that individual protein chains can form confederations with definite shapes, the quaternary structure of a protein assembly. Another possibility is the construction of a *linear* array of polypeptide subunits, extending for an arbitrarily long distance. Such arrays can be thought of as polymers made up of monomers that are themselves proteins. Two examples will be of particular interest in Chapter 10: microtubules and F-actin.

The organelles mentioned in Section 2.1.1 are suspended within the eukaryotic cell's cytosol. The cytosol is far from being a structureless, fluid soup. Instead, a host of structural elements pervade it, both anchoring the organelles in place and conferring mechanical integrity upon the cell itself. These elements are all long, polymeric structures; collectively, they are called the **cytoskeleton.**

The most rigid of the cytoskeletal elements are the **microtubules** (Figure 2.18). Microtubules are 25 nm in diameter and can grow to be as long as the entire cell. They form an interior network of girders, helping the cell to resist overall deformation (Color Figure 1). Another function of microtubules is to serve as highways for the transport of cell products from one place to another (see Figure 2.19 and Section 2.3.2).

Actin filaments (also called "filamentous" actin, or **F-actin**) form a second class of cytoskeletal elements. F-actin fibers are only 7 nm in diameter; they can be several micrometers long (Figure 2.4k). A thin meshwork of these filaments underlies the surface of the cell, forming the cell's **actin cortex**. Filopodia, lamellipodia, and microvilli are all full of actin fibers, which cross-link to one another to form stiff bundles that help to push these projections out of the cell. Finally, actin filaments furnish the "tracks" along which single-molecule motors walk to generate muscle contraction (Chapter 10).

Examples of even more elaborate protein assemblies include the shells surrounding viruses and the whiplike bacterial flagellum (see Figure 2.3 on page 37).

2.3 BRIDGING THE GAP: MOLECULAR DEVICES

We now have a catalog of beautiful *structures* in cells, but little has been said about how they form from the molecules in Section 2.2, nor, indeed, about how cells carry out the many other *activities* characteristic of life. To begin bridging this gap, this section will sketch a few of the molecular devices cells use. The unity of living things becomes very apparent when we study molecular devices: All cells are somewhat similar at the level of physiology, but they are *very* similar at the molecular level. Today's routine use of bacteria as factories for the expression of human genes testifies to this unity.

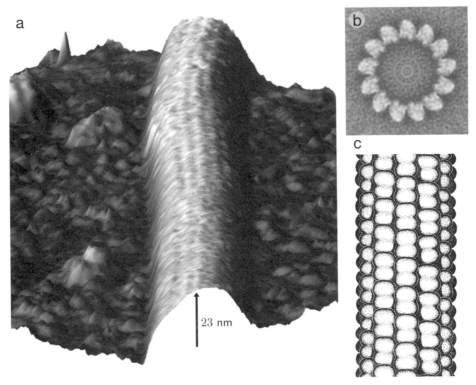

Figure 2.18: (Scanning force micrograph; reconstruction from electron microscopy; drawing based on structural data.) Structure of microtubules. (a) To make this image, a fine probe was scanned over the microtubule and repeatedly brought down to touch it, mapping out its three-dimensional structure. The protofilaments making up the microtubule are visible as longitudinal lines on its surface. (b) Cross-section, again showing the protofilaments. (c) The drawing shows how the subunits line up to form a parallel arrangement of protofilaments. Tubulin monomers, called α and β, first link in $\alpha\beta$ pairs to form the dumbbell-shaped subunits shown in the drawing; the dumbbells then assemble to form the microtubule. The vertical distance between adjacent β subunits is 8 nm. [(a,b) Digital images kindly supplied by I. Schaap and C. Schmidt, and by K. Downing; see also de Pablo et al., 2003 and Li et al., 2002. (c) From Goodsell, 1996.]

2.3.1 The plasma membrane

To maintain its identity (for example, to control its composition), every cell must be surrounded by some sort of envelope. Similarly, every organelle and vesicle must somehow be packaged. Remarkably, all cells have met all of these challenges with a *single* molecular construction: the bilayer membrane (Color Figure 2). For example, the plasma membrane surrounding any cell is a bilayer of this type and so looks like a double layer under the electron microscope. All bilayer membranes have roughly similar chemical composition, electrical capacitance, and other physical properties.

As its name implies, a bilayer membrane consists of two layers of molecules, primarily the phospholipids shown in Color Figure 2. Even though it's only about 4 nm thick, the plasma membrane nevertheless covers the entire exterior of a cell, often a billion or more square nanometers! To be effective, this fragile-looking structure

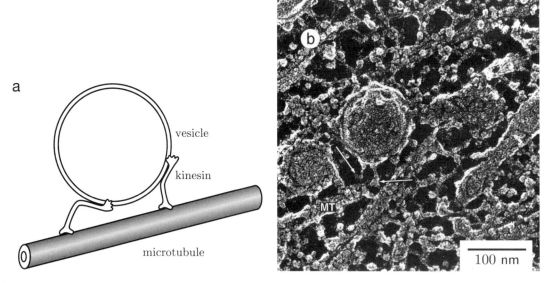

Figure 2.19: (Schematic; electron micrograph.) (a) Model showing how kinesin drags a vesicle along a microtubule. Chapter 10 will discuss the action of this single-molecule motor. (b) Micrograph appearing to show the situation sketched in (a). Arrows show the attachment points. Neurons from rat spinal cord were flash-frozen and deep-etched to create the sample. [(a) Adapted from Kandel et al., 2000. (b) Image kindly supplied by N. Hirokawa; see Hirokawa et al., 1989.]

must not rip; yet it must also be fluid enough to let the cell crawl, endocytose, and divide. We will study the remarkable properties of phospholipid molecules that reconcile these constraints in Chapter 8.

We get another surprise when we mix phospholipid molecules with water: Even without any cellular machinery, *bilayer membranes self-assemble spontaneously.* Chapter 8 will show that this phenomenon is driven by the same interactions that cause salad dressing to separate spontaneously into oil and water. Similarly, microtubules and F-actin can self-assemble from their subunits, without the intervention of any special machinery (see Figure 10.4 on page 408).

Bilayer membranes do far more than partition cells. They also carry a rich variety of molecular devices (see Figure 2.20):

- Integral membrane proteins span the membrane, projecting on both the inner and outer sides. Examples include the **channels**, which allow the passage of specified molecules under specified conditions; receptors, which sense exterior conditions; and **pumps**, which actively pull ions and other material across a membrane (see Figure 2.21).

- Receptors can, in turn, connect to peripheral membrane proteins, which communicate information to the interior of the cell.

- Still other integral membrane proteins anchor the cell's membrane to its underlying actin cortex, helping the cell maintain its required shape. A related example

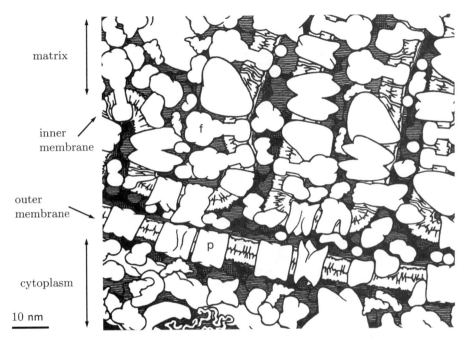

matrix

inner
membrane

outer
membrane

cytoplasm

10 nm

Figure 2.20: (Drawing based on structural data.) Cross section of a part of a mitochondrion (Figure 2.6), showing its two membranes. Each membrane consists of a lipid bilayer (Color Figure 2) with proteins embedded in (or attached to) it. The surrounding cell's cytoplasm appears at the bottom of the figure. (Its own plasma membrane is similarly crowded with embedded proteins.) The mitochondrion's outer membrane is pierced by channel-forming integral membrane proteins (labeled p). The folded inner membrane of the mitochondrion above it is embedded with protein complexes involved in making ATP. Chapter 11 will discuss one of these, the F0–F1 complex (labeled f). A part of the mitochondrial matrix appears at upper left. [From Goodsell, 1993.]

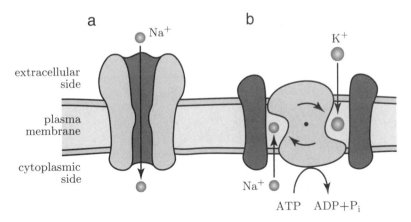

a   Na$^+$ b   K$^+$

extracellular
side

plasma
membrane

cytoplasmic
side

Na$^+$

ATP   ADP+P$_i$

Figure 2.21: (Schematic.) (a) Passive ion channel, like the ones giving rise to the Ohmic part of membrane conductances (see Chapter 11). (b) The sodium–potassium pump (also discussed in Chapter 11). The sketch has been simplified; actually, the pump is believed to bind three Na$^+$ ions and an ATP before its main conformational change, which expels the Na$^+$'s. Then it binds two K$^+$ ions, releases ADP and phosphate, pulls the K$^+$'s inward, and releases them. At this point, the pump is ready to begin its cycle anew. [Adapted from Kandel et al., 2000.]

57

concerns the membrane of the human red blood cell. A network of elastic protein strands (in this case, spectrin) is anchored to the membrane by integral membrane proteins. This network deforms as the red cell squeezes through the body's capillaries, then pops the cell back to its normal shape after passage into a vein.

2.3.2 Molecular motors

As mentioned earlier, actin filaments form the "tracks" along which protein motors walk, thereby generating muscle contraction (see Chapter 10). Many other examples of walking motors are known in cells. Figure 2.19 shows a vesicle being dragged along a microtubule to its destination at an axon terminal. This axonal transport brings needed proteins to the axon terminal, as well as the ingredients from which synaptic vesicles will be built. A family of single-protein motors called kinesins supply the motive force for this and other motions, for example, the dragging of chromosomes to the two halves of a dividing cell. Indeed, selectively staining both the microtubules and the kinesin (by attaching fluorescent markers to each) shows that they are generally found together in the cell (Color Figure 3). It is even possible to follow the progress of individual kinesin molecules as they walk along individual microtubules (Color Figure 4). In such experiments, the kinesin molecules begin to walk as soon as a supply of ATP molecules is added; they stop when the ATP is used up or washed away.

The cilia mentioned in Section 2.1.2 are also powered by walking motors. Each cilium contains a bundle of microtubules. A motor molecule called dynein attaches to one microtubule and walks along its neighbor, inducing a relative motion. Coordinated waves of dynein activity create traveling waves of bending in the cilium, making it beat rhythmically.

Other motors generate *rotary* motion. Examples include the motor that drives the bacterial flagellum (Figure 2.3b; see Chapters 5 and 11), and the one that drives the synthesis of ATP in mitochondria (Chapter 11). Rather than being driven directly by ATP, both of these motors use as their "fuel" a chemical imbalance between the sides of the membrane they span. Ultimately, the imbalance comes from the cell's metabolic activity.

2.3.3 Enzymes and regulatory proteins

Enzymes are molecular devices whose job is to bind particular molecules, under particular conditions, and promote particular chemical changes. The enzyme molecule itself is not modified or used up in this process—it is a **catalyst**, or assistant, for a process that could in principle happen on its own. Enzymes may break down large molecules, as in digestion, or build small molecules into big ones. One feature of enzymes immediately apparent from their structures is their complicated and well-defined *shape* (Color Figure 5). Chapter 7 will begin a discussion of the role of shape in conferring specificity to enzymes; Chapter 9 will look more deeply into how the shapes actually arise and how an enzyme maintains them despite random thermal motion.

Another context where binding specificity is crucial concerns control and feedback. Nearly every cell in your body contains the same collection of chromosomes,[6] and yet only pancreas cells secrete insulin, only hair cells grow hairs, and so on. Each cell type has a characteristic arrangement of genes that are active ("switched on") and inactive ("switched off"). Moreover, individual cells can modulate their gene activities depending on external circumstances: If we deny a bacterium its favorite food molecule but supply an alternative food, the cell will suddenly start synthesizing the chemicals needed to metabolize what's available. The secret to gene switching is a class of regulatory proteins, which recognize and bind specifically to the beginning of the genes they control (Color Figure 6). One subclass, the repressors, can block the start of their gene, thereby preventing transcription. Other regulatory proteins help with the assembly of the transcriptional apparatus and have the opposite effect. Eukaryotic cells have a more elaborate implementation of the same general idea.

Finally, the pumps and channels embedded in cell membranes are also quite specific. For example, a remarkable pump to be studied in Chapter 11 has an operating cycle in which it binds only sodium ions, ferries them to the other side of the membrane, then binds only potassium ions and ferries them in the other direction! As shown in Figure 2.21b, this pump also consumes ATP, in part because the sodium ions are being pulled from a region of negative electrostatic potential (the cell's interior) to a positive region, thereby increasing their potential energy. According to the First Law (Section 1.1.2 on page 6), such a transaction requires a source of energy. (The Example on page 484 will explore the energy budget of this pump in greater detail.)

2.3.4 The overall flow of information in cells

Section 2.3.3 hinted that the cell's genetic message (the genome) should not be regarded as a "blueprint," or literal representation, of the cell, but rather as specifying an algorithm, or set of instructions, for creating and maintaining the entire organism containing the cell. Gene regulatory proteins supply some of the switches turning parts of the algorithm on and off.

We can now describe a simplified version of the flow of information in cells (Figure 2.22).[7]

1. The DNA in the cell nucleus contains the master copy of the software, in duplicate. Under ordinary circumstances, the DNA is not modified but only copied (replicated) during cell division. A molecular machine called **DNA polymerase** accomplishes the replication. Like the machines mentioned in Section 2.3.2, DNA polymerase is made of proteins. The DNA contains genes, which consist of regulatory regions and coding regions that specify the amino acid sequences of various

---

[6]Exceptions include germ cells (genes not present in duplicate) and human red blood cells (no nucleus at all).

[7]Some authors refer to this scheme as the "central dogma" of molecular biology, a playful but unfortunate phrase coined by F. Crick. Several amendments to this scheme are discussed in Section 2.3.4′ on page 63.

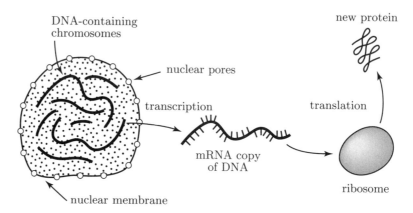

Figure 2.22: (Schematic.) The flow of information in a cell. Sometimes the product of translation is a regulatory protein, which interacts with the cell's genome, thereby creating a feedback loop. [Adapted from Calladine & Drew, 1997.]

needed proteins. A complex organism may have tens of thousands of distinct genes, whereas *E. coli* has fewer than 5000. (The simplest known organism, *Mycoplasma genitalium*, has fewer than 500!) In addition to the genes, the DNA contains a rich array of regulatory sequences for the binding of regulatory proteins, along with immense stretches with no known function.

2. Another molecular machine called **RNA polymerase** reads the master copy in a process called **transcription** (Figure 2.23). RNA polymerase is a combination of walking motor and enzyme; it attaches to the DNA near the start of a gene, then pulls the polymer chain through a slot, simultaneously adding successive monomers to a growing "transcript" made of RNA (Section 2.2.3). The transcript is also called **messenger RNA**, or mRNA. In eukaryotic cells, mRNA leaves the nucleus through pores in the nuclear membrane (see Figure 2.5) and enters the

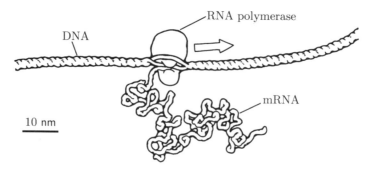

Figure 2.23: (Drawing, based on structural data.) Transcription of DNA to messenger RNA by RNA polymerase, a walking motor. The polymerase reads the DNA as it walks along the DNA strand, synthesizing a mRNA transcript as it moves. [From Goodsell, 1993.]

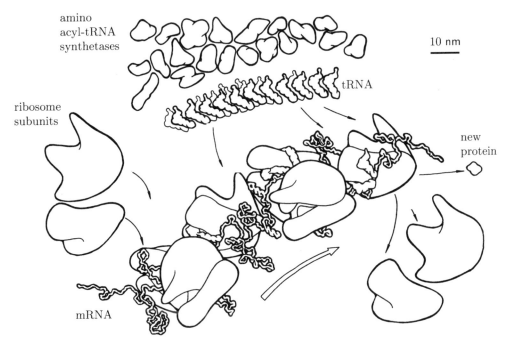

amino
acyl-tRNA
synthetases

10 nm

tRNA

ribosome
subunits

new
protein

mRNA

Figure 2.24: (Drawing, based on structural data.) The information in messenger RNA is translated into a sequence of amino acids making up a new protein by the combined action of over 50 molecular machines. In particular, amino acyl-tRNA synthetases supply transfer RNAs loaded with amino acids to the ribosomes, which construct the new protein as they read the messenger RNA. Not shown are some smaller auxiliary proteins, the initiation, elongation, and transcription factors, that help the ribosomes do their job. [From Goodsell, 1993.]

cytosol. The energy needed to drive RNA polymerase comes from the added nucleotides themselves, which arrive in the high-energy NTP form (Section 2.2.1); the polymerase clips off two of the three phosphate groups from each NTP as it incorporates the nucleotide into the growing transcript (Figure 2.12).

3. In the cytosol, a complex of devices collectively called the **ribosome** binds the transcript and again walks along it, successively building up a polypeptide on the basis of instructions encoded in the transcript. The ribosome accomplishes this **translation** by orchestrating the sequential attachment of **transfer RNA** (or tRNA) molecules (see Figure 2.16), each binding to a particular triplet of monomers (bases) in the transcript and each carrying the corresponding amino acid monomer (residue) to be added to the growing polypeptide chain (Figure 2.24).

4. The polypeptide may spontaneously fold into a functioning protein, or it may fold with the help of other auxiliary devices picturesquely called **chaperones**. Additional chemical bonds (disulfide bonds between residues containing sulfur atoms) can form to cross-link monomers distant from each other along the chain, or even in another chain.

5. The folded protein may then form part of the cell's architecture. It may become a functioning device, for example, one of those shown in Figure 2.24. Or it may be a regulatory protein, helping close a feedback loop. This last option creates a mechanism for orchestrating the development of the cell (or indeed of a multicellular organism).

$\boxed{T_2}$ *Section 2.3.4′ on page 63 mentions some modifications to the simplified scheme given above.*

THE BIG PICTURE

Returning to the Focus Question, we see that we have a lot of work to do: The following chapters will need to shed physical light on the key phenomena of specificity, self-assembly, and active transport. As indicated throughout the chapter, many specific structures and processes will be discussed again later, including flagellar propulsion, RNA folding, the material properties of bilayer membranes and of individual DNA and protein molecules, the structure and function of hemoglobin, the operation of the kinesin motor, the synthesis of ATP in mitochondria, and the transmission of nerve impulses.

It should be clear that the complete descriptions of these processes will occupy whole shelves full of books, at some future date when all the details are known! The purpose of this book is not to give the complete details, but to address the more elementary question: Faced with all these miraculous processes, we will only ask, *"How could anything like that happen at all?"* We will find that simple physical ideas do help with this more modest goal.

FURTHER READING

Semipopular:
Structure and function in cells: Goodsell, 1993; Hoagland & Dodson, 1995.

Intermediate:
General reference: Lackie & Dow, 1999; Smith et al., 2000.
Texts: Cooper, 2000; Alberts et al., 1997; Karp, 2002; Pollard & Earnshaw, 2002.

Technical:
Texts: Alberts et al., 2002; Lodish et al., 2000.
Proteins: Branden & Tooze, 1999.

2.3.4' Track 2

Since its enunciation in the 1950s, several amendments to the simplified picture of information flow given in Section 2.3.4 have been found. (Others were known even at the time.) Just a few examples include

1'. It is an overstatement to claim that all the cell's heritable characteristics are determined solely by its DNA sequence. A cell's entire state, including all the proteins and other macromolecules in its cytoplasm, can potentially affect its descendants. The study of such effects has come to be called epigenetics. One example is cell differentiation: Once a liver cell forms, its descendants will be liver cells. A cell can also give its daughters misfolded proteins, or prions, transmitting a pathology in this way. Even multiple clones of the same animal are generally not identical.[8]

Moreover, the cell's DNA can itself be modified, either permanently or temporarily. Examples of permanent modification include random point mutations (see Chapter 3), random duplication, deletion, and rearrangement of large stretches of the genome from errors in crossing-over (Chapter 3), and insertion of foreign DNA by retroviruses such as HIV. Temporary, reversible changes include chemical modification, for example, methylation.

2'. Other operations, such as RNA editing, may intervene between mRNA synthesis and translation.

3'. A polypeptide can be modified after translation: Additional chemical groups may need to be added, and so on, before the finished protein is functional.

4'. Besides chaperones, cells also have special enzymes to destroy polypeptides that have improperly folded.

---

[8]Identical twins are more similar, but they share more than DNA—they come from a common egg and thus share its cytoplasm.

PROBLEMS

2.1   *All Greek to me*

Now's the time to learn the Greek alphabet. Here are the letters most often used by scientists. The following list gives both lowercase and uppercase (but omits the uppercase when it looks just like a Roman letter):

$$\alpha, \ \beta, \ \gamma/\Gamma, \ \delta/\Delta, \ \epsilon, \ \zeta, \ \eta, \ \theta/\Theta, \ \kappa, \ \lambda/\Lambda, \ \mu, \ \nu, \ \xi/\Xi, \ \pi/\Pi,$$

$$\rho, \ \sigma/\Sigma, \ \tau, \ \upsilon/\Upsilon, \ \phi/\Phi, \ \chi, \ \psi/\Psi, \ \omega/\Omega$$

When reading aloud we call them alpha, beta, gamma, delta, epsilon, zeta, eta, theta, kappa, lambda, mu, nu, xi (pronounced "k'see"), pi, rho, sigma, tau, upsilon, phi, chi (pronounced "ky"), psi, omega. Don't call them all "squiggle."

Practice by examining the quote given in Chapter 1 from D'Arcy Thompson, which in its entirety reads: "Cell and tissue, shell and bone, leaf and flower, are so many portions of matter, and it is in obedience to the laws of physics that their particles have been moved, moulded, and conformed. They are no exception to the rule that $\Theta\epsilon\grave{o}\varsigma \ \grave{\alpha}\epsilon\grave{\iota} \ \gamma\epsilon\omega\mu\epsilon\tau\rho\epsilon\hat{\iota}$." From the sounds made by each letter, can you guess what Thompson was trying to say? [*Hint:* ς is an alternate form of σ.]

2.2   *Do-it-yourself proteins*

This book contains some molecular structure pictures; you can easily make many more yourself. Download RasMol from `http://www.umass.edu/microbio/rasmol/index.html` (or `http://openrasmol.org`), or get some other free molecular viewing application.[9] Now go to the Protein Data Bank,[10] `http://www.rcsb.org/pdb/`. On the main page, try searching for and viewing molecules (see also the "molecule of the month" department, from which the examples below were taken). Once you get the molecule's main entry, click "explore" on the right, then "view" and download in RasMol format. Play with the many RasMol options. Alternatively, you can just click `quickpdb` for a viewer that requires no separate application. Here are some examples; several are discussed in this and later chapters:

a. thrombin, a blood-clotting protein (code `1ppb`).

b. insulin, a hormone (code `4ins`).

c. myosin, a molecular motor (code `1b7t`).

d. the actin-myosin complex (code `1alm`). This entry shows a model of one myosin motor bound to a short actin filament formed of five molecules, based on data from electron microscopy. The file contains only alpha carbon positions for the proteins, so you'll need to use backbone diagrams when you look at it.

e. rhinovirus, responsible for the common cold (code `4rhv`).

---

[9]Protein Explorer, also available at `http://www.umass.edu/microbio/rasmol/index.html` requires installation of additional software. Other popular packages include PyMol (`http://pymol.sourceforge.net`) and VMD (`http://www.ks.uiuc.edu/Research/vmd/`).

[10]The PDB is operated by the Research Collaboratory for Structural Bioinformatics (RCSB). You can also find RasMol there under "software."

f. myoglobin, an oxygen-storing molecule found in muscles (code 1mbn). Myoglobin was the first protein structure ever determined.

g. DNA polymerase (code 1tau).

h. the nucleosome (code 1aoi).

Use your mouse to rotate the pictures. Use the measurement feature of RasMol to find the physical size of each object. Selectively color only the hydrophobic residues. Try the "stereo" option. Print the ones you like.

2.3   *Do-it-yourself nucleic acids*

Go to the Nucleic Acid Database, http://ndbserver.rutgers.edu/. Download coordinates and view, using RasMol or another software:

a. the B-form of DNA (code bd0001). Choose the space-filling representation and rotate the molecule to see its helical structure.

b. transfer RNA (code trna12).

c. RNA hammerhead enzyme, a ribozyme (code urx067).

d. the complex of integration host factor bound to DNA (code pdt040). Try the cartoon display option.

2.4   *Do-it-yourself small molecules*

Go to http://molbio.info.nih.gov/cgi-bin/pdb and search for some small molecule mentioned in this chapter. You'll probably find PDB files for larger molecules binding the one you chose. Look around.

2.5   *Do-it-yourself micelles and bilayers*

Go to http://moose.bio.ucalgary.ca/, http://persweb.wabash.edu/ facstaff/fellers/, http://www.umass.edu/microbio/rasmol/bilayers.htm, or some other database with lipid structures.

a. Go to "downloads" at the first site mentioned and look at the file m65.pdb, which shows a micelle containing 65 molecules of the surfactant. This picture is the output of a molecular simulation. Tell RasMol to remove the thousands of water molecules surrounding the micelle (uncheck "hydrogen" and "hetero atoms"), so you can see it.

b. At the second site mentioned, get the coordinates of the dipalmitoyl phosphatidylcholine bilayer and view it. Again remove the surrounding water. Rotate it to see the layer structure.

PART II

Diffusion, Dissipation, Drive

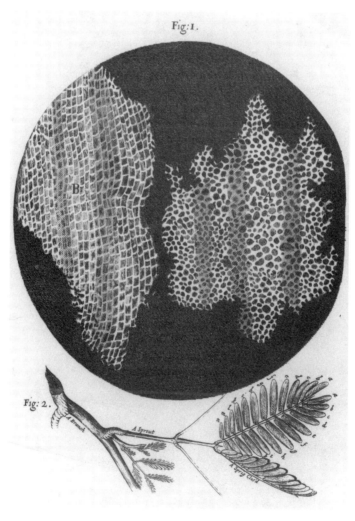

Robert Hooke's original drawing of cork cells (1665). [Hooke, *Micrographia*, 1665]

CHAPTER 3

The Molecular Dance

Who will lead me into that still more hidden and dimmer
region where Thought weds Fact, where the mental operation
of the mathematician and the physical action of the molecules
are seen in their true relation? Does not the way pass through
the very den of the metaphysician, strewed with the remains of
former explorers?
— James Clerk Maxwell, 1870

Chapter 2 made clear that living cells are full of fantastically ordered structures, all the way down to the molecular scale. But Chapter 1 proposed that heat is disorganized molecular motion and tends to destroy order. Does that imply that cells work best at the coldest temperatures? No, life processes *stop* at low temperature.

To work our way out of this paradox, and ultimately own the concept of free energy sketched in Chapter 1, we must first understand more precisely the sense in which heat is a form of motion. This chapter will begin to explain and justify that claim. We will see how the idea of random molecular motion quantitatively explains the ideal gas law (Section 1.5.4), as well as many common observations, from the evaporation of water to the speeding up of chemical reactions when we add heat.

These physical ideas have an immediate biological application: As soon as we appreciate the nanoworld as a violent place, full of incessant thermal motion, we also realize just how miraculous it is that the tiny cell nucleus can maintain a huge database—your genome—without serious loss of information over many genera-tions. Section 3.3 will see how physical reasoning led the founders of molecular bi-ology to infer the existence of a polymer carrying the database, decades before the actual discovery of DNA.

Here is a question to focus our thoughts:

Biological question: Why is the nanoworld so different from the macroworld?
Physical idea: Everything is (thermally) dancing.

3.1 THE PROBABILISTIC FACTS OF LIFE

We want to explore the idea that heat is nothing but random motion of molecules. First, though, we need a closer look at that slippery word *random*. Selecting a person at random on the street, you cannot predict that person's IQ before measuring it. But, on the other hand, you *can* be virtually certain that her IQ is less than 300! In fact, whenever we say that a measured quantity is random, we really implicitly have *some* prior knowledge of the limits its value may take and, more specifically, of the overall distribution that many measurements of that quantity will give, even though we can say little about the result of any one measurement. This observation is the starting point of statistical physics.

Scientists once found it hard to swallow the idea that sometimes physics gives only the expected distribution of measurements and cannot predict the actual measured value of, say, a particle's momentum. Actually, this limitation is a blessing in disguise. Suppose we idealize the air molecules in the room as tiny billiard balls. To specify the "state" of the system at an instant of time, we would list the positions and velocity vectors of every one of these balls. Eighteenth-century physicists believed that if they knew the initial state of a system perfectly, they could, in principle, find its final state perfectly, too. But it's absurd—the initial state of the air in this room consists of the positions and velocities of all 10^{25} or so gas molecules. Nobody has that much initial information, and *nobody wants* that much final information! Rather, we deal in aggregate quantities, such as "how much momentum do the molecules transfer to the floor in one second?" That question relates to the pressure, which we *can* easily measure.

The beautiful discovery made by physicists in the late nineteenth century is that in situations where only probabilistic information is available and only probabilistic information is desired, physics can sometimes make incredibly precise predictions. Physics won't tell you what any one molecule will do, nor will it tell you precisely when a molecule will hit the floor. But it *can* tell you the precise probability distribution of gas molecule velocities in the room, as long as there are lots of them. The following sections introduce some of the terminology we'll need to discuss probability distributions precisely.

3.1.1 Discrete distributions

Suppose some measurable variable x can take only certain discrete values x_1, x_2, ... (see Figure 3.1). Suppose we have measured x on N unrelated occasions, finding $x = x_1$ on N_1 occasions, $x = x_2$ on N_2 occasions, and so on. If we start all over with another N measurements, we'll get different numbers N_i'; but for large enough N, they should be about the same. Then we say that the **probability** of observing x_i is $P(x_i)$, where

$$N_i/N \to P(x_i) \text{ for large } N. \tag{3.1}$$

Thus, $P(x_i)$ is always a number between 0 and 1.

The probability that any given observation will yield *either* x_5 or x_{12} (say) is just $(N_5 + N_{12})/N$, or $P(x_5) + P(x_{12})$. Because the probability of observing *some* value of x is 100% (that is, 1), we must have

Figure 3.1: (Metaphor.) Examples of intermediate outcomes not allowed in a discrete probability distribution. [Cartoon by Larry Gonick, from Gonick & Smith, 1993.]

$$\sum_i P(x_i) = (N_1 + N_2 + \cdots)/N = N/N = 1. \qquad \text{normalization condition}$$

(3.2)

Equation 3.2 is sometimes expressed in the words "the probability distribution P is properly normalized."

3.1.2 Continuous distributions

More often, x can take on any value in a continuous interval. In this case, we partition the interval into **bins** of width dx. Again we imagine making many measurements and drawing a histogram, finding that $dN(x_0)$ of the measurements yield a value for x somewhere between x_0 and $x_0 + dx$. We then say that the probability of observing x in this interval is $P(x_0)\,dx$, where

$$dN(x_0)/N \rightarrow P(x_0)\,dx \text{ for large } N. \tag{3.3}$$

Strictly speaking, $P(x)$ is only defined for the discrete values of x defined by the bins. But if we make enough measurements, we can take the bin widths dx to be as small as we like and still have a lot of measurements in each bin. Thus we suppose $dN(x)$ is much greater than 1, or in symbols $dN(x) \gg 1$. If $P(x)$ approaches a smooth limiting function as we do this, then we say $P(x)$ is the probability distribution (or probability density) for x. Once again, $P(x)$ must always be nonnegative.

Equation 3.3 implies that a continuous probability distribution has dimensions inverse to those of x. A discrete distribution, in contrast, is dimensionless (see Equation 3.1). The reason for this difference is that the actual number of times we land in a small bin depends on the bin width dx. To get a quantity $P(x)$ that is independent of bin width, we must divide $dN(x_0)/N$ by dx in Equation 3.3; this operation introduces dimensions.

What if the interval isn't small? The probability of observing a value of x between x_1 and x_2 is the sum of all the bin probabilities making up that interval, or $\int_{x_1}^{x_2} dx\, P(x)$. The analog of Equation 3.2 is the normalization condition for a continuous distribution:

$$\int dx\, P(x) = 1. \tag{3.4}$$

Dull Example: The **uniform distribution** is a constant from 0 to a:

$$P(x) = \begin{cases} (1/a), & \text{if } 0 \le x \le a; \\ 0, & \text{otherwise.} \end{cases} \tag{3.5}$$

Interesting Example: The famous **Gaussian distribution** (also called the Gaussian, the bell curve, or the normal distribution) is

$$P(x) = Ae^{-(x-x_0)^2/(2\sigma^2)}, \tag{3.6}$$

where A and σ are positive constants and x_0 is some other constant.

Your Turn 3A

You can quickly see what a function looks like with your favorite graphing software. For example, in Maple writing `plot(exp(-(x-1)^2),x=-1..3);` gives Figure 3.2. Try it, then play with the constants A and σ to see how the figure changes.

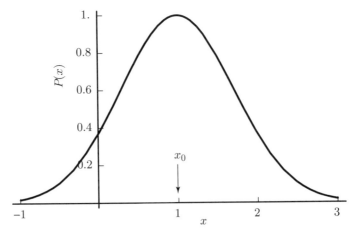

Figure 3.2: (Mathematical function.) Unnormalized Gaussian distribution centered at $x_0 = 1$ with $\sigma = 1/\sqrt{2}$ and $A = 1$ (see Equation 3.6).

The constant A isn't arbitrary; it's fixed by the normalization condition. This derivation is so important and useful that we should see how it works in detail.

Example: Find the value of A required to normalize the Gaussian distribution.
Solution: First we need to know that

$$\int_{-\infty}^{\infty} \mathrm{d}y\, e^{-y^2} = \sqrt{\pi}. \tag{3.7}$$

You can think of this expression as merely a mathematical fact to be looked up in an integral table (or see the derivation in Section 6.2.2' on page 233). What's more important are a couple of easy steps from calculus. Equation 3.4 requires that we choose the value of A in such a way that

$$1 = A \int_{-\infty}^{\infty} \mathrm{d}x\, e^{-(x-x_0)^2/(2\sigma^2)}.$$

Change variables to $y = (x - x_0)/(\sqrt{2}\sigma)$, so $\mathrm{d}y = \mathrm{d}x/(\sqrt{2}\sigma)$. Then Equation 3.7 gives $A = 1/(\sigma\sqrt{2\pi})$.

In short, the Gaussian distribution is

$$P(x) = \frac{1}{\sigma\sqrt{2\pi}} e^{-(x-x_0)^2/(2\sigma^2)}. \qquad \text{Gaussian distribution} \tag{3.8}$$

Looking at Figure 3.2, we see that it's a bump function centered at x_0 (that is, maximum there). The bump has a width controlled by σ. The larger σ is, the fatter the bump, because one can go farther away from x_0 before the factor $e^{-(x-x_0)^2/(2\sigma^2)}$ begins to hurt. Remembering that $P(x)$ is a probability distribution, this observation means that, for bigger σ, you're likely to find measurements with bigger deviations from the most likely value x_0. The prefactor of $1/\sigma$ in front of Equation 3.8 arises because a wider bump (larger σ) needs to be lower to maintain a fixed area. Let's make all these ideas more precise, for any kind of distribution.

3.1.3 Mean and variance

The **average** (or **mean** or **expectation value**) of x for any distribution is written $\langle x \rangle$ and defined by

$$\langle x \rangle = \begin{cases} \sum_i x_i P(x_i), & \text{discrete} \\ \int \mathrm{d}x\, x P(x), & \text{continuous.} \end{cases} \tag{3.9}$$

For the uniform and Gaussian distributions, the mean is the center point, because these distributions are symmetrical: There are exactly as many observations a distance d to the right of the center as there are a distance d to the left of center. For a general distribution, however, the mean needn't equal the center value, nor in general will it equal the **most probable value**, which is the place where $P(x)$ is maximum.

More generally, we may instead want the mean value $\langle f \rangle$ of some other quantity $f(x)$ depending on x. We can find $\langle f \rangle$ via

$$\langle f \rangle = \begin{cases} \sum_i f(x_i)P(x_i), & \text{discrete} \\ \int \mathrm{d}x\, f(x)P(x), & \text{continuous.} \end{cases} \tag{3.10}$$

If you go out and measure x just once, you won't necessarily get $\langle x \rangle$ right on the nose. There is some spread, which we measure by using the **root-mean-square deviation** (or **RMS deviation**, or **standard deviation**):

$$\text{RMS deviation} = \sqrt{\langle (x - \langle x \rangle)^2 \rangle}. \tag{3.11}$$

Example:

a. Show that $\langle \langle \langle f \rangle \rangle \rangle = \langle f \rangle$ for any function f of x. That is, find the average of $\langle f \rangle$.

b. Show that, if the RMS deviation equals zero, then every measurement of x really does give exactly $\langle x \rangle$.

Solution:

a. We note that $\langle f \rangle$ is a constant (that is, a number), independent of x. The average of a constant is just that constant.

b. In the formula $0 = \langle (x - \langle x \rangle)^2 \rangle = \sum_i P(x_i)(x_i - \langle x \rangle)^2$, the right-hand side doesn't have any negative terms. The only way this sum could equal zero is for every term to be zero separately, which in turn requires that $P(x_i) = 0$ unless $x_i = \langle x \rangle$.

Note that it's crucial to square the quantity $(x - \langle x \rangle)$ when defining the RMS deviation; otherwise, we'd trivially get zero for the average value $\langle (x - \langle x \rangle) \rangle$. Then we take the square root to give Equation 3.11 the same dimensions as x. We'll refer to $\langle (x - \langle x \rangle)^2 \rangle$ as the **variance** of x, or variance(x).

Your Turn 3B

a. Show that variance(x) $= \langle x^2 \rangle - (\langle x \rangle)^2$.

b. Show for the uniform distribution (Equation 3.5) that variance(x) $= a^2/12$.

Let's work out the variance of the Gaussian distribution, Equation 3.8. Changing variables as in the Example on normalization (page 73), we see that we need to compute

$$\text{variance}(x) = \frac{2\sigma^2}{\sqrt{\pi}} \int_{-\infty}^{\infty} \mathrm{d}y\, y^2 e^{-y^2}. \tag{3.12}$$

To do this calculation we need a trick, which we'll use again later: Define a function $I(b)$ by

$$I(b) = \int_{-\infty}^{\infty} dy \, e^{-by^2}.$$

Again changing variables gives $I(b) = \sqrt{\pi/b}$. Now consider the derivative dI/db. On one hand, it's

$$dI/db = -\frac{1}{2}\sqrt{\frac{\pi}{b^3}}. \tag{3.13}$$

On the other hand,

$$dI/db = \int_{-\infty}^{\infty} dy \, \frac{d}{db}e^{-by^2} = -\int_{-\infty}^{\infty} dy \, y^2 e^{-by^2}. \tag{3.14}$$

Setting $b = 1$, we see that the last integral in Equation 3.14 is the one we needed (see Equation 3.12). Combining Equations 3.13, 3.14, and 3.12 gives[1]

$$\text{variance}(x) = \frac{2\sigma^2}{\sqrt{\pi}}\left(-\frac{dI}{db}\bigg|_{b=1}\right) = \frac{2\sigma^2}{\sqrt{\pi}} \times \frac{\sqrt{\pi}}{2}.$$

Thus, the RMS deviation of the Gaussian distribution just equals the parameter σ appearing in Equation 3.8.

3.1.4 Addition and multiplication rules

Addition rule Section 3.1.1 noted that, for a discrete distribution, the probability that the next measured value of x is either x_i or x_j equals $P(x_i) + P(x_j)$, unless $i = j$. The key point is that x can't equal *both* x_i and x_j; we say that the alternative values are **exclusive**. More generally, the probability that a person is either taller than 2 m or shorter than 1.9 m is obtained by addition, whereas the probability of being either taller than 2 m or nearsighted cannot be obtained in this way.

For a continuous distribution, the probability that the next measured value of x is either between a and b or between c and d equals the sum, $\int_a^b dx \, P(x) + \int_c^d dx \, P(x)$, provided the two intervals don't overlap. This result follows because the two probabilities (to be between a and b or between c and d) are exclusive in this case.

Multiplication rule Now suppose that we measure two independent quantities, for example, tossing a coin and rolling a die. What is the probability that we get heads *and* roll a 6? To find out, just list all $2 \times 6 = 12$ possibilities. Each is equally probable, so the chance of getting the specified outcome is $\frac{1}{12}$. This example shows that the joint probability distribution for two independent events is the *product* of the two simpler distributions. Let $P_{\text{joint}}(x_i, y_K)$ be the joint distribution, where $i = 1$ or 2 and $x_1 =$(heads), $x_2 =$(tails); similarly, $y_K = K$, the number on the die. Then the

---

[1] The notation $\frac{dI}{db}\big|_{b=1}$ means the derivative of $I(b)$ with respect to b, evaluated at the point $b = 1$. See Appendix A for more on mathematical notation.

multiplication rule says

$$P_{\text{joint}}(x_i, y_K) = P_{\text{coin}}(x_i) \times P_{\text{die}}(y_K).\tag{3.15}$$

Equation 3.15 is correct even for loaded dice (the $P_{\text{die}}(y_K)$ aren't all equal to $\frac{1}{6}$) or a two-headed coin ($P_{\text{coin}}(x_1) = 1$, $P_{\text{coin}}(x_2) = 0$). On the other hand, for two connected events (for example, the chance of rain versus the chance of hail), we don't get such a simple relation.

> **Your Turn 3C**
>
> Show that if P_{coin} and P_{die} are correctly normalized, then so will be P_{joint}.

> **Your Turn 3D**
>
> Suppose we roll *two* dice. What's the probability that the numbers on the dice add up to 2? To 6? To 12? Think about how you used both the addition and the multiplication rule for this.

Here's a more complicated example. Suppose you are shooting arrows into a distant target. Wind currents give random shifts to the x component of your arrows' arrival locations, and independent random shifts to the y component. Suppose that the probability distribution $P_x(x)$ is Gaussian with variance σ^2, and that the same is true for $P_y(y)$.

Example: Find the probability, $P(r)\,dr$, that an arrow lands a distance between r and $r + dr$ from the bull's-eye.

Solution: We must use both the rules discussed earlier. r is the length of the displacement vector: $r \equiv |\mathbf{r}| \equiv \sqrt{x^2 + y^2}$. First, we find the joint distribution, the probability that the x-component lies between x and $x + dx$ *and* the y-component lies between y and $y + dy$. The multiplication rule gives this probability as

$$
\begin{aligned}
P_{xy}(x, y)\mathrm{d}x\mathrm{d}y &= P_x(x)\mathrm{d}x \times P_y(y)\mathrm{d}y \\
&= \left(2\pi\sigma^2\right)^{-2/2} e^{-(x^2+y^2)/(2\sigma^2)} \times \mathrm{d}x\mathrm{d}y \\
&\equiv \left(2\pi\sigma^2\right)^{-1} e^{-r^2/(2\sigma^2)}\mathrm{d}^2\mathbf{r}.
\end{aligned}\tag{3.16}
$$

The two Gaussians combine into a single exponential involving only the distance r.

We're not done. Many different displacement vectors $\mathbf{r}$ all have the same r; to find the total probability that $\mathbf{r}$ has any of these values, we must now use the addition rule. Think about all the $\mathbf{r}$ vectors with length lying between r and $r + dr$. They form a thin ring of width dr. The joint probability distribution $P_{xy}(\mathbf{r})$ is the same for all these $\mathbf{r}$, because it depends only on the length of $\mathbf{r}$. So, to sum all the probabilities, we multiply P_{xy} by the total area of the ring, which is its circumference times its

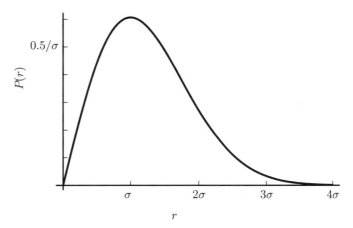

Figure 3.3: (Mathematical function.) The probability distribution $P(r)$ for the distance r from the origin, when both x and y are independent Gaussian distributions with variance σ^2.

thickness: $2\pi r\,dr$. We thus get

$$P(r)\,dr = \left(\frac{1}{2\pi\sigma^2}\right) e^{-r^2/(2\sigma^2)} \times 2\pi r\,dr. \tag{3.17}$$

Figure 3.3 shows this distribution.

Notice two notational conventions used in this Example (see also Appendix A). First, the symbol $\equiv$ is a special form of the equal sign that alerts us to the fact that $r \equiv |\mathbf{r}|$ is a *definition*: it defines the number r in terms of the vector $\mathbf{r}$. We pronounce this symbol "is defined as" or "equals by definition." Second, the symbol $d^2\mathbf{r}$ denotes the *area* of a little box in position space; it is not itself a vector. The integral of $d^2\mathbf{r}$ over a region of space equals that region's area.

Your Turn 3E

Find the fraction of all the arrows you shoot that land outside a circle of some radius R_0, as a function of R_0.

Your Turn 3F

a. Repeat the calculation in the Example just given, for a *three*-component vector $\mathbf{v}$, each of whose components is an independent, random variable distributed as a Gaussian distribution with variance σ^2. That is, let u denote the length of $\mathbf{v}$ and find $P(u)\,du$. [*Hint:* Examine Figure 3.4.]

b. Graph your answer to (a) with a computer math package. Again try various values of σ.

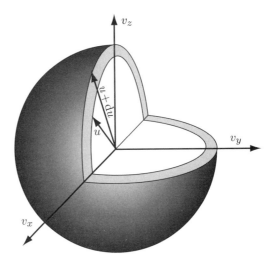

Figure 3.4: (Sketch.) The endpoints of all the vectors $\mathbf{v} = (v_x, v_y, v_z)$ having length u form a sphere. The endpoints of all the vectors with length between u and $u + \mathrm{d}u$ form a spherical shell.

3.2 DECODING THE IDEAL GAS LAW

Let's try to interpret the ideal gas law (Equation 1.11 on page 27), and its universal constant k_B, in the light of the working hypothesis that heat is random motion. Once we make this hypothesis precise, and confirm it, we'll be in a position to understand many physical aspects of the nanoworld.

3.2.1 Temperature reflects the average kinetic energy of thermal motion

When faced with a mysterious new formula, our first impulse should be to think about it in the light of dimensional analysis.

> ***Your Turn 3G***
>
> Examine the left side of the ideal gas law (Equation 1.11 on page 27) and show that the product $k_B T$ has the units of energy, consistent with the numerical value given in Equation 1.12.

So we have a law of Nature, and it contains a fundamental, universal constant with units of energy. We still haven't interpreted the meaning of that constant, but we will in a moment; knowing its units will help us.

Let's think some more about the box of gas introduced in Section 1.5.4 on page 27. If the density is low enough (an ideal gas), the molecules don't hit one another very often.[2] But certainly each one does hit the *walls* of the box. We now ask whether

---

[2] $\boxed{T_2}$ The precise way to say this is that we define an ideal gas to be one for which the time-averaged potential energy of each molecule in its neighbors' potential fields is negligible relative to its kinetic energy.

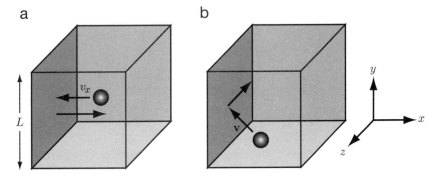

Figure 3.5: (Schematic.) Origin of gas pressure in a cubical box of length L. (a) A molecule traveling parallel to an edge with velocity v_x bounces elastically off a wall of its container. The effect of the collision is to reverse the direction of the molecule, transferring momentum $2mv_x$ to the wall. (b) A molecule traveling with arbitrary velocity **v**. If its next collision is with a wall parallel to the yz-plane, the effect of the collision is to reverse the x-component of the molecule's momentum, again transferring momentum $2mv_x$ to the wall.

that constant hitting of the walls can explain the phenomenon of pressure. Suppose that a gas molecule of mass m is traveling parallel to one edge of the box (say in the x direction) with speed v_x, and the box is a cube of length L (see Figure 3.5a).

Every time the molecule hits the wall, the molecule's momentum changes from mv_x to $-mv_x$; it delivers $2mv_x$ to the wall. This event happens every time the molecule makes a round trip, which takes a time $\Delta t = 2L/v_x$. If there are N molecules, all with this velocity, then the total rate at which they deliver momentum to the wall is $(2mv_x)(v_x/2L)N$. But you learned in first-year physics that the *rate of delivery* of momentum is precisely the force on the box's wall.

Your Turn 3H

Check the dimensions of the formula $f = (2mv_x)(v_x/2L)N$ to make sure they are appropriate for a force.

In reality, every molecule has its own, individual velocity v_x. So what we need is not N times one molecule's velocity-squared, but instead the sum over all molecules, or equivalently, N times the *average* velocity-squared. As in Equation 3.9, we use the shorthand notation $\langle v_x^2 \rangle$ for this quantity.

The force per unit area on the wall is called **pressure**, so we have just found that

$$p = m\langle v_x^2 \rangle N/V. \tag{3.18}$$

Our simple formula Equation 3.18, which embodies the idea that a gas consists of molecules in motion, has already explained two key features of the experimentally observed ideal gas law (Equation 1.11), namely, the facts that the pressure is proportional to N and to $1/V$.

Skeptics may say, "Wait a minute. In a real gas, the molecules aren't all traveling in the x direction!" It's true. Still, it's not hard to do a better job. Figure 3.5b shows the situation. Each individual molecule has a velocity vector **v**. When it hits the wall at $x = L$, its component v_x changes sign, but v_y and v_z don't. So, the momentum delivered to the wall is again $2mv_x$. Also, the time between bounces off this particular wall is once again $2L/v_x$, even though in the meantime the molecule may bounce off other walls as well, as a result of its motion along y and z. Repeating the argument leading to Equation 3.18 in this more general situation, we find that it needs no modifications.

Combining the ideal gas law with Equation 3.18 gives

$$m\langle v_x^2 \rangle = k_{\mathrm{B}}T. \tag{3.19}$$

The gas molecules are flying around at random. So the average, $\langle v_x \rangle$, is zero: There are just as many molecules traveling left as there are traveling right, so their contributions to $\langle v_x \rangle$ cancel. But the *square* of the velocity can have a nonzero average, $\langle v_x^2 \rangle$. Just as in the discussion of Equation 3.11, both the left-movers and right-movers have positive values of v_x^2; so instead of canceling, they add.

In fact, there's nothing special about the x direction. The averages $\langle v_x^2 \rangle$, $\langle v_y^2 \rangle$, and $\langle v_z^2 \rangle$ are all equal. So, their sum is three times as big as any individual term. The sum $v_x^2 + v_y^2 + v_z^2$ is the total length of the velocity vector, so $\langle \mathbf{v}^2 \rangle = 3\langle v_x^2 \rangle$. Thus, we can rewrite Equation 3.19 as

$$\tfrac{1}{2} \times \tfrac{1}{3}m\langle \mathbf{v}^2 \rangle = \tfrac{1}{2}k_{\mathrm{B}}T. \tag{3.20}$$

We now rephrase Equation 3.20, using the fact that the kinetic energy of a particle is $\tfrac{1}{2}mu^2$, to find that

The average kinetic energy of a molecule in an ideal gas is $\tfrac{3}{2}k_{\mathrm{B}}T$, (3.21)

regardless of what kind of gas we have. Even in a mixture of gases, the molecules of each type must separately obey Idea 3.21.

The analysis leading to Idea 3.21 was given by Rudolph Clausius in 1857; it supplies the deep molecular meaning of the ideal gas law. Alternatively, we can regard Idea 3.21 as explaining the concept of temperature itself, in the special case of an ideal gas.

Let's work out some numbers to get a feeling for what our results mean. A mole of air occupies 24 L (that's $0.024\,\mathrm{m}^3$) at atmospheric pressure and room temperature. What's atmospheric pressure? It's a pressure big enough to lift a column of water about ten meters (you can't sip water through a straw taller than this). A 10 m column of water presses down with a force per area (pressure) equal to the height times the mass density of water times the acceleration of gravity, or $z\rho_{\mathrm{m,w}}g$. Thus, atmospheric pressure is

$$p \approx 10\,\mathrm{m} \times \left(10^3\,\frac{\mathrm{kg}}{\mathrm{m}^3}\right) \times \left(9.8\,\frac{\mathrm{m}}{\mathrm{s}^2}\right) \approx 10^5\,\frac{\mathrm{kg}}{\mathrm{m\,s}^2} = 10^5\,\mathrm{Pa}. \tag{3.22}$$

Here $\approx$ means "equals approximately" and Pa stands for **pascal**, the SI unit of pressure. Substituting $V = 0.024\,\mathrm{m}^3$, $p \approx 10^5\,\mathrm{kg\,m^{-1}s^{-2}}$, and $N = N_{\mathrm{mole}}$ into the ideal gas law (Equation 1.11 on page 27) shows that, indeed, it is approximately satisfied:

$$\left(10^5 \frac{\text{kg}}{\text{m s}^2}\right) \times \left(0.024\,\text{m}^3\right) \approx \left(6.0 \cdot 10^{23}\right) \times \left(4.1 \cdot 10^{-21}\,\text{J}\right).$$

We can go further. Air consists mostly of nitrogen molecules. The molar mass of atomic nitrogen is about $14\,\text{g mole}^{-1}$, so a mole of nitrogen molecules, N_2, has mass about 28 g. Thus, the mass of *one* nitrogen molecule is $m = 0.028\,\text{kg}/N_{\text{mole}} = 4.7 \cdot 10^{-26}\,\text{kg}$.

Your Turn 3I

Using Idea 3.21, show that the typical speed of air molecules in the room where you're sitting is about $\sqrt{\langle v^2 \rangle} \approx 500\,\text{m s}^{-1}$. Convert to miles/hour (or km/hour) to see whether you should drive that fast (maybe in the space shuttle).

So the air molecules in your room are pretty frisky. Can we get some independent confirmation to see if this result is reasonable? Well, one thing we know about air is ... there's less of it on top of Mt. Everest. This density difference arises because gravity exerts a tiny pull on every air molecule. On the other hand, the air density in your room is quite uniform from top to bottom. Apparently, the typical kinetic energy of air molecules, $\frac{3}{2}k_B T_r$, is so high that the difference in gravitational potential energy, ΔU, from the top to the bottom of a room is negligible, whereas the difference from sea level to Mt. Everest is not so negligible. Let's make the very rough estimate that Everest is $z = 9\,\text{km}$ high and that the resulting ΔU is roughly equal to the mean kinetic energy:

$$\Delta U = mg(9\,\text{km}) \approx \tfrac{1}{2}m\langle v^2 \rangle. \tag{3.23}$$

Your Turn 3J

Show that the typical speed is about $u = 420\,\text{m s}^{-1}$, or reasonably close to what you just found in Your Turn 3I. (Neglect the temperature difference between sea level and mountaintop.)

This new estimate is completely independent of the one we got from the ideal gas law, so the fact that it gives the same typical u is evidence that we're on the right track.

Your Turn 3K

a. Compare the average kinetic energy $\frac{3}{2}k_B T_r$ of air molecules to the difference in gravitational potential energies ΔU at the top and the bottom of a room. Assume that the height of the ceiling is $z = 3\,\text{m}$. Why doesn't the air in the room fall to the floor? What could you do to *make* it fall?

b. Repeat (a), but this time calculate the appropriate energies for a dirt particle. Suppose that the particle weighs about as much as a 50 μm cube of water. Why does dirt fall to the floor?

In this section, we have seen how the hypothesis of random molecular motion, with an average kinetic energy proportional to the absolute temperature, explains the ideal gas law and a number of other facts. Other questions, however, come to mind. For example, if heating a pan of water raises the kinetic energy of the water molecules, why don't they all suddenly fly away when the temperature gets to some critical value, the one giving them enough energy to escape? To understand questions like this one, we need to keep in mind that the average kinetic energy is far from the whole story. We also want to know about the full *distribution* of molecular velocities, not just its mean-square value.

3.2.2 The complete distribution of molecular velocities is experimentally measurable

The logic in the previous subsection was a bit informal, in keeping with the exploratory character of the discussion. But we ended with a precise question: What is the full distribution of molecular velocities? In other words, how many molecules are moving at $1000\,\mathrm{m\,s^{-1}}$; how many at $10\,\mathrm{m\,s^{-1}}$? The ideal gas law implies that $\langle \mathbf{v}^2 \rangle$ changes in a very simple way with temperature (Idea 3.21), but what about the complete distribution?

These are not just theoretical questions. One can measure directly the distribution of speeds of gas molecules. Imagine taking a box full of gas (in practice, the experiment is done using a vaporized metal) with a pinhole that lets gas molecules emerge into a region of vacuum (Figure 3.6). The pinhole is made so small that the escaping gas molecules do not disturb the state of the others inside the box. The emerging molecules pass through an obstacle course, which only allows those with

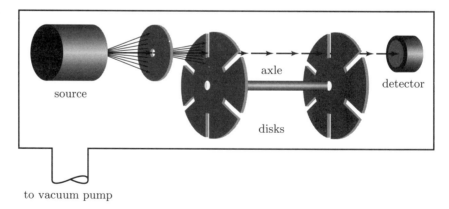

to vacuum pump

Figure 3.6: (Schematic.) An experimental apparatus to measure the distribution of molecular speeds by using a velocity filter consisting of two rotating slotted disks. To pass through the filter, a gas molecule must arrive at the left disk when a slot is in the proper position, then also arrive at the right disk exactly when another slot arrives at the proper position. Thus, only molecules with one selected speed pass through to the detector; the selected speed can be set by adjusting how fast the disks spin. [Adapted from Reif, 1965.]

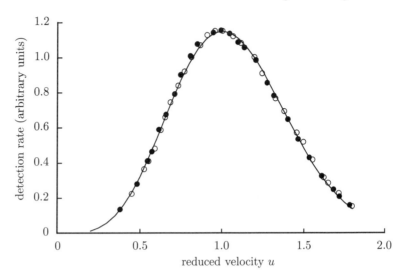

Figure 3.7: (Experimental data with fit.) Speeds of atoms emerging from a box of thallium vapor, at two different temperatures. *Open circles:* $T = 944\,$K. *Solid circles:* $T = 870\,$K. The quantity $\bar{u}$ on the horizontal axis equals $u\sqrt{m/4k_B T}$; both distributions have the same most probable value, $\bar{u}_{max} = 1$. Thus u_{max} is larger for higher temperatures, as implied by Idea 3.21. The vertical axis shows the rate at which atoms hit a detector after passing through a filter like the one sketched in Figure 3.6 (times an arbitrary rescaling factor). *Solid line:* Theoretical prediction (see Problem 3.5). This curve fits the experimental data with *no* adjustable parameters. [Data from Miller & Kusch, 1955.]

speeds in a particular range to pass. The successful molecules then land on a detector, which measures the total number arriving per unit time.

Figure 3.7 shows the results of such an experiment. Even though individual molecules have random velocities, clearly the *distribution* of velocities is predictable and smooth. The data also show clearly that a given gas at different temperatures will have closely related velocity distributions; two different data sets lie on the same curve after rescaling the molecular speed u.

3.2.3 The Boltzmann distribution

Let's use the ideas of Section 3.2.1 to understand the experimental data in Figure 3.7. We are exploring the idea that, even though each molecule's velocity cannot be predicted, there is nevertheless a definite prediction for the *distribution* of molecular velocities. One thing we know about that probability distribution is that it must fall off at large velocities: Certainly there won't be any gas molecules in the room moving at a million meters per second! Moreover, the average speed must increase as we make the gas hotter, because we've argued that the average kinetic energy is proportional to T (see Idea 3.21). Finally, the probability of finding a molecule moving to the left at some velocity v_x should be the same as that for finding it moving to the right at $-v_x$.

One probability distribution with these properties is the Gaussian (Equation 3.8), where the spread σ increases with temperature and the mean is zero. (If the mean were nonzero, there'd be a net, directed, motion of the gas, that is, a wind blowing.) Remarkably, this simple distribution really does describe any ideal gas! More precisely, the probability $P(v_x)$ of finding that a given molecule at a given time has its x-component of velocity equal to v_x is a Gaussian, like the form shown in Figure 3.2, but centered on 0. Each molecule is incessantly changing its speed and direction. What's unchanging is not the velocity of any one molecule but the distribution $P(v_x)$.

We can replace the vague idea that the variance σ^2 of v_x increases with temperature by something more precise. Because the mean velocity equals zero, Your Turn 3B on page 74 says that the variance of v_x is $\langle v_x^2 \rangle$. According to Idea 3.21, the mean kinetic energy is $\frac{3}{2} k_B T$. Combining these statements gives

$$\sigma^2 = k_B T / m. \qquad (3.24)$$

Section 1.5.4 on page 27 gave the numerical value of $k_B T$ at room temperature as $k_B T_r \approx 4.1 \cdot 10^{-21}$ J. That's pretty small, but so is the mass m of one gas molecule, so σ need not be small. In fact, you showed in Your Turn 3I that the quantity $\sqrt{k_B T_r / m}$ corresponds to a large velocity.

Now that we have the probability distribution for one component of the velocity, we can follow the approach of Section 3.1.4 to get the three-dimensional distribution, $P(\mathbf{v})$. Your result in Your Turn 3F on page 77 then gives the distribution of molecular speeds, a function similar to the one shown in Figure 3.3.[3]

Your Turn 3L

Find the most probable value of the speed u. Find the mean speed $\langle u \rangle$. Looking at the graph you drew in Your Turn 3F (or the related function in Figure 3.3), explain geometrically why these are/aren't the same.

Still assuming that the molecules move independently and are not subjected to any external force, we can next find the probability that *all* N molecules in the room have specified velocities $\mathbf{v}_1, \ldots, \mathbf{v}_N$, again using the multiplication rule:

$$P(\mathbf{v}_1, \ldots, \mathbf{v}_N) \propto e^{-m\mathbf{v}_1^2/(2k_B T)} \times \cdots \times e^{-m\mathbf{v}_N^2/(2k_B T)} = e^{-\frac{1}{2}m(\mathbf{v}_1^2 + \cdots + \mathbf{v}_N^2)/k_B T}. \qquad (3.25)$$

James Clerk Maxwell derived Equation 3.25 and showed how it explained many properties of gases. The proportionality sign, $\propto$, reminds us that we haven't bothered to write down the appropriate normalization factor.

Equation 3.25 applies only to an ideal gas, free from any external influences. Chapter 6 will generalize this formula. Although we're not ready to prove this generalization, we can at least form some reasonable expectations:

---

[3] $\boxed{T_2}$ The curve fitting the experimental data in Figure 3.7 is almost, but not quite, the one you found in Your Turn 3F(b). You'll find the precise relation in Problem 3.5.

- If we wanted to discuss the whole atmosphere, for example, we'd have to understand why the distribution is spatially nonuniform—air gets thinner at higher altitudes. But Equation 3.25 gives us a hint. Apart from the normalization factor, the distribution given by Equation 3.25 is just $e^{-E/k_B T}$, where E is the kinetic energy. When altitude (potential energy) starts to become important, it's reasonable to guess that we should just replace E by the molecule's *total* (kinetic plus potential) energy. Indeed, we then find the air thinning out, with density proportional to the exponential of minus the altitude (because the potential energy of a molecule is mgz).

- Molecules in a sample of air hardly interact at all—air is nearly an ideal gas. But in more crowded systems, such as liquid water, the molecules interact a lot. There the molecules are not independent and we can't simply use the multiplication rule. But again we can form some reasonable expectations. The statement that "the molecules interact" means that the potential energy isn't just the sum of independent terms $U(x_1) + \cdots + U(x_N)$ but instead is some joint function $U(x_1, \ldots, x_N)$. Calling the corresponding total energy $E \equiv E(x_1, v_1; \ldots ; x_N, v_N)$, let's substitute *that* into our provisional formula:

$$P(\text{state}) \propto e^{-E/k_B T}. \qquad \text{Boltzmann distribution} \tag{3.26}$$

We will refer to this formula as the **Boltzmann distribution**[4] after Ludwig Boltzmann, who found it in the late 1860s.

We should pause to unpack the very condensed notation in Equation 3.26. To describe a state of the system, we must give the location **r** of each particle, as well as its speed **v**. The probability of finding particle a with its first coordinate lying between $x_{1,a}$ and $x_{1,a} + dx_{1,a}$ and so on, and its first velocity lying between $v_{1,a}$ and $v_{1,a} + dv_{1,a}$ and so on, equals

$$dx_{1,a} \times \cdots \times dv_{1,a} \times \cdots \times P(x_{1,a}, \ldots, v_{1,a}, \ldots). \tag{3.27}$$

For K particles, the probability distribution $P(x_{1,a}, \ldots, v_{1,a}, \ldots)$ is a function of $6K$ variables given by Equation 3.26.

Equation 3.26 has some reasonable features: At very low temperatures, or $T \to 0$, the exponential is a very rapidly decreasing function of **v**: The system is overwhelmingly likely to be in the lowest energy state available to it. (In a gas, this state is the one in which all of the molecules are lying on the floor at zero velocity.) As we raise the temperature, thermal agitation begins; the molecules begin to have a range of energies, which gets broader as T increases.

It's almost unbelievable, but the very simple formula Equation 3.26 is exact. It's not simplified; you'll never have to unlearn it and replace it by anything more complicated. (Suitably interpreted, it holds without changes even in quantum mechanics.) Chapter 6 will derive it from very general considerations.

---

[4]Some authors use the synonym "canonical ensemble."

3.2.4 Activation barriers control reaction rates

We are now in a better position to think about a question posed at the end of Section 3.2.1: If heating a pan of water raises the kinetic energy of its molecules, then why doesn't the water in the pan evaporate suddenly, as soon as it reaches a critical temperature? For that matter, why does evaporation cool the remaining water?

To think about this puzzle, imagine that it takes a certain amount of kinetic energy $E_{barrier}$ for a water molecule to break free of its neighbors (because they attract one another). Any water molecule near the surface with at least this much energy can leave the pan; we say that there is an **activation barrier** to escape. Suppose we heat a covered pan of water, then turn off the heat and momentarily remove the lid, allowing the most energetic molecules to escape. The effect of removing the lid is to *clip* the Boltzmann probability distribution, as suggested by the solid line in Figure 3.8a. We now replace the lid of the pan and thermally insulate it. Now the constant jostling of the remaining molecules once again pushes some up to higher energies, regrowing the tail of the distribution as shown by the dashed line of Figure 3.8a. We say that the remaining molecules have **equilibrated**. But the new distribution is not quite the same as it was initially. Because we removed the most energetic molecules, the average energy of those remaining is less than it was to begin with: Evaporation cooled the remaining water. Moreover, rearranging the distribution takes time: Evaporation doesn't happen all at once. If we had assumed the water to be hotter initially, however, its initial distribution of energies would have been farther to the right (Figure 3.8b), and more of the molecules would have been ready to escape. In other words, evaporation proceeds faster at higher temperature.

The idea of activation barriers can help make sense of our experience with chemical reactions, too. When you flip a light switch, or click your computer's mouse, there is a minimal energy, or activation barrier, which your finger must supply. Tapping the switch too lightly may move it a fraction of a millimeter but doesn't click it over to its "on" position. Now imagine drumming your finger lightly on the switch, giving a series of random light taps with some distribution of energies. Given enough time, eventually one tap will be above the activation barrier and the switch will flip.

Similarly, one can imagine that a molecule with a lot of stored energy, say hydrogen peroxide, can only release that energy after a minimal initial kick pushes it over an activation barrier. The molecule constantly gets kicks from the thermal motion of its neighbors. If most of those thermal kicks are much smaller than the barrier, however, it will be a very long time before a big enough kick occurs. Such a molecule is practically stable. We can speed up the reaction by heating the system, just as in evaporation. For example, a candle is stable, but it burns when we touch it with a lighted match. The energy released by burning in turn keeps the candle hot long enough to burn some more, and so on.

We can do better than these simple qualitative remarks. Our argument implies that the rate of a reaction is proportional to the fraction of all molecules whose energy exceeds the threshold. Consulting Figure 3.8, we see that we want the *area* under the part of the original distribution that gets clipped when molecules escape over the barrier. The fraction of molecules represented by this area is small at low temperatures (see Figure 3.8a). In general, the area depends on the temperature with a factor of

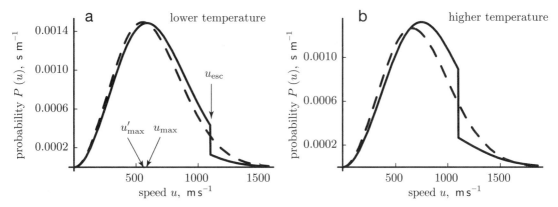

Figure 3.8: (Mathematical functions.) (a) *Solid line:* The distribution of molecular speeds for a sample of water, initially at 100°C, from which some of the most energetic molecules have suddenly been removed. After we reseal the system, molecular collisions bring the distribution of molecular speeds back to the standard form (*dashed line*). The new distribution has regenerated a high-energy tail, but the average kinetic energy did not change; accordingly, the peak has shifted slightly, from u_{max} to u'_{max}. (b) The same system, with the same escape speed; but this time the system starts at a higher temperature. The fraction of the distribution removed is now greater than in (a), and hence the shift in the peak is larger, too.

$e^{-E_{barrier}/k_B T}$. You already found such a result in a simpler situation in Your Turn 3E on page 77: Substituting u_0 for the distance R_0 in that problem, and $k_B T/m$ for σ^2, indeed gives the fraction of molecules over threshold as $e^{-mu_0^2/(2k_B T)}$.

The preceding argument is rather incomplete. For example, it assumes that a chemical reaction consists of a single step, which certainly is not true for many reactions. But there are many elementary reactions between simple molecules for which our conclusion is experimentally true:

> *The rates of simple chemical reactions depend on temperature mainly via a factor of $e^{-E_{barrier}/k_B T}$, where $E_{barrier}$ is some temperature-independent constant characterizing the reaction.* (3.28)

We will refer to Idea 3.28 as the **Arrhenius rate law.** Chapter 10 will discuss it in greater detail.

3.2.5 Relaxation to equilibrium

We are beginning to see the outlines of a big idea: When a gas, or other complicated statistical system, is left to itself under constant external conditions for a long time, it arrives at a situation where the probability distributions of its physical quantities don't change over time. Such a situation is called thermal equilibrium. We will define and explore equilibrium more precisely in Chapter 6, but already something may be troubling you, as it is troubling Gilbert:

Gilbert: Very good, you say the air doesn't fall on the floor at room temperature because of thermal motion. Why then doesn't it slow down and eventually stop (and then fall on the floor), as a result of friction?

Sullivan: Oh, no, that's quite impossible because of the conservation of energy. Each gas molecule makes only elastic collisions with others, just like the billiard balls in first-year physics.

Gilbert: Oh? So then, what *is* friction? If I drop two balls off the Tower of Pisa, the lighter one gets there later, because of friction. Everybody knows that mechanical energy isn't conserved; eventually it winds up as heat.

Sullivan: Uh, um, ...

As you can see, a little knowledge proves to be a dangerous thing for our two fictitious scientists. Suppose that, instead of dropping a ball, we shoot one air molecule into the room with enormous speed, say, 100 times greater than the average molecular speed. (One can actually do this experiment with a particle accelerator.) What happens?

Soon this molecule bangs into one of the ones that was in the room to begin with. There's an overwhelming likelihood that the latter molecule will have kinetic energy much smaller than the injected one and, indeed, probably not much more than the average. When they collide, the fast one transfers a lot of its kinetic energy to the slow one. Even though the collision was elastic, the fast one lost a lot of energy. Now we have two medium-fast molecules; each is closer to the average than it was to begin with. Each one now cruises along till it bangs into another, and so on, until they all blend into the general distribution (Figure 3.9).

Even though the total energy in the system is unchanged after each collision, the original *distribution* (with one molecule way out of line with the others) will settle down to the equilibrium distribution (Equation 3.26), by a process of sharing the

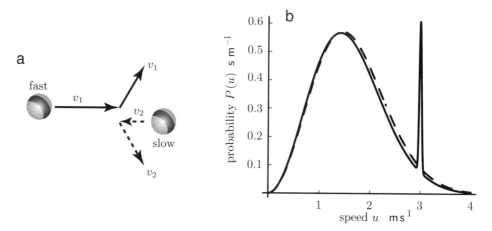

Figure 3.9: (Schematic; sketch graph.) (a) When a fast billiard ball collides with a slow one, in general both move away with a more equal division of their total kinetic energy than before. (b) An initial molecular speed distribution (*solid line*) with one anomalously fast molecule (or a few, creating the bump in the graph) quickly reequilibrates to a Boltzmann distribution at slightly higher temperature (*dashed line*). Compare with Figure 3.8.

energy in the original fast molecule with all the others.[5] What has changed is not energy but the *ordering* of that energy: The one dissident in the crowd has faded into anonymity. Again, the directed motion of the original molecule has been degraded to a tiny increase in the average random motion of its peers. But, average random velocity is just temperature, according to Equation 3.26. In other words, *mechanical energy has been converted to thermal energy* in the process of reaching equilibrium. **Friction** is the name for this conversion.

3.3 EXCURSION: A LESSON FROM HEREDITY

Section 1.2 outlined a broad puzzle about life (the generation of order) and a correspondingly broad outline of a resolution. Many of the points made there were elegantly summarized in a short but enormously influential essay by the physicist Erwin Schrödinger in 1944. Schrödinger then went on to discuss a vexing question from antiquity: the *transmission* of order from one organism to its descendants. Schrödinger noted that this transmission was extremely accurate. Now that we have some concrete ideas about probability and the dance of the molecules, we can better appreciate why Schrödinger found that everyday observation to be so profound. In fact, careful thought about the physical context underlying known biological facts led Schrödinger's contemporary Max Delbrück to an accurate prediction of what the genetic carrier would be like, decades before the discovery of the details of DNA's structure and role in cells. Delbrück's argument rested on ideas from probability theory, as well as on the idea of thermal motion.

3.3.1 Aristotle weighs in

Classical and medieval authors debated long and hard about the material basis of the facts of heredity. Many believed that the only possible solution was that the egg contains somewhere inside itself a tiny but complete chicken, which needed only to grow. In a prescient analysis, Aristotle rejected this view, pointing out, for example, that certain inherited traits can skip a generation entirely. Contrary to Hippocrates, Aristotle argued,

> The male contributes the *plan of development* and the female the
> substrate.... The sperm contributes nothing to the material body of
> the embryo, but only communicates its program of development ...
> just as no part of the carpenter enters into the wood in which he works.

Aristotle missed the fact that the mother also contributes to the "plan of development," but he made crucial progress by insisting on the separate role of an *information carrier* in heredity. The organism uses the carrier in two distinct ways:

---

[5]What if we take one molecule and slow it down to much *smaller* speed than its peers? Now, the molecule tends to *gain* energy by collisions with average molecules, until once again it lies in the Boltzmann distribution.

- It uses the software stored in the carrier to direct its own construction; and

- It *duplicates* the software, and the carrier on which it is stored, for transmission to the offspring.

Today we make this distinction by referring to the collection of physical characteristics of the organism (the output of the software) as the **phenotype**, and the program itself as the **genotype**.

It was Aristotle's misfortune that medieval commentators fastened on his confused ideas about physics, raising them to the level of dogma while ignoring his correct biology. Even Aristotle, however, could not have guessed that the genetic information carrier would turn out to be a single molecule.

3.3.2 Identifying the physical carrier of genetic information

Nobody has ever seen a molecule with their unaided eye. We can nevertheless speak with confidence about molecules, because the molecular hypothesis makes such a tightly interconnected web of falsifiable predictions. A similarly indirect but tight web of evidence drew Schrödinger's contemporaries to their conclusions about the molecular basis of heredity.

To begin, thousands of years' experience in agronomy and animal husbandry had shown that any organism can be inbred to the point where it will breed true for many generations. This statement does not mean that every individual in a purebred lineage will be exactly identical to every other one—certainly there are individual variations. Rather, a purebred stock is one in which there are no *heritable* variations among individuals. To make the distinction clear, suppose we take a purebred population of sheep and make a histogram of, say, femur lengths. A familiar Gaussian-type distribution emerges. Suppose now that we take an unusually big sheep, from the high end of the distribution (see Figure 3.10). Its offspring will not be unusually big; rather, they will lie on the same distribution as the population from which the parent was drawn. Whatever the genetic carrier is, it gets duplicated and transmitted with great accuracy. Indeed, in humans, some characteristic features can be traced through 10 generations.

The significance of this remark may not be immediately obvious. After all, an audio compact disk contains nearly 10^{10} bits of information, duplicated and transmitted with near-perfect fidelity from the factory. But each sheep began with a single cell. A sperm head is only a micrometer or so across, yet it contains roughly the same massive amount of text as that compact disk, in a package around 10^{-13} times the volume! What sort of physical object could lie behind this feat of miniaturization? Nineteenth-century science and technology offered no direct answers to this question. But a remarkable chain of observation and logic broke this impasse, starting with the work of Gregor Mendel, a monk trained in physics and mathematics.

Mendel's chosen model system was the flowering pea plant, *Pisum sativum*. He chose to study seven heritable features (flower position, seed color, seed shape, ripe pod shape, unripe pod color, flower color, and stem length). Each occurred in two clearly identifiable, alternative forms. The distinctness of these features, or traits, endured over many generations, leading Mendel to propose that *sufficiently simple traits*

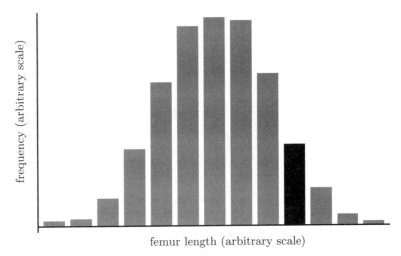

Figure 3.10: (Sketch histogram.) Results of an imaginary experiment measuring the femur lengths of a purebred population of sheep. Selectively breeding sheep from the atypical group shown (*black bar*) doesn't lead to a generation of bigger sheep, but instead to offspring with the same distribution as the one shown.

are inherited in a discrete, yes/no manner. Mendel imagined the genetic code as a collection of switches, which he called factors, each of which could be set to either of two (or more) settings. The various available options for a given factor are now called **alleles** of that factor. Later work would show that other traits, which appear to be continuously variable (for example, hair color), are really the combined effect of so many different factors that the discrete variations from individual factors can't be distinguished.

Painstaking analysis of many pea plants across several generations led Mendel in 1865 to a set of simple conclusions:[6]

- The cells making up most of an individual (somatic cells) each carry two copies of each factor; we say they are diploid. The two copies of a given factor may be "set" to the same allele (the individual is homozygous for that factor) or to different ones (the individual is heterozygous for that factor).
- Germ cells (sperm or pollen, and eggs) are exceptional: They contain only one copy of each factor. Germ cells form from diploid cells by a special form of cell division, in which one copy of each factor gets chosen from the pair in the parent cell. Today, we call this division meiosis and the selection of factors assortment.
- Meiosis chooses each factor randomly and independently of the others, an idea now called the principle of independent assortment.

---

[6]Interestingly, Charles Darwin also did extensive breeding experiments, on snapdragons, and obtained data similar to Mendel's; yet he failed to perceive Mendel's laws. Mendel's advantage was his mathematical background. Later Darwin would express regret that he had not made enough of an effort to know "something of the great leading principles of mathematics," and wrote that persons "thus endowed seem to have an extra sense."

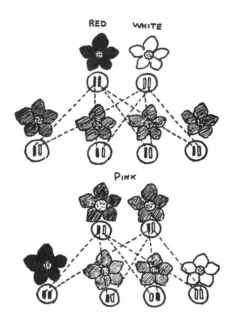

Figure 3.11: (Diagram.) (a) Purebred red and white flowers are cross-pollinated to yield off-spring, each with one chromosome containing the "red" allele and one containing the "white" allele. If neither allele is dominant, the offspring will all be pink. For example, four-o'clocks (a flower) exhibit this "semidominance" behavior. (b) Interbreeding the offspring of the previous generation, we recover pure white flowers in one out of four cases. Even in other species, for which the red allele is dominant, nevertheless one in four of the second-generation offspring will be white. [Cartoon by George Gamow, from Gamow, 1961.]

Thus, each of the four kinds of offspring shown in each generation of Figure 3.11 is equally likely. After the fertilized egg forms, it creates the organism by ordinary division (mitosis), in which both copies of each factor get duplicated. A few of the descendant cells eventually undergo meiosis to form another generation of germ cells, and the process repeats.

If the two copies of the factor corresponding to a given trait represent different alleles, it may be that one allele overrides (or "dominates") the other in determining the organism's phenotype. Nevertheless, both copies persist, with the hidden (or "recessive") one ready to reappear in later generations in a precisely predictable ratio. Verifying such quantitative predictions gave Mendel the conviction that his guesses about the invisible processes of meiosis and mitosis were correct.

Mendel's rules drew attention to the discrete character of inheritance; the image of two alternative alleles as a switch stuck in one of two possible states is physically very appealing. Moreover, Mendel's work showed that most of the apparent variation between generations is simply reassortment of factors, which are themselves extremely stable. Other types of heritable variations do occur spontaneously, but these **mutations** are rare. Moreover, mutations, too, are discrete events, and once formed, a mutation spreads in the population by the same Mendelian rules listed above. Thus, factors are switches that can snap crisply into new positions, but not easily; once changed by mutation, they don't switch back readily.

The history of biology in this period is a beautiful counterpoint between classical genetics and cell biology. Cell biology has a remarkable history of its own; for example, many advances had to await the discovery of staining techniques, without which the various components of cells were invisible. By about the time of Mendel's work, E. Haeckel had identified the nucleus of the cell as the seat of its heritable characters. A recently fertilized egg visibly contained two equal-sized objects called pronuclei, which soon fused. In 1882, W. Flemming noted that the nucleus organized itself into threadlike **chromosomes** just before division. Each chromosome was present in duplicate prior to mitosis, as required by Mendel's rules (see Figure 3.11); and just before cell division, each chromosome appeared to double, after which one copy of each was pulled into each daughter cell. Moreover, E. van Beneden observed that the pronuclei of a fertilized worm egg each had two chromosomes, whereas the ordinary cells had four. van Beneden's result gave visible testimony to Mendel's logical deduction about the mixing of factors from both parents.

By this point, it would have been almost irresistible to conclude that the physical carriers of Mendel's genetic factors were precisely the chromosomes, had anyone been aware of Mendel. Unfortunately, Mendel's results, published in 1865, fell into obscurity, not to be rediscovered until 1900 by H. de Vries, K. Correns, and E. von Tschermak. Immediately upon this rediscovery, W. Sutton and T. Boveri independently proposed that Mendel's genetic factors were physical objects—**genes**—located on the chromosomes. (Sutton was a graduate student at the time.) But what *were* chromosomes, anyway? It seemed impossible to make further progress on this point with the existing cell biology tools.

A surprising quirk of genetics broke the impasse. Although Mendel's rules were approximately correct, later work showed that not all traits assorted independently. Instead, W. Bateson and Correns began to notice that certain pairs of traits seemed to be **linked**, a phenomenon already predicted by Sutton. That is, such pairs of traits will almost always be inherited together: The offspring gets either both, or neither. This complication must have seemed at first to be a blemish on Mendel's simple, beautiful rules. Eventually, however, the phenomenon of linkage opened up a new window on the old question of the nature of genetic factors.

The embryologist T. H. Morgan studied the phenomenon of genetic linkage in a series of experiments starting around 1909. Morgan's first insight was that, in order to generate and analyze huge sets of genealogical data, big enough to find subtle statistical patterns, he would need to choose a very rapidly multiplying organism for his model system. Certainly bacteria multiply rapidly, but they were hard to manipulate individually and lacked readily identifiable hereditary traits. Morgan's compromise choice was the fruit fly *Drosophila melanogaster*.

One of Morgan's first discoveries was that some heritable traits in fruit flies (for example, white eyes) were linked to the fly's sex. Because sex was already known to be related to a gross, obvious chromosomal feature (females have two X chromosomes, whereas males have just one), the linkage of a mutable factor to sex lent direct support to Sutton's and Boveri's idea that chromosomes were the physical carriers of Mendel's factors.

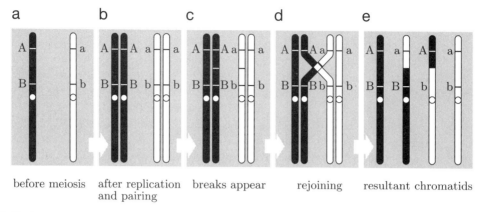

before meiosis   after replication   breaks appear   rejoining   resultant chromatids
                  and pairing

Figure 3.12: (Diagram.) Meiosis with crossing-over. (a) Before meiosis, the cell carries two homologous (similar) copies of a chromosome, carrying genes A, B on one copy and potentially different alleles a, b on the other. (b) Still prior to meiosis, each chromosome is duplicated; the copies are called chromatids. During prophase I of meiosis, the homologous chromatid pairs are brought close together, in register. Recombination may then occur: (c) Two of the four paired chromatids are broken at corresponding locations. (d) The broken ends "cross over," that is, they rejoin with the respective broken ends in the opposite chromatid. (e) The cell now carries new combinations of alleles. The four chromatids then separate into four germ cells by a four-way cell division. [Adapted from Wolfe, 1985.]

But now an even more subtle level of structure in the genetic data was beginning to appear. Two linked traits *almost* always assorted together, but they occasionally would separate. For example, certain body-color and eye-color factors separate in only about 9% of offspring. The rare failure of linkage reminded Morgan that F. Janssens had recently observed chromosome pairs wrapping around each other prior to meiosis and had proposed that this interaction could involve the breakage and exchange of chromosome pieces. Morgan suggested that this **crossing-over** process could explain his observation of incomplete genetic linkage (Figure 3.12). If the carrier object were threadlike, as the chromosomes appeared to be under the microscope, then the genetic factors might be in a fixed sequence, or linear arrangement, along it, like a pattern of knots in a long rope. Some unknown mechanism could bring two corresponding chromosomes together and align them so that each factor was physically next to its partner, then choose a random point at which to break and exchange the two strands. It seemed reasonable to suppose that the chance of two factors on the same chromosome being separated by a physical break should depend on the distance between their fixed positions. After all, when you cut a deck of cards, the chance of two given cards becoming separated is greater if the cards in question are initially far apart in the deck.

Morgan and his undergraduate research student A. Sturtevant analyzed these exceptions in an attempt to confirm the hypothesis of a linear arrangement of genetic factors. They reasoned that it should be possible to list any set of linked traits along a line, in such a way that the probability of two traits' becoming separated in an offspring is related to their distance on the line. Examining the available data, Sturtevant confirmed this deduction, and moreover found that each linked set of traits admitted just one such linear arrangement that fit the data (Figure 3.13). Two years later,

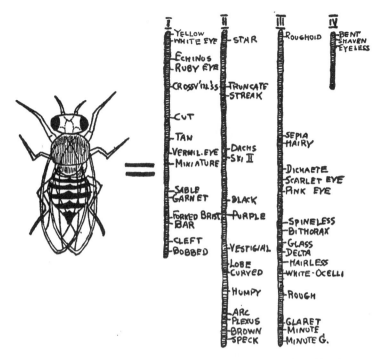

Figure 3.13: (Diagram.) Partial map of the fruit fly genome as deduced by the 1940s from purely genetic experiments. The map is a graphical summary of a large body of statistical information on the degree to which various mutant traits are inherited together. Traits shown on different vertical lines assort independently. Traits appearing near one another on the same line are more tightly linked than those listed as far apart. [Cartoon by George Gamow, from Gamow, 1961.]

the data set had expanded to include 24 different traits, which fell into *exactly four unlinked groups*—the same number as the number of visible chromosome pairs (Figure 3.13)! Now one could hardly doubt that chromosomes were the physical objects carrying genetic factors. The part of a chromosome carrying one factor, the basic unit of heredity, was christened the gene.

Thus, by a tour de force of statistical inference, Morgan and Sturtevant (together with C. Bridges and H. Muller) partially mapped the genome of the fly, concluding that

- The physical carriers of genetic information are indeed the chromosomes; and
- Whatever the chromosomes may be physically, they are chains, one-dimensional "charm bracelets" of subobjects—the genes—in a fixed sequence. Both the individual genes and their sequence are inherited.[7]

---

[7]Later work by Barbara McClintock on maize would show that even the order of the genes along the chromosome is not always fixed: Some genes are transposable elements, that is, they can jump. But this jumping is not caused by simple thermal motion; we now know that it is assisted by special-purpose molecular machines, which cut and splice the otherwise stable DNA molecule.

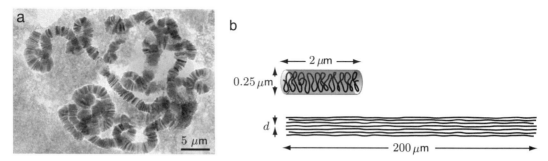

Figure 3.14: (Light micrograph; schematic.) (a) Polytene chromosomes of the fruit fly *Drosophila funebris*. Each chromosome consists of 1000–2000 identical copies of the cell's DNA, all laid parallel and in register. Each visible band is a stretch of DNA about 100 000 basepairs long. (b) Koltzoff's view of the structure of a polytene chromosome (*bottom*) as a bundle of straightened filaments, each of diameter *d*. The normal chromosome seen during mitosis (*top*) consists of just one of these filaments, tightly coiled.

By 1920, Muller could assert confidently that genes were "bound together in a line, in the order of their linkage, by material, solid connections." Like Mendel before them, Morgan's group had applied quantitative, statistical analysis to heredity to obtain insight into the mechanism, and the invisible structural elements, underlying it.

There is a coda to this detective story. One might want to examine the chromosomes directly, to see the genes. Attempts to do this were unsuccessful: Genes are too small to see with ordinary, visible light. Nevertheless, by an almost unbelievable stroke of serendipity, it turned out that salivary-gland cells of *Drosophila* have enormous chromosomes, with details easily visible in the light microscope. N. Koltzoff interpreted these giant (or polytene) chromosomes, arguing that they are really clusters of over a thousand copies of the fly's usual chromosome, all laid side by side in register to form a wide, optically resolvable object (Figure 3.14a). After treatment with an appropriate stain, each polytene chromosome shows a characteristic pattern of dark bands. T. Painter managed to discern differences in these patterns among different individuals and to show that these were inherited and in some cases correlated with observable mutant features. That is, at least some different versions of a chromosome actually *look* different. Moreover, the observed linear sequence of bands associated with known traits matched the sequence of the corresponding genes deduced by genetic mapping. The observed bands are not individual genes (these are still too small to see under the light microscope). Nevertheless, there could be no doubt that genes were physical objects located on chromosomes. Genetic factors, originally a logical construct, had become *things*, the genes.

T_2 *Section 3.3.2′ on page 104 mentions the role of double crossing-over.*

3.3.3 Schrödinger's summary: Genetic information is structural

For some time, it seemed as though the techniques of classical genetics and cell biology, powerful though they were, could shed no further light on the nature of the chromosomal charm bracelet. Even the physical size of a gene remained open for dispute. But by the mid-twentieth century, new experimental techniques and theoretical ideas from physics were opening new windows on cells. Schrödinger's brief summary of the situation in 1944 drew attention to a few of the emerging facts.

To Schrödinger, the biggest question about genes concerned the nearly perfect fidelity of their information storage despite their minute size. To see how serious this problem is, we first need to know just how small a gene is. One crude way to estimate this size is to guess how many genes there are, and note that they must all fit into a sperm head. Muller gave a somewhat better estimate in 1935 by noting that a fruit fly chromosome condenses during mitosis into roughly a cylinder of length $2\,\mu$m and diameter $0.25\,\mu$m (see Figure 3.14b). The total volume of the genetic material in a chromosome is thus no larger than $2\,\mu$m $\times\,\pi(0.25\,\mu$m$/2)^2$. When the same chromosome is stretched out in the polytene form mentioned earlier, however, its length is more like $200\,\mu$m. Suppose that a single thread of the genetic charm bracelet, stretched out straight, has a diameter d. Then its volume equals $200\,\mu$m $\times\,\pi(d/2)^2$. Equating these two expressions for the volume yields the estimate $d \leq 0.025\,\mu$m for the diameter of the genetic information carrier. Although we now know that a strand of DNA is really less than a tenth this wide, still Muller's upper bound on d showed that the genetic carrier is an object of molecular scale. Even the tiny pits encoding the information on an audio compact disk are thousands of times larger than this, just as the disk itself occupies a far larger volume than a sperm cell.

To see what Schrödinger found so shocking about this conclusion, we must again remember that molecules are in constant, random thermal motion (Section 3.2). The words on this page may be stable for many years; but, if we could write them in letters only a few nanometers high, then random motions of the ink molecules constituting them would quickly obliterate them. Random thermal motion becomes more and more destructive of order on shorter length scales, a point to which we will return in Chapter 4. How can genes be so tiny and yet so stable?

Muller and others argued that the only known stable arrangements of just a few atoms are single molecules. Quantum physics was just beginning to explain this phenomenal stability, as the nature of the chemical bond became understood. (As one of the architects of quantum theory, Schrödinger himself had laid the foundations for this understanding.) A molecule derives its enormous stability from the fact that a large activation barrier must be momentarily overcome in order to break the bonds between its constituent atoms. More precisely, Section 1.5.3 pointed out that a typical chemical bond energy is $E_{\text{bond}} \approx 2.4 \cdot 10^{-19}$ J, about 60 times bigger than the typical thermal energy E_{thermal}. Muller argued that this large activation barrier to conversion was the reason why spontaneous thermally induced mutations are so rare, following the ideas of Section 3.2.4.[8]

The hypothesis that the chromosome is a single molecule may appear satisfying, even obvious, today. But to be convinced that it is really true, we must require that a model generate some quantitative, falsifiable predictions. Fortunately, Muller had a powerful new tool in hand: In 1927, he had found that exposure to X-rays could induce mutations in fruit flies. This **X-ray mutagenesis** occurred at a much greater

---

[8]Today we know that cells enhance their genome stability still further with special-purpose molecular machines for the detection and repair of damaged DNA.

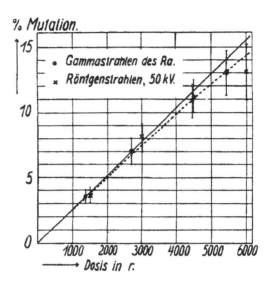

Figure 3.15: (Experimental data.) Some of Timoféeff's original data on X-ray mutagenesis. Cultures of fruit flies were exposed either to gamma rays (*solid circles*) or to X-rays (*crosses*). In each case, the total radiation exposure is given in r units, with 1 r corresponding to about $2 \cdot 10^{12}$ ion pairs created per cubic centimeter of tissue. The vertical axis is the fraction of cultures developing a particular mutant fly (in this case one with abnormal eye color). Both kinds of radiation proved equally effective when their exposures were measured in r units. [From Timoféeff-Ressovsky et al., 1935.]

rate than natural, or spontaneous, mutation. Muller enthusiastically urged the application of modern physics ideas to analyze genes, even going so far as to call for a new science of "gene physics."

Working in Berlin with the geneticist Nikolai Timoféeff, Muller learned how to make precise quantitative studies of the frequency of mutations following different radiation doses. Remarkably, they and others found that, in many instances, *the frequency with which a specific mutation occurred rose linearly with the total X-ray exposure* given to the sample. This linear relation persisted over a wide range of exposures (Figure 3.15). Thus, doubling the exposure simply doubled the number of mutants in a given culture. Prior irradiation had no effect on those individuals not mutated (or killed outright); it neither weakened nor toughened them to further exposure.

Timoféeff went on to find an even more remarkable regularity in his data: *Different kinds of radiation proved equivalent* for inducing mutations. More precisely, cultures of fruit flies were subjected to X-rays produced at various voltages, and even gamma rays from nuclear radioactivity. In each case the exposure was expressed by giving the number of electrically charged molecules (or **ions**) per volume produced by the radiation (Figure 3.15). When the exposures to the various forms of radiation were equal, each was equally effective at producing a particular mutation.

At this point a young physicist named Max Delbrück entered the scene. Delbrück had arrived in the physics world just a few years too late to participate in the feverish discovery days of quantum mechanics. His 1929 thesis (which he later termed "acceptable but dull") nevertheless gave him a thorough understanding of the recently discovered theory of the chemical bond, an understanding that experimentalists like Muller and Timoféeff needed. Updated slightly, Delbrück's analysis of the two key observations (linear response to exposure and equivalency of radiation types) ran as follows: When X-rays pass through any sort of matter, living or not, they knock electrons out of a few of the molecules they encounter. The ions thus formed can in turn react with other molecules, creating highly reactive fragments generically called free radicals. The density c_{ion} of ions created per volume is a convenient, physically measurable index of total radiation exposure; it also reflects the density of free radicals formed.

The reactive molecular fragments generated by the radiation can in turn encounter and damage other nearby molecules. We assume that the density c_* of these damage-inducing fragments equals a constant times the measured ionization: $c_* = Kc_{ion}$. Delbrück argued that if the gene were a single molecule, then a single encounter with a reactive fragment could induce a permanent change in its structure, and so cause a heritable mutation. Suppose that a free radical can wander through a volume v before reacting with something and that a particular gene (for example, the one for eye color) has a chance P_1 of suffering a particular mutation if it is located in this volume (and zero chance otherwise). Then the total chance that a particular egg or sperm cell will undergo the mutation is (see Figure 3.16)

$$\text{probability of mutation} = P_1 c_* v = (P_1 K v) \times c_{ion}. \qquad (3.29)$$

Delbrück did not know the actual numerical values of any of the constants P_1, K, and v appearing in this formula. Nevertheless, his argument implied that

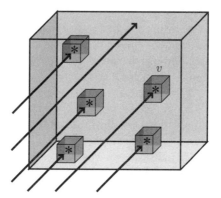

Figure 3.16: (Schematic.) Max Delbrück's simplified model for X-ray induced mutagenesis. Incoming X-rays (*diagonal arrows*) occasionally interact with tissue to create free radicals (*stars*) with number density c_* depending on the X-ray intensity, the wavelength of the X-rays, and the duration of exposure. The chance that the gene of interest lies within a box of volume v centered on one of the radicals, and so has a chance of being altered, is the fraction of all space occupied by the boxes, or $c_* v$.

> *The hypothesis that the gene is a single molecule suggests that a single molecular encounter can break it and, hence, that the probability of mutation equals a constant times the exposure measured in ionizations per volume,* (3.30)

as found in Muller's and Timoféeff's experiments.

Equation 3.29 tells a remarkable story. On the left-hand side, we have a *biological* quantity, which we measure by irradiating a lot of normal flies and seeing how many have offspring with, for example, white eyes. On the right-hand side, we have a purely *physical* quantity, c_{ion}. The formula says that the biological and the physical quantities are linked in a simple way by the hypothesis that the gene is a molecule. Data like those in Figure 3.15 agreed with this prediction, and hence supported the picture of the gene as a single molecule. Combining this idea with the linear arrangement of genes found from Sturtevant's linkage mapping (Section 3.3.2) led Delbrück to his main conclusion:

> *The physical object carrying genetic factors must be a single long-chain molecule, or polymer. The genetic information is carried in the exact identities, and sequence, of the links in this chain. This information is long-lived because the chemical bonds holding the molecule together require a large activation energy to break.* (3.31)

To appreciate the boldness of this proposal, we need to remember that the very idea of a long-chain molecule was quite young and still controversial at the time. Despite the enormous development of organic chemistry in the nineteenth century, the idea that long chains of atoms could retain their structural integrity still seemed like science fiction. Eventually, careful experiments by H. Staudinger around 1922 showed how to synthesize polymer solutions from well-understood small precursor molecules by standard chemical techniques. Staudinger coined the word **macromolecule** to describe the objects he had discovered. These synthesized polymers turned out to mimic their natural analogs: For example, suspensions of synthetic latex behave much like natural rubber-tree sap.

In a sense, Delbrück had again followed the physicist's strategy of thinking about a simple model system. A humble sugar molecule stores some energy through its configuration of chemical bonds. In the language of Section 1.2, this energy is of high quality, or low disorder; and, in isolation, the sugar molecule can retain this energy practically forever. The individual units, or **monomers**, of the genetic polymer also store some chemical energy. But, far more important, they store the entire software needed to direct the construction of the redwood tree from atmospheric CO_2, water with dissolved nitrates, and a source of high-quality energy. Section 1.2.2 proposed that the construction itself is an act of free energy transduction, as is the duplication of the software.

The idea of an enormous molecule with permanent structural arrangements of its constituent atoms was certainly not new. A diamond is an example of such a huge molecule. But nobody (yet) uses diamonds to store and transmit information.

Why not? Because the arrangement of atoms in a diamond, although permanent, is *boring*. We could summarize it by drawing a handful of atoms, then adding the words *et cetera*. A diamond is a periodic structure. Schrödinger's point was that huge molecules need not be so dull: We can equally well imagine a *nonperiodic* string of monomers, just like the words in this book.

Today we know that Nature uses polymers for an enormous variety of tasks. Humans, too, eventually caught on to the versatility of polymers, which now enter technology everywhere from hair conditioner to bulletproof vests. Though we will add little to Schrödinger's remarks on the information storage potential of polymers, the following chapters will return to them over and over as we explore how they carry out the many tasks assigned to them in cells.

Schrödinger's summary of the state of knowledge focused the world's attention on the deepest, most pressing questions: If the gene is a molecule, then which of the many big molecules in the nucleus is it? If mitosis involves duplication of this molecule, then how does such duplication work? Many young scientists heard the call of these questions, including the geneticist James Watson. By this time, further advances in biochemistry had pinpointed DNA as the genetic information carrier: It was the only molecule that, when purified, was capable of permanently transforming cells and their progeny. But how did it work? Watson joined the physicist Francis Crick to attack this problem. Integrating recent physics results (the discovery of the DNA molecule's helical geometry by Rosalind Franklin, Raymond Gosling, and Alec Stokes) with biochemical facts (the base-composition rules observed by Erwin Chargaff), they deduced their now-famous basepaired model for the structure of DNA in 1953. The molecular biology revolution then began in earnest.

$\boxed{T_2}$ *Section 3.3.3' on page 104 mentions more modern views of genetic damage induced by radiation.*

THE BIG PICTURE

Let's return to the Focus Question. This chapter has explored the idea that random thermal motion dominates the molecular world. We found that this idea explains quantitatively some of the behavior of low-density gases. Gas theory may seem remote from the living systems we wish to study, but in fact, it turned out to be a good playing field to develop some themes that transcend this setting. Thus,

- Section 3.1 developed many concepts from probability that will be needed later.

- Sections 3.2.3–3.2.5 used the study of ideal gases to motivate three crucial ideas, namely, the Boltzmann distribution, the Arrhenius rate law, and the origin of friction, all of which will turn out to be general.

- Section 3.3 also showed how the concept of activation barrier, on which the Arrhenius law rests, led to the correct hypothesis that a long-chain molecule was the carrier of genetic information.

Chapters 7 and 8 will develop the general concept of entropic forces, again starting with ideas from gas theory. Even when we cannot neglect the interactions between particles, for example, when studying electrostatic interactions in solution, Chapter 7 will show that sometimes the noninteracting framework of ideal-gas theory can still be used.

KEY FORMULAS

- *Probability:* The mean value of any quantity f is $\langle f \rangle = \int \mathrm{d}x\, f(x)P(x)$ (Equation 3.10). The variance is the mean-square deviation, variance$(f) = \langle (f - \langle f \rangle)^2 \rangle$. Addition rule: The probability of getting either of two mutually exclusive outcomes is the sum of the individual probabilities.
  Multiplication rule: The probability of getting particular outcomes in each of two independent random strings is the product of the individual probabilities (Equation 3.15).
  Gaussian distribution: $P(x) = (2\pi\sigma^2)^{-1/2}\mathrm{e}^{-(x-x_0)^2/(2\sigma^2)}$ (Equation 3.8). The root-mean-square deviation of this distribution equals σ.

- *Thermal energy:* The average kinetic energy of an ideal gas molecule at temperature T is $\frac{3}{2}k_\mathrm{B}T$ (Idea 3.21).

- *Boltzmann distribution:* In a free, ideal gas, the probability distribution for a molecule to have x-component of velocity between v_x and $v_x + \mathrm{d}v_x$ is a constant times $\mathrm{e}^{-m(v_x)^2/(2k_\mathrm{B}T)}\,\mathrm{d}v_x$. The total distribution for all three components is then the product, namely, another constant times $\mathrm{e}^{-m\mathbf{v}^2/(2k_\mathrm{B}T)}\,\mathrm{d}^3\mathbf{v}$. Equation 3.25 generalizes this statement for the case of many particles.
  In an ideal gas on which forces act, the probability that one molecule has given position and momentum is a constant times $\mathrm{e}^{-E/k_\mathrm{B}T}\,\mathrm{d}^3\mathbf{v}\mathrm{d}^3\mathbf{x}$, where the total energy E of the molecule (kinetic plus potential) depends on position and velocity. In the special case where the potential energy is a constant, this formula reduces to Maxwell's result (Equation 3.25). More generally, for many *interacting* molecules in equilibrium, the probability for molecule *1* to have velocity $\mathbf{v}_1$ and position $\mathbf{x}_1$, and so on, equals a constant times $\mathrm{e}^{-E/k_\mathrm{B}T}\,\mathrm{d}^3\mathbf{v}_1\,\mathrm{d}^3\mathbf{v}_2\ldots\mathrm{d}^3\mathbf{x}_1\mathrm{d}^3\mathbf{x}_2\ldots$ (Equations 3.26 and 3.27), where now E is the total energy for *all* the molecules.

- *Rates:* The rates of many chemical reactions depend on temperature mainly via the Arrhenius exponential factor, $\mathrm{e}^{-E_\mathrm{barrier}/k_\mathrm{B}T}$ (Idea 3.28).

FURTHER READING

Semipopular:
Probability: Gonick & Smith, 1993.
Genetics: Gonick & Wheelis, 1991.
Schrödinger's and Gamow's reviews: Schrödinger, 1967; Gamow, 1961
Polymers: deGennes & Badoz, 1996.

Intermediate:
Probability: Ambegaokar, 1996.
Molecular theory of heat: Feynman et al., 1963a, §39.4.
History of genetics: Judson, 1995.

3.3.2′ **Track 2**

Sturtevant's genetic map (Figure 3.13) also has a more subtle, and remarkable, property. If we choose any three traits A, B, and C appearing in the map in that order on the same linkage group, we find that the probability P_{AC} that A and C will be separated in a single meiosis is less than or equal to the sum $P_{AB} + P_{BC}$ of the corresponding probabilities of separation of AB and BC. There was initially some confusion on this point. Requiring that P_{AC} be *equal* to $P_{AB} + P_{BC}$ led W. Castle to propose a three-dimensional arrangement of the fly genes. Muller later pointed out that requiring strict equality amounted to neglecting the possibility of *double* crossing-over. Revising his model to incorporate this effect, and including later data, Castle soon found that the data actually required that the genes be linearly arranged, as Morgan and Sturtevant had assumed all along.

3.3.3′ **Track 2**

Delbrück's picture of genetic damage by ionizing radiation was rather incomplete. DNA repair mechanisms in eukaryotic cells can usually fix the harm done when a free radical damages only one strand of the double helix. For many more details see Hobbie, 1997, §15.10.

3.1 *White-collar crime*

a. You are a city inspector. You go undercover to a bakery and buy 30 loaves of bread marked 500 g. Back at the lab you weigh them and find their masses to be 493, 503, 486, 489, 501, 498, 507, 504, 493, 487, 495, 498, 494, 490, 494, 490, 497, 503, 498, 495, 503, 496, 492, 492, 495, 498, 490, 490, 497, and 482 g. You go back to the bakery and issue a warning. Why?

b. Later you return to the bakery (this time, they know you). They sell you 30 more loaves of bread. You take them home, weigh them, and find their masses to be 504, 503, 503, 503, 501, 500, 500, 501, 505, 501, 501, 500, 508, 503, 503, 500, 503, 501, 500, 502, 502, 501, 503, 501, 501, 502, 503, 501, 502, and 500 g. You're satisfied, because all the loaves weigh at least 500 g. But your boss reads your report and tells you to go back and close the shop down. What did she notice that you missed?

3.2 *Relative concentration versus altitude*

Earth's atmosphere has roughly four molecules of nitrogen for every oxygen molecule at sea level; more precisely, the ratio is 78:21. Assuming a constant temperature at all altitudes (not really very accurate), what is the ratio at an altitude of 10 km? Explain why your result is qualitatively reasonable. [*Hint:* This problem concerns the number density of oxygen molecules as a function of height. The density is related in a simple way to the *probability* that a given oxygen molecule will be found at a particular height. You know how to calculate such probabilities.]
[*Remark:* Your result is also applicable to the sorting of macromolecules by sedimentation to equilibrium (see Problem 5.2).]

3.3 *Stop the dance*

A suspension of virus particles is flash-frozen and chilled to a temperature of nearly absolute zero. When the suspension is gently thawed, it is found to be still virulent. What conclusion do we draw about the nature of hereditary information?

3.4 *Photons*

Section 3.3.3 reviewed Muller's and Timoféeff's empirical results that the rate of induced mutations is proportional to the radiation exposure. Not only X-rays can induce mutations; even ultraviolet light will work (that's why you wear sunblock). To get a feeling for what is so shocking about these results, notice that they imply that there's no "safe," or threshold, dose level. The amount of damage (probability of damaging a gene) is directly proportional to the total radiation exposure. Extrapolating to the smallest possible dose, we must conclude that even a single photon of UV light has the ability to cause permanent genetic damage to a skin cell and its progeny. (Photons are the packets of light mentioned in Section 1.5.3.)

a. Somebody tells you that a single ultraviolet photon carries an energy equivalent of about 10 electron volts (eV, see Appendix B). You propose a damage mechanism: A photon delivers that energy into a volume the size of the cell nucleus and heats it up; then the increased thermal motion knocks the chromosomes apart in some way. Is this a reasonable proposal? Why or why not? [*Hint:* Use Equation 1.2, and the definition of calorie found just below it, to calculate the temperature change.]

b. Turning the result around, suppose that that photon's energy is delivered to a small volume L^3 and heats it up. We might suspect that if it heats up the region to boiling, this change could disrupt any genetic message contained in that volume. How small must L be for this amount of energy to heat that volume up to boiling (from 30°C to 100°C)? What could we conclude about the size of a gene if this proposal were correct?

3.5 $\boxed{T_2}$ *Effusion*

Figure 3.6 shows how to check the Boltzmann distribution of molecular speeds experimentally. Interpreting the data, however, requires some analysis.

Figure 3.17 shows a box full of gas with a tiny pinhole of area A, which slowly allows gas molecules to escape into a region of vacuum. You can assume that the gas molecules have a nearly equilibrium distribution inside the box; the disturbance caused by the pinhole is small. The gas molecules have a known mass m. The number density of gas in the box is c. The emerging gas molecules pass through a velocity selector, which admits only those with speed in a particular range, from u to $u + du$. A detector measures the total number of molecules arriving per unit time. It is located a distance d from the pinhole, on a line perpendicular to the hole, and its sensitive region is of area A_*.

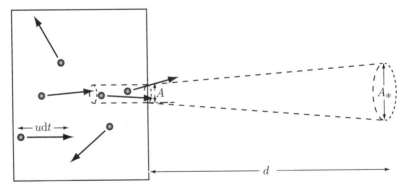

Figure 3.17: (Schematic.) Gas escaping from a pinhole of area A in the wall of a box. The number density of gas molecules is c inside the box and zero outside. A detector counts the rate at which molecules land on a sensitive region of area A_*. The six arrows in the box depict schematically six molecules, all with one particular speed $u = |\mathbf{v}|$. Of these, only two will emerge from the box in time dt, and of those two, only one will arrive at the detector a distance d away.

a. The detector catches only those molecules emitted in certain directions. If we imagine a sphere of radius d centered on the pinhole, then the detector covers only a fraction α of the full sphere. Find α.

Thus, the fraction of all gas molecules whose $\mathbf{v}$ makes them candidates for detection is $P(\mathbf{v})d^3\mathbf{v}$, where $\mathbf{v}$ points perpendicular to the pinhole and has magnitude u and

$d^3\mathbf{v} = 4\pi\alpha u^2 du$. Of these, the molecules that actually emerge from the box in time dt will be those initially located within a cylinder of cross-sectional area A and length $u dt$ (see the dashed cylinder in the figure).

b. Find the total number of gas molecules per unit time arriving at the detector.

c. Some authors report their results in terms of the transit time $\tau = d/u$ instead of u. Rephrase your answer to (b) in terms of τ and $d\tau$, not u and du.

[*Note:* In practice, the selected velocity range du depends on the width of the slots in Figure 3.6, *and* on the value of u selected. For thin slots, du is roughly a constant times u. Thus, the solid curve drawn in Figure 3.7 consists of your answer to (b), multiplied by another factor of u, and normalized; the experimental points reflect the detector response, similarly normalized.]

CHAPTER **4**

Random Walks, Friction, and Diffusion

> *It behoves us always to remember that in physics it has taken*
> *great minds to discover simple things. They are very great*
> *names indeed which we couple with the explanation of the*
> *path of a stone, the droop of a chain, the tints of a bubble, the*
> *shadows in a cup.*
> — D'Arcy Thompson, 1917

Section 3.2.5 argued that the origin of friction was the conversion of organized motion to disordered motion by collisions with a surrounding, disordered medium. In this picture, the First Law of thermodynamics is just a restatement of the conservation of energy. To justify such a unifying conclusion, we'll continue to look for nontrivial, testable, quantitative predictions from the model.

This process is not just an exercise in retracing others' historical footsteps. Once we understand the origin of friction, a wide variety of other **dissipative processes**— those that irreversibly turn order into disorder—will make sense, too:

- The diffusion of ink molecules in water erases order; for example, any pattern initially present disappears (Section 4.4.2).
- Friction erases order in the initial directed *motion* of an object (Section 4.1.4).
- Electrical resistance runs down your flashlight batteries, making heat (Section 4.6.4).
- The conduction of heat erases the initial separation into hot and cold regions (Section 4.4.2′).

In every case just listed, organized kinetic or potential energy gets degraded into disorganized motion, by collisions with a large, random environment. The paradigm we will study for all these processes will be the physics of the random walk (Section 4.1.2).

None of the dissipative processes listed in the preceding paragraph matters much for the Newtonian questions of celestial mechanics. But all will turn out to be of supreme importance in understanding the physical world of cells. The difference is that, in cells, the key actors are single molecules or perhaps structures of at most a few thousand molecules. In this nanoworld, the tiny energy $k_B T_r$ is *not* so tiny; the randomizing kicks of neighboring molecules can quickly degrade any concerted motion. For example,

- Diffusion turns out to be the dominant form of material transport on submicrometer scales (Section 4.4.1).
- The mathematics of random walks is also the appropriate language to understand the conformations of many biological macromolecules (Section 4.3.1).
- Diffusion ideas will give us a quantitative account of the permeability of bilayer membranes (Section 4.6.1) and the electrical potentials across them (Section 4.6.3), two topics of great importance in cell physiology.

The Focus Question for this chapter is
Biological question: If everything is so random in the nanoworld of cells, how can we say anything predictive about what's going on there?
Physical idea: The collective activity of many randomly moving actors can be effectively predictable, even if the individual motions are not.

4.1 BROWNIAN MOTION

4.1.1 Just a little more history

Even up to the end of the nineteenth century, influential scientists were criticizing, even ridiculing, the hypothesis that matter consisted of discrete, unchangeable, real particles. The idea seemed to them philosophically repugnant. Many physicists, however, had by this time long concluded that the atomic hypothesis was indispensable for explaining the ideal gas law and a host of other phenomena. Nevertheless, doubts and controversies swirled. For one thing, the ideal gas law doesn't actually tell us how big molecules are. We can take 2 g of molecular hydrogen (one mole) and measure its pressure, volume, and temperature, but all we get from the gas law is the product $k_B N_{mole}$, not the separate values of k_B and N_{mole}; thus we don't actually find how *many* molecules were in that mole. Similarly, in Section 3.2 on page 78, the decrease of atmospheric density on Mt. Everest told us that $mg \times 10\,\mathrm{km} \approx \frac{1}{2}mv^2$, but we can't use this to find the mass m of a single molecule—m drops out.

If only it were possible to *see* molecules and their motion! But this dream seemed hopeless. The many improved estimates of Avogadro's number deduced in the century since Franklin all pointed to an impossibly small size for molecules, far below what could ever be seen with a microscope. But there was one ray of hope.

In 1828, a botanist named Robert Brown had noticed that pollen grains suspended in water do a peculiar incessant dance, visible with his microscope. At roughly 1 μm in diameter, pollen grains seem tiny to us. But they're enormous on the scale of atoms, and big enough to see under the microscopes of Brown's time (the wavelength of visible light is around half a micrometer). We will generically call such objects **colloidal particles**. Brown naturally assumed that what he was observing was some life process, but being a careful observer, he proceeded to check this assumption. What he found was that:

- The motion of the pollen never stopped, even after the grains were kept for a long time in a sealed container. If the motion were a life process, the grains would run out of food eventually and stop moving. They didn't.

- Totally lifeless particles do exactly the same thing. Brown tried using soot ("deposited in such Quantities on all Bodies, especially in London") and other materials, eventually getting to the most exotic material available in his day: ground-up bits of the Sphinx. The motion was always the same for similar-size particles in water at the same temperature.

Brown reluctantly concluded that his phenomenon had nothing to do with life.

By the 1860s several people had proposed that the dance Brown observed was caused by the constant collisions between the pollen grains and the molecules of water agitated by their thermal motion. Experiments by several scientists confirmed that this **Brownian motion** was more vigorous at higher temperature, as expected from the relation (average kinetic energy)$=\frac{3}{2}k_B T$ (Idea 3.21). (Other experiments had ruled out other, more prosaic, explanations for the motion, such as convection currents.) It looked as though Brownian motion could be the long-awaited missing link between the macroscopic world of bicycle pumps (the ideal gas law) and the nanoworld (individual molecules). Missing from these proposals, however, was any precise quantitative test.

But the molecular-motion explanation of Brownian motion seems, on the face of it, absurd, as others were quick to point out. The critique hinged on two points:

1. If molecules are tiny, then how can a molecular collision with a comparatively enormous pollen grain make the grain move appreciably? The grain takes steps that are visible in light microscopy and hence are enormous relative to the size of a molecule.

2. Section 3.2 argued that molecules are moving at high speeds, around 10^3 m s^{-1}. If water molecules are about a nanometer in size and closely packed, then each one moves less than a nanometer before colliding with a neighbor. The collision *rate* is then at least $(10^3$ m s$^{-1})/(10^{-9}$ m$)$, or about 10^{12} collisions per second. Our eyes can resolve events at rates no faster than 30 s^{-1}. How could we see these hypothetical dance steps?

This is where matters stood when a graduate student was finishing his thesis in 1905. The student was Albert Einstein. The thesis kept getting delayed because Einstein had other things on his mind that year. But everything turned out all right in the end. One of Einstein's distractions was Brownian motion.

4.1.2 Random walks lead to diffusive behavior

Random walks Einstein's beautiful resolution to the two paradoxes just mentioned was that *the two problems cancel each other*. To understand his logic, imagine moving a marker on the sidewalk below a skyscraper. Once per second, you toss a coin. Each time you get heads, you move the marker one step to the east; for tails, one step to the west. You have a friend looking down from the top of the building. She cannot resolve the individual squares on the sidewalk; they are too distant for that. Nevertheless, *once in a while* you will flip 100 heads in a row, thus producing a step clearly visible from

Figure 4.1: (Metaphor.) A random (or "drunkard's") walk. [Cartoon by George Gamow, from Gamow, 1961.]

afar. Certainly such events are rare; your friend can check up on your game only every hour or so and still not miss them.

In just the same way, Einstein said, although we cannot see the small, rapid jerks of the pollen grain due to individual molecular collisions, still we can and will see the rare large displacements.[1]

The fact that rare large displacements exist is sometimes expressed by the statement that *a random walk has structure on all length scales*, not just on the scale of a single step. Moreover, studying only the rare large displacements will not only confirm that the picture is correct but will also tell us something quantitative about the invisible molecular motion (namely, the value of the Boltzmann constant). The motion of pollen grains may not seem to be very significant for biology, but Section 4.4.1 will argue that thermal motion becomes more and more important as we look at smaller objects—and biological macromolecules are much smaller than pollen grains.

It's easy to adapt this logic to more realistic motions, in two or three dimensions. For two dimensions, place the marker on a checkerboard and flip *two* coins each second, a penny and a nickel. Use the penny to move the marker east/west as before. Use the nickel to move the marker north/south. The path traced by the marker is then a two-dimensional **random walk** (Figures 4.1 and 4.2); each step is a diagonal across a square of the checkerboard. We can similarly extend our procedure to three dimensions. But to keep the formulas simple, the rest of this section will only discuss the one-dimensional case.

---

[1] $\boxed{T_2}$ What follows is a simplified version of Einstein's argument. Track-2 readers will have little difficulty following his original paper (see Einstein, 1956) after reading Chapter 6 of this book.

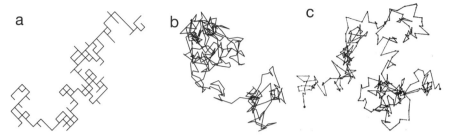

Figure 4.2: (Mathematical functions; experimental data.) (a) Computer simulation of a two-dimensional random walk with 300 steps. Each step lies on a diagonal as discussed in the text. (b) The same with 7500 steps, each 1/5 the size of the steps in (a). The walk has been sampled every 25 steps, giving a mean step size similar to that in (a). The figure has both fine detail and an overall structure: We say there is structure on all length scales. (c) Jean Perrin's actual experimental data from 1908. Perrin periodically observed the location of a single particle, then plotted these locations joined by straight lines, a procedure similar to the periodic sampling used to generate the mathematical graph (b). The field of view is about 75 μm wide. [Simulations kindly supplied by P. Biancaniello; experimental data from Perrin, 1948.]

Suppose our friend looks away for 10 000 s (about three hours). When she looks back, it's quite unlikely that our marker will be exactly where it was originally. For that to happen, we would have to have taken exactly 5000 steps right and 5000 steps left. Just how improbable is this outcome? For a walk of *two* steps, there are two possible outcomes that end where we started (HT and TH), out of a total of $2^2 = 4$ possibilities; thus the probability to return to the starting point is $P_0 = 2/2^2$ or 0.5. For a walk of four steps, there are six ways to end at the starting point, so $P_0 = 6/2^4 = 0.375$. For a walk of 10 000 steps, we again need to find M_0, the number of different outcomes that land us at the starting point, then divide by $M = 2^{10\,000}$.

Example: Finish the calculation.

Solution: Of the M possible outcomes, we can describe the ones with exactly 5000 heads as follows: To describe a particular sequence of coin tosses, we make a list of which tosses came out heads. This list contains 5000 different integers, $(n_1, \ldots, n_{5000})$, each less than 10 000. We want to know how many such distinct lists there are.

We can take n_1 to be any number between 1 and 10 000, n_2 to be any of the 9999 remaining choices, and so on, for a total of $10\,000 \times 9999 \times \cdots \times 5001$ lists. We can rewrite this quantity as $(10\,000!)/(5000!)$, where the exclamation point denotes the factorial function. But any two lists differing by exchange (or permutation) of the n_i's are not really different, so we must divide our answer by the total number of possible permutations, which is $5000 \times 4999 \times \cdots \times 1$. Altogether, then, we have

$$M_0 = \frac{10\,000!}{5000! \times 5000!} \tag{4.1}$$

distinct lists.

Dividing by the total number of possible outcomes gives the probability of landing exactly where you started as $P_0 = M_0/M \approx 0.008$. It's less than a 1% chance.

The probability distribution found in the Example is called the **binomial distribution**. (Some authors abbreviate Equation 4.1 as $M_0 = \binom{10\,000}{5000}$, pronounced "ten thousand choose five thousand.")

Your Turn 4A

You can't do the preceding calculation on a calculator. You could do it with a computer-algebra package, but now is a good time to learn a handy tool: **Stirling's formula** gives an approximation for the factorial $M!$ of a large number M as

$$\ln M! \approx M \ln M - M + \tfrac{1}{2}\ln(2\pi M). \qquad (4.2)$$

Work out for yourself the result for P_0 just quoted, using this formula.

The preceding discussion shows that it's quite unlikely that you will end up exactly where you started. But you're even less likely to end up 10 000 steps to the left of your starting point, a movement requiring that you flip 10 000 consecutive tails, with $P \approx 5 \cdot 10^{-3011}$. Instead, you're likely to end up somewhere in the middle. Figure 4.3 illustrates these ideas with some shorter walks.

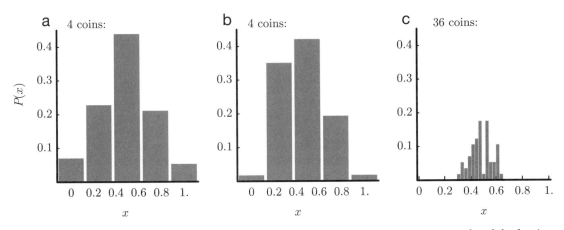

Figure 4.3: (Experimental data.) Behavior of the binomial distribution. (a) Four coins were tossed, and the fraction x that came up heads was recorded. The histogram shows the result for a sample of 57 such trials. Because this is a discrete distribution, the bars have been normalized so that the sum of their heights equals 1. (b) Another sample of 57 tosses of 4 coins. (c) This time, 36 coins were tossed, again 57 times. The resulting distribution is much narrower than (a,b); we can say with greater certainty that "about half" our coin tosses will come up heads if the total number of tosses is large. The bars are not as tall as in (a,b) because the same number of tosses (57) is now being divided among a larger number of bins (37 rather than 5). [Data kindly supplied by R. Nelson.]

The diffusion law One way to find how far you're likely to go in a random walk would be to list explicitly all the possible outcomes for a 10 000-toss sequence, then find the average over all outcomes of $(x_{10\,000})^2$, the mean-square position after step 10 000. Luckily, there is an easier way.

Suppose each step is of length L. Thus the displacement of step j is $k_j L$, where k_j is equally likely to be ± 1. Call the position after j steps x_j; the initial position is $x_0 = 0$ (see Figure 4.4a). Then $x_1 = k_1 L$, and similarly the position after j steps is $x_j = x_{j-1} + k_j L$.

We can't say anything about x_j because each walk is random. We *can*, however, make definite statements about the *average* of x_j over many different trials: For example, Figure 4.4b shows that $\langle x_3 \rangle = 0$. The diagram makes it clear why we got this result: In the average over all possible outcomes, those with net displacement to the left will cancel the contributions of their equally likely analogs with net displacement to the right.

Thus the mean displacement of a random walk is zero. But this doesn't imply we won't go anywhere! The preceding Example showed that the probability of ending right where we started is *small* for large N. To get a meaningful result, recall the discussion in Section 3.2.1: For an ideal gas, $\langle v_x \rangle = 0$ but $\langle v_x^2 \rangle \neq 0$. Following that hint, let's compute $\langle x_N^2 \rangle$ in our problem. Figure 4.4 shows such a computation, yielding $\langle x_3^2 \rangle = 3L^2$.

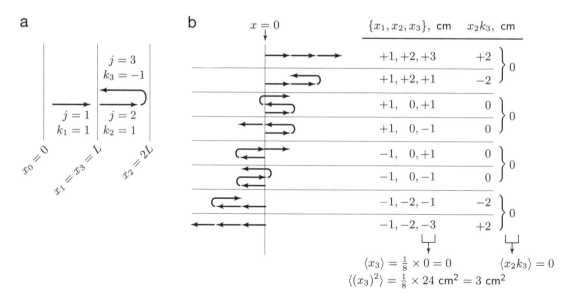

Figure 4.4: (Diagram.) (a) Anatomy of a random walk. Three steps, labeled $j = 1, 2, 3$, are shown. Step j makes a displacement of $k_j = \pm 1$. (b) Complete list of the eight distinct 3-step walks, with step length $L = 1$ cm. Each of these outcomes is equally probable in our simplest model.

<table>
<tr><td>**Your Turn 4B**</td><td>Repeat this calculation for a walk of four steps, just to make sure you understand how it works.</td></tr>
</table>

Admittedly, the math gets tedious. Instead of exhaustively listing all possible outcomes, though, we can note that

$$\langle (x_N)^2 \rangle = \langle (x_{N-1} + k_N L)^2 \rangle = \langle (x_{N-1})^2 \rangle + 2L \langle x_{N-1} k_N \rangle + L^2 \langle (k_N)^2 \rangle. \qquad (4.3)$$

In the last expression, the final term just equals L^2, because $(\pm 1)^2 = 1$. For the middle term, note that we can group all 2^N possible walks into pairs (see the last column of Figure 4.4). Each pair consists of two equally probable walks with the same x_{N-1}, differing only in their last step, so each pair contributes zero to the average of $x_{N-1} k_N$. Think about how this step implicitly makes use of the multiplication rule for probabilities (see page 75) and the assumption that every step was independent of the previous ones.

Thus, Equation 4.3 says that a walk of N steps has mean-square displacement bigger by L^2 than a walk of $N - 1$ steps, which in turn is L^2 bigger than a walk of $N - 2$ steps, and so on. Carrying this logic to its end, we find

$$\langle (x_N)^2 \rangle = NL^2. \qquad (4.4)$$

We can now apply our result to our original problem of moving a marker in one dimension, once per second. If we wait a total time t, the marker makes $N = t/\Delta t$ random steps, where $\Delta t = 1\,\text{s}$. Define the **diffusion constant** of the process as $D = L^2/(2\Delta t)$. Then,[2]

> a. *The mean-square displacement in a one-dimensional random walk increases linearly in time:* $\langle (x_N)^2 \rangle = 2Dt$, *where*   (4.5)
> b. *The constant D equals* $L^2/(2\Delta t)$.

The first part of Idea 4.5 is called the one-dimensional **diffusion law**. In our example, the time between steps is $\Delta t = 1\,\text{s}$; so if the marker makes 1 cm steps, we get $D = 0.5\,\text{cm}^2\,\text{s}^{-1}$. Figure 4.5 illustrates the fact that the averaging symbol in Idea 4.5a must be taken seriously—any *individual* walk will not conform to the diffusion law, even approximately.

Idea 4.5a makes our expectations about random walks precise. For example, we will observe excursions of any size X, even if X is much longer than the elementary step length L, as long as we are prepared to wait a time on the order of $X^2/(2D)$.

Returning to the physics of Brownian motion, our result means that, even if we cannot see the elementary steps in our microscope, we can nevertheless confirm Idea 4.5a and measure D experimentally: Simply note the initial position of a colloidal particle, wait a time t, note the final position, and calculate $x^2/(2t)$. Repeat the

---

[2]The definition of D in Idea 4.5b contains a factor of $1/2$. We can define D any way we like, as long as we're consistent; the definition we chose results in a compensating factor of 2 in the diffusion law, Idea 4.5a. This convention will be convenient when we derive the diffusion equation in Section 4.4.2.

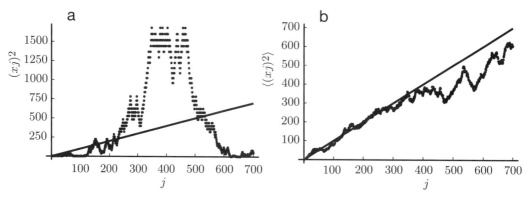

Figure 4.5: (Mathematical functions.) (a) Squared deviation $(x_j)^2$ for a single, one-dimensional random walk of 700 steps. Each step is one unit long. The solid line shows j itself; the graph shows that $(x_j)^2$ is not at all the same as j. (b) Here the the dots represent the *average* $\langle (x_j)^2 \rangle$ for 30 walks, each having 700 steps. Again the solid line shows j. This time $\langle (x_j)^2 \rangle$ does resemble the idealized diffusion law (Equation 4.4).

observation many times; the average of $x^2/2t$ gives D. The content of Idea 4.5a is that the value of D thus found will not depend on the elapsed time t.

We can extend all these ideas to two or more dimensions (Figure 4.2). For a walk on a two-dimensional checkerboard with squares of side L, we still define $D = L^2/(2\Delta t)$. Now, however, each step is a diagonal and hence has length $L\sqrt{2}$. Also, the position $\mathbf{r}_N$ is a vector, with two components x_N and y_N. Thus $\langle (\mathbf{r}_N)^2 \rangle = \langle (x_N)^2 \rangle + \langle (y_N)^2 \rangle = 4Dt$ is twice as large as before, because each term on the right separately obeys Idea 4.5a. Similarly, in three dimensions, we find

$$\langle (\mathbf{r}_N)^2 \rangle = 6Dt. \qquad \text{diffusion in three dimensions} \qquad (4.6)$$

It may seem confusing to keep track of all these different cases. But the important features about the diffusion law are simple: In any number of dimensions, mean-square displacement increases linearly with time, so the constant of proportionality D has dimensions $\mathbb{L}^2\mathbb{T}^{-1}$. Remember this, and many other formulas will be easy to remember.

From macro to micro Section 4.1.1 introduced a puzzle: How can we learn things about the molecular-scale (or "microscopic") world, when we can't see molecules? This section has explored the idea that Brownian motion supplies the link between the microscopic world and the "macroscopic" world (things we can see with light). Ultimately, we'd like to find that observations of Brownian motion, a macroscopic phenomenon, not only support the molecular theory of heat qualitatively but also test some quantitative prediction of that theory. We're not ready to get this prediction yet (it's Equation 4.16). But at least we have found one relation between the microscopic parameters of Brownian motion (the step size L and step time Δt) and a quantity observable in macroscopic experiments (the diffusion constant D), namely, Idea 4.5b.

Unfortunately, we cannot solve one equation for two unknowns: Just measuring D is not enough to find specific values for either one of these parameters. We need a second formula relating L and Δt to some macroscopic observation, so that we can solve two equations for the two unknowns. Section 4.1.4 will provide the required additional formula.

4.1.3 The diffusion law is model independent

Our mathematical treatment of the random walk made some drastic simplifying assumptions. One might well worry that our simple result, Idea 4.5, may not survive in a more realistic model. This subsection will show that, on the contrary, *the diffusion law is universal*—it's independent of the model, as long as we have some distribution of random, independent steps.

For simplicity, we'll continue to work in one dimension. (Besides being mathematically simpler than three dimensions, the one-dimensional case will be of great interest in Section 10.4.4.) Suppose that our marker makes steps of various lengths. We are given a set of numbers P_k, the probabilities of taking steps of length kL, where k is an integer. The length k_j of step j can be positive or negative, for forward or backward steps. We assume that the relative probabilities of the various step sizes are all the same for each step (that is, each value of j). Let u be the mean value of k_j:

$$u = \langle k_j \rangle = \sum_k k P_k. \tag{4.7}$$

u describes average drift motion superimposed on the random walk. (The analysis of the preceding subsection corresponds to the special case $P_{\pm 1} = \frac{1}{2}$, with all the other $P_k = 0$. For that case, $u = 0$.)

The mean position of the walker is now

$$\langle x_N \rangle = \langle x_{N-1} \rangle + L \langle k_N \rangle = \langle x_{N-1} \rangle + uL = NuL. \tag{4.8}$$

To get the last equality, we noticed that a walk of N steps can be built one step at a time; after each step, the mean displacement grows by uL.

The mean displacement is not the whole story: We know from our earlier experience that diffusion concerns the *fluctuations* about the mean. Accordingly, let's now compute the variance (or mean-square deviation, Equation 3.11) of the actual position about its mean. Repeating the analysis leading to Equation 4.3 gives

$$\text{variance}(x_N) \equiv \langle (x_N - \langle x_N \rangle)^2 \rangle = \langle (x_{N-1} + k_N L - NuL)^2 \rangle$$
$$= \langle \left((x_{N-1} - u(N-1)L) + (k_N L - uL) \right)^2 \rangle$$
$$= \langle (x_{N-1} - u(N-1)L)^2 \rangle + 2\langle (x_{N-1} - u(N-1)L)(k_N L - uL) \rangle$$
$$+ L^2 \langle (k_N - u)^2 \rangle. \tag{4.9}$$

As before, we now recall that kL, the length of the Nth step, was assumed to be a random variable, statistically independent of all the previous steps. Thus the middle term of the last formula becomes $2L\langle x_{N-1} - u(N-1)L \rangle \langle k_N - u \rangle$, which is zero by the definition of u (Equation 4.7). Thus Equation 4.9 says that the variance of x_N increases by a fixed amount on every step, or

$$\text{variance}(x_N) = \langle(x_{N-1} - \langle x_{N-1}\rangle)^2\rangle + L^2\langle(k_N - \langle k_N\rangle)^2\rangle$$

$$= \text{variance}(x_{N-1}) + L^2 \times \text{variance}(k).$$

After N steps, the variance is then $NL^2 \times \text{variance}(k)$. Suppose the steps come every Δt, so that $N = t/\Delta t$. Then

$$\text{variance}(x_N) = 2Dt, \quad \text{where } D = \frac{L^2}{2\Delta t} \times \text{variance}(k). \tag{4.10}$$

In the special case where $u = 0$ (no drift), Equation 4.10 just reduces to our earlier result, Idea 4.5a!

Thus the diffusion law (Idea 4.5a) is model independent. Only the detailed formula for the diffusion constant depends on the microscopic details of the model (compare Idea 4.5b to Equation 4.10).[3] Such universality, whenever we find it, gives a result great power and wide applicability.

4.1.4 Friction is quantitatively related to diffusion

Diffusion is essentially a question of random fluctuations: Knowing where a particle is now, we seek the spread in its expected position at a later time t. Section 3.2.5 argued qualitatively that the *same* random collisions responsible for this spread also give rise to friction. So we should be able to relate the microscopic quantities L and Δt to friction, another experimentally measurable, macroscopic quantity. As usual, we'll make some simplifications to get to the point quickly. For example, we again consider an imaginary world where everything moves only in one dimension.

To study friction, we want to consider a particle pulled by a constant external force f in the $\hat{x}$ direction. For example, f could be the force mg of gravity, or the artificial gravity inside a centrifuge. We want to know the average motion of each particle as it falls in the direction of the force. In first-year physics, you probably learned that a falling body eventually comes to a "terminal velocity" determined by friction. Let's investigate the origin of friction, in the case of a small body suspended in fluid.

In the same spirit as Section 4.1.2, suppose that the collisions occur exactly once per Δt (although really there is a distribution of times between collisions). In between kicks, the particle is free of random influences, so it is subject to Newton's Law of motion, $dv_x/dt = f/m$; its velocity accordingly changes with time as $v_x(t) = v_{0,x} + ft/m$, where $v_{0,x}$ is the starting value just after a kick and m is the mass of the particle. The resulting uniformly accelerated motion of the particle is then

$$\Delta x = v_{0,x}\Delta t + \frac{1}{2}\frac{f}{m}(\Delta t)^2. \tag{4.11}$$

---

[3] $\boxed{T_2}$ Section 9.2.2' on page 389 will show that, similarly, the structure of the three-dimensional diffusion law (Equation 4.6) does not change if we replace our simple model (diagonal steps on a cubic lattice) by something more realistic (steps in any direction).

Following Section 4.1.1, we assume that each collision obliterates all memory of the previous step. Thus, after each step, $v_{0,x}$ is randomly pointing left or right, so its average value, $\langle v_{0,x} \rangle$, equals zero. Taking the average of Equation 4.11 thus gives $\langle \Delta x \rangle = (f/2m)(\Delta t)^2$. In other words, the particle, although buffeted about by random collisions, nevertheless acquires a net **drift velocity** equal to $\langle \Delta x \rangle / \Delta t$, or

$$v_{\text{drift}} = f/\zeta, \qquad (4.12)$$

where

$$\zeta = 2m/\Delta t. \qquad (4.13)$$

Equation 4.12 shows that, under the assumptions made, a particle under a constant force indeed comes to a terminal velocity proportional to the force. The **viscous friction coefficient** ζ, like the diffusion constant, is experimentally measurable—we just look through a microscope and see how fast a particle settles under the influence of gravity, for example.

Recovering the familiar friction law (Equation 4.12) strengthens the idea that friction originates in randomizing collisions of a body with the thermally disorganized surrounding fluid. Our result goes well beyond the motion of Robert Brown's pollen grains: *Any* macromolecule, small dissolved solute molecule, or even the molecules of water itself are subject to Equations 4.12 and 4.13. Each type of particle, in each type of solvent, has its own characteristic values of D and ζ.

Returning to colloidal particles, in practice it's often not necessary to measure ζ directly. The viscous friction coefficient for a spherical object is related to its *size* by a simple relation:

$$\boxed{\zeta = 6\pi\eta R. \qquad \textbf{Stokes formula}} \qquad (4.14)$$

In this expression, R is the radius of the particle and η is a constant called the viscosity of the fluid. Chapter 5 will discuss viscosity in greater detail; for now, we only need to know that the viscosity of water at room temperature is about $10^{-3}\,\text{kg}\,\text{m}^{-1}\text{s}^{-1}$. Equation 4.14 gives us ζ once we measure the size of a colloidal particle (for example, by looking at it). If we also know the density of the particle (for example, by weighing a bulk sample of soot), then knowing its size also lets us determine its mass m.

Summarizing, we have found that ζ and m are experimentally measurable properties of a macroscopic colloidal particle. Equation 4.13 connects them to a molecular-scale quantity, the collision time Δt. We can also substitute this value back into Idea 4.5b and use the particle's diffusion constant D to find another molecular-scale quantity, the effective step size L.

Unfortunately, however, our theory has *not* made a falsifiable, quantitative prediction yet. It lets us compute the molecular-scale parameters L and Δt of the random walk's steps, but these are unobservable! To test the idea that diffusion and friction are merely two faces of thermal motion, we must take one further step.

Einstein noticed that there's a *third* relation involving L and Δt. To find it, note that $(L/\Delta t)^2 = (v_{0,x})^2$. Our discussion leading to the ideal gas law concluded that

$$\langle (v_{0,x})^2 \rangle = k_B T/m. \tag{4.15}$$

(Unlike Idea 3.21 on page 80, there's no factor of 3: We need only one component of the velocity.)

Combining Equation 4.15 with our earlier results (Idea 4.5b and Equation 4.13) *overdetermines* L and Δt. That is, these three relations in two unknowns can only hold if D and ζ themselves satisfy a particular relation. This relation between experimentally measurable quantities is the prediction we were seeking. To find it, consider the product ζD.

Your Turn 4C

Put all the pieces together: Use Equations 4.5b and 4.13 to express ζD in terms of m, L, and Δt. Then use the definition $v_{0,x} = L/\Delta t$, and Equation 4.15, to show that

$$\boxed{\zeta D = k_B T.} \quad \textbf{Einstein relation} \tag{4.16}$$

Equation 4.16 is Einstein's 1905 result. It states that the *fluctuations* in a particle's position are linked to the *dissipation* (or frictional drag) that it is subject to.

The Einstein relation is remarkable in a number of ways. For one thing, it tells us how to find k_B by making macroscopic measurements. Einstein was then able to find Avogadro's number by dividing the ideal gas law constant, $N_{mole}k_B$, by k_B. That is, he found how many molecules are in a mole, and hence how small molecules are—without seeing molecules.

The Einstein relation is quantitative and *universal*: It always yields the same value for $k_B T$, no matter what sort of particle and solvent we study. For example, the right-hand side of Equation 4.16 does not depend on the mass m of the particle. Smaller particles will feel less drag (smaller ζ), but will diffuse more readily (bigger D), in such a way that all particles obey Equation 4.16. Also, although both ζ and D generally depend on temperature in a complicated way, Equation 4.16 says their *product* depends on T in a very simple way.

The universality of ζD is a falsifiable prediction of the hypothesis that heat is disordered molecular motion: We can check whether various kinds of particles, of various sizes, at various temperatures, all give the same value of k_B. (They do; you'll see one example in Problem 4.5.)

Einstein also checked whether the experiment he was proposing was *actually doable*. He reasoned that, to see a measurable displacement of a single 1 μm colloidal particle, we'd have to wait until it had moved several micrometers. If the waiting time for such a motion were impracticably long, then the experiment itself would be impractical. Using existing estimates of k_B, Einstein estimated that a 1 μm sphere in water would take about a minute, a convenient waiting time, to wander a mean-square distance of 5 μm. Einstein concluded that colloidal particles occupy a window of experimental opportunity: They are large enough to resolve optically, yet not so large as to render their Brownian motion unobservably sluggish. Very soon after his prediction, Jean Perrin and others did the experiments and confirmed the predictions. As Einstein put it later, "Suddenly all doubts vanished about the foundations of Boltzmann's theory [of heat]."

T_2 *Section 4.1.4' on page 147 mentions several finer points about random walks.*

4.2 EXCURSION: EINSTEIN'S ROLE

Einstein was not the first to suggest that the origin of Brownian motion was thermal agitation. What did he do that was so great?

First of all, Einstein had exquisite taste in realizing what problems were important. At a time when others were pottering with acoustics and such, he realized that the pressing questions of the day were the reality of molecules, the structure of Maxwell's theory of light, the apparent breakdown of statistical physics in the radiation of hot bodies, and radioactivity. His three articles from 1905 practically form a syllabus for all of twentieth-century physics.

Einstein's interests were also interdisciplinary. Most scientists at that time could hardly comprehend that these problems even belonged to the same field of inquiry, and certainly no one guessed that they would all interlock as they did in Einstein's hands.

Third, Einstein grasped that the way to take the molecular theory out of its disreputable state was to find new, testable, quantitative predictions. Thus Section 4.1.4 discussed how the study of Brownian motion gives a numerical value for the constant k_B, and hence, for N_{mole}. The molecular theory of heat says that the value obtained in this way should agree with earlier, approximate, determinations—and it did.

Nor did Einstein stop there. His doctoral thesis gave yet another independent determination of N_{mole} (and hence of k_B), again making use of Equation 4.16. Over the next few years, he published *four more* independent determinations of N_{mole}! Einstein was making a point: If molecules are real, then they have a real, finite size, which manifests itself in many different ways. If they were not real, it would be an absurd coincidence that all these independent measurements pointed to the *same* size scale.

These theoretical results had technological implications. Einstein's thesis work, on the viscosity of suspensions, remains his most heavily cited work today. At the same time, Einstein was also sharpening his tools for a bigger project: Showing that matter consisted of discrete particles prepared his mind to show that *light* does as well (see Section 1.5.3 on page 26). It is no accident that the Brownian motion work immediately preceded the light-quantum paper.

T_2 *Section 4.2' on page 148 views some of Einstein's other early work in the light of the preceding discussion.*

4.3 OTHER RANDOM WALKS

4.3.1 The conformation of polymers

Up to this point, we have been thinking of Figure 4.2 as a time-lapse photo of the *motion* of a *point* particle. Here is another application of exactly the same mathematics to a totally different physical problem, with biological relevance: the conformation of a polymer.

To describe the exact state of a polymer, we'd need an enormous number of geometrical parameters, for example, the angles of every chemical bond. It's hopeless to *predict* this state, because the polymer is constantly being knocked about by the thermal motion of the surrounding fluid. But, here again, we may turn frustration into opportunity. Are there some *overall, average* properties of the whole polymer's shape that we could try to predict?

Let's imagine that the polymer can be regarded as a string of N units. Each unit is joined to the next by a perfectly flexible joint, like a string of paperclips.[4] In thermal equilibrium, the joints will all be at random angles. An instantaneous snapshot of the polymer will be different at each instant of time, but there will be a family resemblance in a series of such snapshots: Each one will be a random walk. Following the approach of Section 4.1.2, we will simplify the problem by supposing that each joint of the chain sits at one of the eight corners of a cube centered on the previous joint (Figure 4.6). Taking the length of the cube's sides to be $2L$, then the length of one link is $\sqrt{3}L$. We can now apply our results from Section 4.1.2. For instance, the polymer is extremely unlikely to be stretched out straight, just as in our imagined checker game

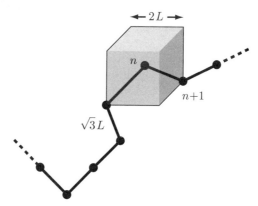

Figure 4.6: (Schematic.) A fragment of a three-dimensional random walk, simplified so that every joint can make any of eight possible bends. In the configuration shown, the step from joint n to joint $n + 1$ is the vector sum of one step to the right, one step down, and one step into the page.

---

[4]In a real polymer, the joints will not be perfectly flexible. Chapter 9 will show that, even in this case, the freely jointed chain model has some validity, as long as we understand that each of the "units" just mentioned may actually consist of many monomers.

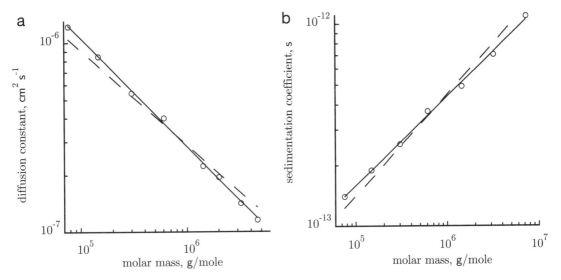

Figure 4.7: (Experimental data with fits.) (a) Log-log plot of the diffusion constant D of polymethyl methacrylate in acetone, as a function of the polymer's molar mass M. The *solid line* corresponds to the function $D \propto M^{-0.57}$. For comparison, the *dashed line* shows the best fit with scaling exponent fixed to $-1/2$, which is the prediction of the simplified analysis in this chapter. (b) The sedimentation coefficient s of the same polymer, to be discussed in Chapter 5. The solid line corresponds to the function $s \propto m^{0.44}$. For comparison, the dashed line shows the best fit with scaling exponent fixed to $1/2$. [Data from Meyerhoff & Schultz, 1952.]

we're unlikely to take every step to the right. Instead, the polymer is likely to be a blob, or **random coil**.

From Equation 4.4 on page 115, we find that the root-mean-square distance between the ends of the random coil is $\sqrt{\langle \mathbf{r}_N{}^2 \rangle} = \sqrt{\langle x_N{}^2 \rangle + \langle y_N{}^2 \rangle + \langle z_N{}^2 \rangle} = \sqrt{3L^2 N} = L\sqrt{3N}$. This is an experimentally testable prediction. The molar mass of the polymer equals the number of units, N, times the molar mass of each unit, so we predict that

> *If we synthesize polymers made from various numbers of the same units, then the coil size increases proportionally as the square root of the molar mass.* (4.17)

Figure 4.7a shows the results of an experiment in which eight batches of polymer, each with a different chain length, were synthesized. The diffusion constants of dilute solutions of these polymers were measured. The Stokes and Einstein relations (Equations 4.14 and 4.16) imply that D is a constant divided by the radius of the polymer blob. Idea 4.17 then leads us to expect that D should be proportional to $M^{-1/2}$, roughly as seen in the experimental data.[5]

Figure 4.7 also illustrates an important graphical tool. If we wish to show that D is a constant times $M^{-1/2}$, we could try graphing the data and superimposing the

---

[5]See Section 5.1.2 and Problem 5.8 for more about random-coil sizes.

curves $D = AM^{-1/2}$ for various values of the constant A and seeing whether any of them fit. A far more transparent approach is to plot instead $(\log D)$ versus $(\log M)$. Now the different predicted curves $(\log D) = (\log A) - \frac{1}{2}(\log M)$ are all *straight lines of slope* $-\frac{1}{2}$. We can thus test our hypothesis by laying a ruler along the observed data points, seeing whether they lie on any straight line, and if so, finding the slope of that line.

One consequence of Idea 4.17 is that random-coil polymers are loose structures. To see this, note that, if each unit of a polymer takes up a fixed volume v, then packing N units tightly together would yield a ball of radius $(3Nv/4\pi)^{1/3}$. For large enough polymers (N large enough), this size will be smaller than the random-coil size, because $N^{1/2}$ increases more rapidly than $N^{1/3}$.

We made a number of expedient assumptions to arrive at Idea 4.17. Most important, we assumed that every polymer unit is equally likely to occupy all the spaces adjacent to its neighbor (the eight corners of the cube in the idealization of Figure 4.6). This assumption could fail if the monomers are strongly attracted to one another; in that case, the polymer will not assume a random-walk conformation but will instead pack itself tightly into a ball. Examples of this behavior include globular proteins such as serum albumin. We can crudely classify polymers as "compact" or "extended" by comparing the volume occupied by the polymer with the minimal volume it would occupy if all its monomers were tightly packed together. Most large proteins and nonbiological polymers then fall unambiguously into one or the other category; see Table 4.1.

Even if a polymer does not collapse into a packed coil, its monomers are not really free to sit anywhere: Two monomers cannot occupy the same point of space! Our treatment ignored this self-avoidance phenomenon. Remarkably, introducing the physics of self-avoidance simply ends up changing the scaling exponent in Idea 4.17 from $\frac{1}{2}$ to another, calculable, value. The actual value of this exponent depends on temperature and solvent conditions. For a walk in three dimensions, in "good solvent," the corrected value is about 0.58. The experiment shown in Figure 4.7 is an example of this situation; and, indeed, its scaling exponent is seen to be slightly larger than the simple model's prediction of $\frac{1}{2}$. Whatever the precise value of this exponent, the main point is that simple scaling relations emerge from the complexity of polymer motions.

Table 4.1: Properties of various polymers. R_G is the measured radius of gyration for a few natural and artificial polymers, along with the radius of the ball the polymer would occupy if it were tightly packed, estimated from the molar mass and approximate density.

polymer	molar mass, g/mole	R_G, nm	packed–ball radius, nm	type
serum albumin	$6.6 \cdot 10^4$	3	2	compact
catalase	$2.25 \cdot 10^5$	4	3	compact
bushy stunt virus	$1.1 \cdot 10^7$	12	11	compact
myosin	$4.93 \cdot 10^5$	47	4	extended
polystyrene	$3.2 \cdot 10^6$	49	8	extended
DNA, in vitro	$4.0 \cdot 10^6$	117	7	extended

[From Tanford, 1961.]

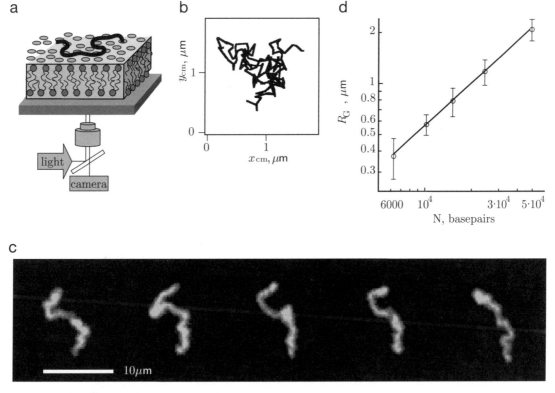

Figure 4.8: (Schematic; experimental data; photomicrograph.) Experimental test of the self-avoiding random walk model of polymer conformation, in two dimensions. (a) Experimental setup. A negatively charged DNA molecule sticks to a positively charged surface. The DNA has been labeled with a fluorescent dye to make it visible in a light microscope. (b) The entire molecule performs a random walk in time. The plot shows the molecule's center of mass on successive observations (compare Figure 4.2b,c on page 112). (c) Successive snapshots of the molecule taken at 2 s intervals. Each one shows a different random conformation. The fine structure of the conformation is not visible, because of the limited resolving power of an optical microscope, but the mean-square distance of the molecule from its center of mass can still be calculated.(d) Log-log plot of the size of a random coil of length N basepairs versus N. For each N, the coil size has been averaged over 30 independent snapshots like the ones in (c) (see Figure 4.5). The average size increases proportionally to $N^{0.79\pm0.04}$, close to the theoretically predicted $N^{3/4}$ behavior (see Problem 7.9). [(c) Digital image kindly supplied by B. Maier; see also Maier & Rädler, 1999.]

Figure 4.8 shows a particularly direct test of a scaling law for a polymer conformation. B. Maier and J. Rädler formed a positively charged surface and let it attract single strands of DNA, which is negatively charged. They then took successive snapshots of the attached DNA's changing conformation (the DNA contained a fluorescent dye to make it visible). The DNA may cross over itself; but each time it does so, there is a cost in binding energy because the negatively charged upper strand does not contact the positive surface at the point of crossing and, instead, is forced to contact another negative strand. Thus we may expect the coil size to follow a scaling relation appropriate to a two-dimensional, self-avoiding random walk. Problem 7.9 will show that the predicted scaling exponent for such a walk is $\frac{3}{4}$.

Once bound to the plate, the strands began to wander (Figure 4.8c). Measuring the fluorescence intensity as a function of position and averaging over many video frames allowed Maier and Rädler to compute the polymer chain's radius of gyration R_G, which is related to the chain's mean-square end-to-end distance. The data in Figure 4.8d show that $R_G \propto M^{0.79}$, close to the $\frac{3}{4}$ power law predicted by theory.

$\boxed{T_2}$ *Section 4.3.1′ on page 148 mentions some finer points about the conformation of random-coil polymers.*

4.3.2 Vista: Random walks on Wall Street

Stock markets are interacting systems of innumerable, independent biological subunits—the investors. Each investor is governed by a personal mixture of prior experience, emotion, and incomplete knowledge. Each bases his decisions on the aggregate of the other investors' decisions, as well as on the totally unpredictable events in the daily news. How could we possibly say anything predictive about such a tremendously complex system?

Indeed, we cannot predict an individual investor's behavior. But remarkably, the very fact that investors are so well informed about one another's aggregate behavior *does* lead to a certain statistical regularity in their behavior: It turns out that over the long term, *stock prices execute a random walk with drift.* The "thermal motion" driving this walk includes the whims of individual investors, along with natural disasters, collapses of large firms, and other unpredictable news items. The overall drift in the walk comes from the fact that, in the long run, investing money in firms does make a profit.

Why is the walk random? Suppose that a technical analyst finds that there was a reliable year-end rally, that is, every year stock prices rise in late December, then

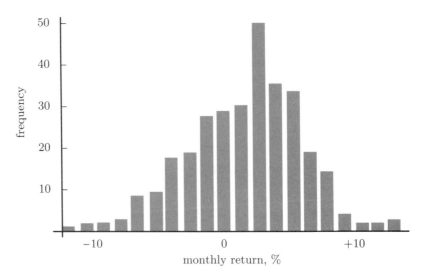

Figure 4.9: (Experimental data.) Ubiquity of random walks. The distribution of monthly returns for a 100-security portfolio, January 1945–June 1970. [Data from Malkiel, 1996.]

fall in early January. The problem is that once such a regularity becomes known to market participants, many people will naturally choose to sell during this period, driving prices down and eliminating the effect in the future. More generally, the past history of stock-price movements, which is public information, contains no useful information that will enable an investor consistently to beat other investors.

If this idea is correct, then some of our results from random-walk theory should show up in financial data. Figure 4.9 shows the distribution of step sizes taken by the market value of a stock portfolio. The value was sampled at one-month intervals, over 306 consecutive periods. The graph indeed bears a strong resemblance to Figure 4.3. In fact, Section 4.6.5 will argue that the distribution of step sizes in a random walk should be a Gaussian, as seen approximately in the figure.

4.4 MORE ABOUT DIFFUSION

4.4.1 Diffusion rules the subcellular world

Cells are full of localized structures; "factory" sites must transport their products to distant "customers." For example, mitochondria synthesize ATP, which then is used throughout the cell. We may speculate that thermal motion, which we have found is a big effect in the nanoworld, somehow causes molecular transport. It's time to put this speculation on a firmer footing.

Suppose we look at one colloidal particle—perhaps a visible pollen grain—every $\frac{1}{30}$ second, the rate at which an ordinary video camera takes pictures. An enormous number of collisions happen in this time, and they lead to some net displacement. Each such displacement is independent of the preceding ones, just like the successive tosses of a coin, because the surrounding fluid is in random motion. It's true that the steps won't be all the same length, but Section 4.1.3 showed that correcting this oversimplification complicates the math but doesn't change the physics.

With enough patience, one can watch a single particle for, say, one minute, note its displacement squared, then repeat the process enough times to get the mean. If we start over, this time using two-minute runs, the diffusion law says that we should get a value of $\langle (x_N)^2 \rangle$ twice as great as before, and we do. The actual value of the diffusion constant D needed to fit the observations to the diffusion law (Idea 4.6) will depend on the size of the particle and the nature of the surrounding fluid.

Moreover, what works for a pollen grain holds equally for the individual molecules in a fluid. They, too, will wander from their positions at any initial instant. We don't need to see individual molecules to confirm this prediction experimentally. Simply release a large number N of ink molecules at one point, for example, with a micropipette. Each begins an independent random walk through the surrounding water. We can come back at time t and examine the solution optically by using a photometer. The solution's *color* gives the number density $c(\mathbf{r}, t)$ of ink molecules, which, in turn, allows us to calculate the mean-square displacement $\langle r(t)^2 \rangle$ as $N^{-1} \int d^3\mathbf{r}\, r^2 c(\mathbf{r}, t)$. By watching the ink spread, we can not only verify that diffusion obeys Idea 4.6, but also find the value of the diffusion constant D. For small molecules, in water, at room temperature, one finds $D \approx 10^{-9}\ \mathrm{m}^2\,\mathrm{s}^{-1}$. A more useful form of this number, and one worth memorizing, is $D \approx 1\ \mu\mathrm{m}^2\,\mathrm{ms}^{-1}$.

Example: Pretend that the interior of a bacterium could be adequately modeled as a sphere of water of radius $1 \, \mu$m. About how long does it take for a sudden supply of sugar molecules at, say, the center of the bacterium to spread uniformly throughout the cell? How long would this diffusion take in a container the size of a eukaryotic cell?

Solution: Rearranging Equation 4.6 slightly and substituting $D = 1 \, \mu\text{m}^2 \, \text{ms}^{-1}$ gives that the time is around $(1 \, \mu\text{m})^2/(6D) \approx 0.2$ ms for the bacterium. It takes a hundred times longer for sugar to spread throughout a "cell" of radius $10 \, \mu$m.

The estimate just made points out an engineering design problem that larger, more complex cells need to address: Although diffusion is very rapid on the micrometer scale, it quickly becomes inadequate as a means of transporting material on long scales. As an extreme example, you have some single cells, the neurons that stretch from your spinal cord to your toes, that are about a meter long! If the specialized proteins needed at the terminus of these nerves had to arrive there from the cell body by diffusion, you'd be in trouble. Indeed, many animal and plant cells (not just neurons) have developed an infrastructure of "highways" and "trucks," all to carry out such transport (see Section 2.2.4). But on the subcellular, $1 \, \mu$m level, diffusion is fast, automatic, and free. And, indeed, bacteria don't have all that transport infrastructure; they don't need it.

4.4.2 Diffusion obeys a simple equation

Although the motion of a colloidal particle is totally unpredictable, Section 4.1.2 showed that a certain *average* property of many random walks obeys a simple law (Equation 4.5a on page 115). But the mean-square displacement is just one of many properties of the full probability distribution $P(x, t)$ of particle displacements after a given time t has passed. Can we find any simple rule governing the *full* distribution?

We could try to use the binomial distribution to answer this question (see the random walk Example on page 112). Instead, however, this section will derive an approximation, valid when there are very many steps between each observation.[6] The approximation is simpler and more flexible than the binomial distribution approach and will lead us to some important intuition about dissipation in general.

It's possible experimentally to observe the initial position of a colloidal particle, watch it wander, log its position at various times t_i, then repeat the experiment and compute the probability distribution $P(x, t)\text{d}x$ by using its definition (Equation 3.3 on page 71). But we have already seen in Section 4.4.1 that an alternative approach is much easier in practice. If we simply release a *trillion* random walkers in some initial distribution $P(x, 0)$, then monitor their *density*, we'll find the later distribution $P(x, t)$, automatically averaged for us over those trillion independent random walks, all in one step.

---

[6] $\boxed{T_2}$ Section 4.6.5' on page 150 explores the validity of this approximation.

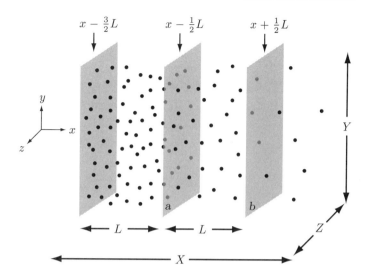

Figure 4.10: (Schematic.) Simultaneous diffusion of many particles in three dimensions. For simplicity, the figure shows a distribution uniform in y and z, but nonuniform in x. Space is subdivided into imaginary bins centered at $x - L, x, x + L, \ldots$ The *planes* labeled a, b represent the (imaginary) boundaries between these bins. X, Y, and Z denote the overall size of the system.

Suppose that the initial distribution is everywhere uniform in the y, z directions but nonuniform in x (Figure 4.10). We again simplify the problem by supposing that, on every time step Δt, every suspended particle moves a distance L either to the right or to the left, at random (see Section 4.1.2). Thus about half of a given bin's population hops to the left, and half to the right. And more will hop from the slot centered on $x - L$ to the one centered on x than will hop in the other direction, simply because there were more at $x - L$ to begin with.

Let $N(x)$ be the total number of particles in the slot centered at x, and Y, Z the widths of the box in the y, z directions. The *net* number of particles crossing the bin boundary a from left to right is then the difference between N evaluated at two nearby points, or $\frac{1}{2}\big(N(x - L) - N(x)\big)$; we count the particles crossing the other way with a minus sign.

We now come to a crucial step: The bins have been imaginary all along, so we can, if we choose, imagine them to be very narrow. But the difference between a function, like $N(x)$, at two nearby points is L times the derivative of N:

$$N(x - L) - N(x) \rightarrow -L\frac{\mathrm{d}N}{\mathrm{d}x}. \tag{4.18}$$

The point of this step is that we can now simplify our formulas by eliminating L altogether, as follows.

The number density of particles, $c(x)$, is just the number $N(x)$ in a slot, divided by the volume LYZ of the slot. Clearly, the future development of the density won't depend on how big the box is (that is, on X, Y, or Z); the important thing is not really the number crossing the boundary a, but rather the number *per unit area* of a. This notion is so useful that the average rate of crossing a surface per unit area has a special name, the number flux, denoted by the letter j (see Section 1.4.4 on page 22). Thus, number flux has dimensions $\mathbb{T}^{-1}\mathbb{L}^{-2}$.

We can restate the result of the three preceding paragraphs in terms of the number density $c = N/(LYZ)$, finding that

$$j = \frac{1}{YZ \times \Delta t} \times \frac{1}{2} \times L \times \left(-\frac{d}{dx}LYZc(x)\right) = -\frac{1}{\Delta t}\frac{L^2}{2} \times \frac{dc}{dx}.$$

We have already given a name to the combination $L^2/(2\Delta t)$, namely, the diffusion constant D (see Equation 4.5b). Thus we have

$$\boxed{\quad j = -D\frac{dc}{dx}. \qquad \textbf{Fick's law} \quad} \tag{4.19}$$

j measures the net number of particles moving from left to right. If there are more on the left than on the right, then c is decreasing, its derivative is negative, so the right-hand side is *positive*. That makes sense intuitively: A net drift to the right ensues, tending to even out the distribution, or make it more uniform. If there's structure (or order) in the original distribution, Fick's law says that diffusion will tend to erase it. The diffusion constant D enters the formula because more-rapidly diffusing particles will erase their order faster.

What "drives" the flux? It's *not* that the particles in the crowded region push against one another, driving one another out. Indeed, we assumed that each particle is moving totally independently of the others; we've neglected any possible interactions among the particles, which is appropriate if they're greatly outnumbered by the surrounding solution molecules. The only thing causing the net flow is simply that, if there are *more* particles in one slot than in the neighboring one and if each particle is equally likely to hop in either direction, then more will hop out of the slot with the higher initial population. *Mere probability seems to be "pushing" the particles.* This simple observation is the conceptual rock upon which we will build the notion of *entropic forces* in later chapters.

Fick's law is still not as useful as we'd like, though. We began this subsection with a very practical question: If all the particles are initially concentrated at a point (that is, the number density $c(\mathbf{r}, 0)$ is sharply peaked at one point), what will we measure for $c(\mathbf{r}, t)$ at a later time t? We'd like an equation we could solve; but all Equation 4.19 does is tell us j, given c. That is, we've found one equation in *two* unknowns, namely, c and j. But to find a solution, we need one equation in *one* unknown, or equivalently a second independent equation in c and j.

Looking again at Figure 4.10, we see that the average number $N(x)$ changes in one time step for two reasons: Particles can cross the imaginary wall a, and they can cross b. Recalling that j refers to the net flux from left to right, we find the net change

$$\frac{d}{dt}N(x) = \left(YZj\left(x - \frac{L}{2}\right) - YZj\left(x + \frac{L}{2}\right)\right).$$

Once again, we may take the bins to be narrow, whereupon the right-hand side of this formula becomes $(-L)$ times a derivative. Dividing by LYZ then gives

$$\frac{dc}{dt} = -\frac{dj}{dx},$$

a result known as the **continuity equation**. That's the second equation we were seeking. We can now combine it with Fick's law to eliminate j altogether. Simply take the derivative of Equation 4.19 and substitute to find[7]

$$\boxed{\frac{dc}{dt} = D\frac{d^2c}{dx^2}. \qquad \textbf{diffusion equation}} \qquad (4.20)$$

In more advanced texts, you will see the diffusion equation written as

$$\frac{\partial c}{\partial t} = D\frac{\partial^2 c}{\partial x^2}.$$

The curly symbols are just a stylized way of writing the letter "d," and they refer to derivatives, as always. The ∂ notation simply emphasizes that there is more than one variable in play, and the derivatives are to be taken by wiggling one variable while holding the others fixed. This book will use the more familiar "d" notation.

$\boxed{T_2}$ *Section 4.4.2' on page 149 casts the diffusion equation in vector notation and identifies thermal conduction as another diffusion problem.*

4.4.3 Precise statistical prediction of random processes

Something magical seems to have happened. Section 4.4.1 started with the hypothesis of random molecular motion. But the diffusion equation (Equation 4.20) is **deterministic**; that is, given the initial profile of concentration $c(x, 0)$, we can solve the equation and *predict the future* profile $c(x, t)$.

Did we get something from nothing? Almost—but it's not magic. Section 4.4.2 started from the assumption that the number of random-walking particles, and in particular, the number in any one slice, was huge. Thus we have a large collection of independent random events, each of which can take either of two equally probable options, just like a sequence of coin flips. Figure 4.3 illustrates how, in this limit, the fraction taking one of the two options will be very nearly equal to $\frac{1}{2}$, as assumed in the derivation of Equation 4.20.

Equivalently, we can consider a smaller number of particles but imagine repeating an observation on them many times and finding the average of the flux over the many trials. Our derivation can be seen as giving this average flux, $\langle j(x)\rangle$, in terms of the average number density, $c(x) = \langle N(x)\rangle/(LYZ)$. The resulting equation for $c(x)$ (the diffusion equation) is deterministic. Similarly, a deterministic formula for the

---

[7]Some authors call Equation 4.20 "Fick's second law."

squared displacement (the diffusion law, Equation 4.5 on page 115) emerged from averaging many individual random walks (see Figure 4.5).

When we don't deal with the ideal world of infinitely repeated observations, we should expect some deviation of actual results from their predicted average values. Thus, for example, the peak in the coin-flipping histogram in Figure 4.3c is narrow, but not infinitely narrow. This deviation from the average is called **statistical fluctuation**. For a more interesting example, we'll see that the diffusion equation predicts that a uniformly mixed solution of ink in water won't spontaneously assemble itself into a series of stripes. Certainly this *could* happen spontaneously, as a statistical fluctuation from the behavior predicted by the diffusion equation. But, for the huge number of molecules in a drop of ink, spontaneous unmixing is so unlikely that we can forget about the possibility. (Section 6.4 on page 206 will give a quantitative estimate.) Nevertheless, in a box containing just *ten* ink molecules, there's a reasonable chance of finding all of them on the left-hand side, a big nonuniformity of density. The probability is $(1/2)^{10}$, or $\approx 0.1\%$. In such a situation, the average behavior predicted by the diffusion equation won't be very useful in predicting what we'd see: The statistical fluctuations will be significant, and the system's evolution really will appear random, not deterministic.

So we need to take fluctuations seriously in the nanoworld of single molecules. Still, there are many cases in which we study collections of molecules large enough for the average behavior to be a good guide to what we'll actually see.

$\boxed{T_2}$ *Section 4.4.3' on page 149 mentions a conceptual parallel to quantum mechanics.*

4.5 FUNCTIONS, DERIVATIVES, AND SNAKES UNDER THE RUG

4.5.1 Functions describe the details of quantitative relationships

Before solving the diffusion equation, it's important to get an intuitive feeling for what the symbols are saying. Even if you already have the technical skills to handle equations of this sort, take some time to see how Equation 4.20 summarizes everyday experience in one terse package.

The simplest possible situation, Figure 4.11a, is a suspension of particles that already has uniform density at time $t = 0$. Because $c(x)$ is a constant, Fick's law says there's zero net flux. The diffusion equation says that c doesn't change: A uniform distribution stays that way. In the language of this book, we can say that it stays uniform because any nonuniformity would increase its order, and order doesn't increase spontaneously.

The next simplest situation, Figure 4.11b, is a uniform concentration gradient. The first derivative dc/dx is the slope of the curve shown, which is a constant. Fick's law then says there's a constant flux j to the right. The second derivative d^2c/dx^2 is the *curvature* of the graph, which is zero for the straight line shown. Thus, the diffusion equation says that once again c is unchanging in time: Diffusion maintains the profile shown. This conclusion may be surprising at first, but it makes sense: Every

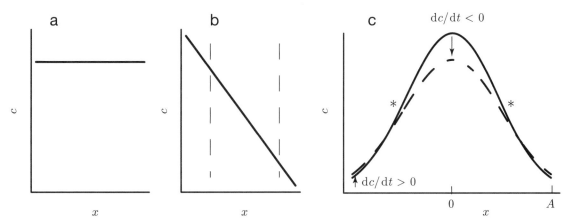

Figure 4.11: (Mathematical functions.) (a) A uniform (well-mixed) solution has constant concentration $c(x)$ of solute. A constant function has $dc/dx = 0$ and $d^2c/dx^2 = 0$; its graph is a horizontal line. (b) A linear function has $d^2c/dx^2 = 0$; its graph is a straight line. If the slope, dc/dx, is not zero, then this function represents a uniform concentration gradient. The *dashed lines* denote two fixed locations; see the text. (c) A lump of dissolved solute centered on $x = 0$. The curvature, d^2c/dx^2, is now negative near the bump, zero at the points labeled $*$, and positive beyond those points. The ensuing flux of particles will be directed outward. This flux will deplete the concentration in the region between the points labeled with stars, while increasing it elsewhere, for example, at the point labeled A. The flux changes the distribution from the *solid curve* at one instant of time to the *dashed curve* at a later time.

second, the net number of particles entering the region bounded by dashed lines in Figure 4.11b from the left is just equal to the net number *leaving* to the right, so c doesn't change.

Figure 4.11c shows a more interesting situation: a *bump* in the initial concentration at 0. For example, at the moment when a synaptic vesicle fuses (Figure 2.7 on page 43), it suddenly releases a large concentration of neurotransmitter at one point, creating such a bump distribution in three dimensions. Looking at the slope of the curve, we see that the flux will be everywhere *away* from 0, indeed tending to erase the bump. More precisely, the curvature of this graph is concave-down between the two starred points. Here the diffusion equation says that dc/dt will be negative: The height of the bump goes down. But outside the two starred points, the curvature is concave-*up*: dc/dt will be positive, and the concentration grows. This conclusion also makes sense: Particles leaving the bump must go somewhere, enhancing the concentration away from the bump. The starred points, where the curvature changes sign, are called **inflection points** of the graph of concentration. We'll soon see that they move apart in time, thereby leading to a wider, lower bump.

Suppose you stand at the point $x = A$ and watch. Initially, the concentration is low. Then it starts to increase, because you're outside the inflection point. Later, as the inflection point moves past you, the concentration again decreases: You've seen a *wave* of diffusing particles pass by. Ultimately, the bump is so small that the concentration is uniform: Diffusion erases the bump and the order it represents.

4.5.2 A function of two variables can be visualized as a landscape

Implicit in all the discussion so far has been the idea that c is a function of *two variables*, space x and time t. All the pictures in Figure 4.11 have been snapshots, graphs of $c(x, t_1)$ at some fixed time $t = t_1$. But the stationary observer just mentioned has a different point of view: She would graph the *time* development by $c(A, t)$ holding $x = A$ fixed. We can visualize both points of view at the same time by drawing a picture of the whole function as a *surface* in space (Figure 4.12). In these figures, points in the horizontal plane correspond to all points in space and time; the height of the surface above this plane represents the concentration at that point and that time. The two derivatives dc/dx and dc/dt are then *both* interpreted as slopes, corresponding to the two directions you could walk away from any point. Sometimes it's useful to be ultraexplicit and indicate both what's being varied *and* what's held fixed. For example, the notation $\frac{dc}{dx}\big|_t$ denotes the derivative holding t fixed. To get the sort of graphs shown in Figure 4.11, we *slice* the surface-graph along a line of constant time; to get the graph made by our stationary observer, we instead slice along a line of constant x (heavy line in Figure 4.12b).

Figure 4.12a shows the behavior we'll find for the solution to the diffusion equation.

Your Turn 4D	Examine Figure 4.12a and convince yourself visually that a stationary observer, for example, one located at $x = -0.7$, indeed sees a transient increase in concentration.

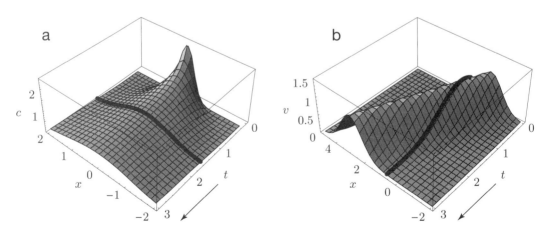

Figure 4.12: (Mathematical functions.) (a) The surface specifies a function $c(x, t)$, describing diffusion as a concentrated lump of solute begins to spread (see Section 4.6.5). Notice that time is drawn as increasing as we move diagonally downward in the page (*arrow*). The *heavy line* is the concentration profile at one particular time, $t = 1.6$. (b) This surface specifies a function $v(x, t)$, describing a hypothetical traveling wave. The diffusion equation has no such solutions, but Chapter 12 will find this behavior in the context of nerve impulses. The *heavy line* is the concentration as seen by an observer fixed at $x = 0.7$.

In contrast, Figure 4.12b depicts a behavior very different from what you just found in Your Turn 4D. This snake-under-the-rug surface shows a localized bump in a function $v(x, t)$, initially centered on $x = 0$, which moves steadily to the left (larger x) as time proceeds, without changing its shape. This function describes a **traveling wave**.

The ability to look at a graph and see at a glance what sort of physical behavior it describes is a key skill, so please don't proceed until you're comfortable with these ideas.

4.6 BIOLOGICAL APPLICATIONS OF DIFFUSION

Up to now, we have admired the diffusion equation but not solved it. This book is not about the elaborate mathematical techniques used to solve differential equations. But it's well worth our while to examine some of the simplest solutions and extract their intuitive content.

4.6.1 The permeability of artificial membranes is diffusive

Imagine a long, thin glass tube (or capillary tube) of length L, full of water. One end sits in a bath of pure water, the other in a solution of ink in water with concentration c_0. Eventually, the containers at both ends will come to equilibrium with the same ink concentration, somewhere between 0 and c_0. But equilibrium will take a long time to achieve if the two containers are both large. Prior to equilibrium, the system will instead come to a nearly steady, or **quasi-steady**, state. That is, all variables describing the system will be nearly unchanging in time: The concentration stays fixed at $c(0) = c_0$ at one end of the tube and $c(L) = 0$ at the other and will take various intermediate values $c(x)$ in between.

To find the quasi-steady state, we look for a solution to the diffusion equation with $dc/dt = 0$. According to Equation 4.20, this condition means that $d^2c/dx^2 = 0$. Thus the graph of $c(x)$ is a straight line (see Figure 4.11b), or $c(x) = c_0(1 - x/L)$. A constant number flux $j_s = Dc_0/L$ of ink molecules then diffuses through the tube. (The subscript "s" reminds us that this is a flux of solute, not of water.) If the concentrations on each side are both nonzero, the same argument gives the flux in the $+\hat{x}$ direction as $j_s = -D(\Delta c)/L$, where $\Delta c = c_L - c_0$ is the concentration difference.

The sketch in Figure 2.21a on page 57 shows cell membranes as having channels even narrower than the membrane thickness. Accordingly, let's try to apply the preceding picture of diffusion through a long, thin channel to membrane transport. Thus we expect that the flux through the membrane will be of the form

$$j_s = -\mathcal{P}_s \Delta c. \qquad (4.21)$$

Here the **permeability** of the membrane to solute, $\mathcal{P}_s$, is a number depending on both the membrane and the molecule whose permeation we're studying. In simple cases, the value of $\mathcal{P}_s$ roughly reflects the width of the pore, the thickness of the membrane (length of the pore), and the diffusion constant for the solute molecules.

Your
Turn
4E

> **a.** Show that the units of P_s are the same as those of velocity.
> **b.** Using this simplified model of the cell membrane, show that P_s is given by D/L times the fraction α of the membrane area covered by pores.

Example: Think of a cell as a spherical bag of radius $R = 10\,\mu m$, bounded by a membrane that passes alcohol with permeability $P_s = 20\,\mu m\,s^{-1}$. *Question:* If, initially, the alcohol concentration is c_{out} outside the cell and $c_{in}(0)$ inside, how does the interior concentration c_{in} change with time?

Solution: The outside world is so immense and the permeation rate so slow that the concentration outside is essentially always the same. The concentration inside is related to the number $N(t)$ of molecules inside by $c_{in}(t) = N(t)/V$, where $V = 4\pi R^3/3$ is the volume of the cell. According to Equation 4.21, the outward flux through the membrane is then $j_s = -P_s(c_{out} - c_{in}(t)) \equiv -P_s \times \Delta c(t)$. Note that j_s can be negative: Alcohol will move inward if there's more outside than inside.

Let $A = 4\pi R^2$ be the area of the cell. From the definition of flux (Section 4.4.2), N changes at the rate $dN/dt = -Aj_s$. Remembering that $c_{in} = N/V$, we find that the concentration jump Δc obeys the equation

$$-\frac{d(\Delta c)}{dt} = \left(\frac{AP_s}{V}\right)\Delta c. \qquad \text{relaxation of a concentration jump} \qquad (4.22)$$

This is an easy differential equation: Its solution is $\Delta c(t) = \Delta c(0)e^{-t/\tau}$, where $\tau = V/(AP_s)$ is the **decay constant** for the concentration difference. Putting in the given numbers shows that $\tau \approx 0.2\,s$. Finally, to answer the question we need c_{in}, which we write in terms of known quantities as $c_{in}(t) = c_{out} - (c_{out} - c_{in}(0))e^{-t/\tau}$.

We say that an initial concentration jump **relaxes** exponentially to its equilibrium value. In one second, the concentration difference drops to about $e^{-5} = 0.7\%$ of its initial value. A smaller cell would have a bigger surface-to-volume ratio, so it would eliminate the concentration difference even faster.

The rather literal model for permeability via membrane pores, used in Your Turn 4E, is certainly oversimplified. Other processes also contribute to permeation. For example, a molecule can dissolve in the membrane material from one side, diffuse to the other side, then leave the membrane. Even artificial membranes, with no pores at all, will pass some solutes in this way. Here, too, a Fick-type law, Equation 4.21, will hold; after all, *some* sort of random walk is still carrying molecules across the membrane.

Because artificial bilayers are quite reproducible in the laboratory, we should be able to test the dissolve→diffuse→undissolve mechanism of permeation by checking a quantitative deduction from the model. Figure 4.13 shows the result of such an

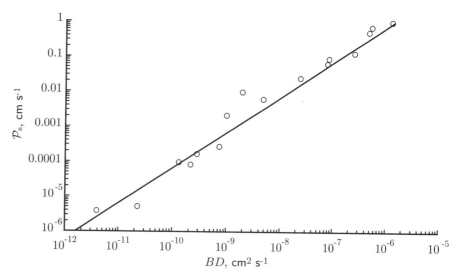

Figure 4.13: (Experimental data with fit.) Log-log plot of the permeability $\mathcal{P}_s$ of artificial bilayer membranes (made of egg phosphatidylcholine) to various small molecules, ranging from urea (*far left point*) to hexanoic acid (*far right point*). The horizontal axis gives the product BD of the diffusion constant D of each solute times its partition coefficient B in oil versus water. The solid line has slope equal to 1, indicating a strict proportionality $\mathcal{P}_s \propto BD$. [Data from Finkelstein, 1987.]

experiment by A. Finkelstein, who measured the permeabilities of a membrane to 16 small molecules. To understand these data, first imagine a simpler situation, a container with a layer of oil floating on a layer of water. If we introduce some sugar, stir well, and wait, eventually we will find that almost, but not all, of the sugar is in the water. The ratio of the concentration of sugar in the oil to that in the water is called the **partition coefficient** B; it characterizes the degree to which sugar molecules prefer one environment to another. We will investigate the reasons for this preference in Chapter 7; for now, we only note that this ratio is some measurable constant.

We will see in Chapter 8 that a bilayer membrane is essentially a thin layer of oil (sandwiched between two layers of head groups). Thus, a membrane separating two watery compartments with sugar concentrations c_1 and c_2 will itself have sugar concentration Bc_1 on one side and Bc_2 on the other, and hence a drop of $\Delta c = B(c_1 - c_2)$ across the membrane. Adapting the model discussed at the start of this section shows that the resulting flux of sugar gives the membrane a permeability $\mathcal{P}_s = BD/L$. Thus, even if we don't know the value of L, we can still assert that

> *The permeability of a pure bilayer membrane is roughly BD times a constant independent of the solute, where B is the partition coefficient* (4.23)
> *of solute and D its diffusion constant in oil.*

The data in Figure 4.13 support this simple conclusion, over a remarkably wide range (six orders of magnitude) of BD.

Typical real values are $\mathcal{P}_s \approx 10^{-3}\,\mu\text{m s}^{-1}$ for glucose diffusing across an artificial lipid bilayer membrane, or three to five orders of magnitude less than this (that is, 0.001 to 0.000 01 times as great) for charged ions like Cl^- or Na^+, respectively.

The bilayer membranes surrounding living cells have much larger values of $\mathcal{P}_s$ than do artificial bilayers. Indeed, Chapter 11 will show that the transport of small molecules across cell membranes is far more complicated than simple diffusion would suggest. Nevertheless, passive diffusion is one important ingredient in the full membrane-transport picture.

4.6.2 Diffusion sets a fundamental limit on bacterial metabolism

Let's idealize a single bacterium as a sphere of radius R. Suppose that the bacterium is suspended in a lake and that it needs oxygen to survive (it's ærobic). The oxygen is all around it, dissolved in the water, with a concentration c_0. But the oxygen nearby gets depleted, as the bacterium uses it up.

The lake is huge, so the bacterium won't affect the lake's overall oxygen level; instead, the environment near the bacterium will come to a steady state, in which the oxygen concentration c doesn't depend on time. In this state, the oxygen concentration $c(r)$ will depend on the distance r from the center of the bacterium. Very far away, we know that $c(\infty) = c_0$. We'll assume that every oxygen molecule reaching the bacterium's surface gets immediately gobbled up. Hence, at the cell surface, $c(R) = 0$. From Fick's law, there must therefore be a flux j of oxygen molecules inward.

Example: Find the full concentration profile $c(r)$ and the maximum number of oxygen molecules per time that the bacterium can consume.

Solution: Imagine drawing a series of concentric spherical shells around the bacterium with radii $r_1, r_2, \ldots$. Oxygen is moving across each shell on its way to the center. Because we're in a steady state, oxygen does not accumulate anywhere: The number of molecules per time crossing each shell equals the number per time crossing the next shell. This condition means that the inward flux $j(r)$ times the surface area of the shell must be a constant, independent of r. Call this constant I. Now we know $j(r)$ in terms of I (but we don't know I yet).

Next, Fick's law says $j = D(dc/dr)$, but we also know $j = I/(4\pi r^2)$. Solving for $c(r)$ gives $c(r) = A - (1/r)(I/4\pi D)$, where A is some constant. We can fix both I and A by imposing $c(\infty) = c_0$ and $c(R) = 0$, thereby finding that $A = c_0$ and $I = 4\pi DRc_0$. Along the way, we also find that the concentration profile itself is $c(r) = c_0(1 - (R/r))$.

Remarkably, *we have just computed the maximum rate at which oxygen molecules can be consumed by any bacterium whatsoever!* We didn't need to use any biochemistry at all, just the fact that living organisms are subject to constraints from the physical world. Notice that the oxygen uptake I increases with increasing bacterial size, but only as the first power of R. We might expect the oxygen *consumption*, however, to increase roughly with an organism's *volume*. Together, these statements imply an upper limit to the size of a bacterium: If R were too large, the bacterium would literally suffocate.

Your
Turn
4F

a. Evaluate the expression for I in the Example, using the illustrative values $R = 1\,\mu\text{m}$ and $c_0 \approx 0.2\,\text{mole}/\text{m}^3$.

b. A convenient measure of an organism's overall metabolic activity is its rate of O_2 consumption divided by its mass. Find the maximum possible metabolic activity of a bacterium of arbitrary radius R, again using $c_0 \approx 0.2\,\text{mole}\,\text{m}^{-3}$.

c. The actual metabolic activity of a bacterium is about $0.02\,\text{mole}\,\text{kg}^{-1}\text{s}^{-1}$. What limit do you then get on the size R of a bacterium? Compare your answer to the size of real bacteria. Can you think of some way for a bacterium to evade this limit?

T_2 *Section 4.6.2′ on page 149 mentions the concept of allometric exponents.*

4.6.3 The Nernst relation sets the scale of membrane potentials

Many of the molecules floating in water carry a net electric charge, unlike the alcohol molecules studied in the concentration decay Example (page 136). When table salt dissolves, for example, the individual sodium and chlorine atoms separate, but the chlorine atom grabs one extra electron from sodium, thereby becoming a negatively charged chloride ion, Cl^-, and leaving the sodium as a positive ion, Na^+. Any electric field $\mathcal{E}$ present in the solution will then exert forces on the individual ions, dragging them just as gravity drags colloidal particles to the bottom of a test tube.

Suppose first that we have a uniform-density solution of charged particles, each of charge q, in a region with electric field $\mathcal{E}$. For example, we could place two parallel plates just outside the solution's container, a distance ℓ apart, and connect them to a battery that maintains a constant electrostatic potential difference ΔV across them. We know from first-year physics that $\mathcal{E} = \Delta V/\ell$ and each charged particle feels a force $q\mathcal{E}$, so it drifts with the net speed we found in Equation 4.12: $v_{\text{drift}} = q\mathcal{E}/\zeta$, where ζ is the viscous friction coefficient.

Imagine a small net of area A stretched out perpendicular to the electric field (that is, parallel to the plates); see Figure 4.14. To find the flux of ions induced by the field, we ask how many ions get caught in the net each second. The average ion drifts a distance $v_{\text{drift}}dt$ in time dt, so, in this time, all the ions contained in a slab of volume $Av_{\text{drift}}dt$ get caught in the net. The number of ions caught equals this volume times the number density c. The flux j is then the number crossing per area per time, or cv_{drift}. (Check to make sure this formula has the proper units.) Substituting the drift velocity gives $j = q\mathcal{E}c/\zeta$, the **electrophoretic flux** of ions.

Now suppose that the density of ions is *not* uniform. For this case, we add the driven (electrophoretic) flux just found to the probabilistic (Fick's law) flux, Equation 4.19, thereby obtaining

$$j(x) = \frac{q\mathcal{E}(x)c(x)}{\zeta} - D\frac{dc}{dx}.$$

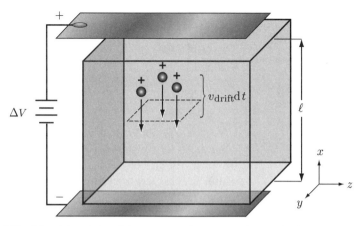

Figure 4.14: (Sketch.) Origin of the Nernst relation. An electric field pointing downward drives positively charged ions down. The system comes to equilibrium with a downward density gradient of positive ions and an upward gradient of negative ions. The flux through the surface element shown (*dashed square*) equals the number density c times v_{drift}.

We next rewrite the viscous friction coefficient in terms of D, using the Einstein relation (Equation 4.16 on page 120) to get[8]

$$j = D\left(-\frac{dc}{dx} + \frac{q}{k_B T}\mathcal{E}c\right). \qquad \textbf{Nernst–Planck formula} \qquad (4.24)$$

The Nernst–Planck formula helps us to answer a fundamental question: What electric field would be needed to get *zero* net flux, that is, to cancel the diffusive tendency to erase nonuniformity? To answer the question, we set $j = 0$ in Equation 4.24. In a planar geometry, where everything is constant in the y, z directions, we get the condition

$$\frac{1}{c}\frac{dc}{dx} = \frac{q}{k_B T}\mathcal{E}. \qquad \text{(in equilibrium)} \qquad (4.25)$$

The left side of this formula can be written as $\frac{d}{dx}(\ln c)$.

To use Equation 4.25, we now integrate both sides from the top plate to the bottom one (see Figure 4.14). The left side is $\int_0^\ell dx \frac{d}{dx} \ln c = \ln c_{\text{bot}} - \ln c_{\text{top}}$, that is, the difference in $\ln c$ from one plate to the other. To understand the right side, we first note that $q\mathcal{E}$ is the force acting on a charged particle, so the particle's potential energy obeys $-dU/dx = q\mathcal{E}$, or $U(x) = -q\mathcal{E}x$. The electrostatic potential V is the

---

[8] $\boxed{T_2}$ In the three-dimensional language introduced in Section 4.4.2′ on page 149, the Nernst–Planck formula becomes $\mathbf{j} = D(-\nabla c + (q/k_B T)\boldsymbol{\mathcal{E}}c)$. The gradient ∇c points in the direction of most steeply increasing concentration.

potential energy per unit charge, so $\Delta V \equiv V_{bot} - V_{top} = -\mathcal{E}\ell$. Writing $\Delta(\ln c)$ for $\ln c_{bot} - \ln c_{top}$ then gives the condition for equilibrium:

$$\Delta(\ln c) = -q\Delta V_{eq}/k_B T. \qquad \textbf{Nernst relation} \qquad (4.26)$$

The subscript on ΔV_{eq} reminds us that this is the voltage needed to maintain a concentration jump *in equilibrium*. (Chapter 11 will consider nonequilibrium situations, where the actual potential difference differs from ΔV_{eq}, thereby driving a net flux of ions.)

Equation 4.26 predicts that positive charges will migrate toward the bottom of Figure 4.14. It makes sense: They're attracted to the negative plate. We have so far been ignoring the corresponding negative charges (for example, the chloride ions in table salt), but the same formula applies to them as well. Because they carry negative charge ($q < 0$), Equation 4.26 says they migrate toward the positive plate.

Substituting some real numbers into Equation 4.26 yields a suggestive result. Consider a singly charged ion like Na$^+$, for which $q = e$. Suppose we have a moderately big concentration jump, $c_{bot}/c_{top} = 10$. Using the fact that

$$\frac{k_B T_r}{e} = \frac{1}{40} \text{ volt}$$

(see Appendix B), we find $\Delta V = +58\,\mathrm{mV}$. What's suggestive about this result is that many living cells, particularly nerve and muscle cells, really do maintain a potential difference across their membranes of a few tens of millivolts! We haven't proven that these potentials are equilibrium Nernst potentials, and indeed Chapter 11 will show that they're not. But the observation *does* show that dimensional arguments successfully predict the scale of membrane potentials with almost no hard work at all.

Something interesting happened on the way from Equation 4.24 to Equation 4.26: When we consider equilibrium only, the value of D drops out. That's reasonable: D controls how *fast* things move in response to a field; its units involve time. But equilibrium is an eternal state; it can't depend on time. In fact, exponentiating the Nernst relation gives that $c(x)$ is a constant times $e^{-qV(x)/k_B T}$. This result is an old friend: It says that the spatial distribution of ions follows the Boltzmann distribution (Equation 3.26 on page 85). A charge q in an electric field has electrostatic potential energy $qV(x)$ at x; its probability to be there is proportional to the exponential of minus its energy, measured in units of the thermal energy $k_B T$. Thus, a positive charge doesn't like to be in a region of large positive potential, and vice versa for negative charges. Our formulas are mutually consistent.[9]

---

[9] $\boxed{T_2}$ Einstein's original derivation of his relation inverted the logic here. Instead of starting with Equation 4.16 and rediscovering the Boltzmann distribution, as we just did, he began with Boltzmann and arrived at Equation 4.16.

4.6.4 The electrical resistance of a solution reflects frictional dissipation

Suppose we place the metal plates in Figure 4.14 *inside* the container of salt water, so that they become electrodes. Then the ions in solution migrate, but they don't accumulate: The positive ones get electrons from the − electrode while the negative ones hand their excess electrons over to the + electrode. The resulting neutral atoms leave the solution; for example, they can electroplate onto the attracting electrode or bubble away as gas.[10] Then, instead of establishing equilibrium, our system continuously *conducts* electricity, at a rate controlled by the steady-state ion fluxes.

The potential drop across our cell is $\Delta V = \mathcal{E}\ell$, where ℓ is the separation of the plates. According to the Nernst–Planck formula (Equation 4.24), this time with uniform c, the electric field is

$$\mathcal{E} = \frac{k_{B}T}{Dqc} j.$$

Recall that j is the number of ions passing per area per time. To convert this expression to the total electric current I, note that each ion deposits charge q when it lands on a plate; thus, $I = qAj$, where A is the plate area. Putting everything together gives

$$\Delta V = \left(\frac{k_{B}T}{Dq^{2}c} \frac{\ell}{A} \right) I. \tag{4.27}$$

This is a familiar result: It's **Ohm's law**, $\Delta V = IR$. Equation 4.27 gives the **electrical resistance** R of the cell as the constant of proportionality between voltage and current. To use this formula, we must remember that each type of ions contributes to the total current; for table salt, we need to add separately the contributions from Na^{+} with $q = e$ and Cl^{-} with $q = -e$, or in other words, double the right-hand side of the formula.

The resistance depends not only on the solution but also on the geometry of the cell. It's customary to eliminate the geometry dependence by defining the **electrical conductivity** of the solution as $\kappa = \ell/(RA)$. Then our result is that each ion species contributes $\kappa = Dq^{2}c/k_{B}T$ to κ. It makes sense: Saltier water conducts better.

$\boxed{T_2}$ *Section 4.6.4′ on page 149 mentions other points about electrical conduction.*

4.6.5 Diffusion from a point gives a spreading, Gaussian profile

Let's return to one dimension, and to the question of time-dependent diffusion processes. Section 4.4.2 on page 128 posed the question of finding the full distribution function of particle positions after an initial density profile $c(x, 0)$ has spread out for time t.

---

[10] $\boxed{T_2}$ Electroplating does not occur with a solution of table salt, nor does chlorine gas bubble away, because sodium metal and chlorine gas are so strongly reactive with water. Nevertheless, the following discussion is valid for the *alternating-current* conductivity of NaCl.

Suppose we release many particles all at one place (a "pulse" of concentration). We expect the resulting distribution to get broader with time. We might, therefore, guess that the solution we seek is a Gaussian; perhaps $c(x, t) \overset{?}{=} Be^{-x^2/(2At)}$, where A and B are some constants. This profile has the desired property that its variance, $\sigma^2 = At$, indeed grows with time. But substituting it into the diffusion equation, we find that it is not a solution, regardless of what we choose for A and B.

Before abandoning our guess, notice that it has a more basic defect: It's not properly normalized (see Section 3.1.1 on page 70). The integral $\int_{-\infty}^{\infty} dx\, c(x, t)$ is the total number of particles and hence cannot change in time. The proposed solution doesn't have that property.

Your Turn 4G

a. Establish that last statement. Then show that the profile

$$c(x, t) = \frac{\text{const}}{\sqrt{t}} e^{-x^2/(4Dt)}$$

does always maintain the same normalization. Find the constant, assuming that N particles are present. [*Hint:* Use the change of variables trick from the Gaussian normalization Example on page 73.]

b. Substitute your expression from (a) into the one-dimensional diffusion equation, take the derivatives, and show that with this correction we do get a solution.

c. Verify that $\langle x^2 \rangle = 2Dt$ for this distribution: It obeys the fundamental diffusion law (Idea 4.5a on page 115).

The solution you just found is the function shown in Figure 4.12 on page 134. You can now find the inflection points, where the concentration switches from increasing to decreasing, and can verify that they move outward in time, as mentioned in Section 4.5.1.

The result of Your Turn 4G pertains to one-dimensional walks, but we can promote it to three dimensions. Let $\mathbf{r} = (x, y, z)$. Because each diffusing particle moves independently in all three dimensions, we can use the multiplication rule for probabilities: The concentration $c(\mathbf{r})$ is the product of three one-dimensional distributions:

$$c(\mathbf{r}, t) = \frac{N}{(4\pi Dt)^{3/2}} e^{-r^2/(4Dt)}. \qquad \text{fundamental pulse solution} \qquad (4.28)$$

In this formula, the symbol r^2 refers to the length-squared of the vector $\mathbf{r}$, that is, $x^2 + y^2 + z^2$. Equation 4.28 has been normalized to make N the total number of particles released at $t = 0$. Applying your result from Your Turn 4G(c) to x, y, and z separately and adding the results recovers the three-dimensional diffusion law (Equation 4.6).

We get another important application of Equation 4.28 when we recall the discussion of polymers. Section 4.3.1 argued that, although a polymer in solution is constantly changing its shape, nevertheless its mean-square end-to-end length is a constant times its length. We can now sharpen that statement to say that the *distribution* of end-to-end vectors **r** will be Gaussian.

$\boxed{T_2}$ *Section 4.6.5′ on page 150 points out that an approximation used in Section 4.4.2 limits the accuracy of our result in the far tail of the distribution.*

THE BIG PICTURE

Returning to the Focus Question, we've seen how large numbers of random, independent actors can collectively behave in a predictable way. For example, we found that the purely random Brownian motion of a single molecule gives rise to a rule of diffusive spreading for *collections* of molecules (Equation 4.5a) that is simple, deterministic, and repeatable. Remarkably, we also found that precisely the same math gives useful results about the sizes of polymer coils, at first sight a completely unrelated problem.

We have already found a number of biological applications of diffusion and its other side, dissipation. Later chapters will carry this theme even further:

- Frictional effects dominate the mechanical world of bacteria and cilia, dictating the strategies they have chosen to do their jobs (Chapter 5).
- Our discussion in Section 4.6.4 about the conduction of electricity in solution will be needed when we discuss nerve impulses (Chapter 12).
- Variants of the random walk help explain the operation of some of the walking motors mentioned in Chapter 2 (see Chapter 10).
- Variants of the diffusion equation also control the rates of enzyme-mediated reactions (Chapter 10) and even the progress of nerve impulses (Chapter 12).

More bluntly, we cannot be satisfied with understanding thermal equilibrium (for example, the Boltzmann distribution found in Chapter 3), because *equilibrium is death.* Chapter 1 emphasized that life prospers on Earth only by virtue of an incoming stream of high-quality energy, which keeps us *far* from thermal equilibrium. The present chapter has provided a framework for understanding the dissipation of order in such situations; later chapters will apply this framework.

KEY FORMULAS

- *Binomial:* The number of ways to choose k objects out of a jar full of n distinct objects is $n!/(k!(n-k)!)$ (Equation 4.1).
- *Stirling:* The formula: $\ln N! \approx N \ln N - N + \frac{1}{2}\ln(2\pi N)$ allows us to approximate $N!$ for large values of N (Equation 4.2).
- *Random walk:* The average location after random-walking N steps of length L in one dimension is $\langle x_N \rangle = 0$. The mean-square distance from the starting point is

$\langle x_N{}^2 \rangle = NL^2$, or $2Dt$, where $D = L^2/(2\Delta t)$ if we take a step every Δt (Idea 4.5). Similarly, taking diagonal steps on a two-dimensional grid gives $\langle (\mathbf{x}_N)^2 \rangle = 4Dt$. D is given by the same formula as before; this time L is the edge of one square of the grid. In three dimensions, the 4 becomes a 6 (Equation 4.6).

- *Einstein:* An imposed force f on a particle in suspension, if small enough, results in a slow net drift with velocity $v_{\text{drift}} = f/\zeta$ (Equation 4.12). Drag and diffusion are related by the Einstein relation, $\zeta D = k_B T$ (Equation 4.16). This relation is not limited to our simplified model.

- *Stokes:* For a macroscopic (many nanometers) sphere of radius R moving slowly through a fluid, the drag coefficient is $\zeta = 6\pi \eta R$ (Equation 4.14), where η is the fluid viscosity.

  (In contrast, at high speed, the drag force on a fixed object in a flow has the form $-Bv^2$ for some constant B characterizing the object and the fluid; see Problem 1.7.)

- *Fick and diffusion:* The flux of particles along $\hat{\mathbf{x}}$ is the net number of particles passing from negative to positive x, per area per time. The flux created by a concentration gradient is $j = -D\, dc/dx$ (Equation 4.19), where $c(x)$ is the number density (concentration) of particles. (In three dimensions, $\mathbf{j} = -D\nabla c$.) The rate of change of $c(x, t)$ is then $dc/dt = D(d^2c/dx^2)$ (Equation 4.20).

- *Membrane permeability:* The flux of solute through a membrane is $j_s = -\mathcal{P}_s \Delta c$ (Equation 4.21), where $\mathcal{P}_s$ is the permeability and Δc is the jump in concentration across the membrane.

- *Relaxation:* The concentration difference of a permeable solute between the inside and outside of a spherical bag decreases in time, following the equation

$$-\frac{d(\Delta c)}{dt} = \left(\frac{A\mathcal{P}_s}{V} \right) \Delta c$$

  (Equation 4.22).

- *Nernst–Planck:* When charged particles diffuse in the presence of an electric field, we must modify Fick's law to include the electrophoretic flux:

$$j = D\left(-\frac{dc}{dx} + \frac{q}{k_B T}\mathcal{E}c \right)$$

  (Equation 4.24).

- *Nernst:* If an electrostatic potential difference ΔV is imposed across a region of fluid, then each dissolved ion species with charge q comes to equilibrium (no net flux) with a concentration change across the region fixed by $\Delta V = -(k_B T/q)\Delta(\ln c)$ (Equation 4.26) or equivalently

$$V_2 - V_1 = -\frac{58\,\text{mV}}{z}\log_{10}(c_2/c_1),$$

  where the valence z is defined by $z = q/e$.

- *Ohm:* The flux of electric current created by an electric field $\mathcal{E}$ is proportional to $\mathcal{E}$, a relation leading to Ohm's law. The resistance of a conductor of length ℓ and cross section A is $R = \ell/(A\kappa)$, where κ is the conductivity of the material. In our simplified model, each ion species contributes $Dq^2 c/k_{\mathrm{B}}T$ to κ (Section 4.6.4).

• *Diffusion from an initial sharp point:* Suppose N molecules all begin at the same location in three-dimensional space at time zero. Later the concentration is

$$c(\mathbf{r}, t) = \frac{N}{(4\pi Dt)^{3/2}} e^{-\mathbf{r}^2/(4Dt)}$$

(Equation 4.28).

FURTHER READING

Semipopular:
Historical: Pais, 1982, §5.
Finance: Malkiel, 1996.

Intermediate:
General: Berg, 1993; Tinoco et al., 2001.
Polymers: Grosberg & Khokhlov, 1997.
Better derivations of the Einstein relation: Benedek & Villars, 2000b, §2.5A–C; Feynman et al., 1963a, §43.

Technical:
Einstein's original discussion: Einstein, 1956.

$\boxed{T_2}$ **4.1.4′ Track 2**

Some fine points:

1. Sections 4.1.2 and 4.1.4 made a number of idealizations, so Equations 4.5b and 4.13 should not be taken too literally. Nevertheless, it turns out that the Einstein relation (Equation 4.16) is both general and accurate. This broad applicability must mean that it actually rests on a more general, although more abstract, argument than the one given here. Indeed, Einstein gave such an argument in his original 1905 paper (Einstein, 1956).

    For example, introducing a realistic distribution of times between collisions does not change our main results, Equations 4.12 and 4.16. See Feynman et al., 1963a, §43 for the analysis of this more detailed model. In it, Equation 4.13 for the viscous friction coefficient ζ expressed in terms of microscopic quantities becomes instead $\zeta = m/\tau$, where τ is the mean time between collisions.

2. The assumption that each collision wipes out all memory of the previous step is also not always valid. A bullet fired into water does not lose all memory of its initial motion after the first molecular collision! Strictly speaking, the derivation given here applies to the case where the particle of interest starts out with momentum comparable to that transferred in each collision, that is, not too far from equilibrium. We must also require that the momentum *imparted* by the external force in each step not be bigger than that transferred in molecular collisions, or, in other words, that the applied force is not too large. Chapter 5 will explore how great the applied force may be before "low Reynolds-number" formulas like Equation 4.12 become invalid, concluding that the results of this chapter are indeed applicable in the world of the cell. Even in this world, however, our analysis can certainly be made more rigorous: Again see Feynman et al., 1963a, §43.

3. Cautious readers may worry that we have applied a result obtained for the case of low-density gases (Idea 3.21, that the mean-square velocity is $\langle (v_x)^2 \rangle = k_B T/m$), to a dense *liquid*, namely, water. But our working hypothesis, the Boltzmann distribution (Equation 3.26 on page 85) assigns probabilities on the basis of the total system energy. This energy contains a complicated potential energy term, plus a simple kinetic energy term, so the probability distribution factors into the product of a complicated function of the positions, times a simple function of the velocities. But we don't care about the positional correlations. Hence we may simply integrate the complicated factor over $d^3x_1 \cdots d^3x_N$, leaving behind a constant times the *same* simple probability distribution function of velocities (Equation 3.25 on page 84) as the one for an ideal gas. Taking the mean-square velocity then leads again to Idea 3.21.

    Thus, in particular, the average kinetic energy of a colloidal particle is the same as that of the water molecules, just as argued in Section 3.2.1 for the different kinds of gas molecule in a mixture. We implicitly used this equality in arriving at Equation 4.16.

4. The Einstein relation, Equation 4.16, was the first of many similar relations between fluctuations and dissipation. In other contexts such relations are generically called fluctuation–dissipation theorems.

4.2′ Track 2

The themes explored in Section 4.2 also pervade the rest of Einstein's early work:

1. Einstein did not originate the idea that energy levels are quantized; Max Planck did, in his approach to thermal radiation. Einstein pointed out that applying this idea directly to light explained another, seemingly unrelated phenomenon, the photoelectric effect. Moreover, if the light-quantum idea was right, then both Planck's thermal radiation and the photoelectric experiments should independently determine a number, which we now call the Planck constant. Einstein showed that both experiments gave the *same numerical value* of this constant.

2. Einstein did not invent the equations for electrodynamics; Maxwell did. Nor was Einstein the first to point out their curious invariances; H. Lorentz did. Einstein did draw attention to a consequence of this invariance: the existence of a fundamental limiting velocity, the speed of light c. Once again, the idea seemed crazy. But Einstein showed that doggedly following it to its logical end point led to a new, quantitative, experimentally testable prediction in an apparently very distant field of research. In his very first relativity paper, also published in 1905, he observed that, if the mass m of a body could change, the transformation would necessarily liberate a definite amount of energy equal to $\Delta E = (\Delta m)c^2$. Yet again, Einstein offered a highly falsifiable prediction to test his seemingly crazy theory: The numerical value of c can be deduced from measuring Δm and ΔE of any nuclear reaction. Later experiments confirmed this prediction, with the same numerical value of c as that measured from light propagation.

3. Einstein said some deep things about the geometry of space and time, but D. Hilbert was saying many similar things at about the same time. Only Einstein, however, realized that measuring an apple's fall yields the numerical value of a physical parameter (Newton's constant), which also controls the fall of a *photon*. His theory thus made quantitative predictions about both the bending of light by the Sun and the gravitational blue-shift of a falling photon. The experimental confirmation of the light-bending prediction catapulted Einstein to international fame.

4.3.1′ Track 2

1. We saw that typically the scaling exponent for a polymer in solvent is not exactly $\frac{1}{2}$. One special condition, called theta solvent, actually does give a scaling exponent of $\frac{1}{2}$, the same as the result of our naïve analysis. Theta conditions roughly correspond to the case where the monomers attract one another just as much as they attract solvent molecules. (Problem 5.8 will explore this situation.) In some cases, theta conditions can be reached simply by adjusting the temperature.

2. The precise definition of the radius of gyration R_G is the root-mean-square distance of the individual monomers from the polymer's center of mass. For long polymer chains, it is related to the end-to-end distance $\mathbf{r}_N$ by the relation $(R_G)^2 = \frac{1}{6}\langle(\mathbf{r}_N)^2\rangle$.

3. Another test for polymer coil size uses light scattering; see Tanford, 1961.

4.4.2′ Track 2

1. What if we don't have everything uniform in the y and z directions? The net flux of particles is really a *vector*, like velocity; our j was just the x component of this vector. Likewise, the derivative dc/dx is just the x component of a vector, the **gradient**, denoted ∇c (and pronounced "grad c"). In this language, the general form of Fick's law is then $\mathbf{j} = -D\nabla c$, and the diffusion equation reads

$$\frac{\partial c}{\partial t} = D\nabla^2 c.$$

2. Actually, *any* conserved quantity carried by random walkers will have a diffusive transport law. We've studied the *number* of particles, which is conserved because we assumed them to be indestructible. But particles also carry *energy*, another conserved quantity. So it shouldn't surprise us that there's also a transfer of *heat* whenever molecular energy is not uniform to begin with, that is, when the temperature is nonuniform. And indeed, the law of heat conduction reads just like another Fick-type law: The flux j_Q of thermal energy is a constant (the thermal conductivity) times minus the gradient of temperature. (Various versions of this law are sometimes called Newton's law of cooling, or Fourier's law of conduction.)

    Section 5.2.1′ on page 187 discusses another important example, the dissipative transport of momentum.

 4.4.3′ Track 2

One can hardly overstate the conceptual importance of the idea that a probability distribution may have deterministic evolution, even if the events it describes are themselves random. The same idea (with different details) underlies quantum mechanics. There is a popular conception that quantum theory says "everything is uncertain; nothing can be predicted." But Schrödinger's equation is deterministic. Its solution, the wave function, when squared yields the *probability* of certain observations being made in any given trial, just as $c(x, t)$ reflects the probability of finding a particle near x at time t.

 4.6.2′ Track 2

Actually, a wide range of organisms have basal metabolic rates scaling with a power of body size that is less than three. All that matters for the structure of our argument is that this "allometric scaling exponent" is bigger than 1.

4.6.4′ Track 2

1. Section 3.2.5 on page 87 mentioned that frictional drag must generate *heat*. Indeed, it's well known that electrical resistance creates heat, for example, in your

toaster. Using the First Law, we can calculate the heat: Each ion passed between the plates falls down a potential hill, losing potential energy $q \times \Delta V$. The total number of ions per time making the trip is I/q, so the power (energy per time) expended by the external battery is $\Delta V \times I$. Using Ohm's law gives the familiar formula: power $= I^2 R$.

2. The conduction of electricity through a copper wire is also a diffusive transport process and also obeys Ohm's law. But the charge carriers are electrons, not ions; and the nature of the collisions is quite different from that in salt solution. In fact, the electrons could pass perfectly freely through a perfect single crystal of copper; they only bounce off imperfections (or thermally induced distortions) in the crystal lattice. Figuring out this story required the invention of quantum theory. Luckily, your body doesn't contain any copper wires; the picture developed in Section 4.6.4 is adequate for our purposes.

 4.6.5′ Track 2

1. *Gilbert says:*   Something is bothering me about the pulse solution (Equation 4.28 on page 143). For simplicity, let's work in just one dimension. Recall the setup (Section 4.1.2): At time $t = 0$, I release some random walkers at the origin, $x = 0$. A short time t later, the walkers have taken N steps of length L, where $N = t/\Delta t$. Then *none* of the walkers can be found farther away than $x_{max} = \pm NL = tL/\Delta t$. And yet, the solution (Equation 4.28) says that the density $c(x, t)$ of walkers is nonzero for any x, no matter how large! Did we make some error or approximation when solving the diffusion equation?

Sullivan: No, Your Turn 4G showed that it was an exact solution. But let's look more closely at the derivation of the diffusion equation itself—maybe what we've got is an exact solution to an approximate *equation*. Indeed, it's suspicious that we don't see the step size L, nor the time step Δt, anywhere in Equation 4.20.

Gilbert: Now that you mention it, I see that Equation 4.18 replaced the discrete difference of the populations N in adjacent bins by a *derivative*, remarking that this was legitimate in the limit of small L.

Sullivan: That's right. But we took this limit *holding D fixed*, where $D = L^2/(2\Delta t)$. So we're also taking $\Delta t \to 0$ as well. At any fixed time t, then, we're taking a limit where the number of steps is becoming infinite. So the diffusion equation is an approximate, limiting representation of a discrete random walk. In this limit, the maximum distance $x_{max} = tL/\Delta t = 2Dt/L$ really does become infinite, as implied by Equation 4.28.

Gilbert: Should we trust this approximation?

Let's help Gilbert out by comparing the exact, discrete probabilities for a walk of N steps to Equation 4.28 and seeing how fast they converge with increasing N. We seek the probability that a random walker will end up at a position x after a fixed amount of time t. We want to explore walks of various step sizes, while holding fixed the macroscopically observable quantity D.

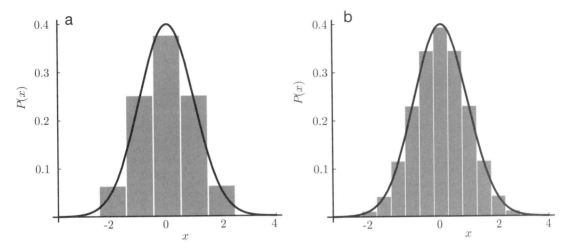

Figure 4.15: (*Mathematical functions.*) The discrete binomial distribution for N steps (*bars*), versus the corresponding solution to the diffusion equation (*curve*). In each case, the random walk under consideration had $2Dt = 1$ in the arbitrary units used to express x; thus, the curve is given by $(2\pi)^{-1/2}e^{-x^2/2}$. The discrete distribution (Equation 4.29) has been rescaled so that the area under the bars equals 1, for easier comparison to the curves. (a) $N = 4$. (b) $N = 14$.

Suppose that N is even. An N-step random walk can end up at one of the points $(-N), (-N + 2), \ldots, +N$. Extending the random walk Example (page 112) shows that the probability of taking $(N + j)/2$ steps to the right (and hence $(N - j)/2$ steps left), ending up j steps from the origin, is

$$P_j = \frac{N!}{2^N \left(\frac{N+j}{2}\right)! \left(\frac{N-j}{2}\right)!}. \tag{4.29}$$

Such a walk ends up at position $x = jL$. We set the step size L by requiring a fixed, given D: Noting that $\Delta t = t/N$ and $D = L^2/(2\Delta t)$ gives $L = \sqrt{2Dt/N}$. Thus, if we plot a bar of width $2L$ and height $P_j/(2L)$, centered on $x = jL$, then the area of the bar represents the probability that a walker will end up at x. Repeating for all even integers j between $-N$ and $+N$ gives a bar chart to be compared with Equation 4.28. Figure 4.15 shows that the approximate solution is quite accurate even for small values of N.

Strictly speaking, Gilbert is right to note that the true probability must be zero beyond x_{max}, whereas the approximate solution (Equation 4.28) instead equals $(4\pi Dt)^{-1/2}e^{-(x_{max})^2/(4Dt)}$. But the ratio of this error to the peak value of P, $(4\pi Dt)^{-1/2}$, is $e^{-N/2}$, which is already less than 1% when $N = 10$.

Similar remarks apply to polymers: The **Gaussian model** of a polymer mentioned at the end of Section 4.6.5 gives an excellent account of many polymer properties. We do need to be cautious, however, about using it to study any property that depends sensitively on the part of the distribution representing highly extended molecular conformations.

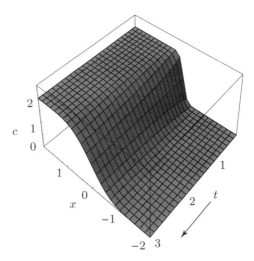

Figure 4.16: (Mathematical functions.) Diffusion from an initial concentration step. Time increases as we move diagonally downward (*arrow*). The sharp step gradually smooths out, starting from its edges.

> **Your Turn 4H**
>
> Instead of graphing the explicit formula, use Stirling's approximation (Equation 4.2 on page 113) to find the limiting behavior of the logarithm of Equation 4.29 when $N \to \infty$, holding x, t, and D fixed. Express your answer as a probability distribution $P(x, t)\mathrm{d}x$ and compare it with the diffusion solution.

2. Once we've found one solution to the diffusion equation, we can manufacture others. For example, if $c_1(x, t)$ is one solution, then so is $c_2(x, t) = \mathrm{d}c_1/\mathrm{d}t$, as we see by differentiating both sides of the diffusion equation. Similarly, the antiderivative $c_2(x, t) = \int^x \mathrm{d}x' \, c_1(x', t)$ yields a solution. The latter procedure, applied to the fundamental pulse solution in Your Turn 4G on page 143, gives a new solution describing the gradual smoothing-out of a sharp concentration *step*; see Figure 4.16. Mathematicians give the function $2/\sqrt{\pi} \int_0^x \mathrm{d}x' \, \mathrm{e}^{-(x')^2}$ the name Erf(x), the **error function**.

PROBLEMS*

4.1   *Bad luck*

a. You go to a casino with a dishonest coin, which you have filed down in such a way that it comes up heads 51% of the time. You find a credulous rube willing to bet $1 on tails for 1000 consecutive throws. He merely insists in advance that if after 1000 throws you're exactly even, then he'll take your shirt. You figure that you'll win about $20 from this sucker, but instead you lose your shirt. How could this happen? You come back every weekend with the same proposition, and indeed, usually you do win. How often on average do you lose your shirt?

b. You release a billion protein molecules at position $x = 0$ in the middle of a narrow capillary test tube. The molecules' diffusion constant is 10^{-6} cm^2 s^{-1}. An electric field pulls the molecules to the right (larger x) with a drift velocity of $1\,\mu$m s^{-1}. Nevertheless, after 80 s you see that a few protein molecules are actually to the *left* of where you released them. How could this happen? What is the ending number density right at $x = 0$? [*Note:* This is a one-dimensional problem, so you should express your answer in terms of the number density integrated over the cross-sectional area of the tube, a quantity with dimensions $\mathbb{L}^{-1}$.]

c. $\boxed{T_2}$ Explain why (a) and (b) are essentially, but not exactly, the same mathematical situation.

4.2   *Binomial distribution*

The genome of the HIV–1 virus, like any genome, is a string of "letters" (basepairs) in an "alphabet" containing only four letters. The message for HIV is rather short, just $n \approx 10^4$ letters in all. Because any of the letters can mutate to any of the three other choices, there's a total of 30 000 possible distinct one-letter mutations.

In 1995, A. Perelson and D. Ho found that every day about 10^{10} new virus particles are formed in an asymptomatic HIV patient. They further estimated that about 1% of these virus particles proceed to infect new white blood cells. It was already known that the error rate in duplicating the HIV genome was about one error for every $3 \cdot 10^4$ "letters" copied. Thus the number of newly infected white cells receiving a copy of the viral genome with one mutation is roughly

$$10^{10} \times 0.01 \times (10^4/(3 \cdot 10^4)) \approx 3 \cdot 10^7$$

per day. This number is much larger than the total 30 000 possible 1-letter mutations, so every possible mutation will be generated many times per day.

a. How many distinct *two*-base mutations are there?

b. You can work out the probability P_2 that a given viral particle has *two* bases copied inaccurately from the previous generation by using the sum and product rules of probability. Let $P = 1/(3 \cdot 10^4)$ be the probability that any given base is copied incorrectly. Then the probability of exactly two errors is P^2, times the probability

---

*Problem 4.7 is adapted with permission from Benedek & Villars, 2000b.

that the remaining 9998 letters *don't* get copied inaccurately, times the number of distinct ways to choose *which* two letters get copied inaccurately. Find P_2.

c. Find the expected number of two-letter mutant viruses infecting new white cells per day and compare to your answer to (a).

d. Repeat (a–c) for *three* independent mutations.

e. Suppose that an antiviral drug attacks some part of HIV but that the virus can evade the drug's effects by making one particular, single-base mutation. According to the preceding information, the virus will very quickly stumble upon the right mutation—the drug isn't effective for very long. Why do you suppose an effective HIV therapy involves a combination of *three* different antiviral drugs administered simultaneously?

4.3 *Limitations of passive transport*

Most eukaryotic cells are about $10\,\mu$m in diameter, but a few cells in your body are about a meter long. These are the neurons running from your spinal cord to your feet. They have a normal-sized cell body, with various bits sticking out, notably a very long axon (see Section 2.1.2 on page 43).

Many molecules needed at the tip of the axon, for example proteins, are synthesized in the cell body and packaged into vesicles or other particles. Even entire organelles, like mitochondria, need to be transported from their construction sites in the cell body to the periphery. Section 2.3.2 asserted that these objects are all transported along the axon by molecular motors. It might seem that an attractive alternative would be for them to arrive by simple diffusion, but Section 4.4.1 claimed that this mechanism is too slow. Let's see.

Model the axon as a tube 1 m long. At one end of the axon, some synthetic process creates objects similar to those seen in Figure 2.19 on page 56, maintaining them at a number density c_0 (we won't need the numerical value of c_0). Objects arriving at the axon terminal are immediately gobbled up by some other process, and so the number density at this end is zero.

a. Use the Stokes and Einstein relations to estimate the diffusion constant D for an object the size of the vesicle in Figure 2.19b.

b. What is the diffusive number flux j_{diffus} of these objects along the axon?

c. In the microscope one sees organelles and other objects moving at about 400 mm per day. Convert this speed to another number flux j_{obs}, again assuming a number density of c_0.

d. Find the ratio $j_{\mathrm{diffus}}/j_{obs}$ and comment.

4.4 *Diffusion versus size*

Table 4.2 lists the diffusion constants D and radii r of various biologically interesting molecules in water. Consider the last four entries. Interpret these data in light of the diffusion law. [*Hint:* Plot D versus $1/R$, and remember Equation 4.14.]

4.5 *Perrin's experiment*

Figure 4.17 shows some experimental data on Brownian motion taken by Jean Perrin. Perrin took colloidal particles of gutta-percha (natural rubber), with radius $0.37\,\mu$m. He watched their projections into the xy plane, so the two-dimensional random walk

Table 4.2: Sizes and diffusion constants of some molecules in water at 20°C.

molecule	molar mass, g/mole	radius, nm	$D \times 10^9$, m^2 s^{-1}
water	18	0.15	2.0
oxygen	32	0.2	1.0
urea	60	0.4	1.1
glucose	180	0.5	0.7
ribonuclease	13 683	1.8	0.1
β-lactoglobulin	35 000	2.7	0.08
hemoglobin	68 000	3.1	0.07
collagen	345 000	31	0.007

[From Tanford, 1961.]

should describe their motions. Following a suggestion of his colleague P. Langevin, Perrin observed the location of a particle, waited 30 s, then observed again and plotted the net displacement in that time interval. He collected 508 data points in this way and calculated the root-mean-square displacement to be $d = 7.84\,\mu$m. The concentric circles drawn on the figure have radii $d/4, 2d/4, 3d/4, \dots.$

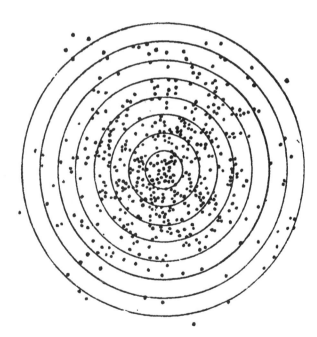

Figure 4.17: (Experimental data.) See Problem 4.5. [From Perrin, 1948.]

a. Find the expected coefficient of friction for a sphere of radius $0.37\,\mu$m, using the Stokes formula (Equation 4.14). Then work out the predicted value of d, using the Einstein relation (Equation 4.16) and compare with the measured value.

b. $\boxed{T_2}$ How many dots do you expect to find in each of the rings? How do your expectations compare with the actual numbers?

4.6 *Permeability versus thickness*
Look at Figure 4.13 on page 137 again. Find the thickness of the bilayer membrane used in Finkelstein's experiments.

4.7 *Vascular design*
Blood carries oxygen to your body's tissues. For this problem, you may neglect the role of the red cells: Just suppose that the oxygen is dissolved in the blood and diffuses out through the capillary wall because of a concentration difference. Model a capillary as a cylinder of length L and radius r, and describe its oxygen transport by a permeability $\mathcal{P}$.

a. If the blood did not flow, the interior oxygen concentration would approach that of the exterior as an exponential, similarly to the concentration decay Example (page 136). Show that the corresponding time constant would be $\tau = r/(2\mathcal{P})$.

b. But blood *does* flow. For efficient transport, the time that the flowing blood remains in the capillary should be at least $\approx \tau$; otherwise the blood would carry its incoming oxygen right back out of the tissue after entering the capillary. Using this constraint, derive a formula for the maximum speed of blood flow in the capillary. Evaluate your formula numerically, using $L \approx 0.1\,\mathrm{cm}$, $r = 4\,\mu\mathrm{m}$, $\mathcal{P} = 3\,\mu\mathrm{m\,s^{-1}}$. Compare with the actual speed $v \approx 400\,\mu\mathrm{m\,s^{-1}}$.

4.8 *Spreading burst*
Your Turn 4D on page 134 claimed that, in one-dimensional diffusion, an observer sitting at a fixed point sees a transient pulse of concentration pass by. Make this statement more useful, as follows: Write the explicit solution of the diffusion equation for release of a million particles from a point source in *three* dimensions. Then show that the concentration measured by an observer at fixed distance r from the initial release point peaks at a certain time.

a. Find that time, in terms of r and D.

b. Show that the value of concentration at that time is a constant times r^{-3} and evaluate the constant numerically.

4.9 $\boxed{T_2}$ *Rotational random walk*
A particle in fluid will wander: Its center does a random walk. But the same particle will also *rotate* randomly, leading to diffusion in its orientation. Rotational diffusion affects the precision with which a microorganism can swim in a straight line. We can estimate this effect as follows.

a. You look up in a book that a sphere of radius R can be twisted in a viscous fluid by applying a torque $\tau = \zeta_r \omega$, where ω is the speed in radians/s and $\zeta_r = 8\pi\eta \times (??)$ is the rotational friction coefficient. Unfortunately, the dog has chewed your copy of the book and you can't read the last factor. What is it?

b. But you didn't want to know about friction—you wanted to know about diffusion. After time t, a sphere will reorient with its axis at an angle θ to its original direction. Not surprisingly, rotational diffusion obeys $\langle \theta^2 \rangle = 4D_r t$, where D_r is a rotational diffusion constant. (This formula is valid as long as t is short enough that this quantity stays small.) Find the dimensions of D_r.

c. Use your answer to (a) to obtain a numerical value for D_r. Model the bacterium as a sphere of radius $1\ \mu m$ in water at room temperature.

d. If this bacterium is swimming, about how long will it take to wander significantly (say, $30°$) off its original direction?

4.10 $\boxed{T_2}$ *Spontaneous versus driven permeation*

This chapter discussed the permeability $\mathcal{P}_s$ of a membrane to dissolved solute. But membranes also let water pass. The permeability $\mathcal{P}_w$ of a membrane to water may be measured as follows. Heavy water, HTO, is prepared with tritium in place of one of the hydrogens; it's chemically identical to water but radioactive. We take a membrane patch of area A. Initially, one side is pure HTO, the other pure H_2O. After a short time dt, we measure some radioactivity on the other side, corresponding to a net passage of $(2.9\ \text{mole s}^{-1}\text{m}^{-2})) \times A\,dt$ radioactive water molecules.

a. Rephrase this result as a Fick-type formula for the diffusive flux of water molecules. Find the constant $\mathcal{P}_w$ appearing in that formula. [*Hint:* Your answer will contain the number density of water molecules in liquid water, about 55 mole/L.]

Next suppose that we have ordinary water, H_2O, on both sides, but we *push* the fluid across the membrane with a pressure difference Δp. The pressure results in a flow of water, which we can express as a flux of volume j_v (see the general discussion of fluxes in Section 1.4.4 on page 22). The volume flux will be proportional to the mechanical driving force: $j_v = -L_p\,\Delta p$. The constant L_p is called the membrane's filtration coefficient.

b. There should be a simple relation between L_p and $\mathcal{P}_w$. Guess it, remembering to check your guess with dimensional analysis. Using your guess, estimate L_p, using your answer to (a). Express your answer both in SI units and in the traditional units $\text{cm s}^{-1}\text{atm}^{-1}$ (see Appendix A). What will be the net volume flux of water if $\Delta p = 1\ \text{atm}$?

c. Human red blood cell membranes have water permeability corresponding to the value you found in (a). Compare your result in (b) to the measured value of the filtration coefficient for this membrane, $9.1 \cdot 10^{-6}\ \text{cm s}^{-1}\text{atm}^{-1}$.

CHAPTER 5

Life in the Slow Lane: The Low Reynolds-Number World

Nobody is silly enough to think that an elephant will only fall under gravity if its genes tell it to do so, but the same underlying error can easily be made in less obvious circumstances. So [we must] distinguish between how much behavior, and what part, has a genetic origin, and how much comes solely because an organism lives in the physical universe and is therefore bound by physical laws.

— Ian Stewart, *Life's Other Secret*

Before our final assault on the citadel of statistical physics in Chapter 6, this chapter will show how the ideas we have already developed give some simple but powerful conclusions about cellular, subcellular, and physiological processes, as well as helping us understand some important laboratory techniques. One key example will be the propulsion of bacteria by their flagella (see Figure 2.3b on page 37).

Section 4.4.1 described how diffusion dominates transport of molecules in the nanoworld. Diffusion is a dissipative process: It tends to erase ordered arrangements of molecules. Similarly, this chapter will outline how viscous friction dominates *mechanics* in the nanoworld. Friction, too, is dissipative: It tends to erase ordered *motion*, converting it to thermal energy. The physical concept of symmetry will help us to understand and unify the sometimes surprising ramifications of this statement. The Focus Question for this chapter is

Biological question: Why don't bacteria swim like fish?

Physical idea: The equations of motion appropriate to the nanoworld behave differently under *time reversal* than do those of the macroworld.

5.1 FRICTION IN FLUIDS

First let's see how the friction formula $v_{\text{drift}} = f/\zeta$ (Equation 4.12 on page 119) tells us how to sort particles by their weight or electric charge, an eminently practical laboratory technique. Then we'll look at some odd but suggestive phenomena in viscous liquids like honey. Section 5.2 will argue that, in the nanoworld, water itself acts as a very viscous liquid; so these phenomena are actually representative of the physical world of cells.

5.1.1 Sufficiently small particles can remain in suspension indefinitely

If we suspend a mixture of several particle types (for example, several proteins) in water, then gravity pulls on each particle with a force mg proportional to its mass. (If we prefer, we can put our mixture in a centrifuge, where the centrifugal "force" mg' is again proportional to the particle mass, although g' can be much greater than the ordinary acceleration of gravity.)

The *net* force propelling the particle downward is less than mg, because for the particle to go down, an equal volume of water must move *up*. Gravity pulls on the water, too, with a force $(V\rho_m)g$, where ρ_m is the mass density of water and V the volume of the particle. Let z denote the particle's height. Thus, when the particle moves downward a distance $|\Delta z|$, displacing an equal volume of water up a distance $|\Delta z|$, the total change in gravitational potential energy is $\Delta U = (mg)\Delta z - (V\rho_m g)\Delta z$. The net force driving sedimentation is then the derivative $f = -dU/dz = -(m - V\rho_m)g$, which we'll abbreviate as $-m_{net}g$. All we have done so far is to derive Archimedes' principle: The net weight of an object under water gets reduced by a **buoyant force** equal to the weight of the water displaced by the object.

What happens after we let a suspension settle for a very long time? Won't all the particles just fall to the bottom? Pebbles would, but colloidal particles smaller than a certain size won't, for the same reason that the air in the room around you doesn't: Thermal agitation creates an equilibrium distribution in which some particles are constantly off the bottom. To make this idea precise, consider a test tube filled to a height h with a suspension. In equilibrium, the profile of particle density $c(z)$ has stopped changing, so we can apply the argument that led to the Nernst relation (Equation 4.26 on page 141), replacing the electrostatic force by the net gravitational force $= m_{net}g$. Thus the density of particles in equilibrium is

$$c(z) \propto e^{-m_{net}gz/k_B T}. \qquad \text{(sedimentation equilibrium, Earth's gravity)} \qquad (5.1)$$

Here are some typical numbers. Myoglobin is a globular protein, with molar mass $m \approx 17\,000\,\text{g mole}^{-1}$. The buoyant correction typically reduces m to $m_{net} \approx 0.25m$. Defining the **scale height** as $z_* \equiv k_B T_r/(m_{net}g) \approx 59\,\text{m}$, we expect $c(z) \propto e^{-z/z_*}$. Thus, in a 4 cm test tube, in equilibrium, the concentration at the top equals $c(0)e^{-0.04\,\text{m}/59\,\text{m}}$, or 99.9% as great as at the bottom. In other words, *the suspension never settles out*. In that case, we call it an equilibrium **colloidal suspension**, or just a **colloid**. Macromolecules like DNA or soluble proteins form colloidal suspensions in water; another example is Robert Brown's pollen grains in water. On the other hand, if m_{net} is big (as it would be for sand grains), then the density at the top will be essentially zero: The suspension settles. How big is "big"? Looking at Equation 5.1 shows that, for settling to occur, the gravitational potential energy difference $m_{net}gh$ between the top and bottom must be bigger than the thermal energy.

Your Turn 5A

Here is another example. Suppose that the container is a carton of milk, with $h = 25\,\text{cm}$. We idealize homogenized milk as a suspension of fat droplets (spheres of diameter up to about a micrometer) in water. The *Handbook of Chemistry and Physics* lists the mass density of butterfat as $\rho_{m,fat} = 0.91\,\text{g cm}^{-3}$ (the density of water is $1\,\text{g cm}^{-3}$). Find $c(h)/c(0)$ in equilibrium. Is homogenized milk an equilibrium colloidal suspension?

Returning to myoglobin, it may seem as though sedimentation is not a very useful tool for protein analysis. But the scale height depends not only on properties of the protein and solvent but also on the acceleration of gravity, g. Artificially increasing g with a centrifuge can reduce z_* to a manageably small value; indeed, laboratory centrifuges can attain values of g' up to around 10^6 m s^{-2}, making protein separation feasible.

To make these remarks precise, first note that, when a particle gets whirled about at angular frequency ω, a first-year physics formula gives its centripetal acceleration as $r\omega^2$, where r is the distance from the center.

Your Turn 5B

Suppose you didn't remember this formula. Show how to guess it by dimensional analysis, knowing that angular frequency is measured in radians/s.

Suppose that the sample is in a tube lying in the plane of rotation, so that its long axis points radially. The centripetal acceleration points inward, toward the axis of rotation, so there must be an inward-pointing force, $f = -m_{net}r\omega^2$, causing it. This force can only come from the frictional drag of the surrounding fluid as the particle drifts slowly outward. Thus, the drift velocity is given by $m_{net}r\omega^2/\zeta$ (see Equation 4.12 on page 119). Repeating the argument that led to the Nernst relation (Section 4.6.3 on page 139) now gives the drift flux as $cv_{drift} = cm_{net}r\omega^2 D/k_B T$, where $c(r)$ is the number density. In equilibrium, this drift flux is canceled by a diffusive flux, given by Fick's law. We thus find that, in equilibrium,

$$j = 0 = D\left(-\frac{dc}{dr} + \frac{r\omega^2 m_{net}}{k_B T}c\right),$$

a result analogous to the Nernst–Planck formula (Equation 4.24 on page 140). To solve this differential equation, divide by $c(r)$ and integrate:

$$c = \text{const} \times e^{m_{net}\omega^2 r^2/(2k_B T)}. \qquad \text{(sedimentation equilibrium, centrifuge)} \qquad (5.2)$$

5.1.2 The rate of sedimentation depends on solvent viscosity

The drift velocity $v_{drift} = m_{net}g/\zeta$ isn't an intrinsic property of the particle, because it depends on the strength of gravity, g. To get a quantity that we can tabulate for various particle types (in given solvents), we instead define the **sedimentation coefficient**

$$s = v_{drift}/g \equiv m_{net}/\zeta. \qquad (5.3)$$

Measuring s and looking in a table thus gives a rough-and-ready particle identification. The quantity s can be interpreted as the time required for a particle to come to terminal velocity. It is sometimes expressed in units of **svedbergs**; a svedberg by definition equals 10^{-13} s.

What determines the sedimentation coefficient s? Surely sedimentation will be slower in a "thick" liquid like honey than in a "thin" one like water. That is, we expect the viscous friction coefficient ζ for a single particle in a fluid to depend not only on the size of the particle but also on some intrinsic property of the fluid, called the viscosity. In fact, Section 4.1.4 already quoted an expression for ζ, namely, the Stokes formula, $\zeta = 6\pi\eta R$, for an isolated, spherical particle of radius R.

> **Your Turn 5C**
>
> **a.** Work out the dimensions of η from the Stokes formula. Show that they can be regarded as those of pressure times time and that, hence, the SI units for viscosity are Pa s.
>
> **b.** Your Turn 5A raised a paradox: The equilibrium formula you found suggested that milk should separate, and yet we don't normally observe this happening. Use the Stokes formula to estimate how long it takes for fat globules in homogenized milk to drift a distance equal to the size of a milk bottle. Then compare homogenized milk with raw milk (which has fat droplets up to about 5 μm in diameter), and comment.

It's worth memorizing the value of η for water at room temperature:[1] $\eta_w \approx 10^{-3}\,\mathrm{kg\,m^{-1}\,s^{-1}} = 10^{-3}\,\mathrm{Pa\,s}$.

We can use the preceding remarks to look once again at the sizes of polymer coils. Let's suppose that a particular type of polymer forms random coils, with radius given by a constant times some power of the molecular mass: $R \propto m^p$. We'd like to verify this claim, and extract the value of the scaling exponent p, from an experiment. Then we'll compare the result to the prediction from random-walk theory, which is that $p = \frac{1}{2}$ (Idea 4.17 on page 123).

Combining Equation 5.3 with the Stokes formula gives $s = (m - V\rho_m)/(6\pi\eta R)$. Assuming that the polymer displaces a volume of water proportional to the number of monomers yields $s \propto m^{1-p}$. Figure 4.7b on page 123 shows that our prediction $p = \frac{1}{2}$ indeed is roughly true. (More precisely, for one particular polymer/solvent combination Figure 4.7a gives the scaling exponent for R as $p = 0.57$. Figure 4.7b gives the exponent for s as 0.44, which is quite close to $1 - p$.)

5.1.3 It's hard to mix a viscous liquid

Section 5.2 will argue that, in the nanoworld of cells, ordinary water behaves as a very viscous liquid. Because most people have made only limited observations in this world, it's worthwhile to pause first and notice some of the spooky phenomena that happen there.

Pour a few centimeters of clear corn syrup into a clear cylindrical beaker or wide cup. Set aside some of the syrup and mix it with a small amount of ink to serve as a marker. Put a stirring rod in the beaker, then inject a small blob of marked syrup

---

[1]Some authors express this result in units of **poise**, defined as erg s/cm^3 = 0.1 Pa s; thus η_w is about one centipoise. Values of η for other biologically relevant fluids appear in Table 5.1 on page 165.

a

b

c

Figure 5.1: (Photographs.) An experiment showing the peculiar character of low Reynolds-number flow. (a) A small blob of colored glycerine is injected into clear glycerine in the space between two concentric cylinders. (b) The inner cylinder is turned through four full revolutions, apparently mixing the blob into a thin smear. (c) Upon turning the inner cylinder back exactly four revolutions, the blob reassembles, only slightly blurred by diffusion. The finger belongs to Sir Geoffrey Taylor. [From Shapiro, 1972.]

somewhere below the surface, far from both the rod and the walls of the beaker. (A syringe with a long needle helps with this step, but a medicine dropper will do; remove it gently to avoid disturbing the blob.) Now try moving the stirring rod slowly. One particularly revealing experiment is to hold the rod against the wall of the beaker, slowly run it around the wall once clockwise, then slowly reverse your first motion, running it counterclockwise to its starting position.

You'll note several phenomena:

- It's very hard to mix the marked blob into the bulk.
- The marked blob actually seems to take evasive action when the stirring rod approaches.
- In the clockwise–counterclockwise experiment, the blob will smear out in the first step. But if you're careful in the second step to retrace the first step exactly, you'll see the blob magically reassemble itself into nearly its original position and shape! That's not what happens when you stir cream into your coffee.

Figure 5.1 shows the result of a more controlled experiment. A viscous liquid sits between two concentric cylinders. One cylinder is rotated through several full turns, smearing out the marker blob as shown (Figure 5.1b). Upon rotation through an equal and opposite angle, the blob reassembles itself (Figure 5.1c).

What's going on? Have we stumbled onto some violation of the Second Law? Not necessarily. If you just leave the marked blob alone, it does diffuse away, but extremely slowly, because the viscosity η is large, and the Einstein and Stokes relations give $D = k_B T/\zeta \propto \eta^{-1}$ (Equations 4.16 and 4.14). Moreover, diffusion initially only changes the density of ink near the *edges* of the blob (see Figure 4.16 on page 152), so a compact blob cannot change much in a short time. One could imagine

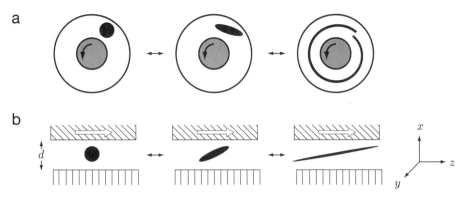

Figure 5.2: (Schematics.) Shearing motion of a fluid in laminar flow, in two geometries. (a) Cylindrical (ice-cream maker) geometry, viewed from above. The central cylinder rotates while the outer one is held fixed. (b) Planar (sliding plates) geometry. The top plate is pushed to the right while the bottom one is held fixed. The plates have area A and are separated by distance d.

that stirring causes an organized motion, in which successive layers of fluid simply slide over one another and stop as soon as the stirring rod stops (Figure 5.2). Such a stately fluid motion is called **laminar flow**. Then the motion of the stirring rod, or of the container walls, would just stretch out the blob, leaving it still many billions of molecules thick. The ink molecules are spread out but are still not random, because diffusion hasn't yet had enough time to randomize them fully. When we slide the walls back to their original configuration, the fluid layers could then each slide right back and reassemble the blob. In short, we could explain the reassembly of the blob by arguing that it never "mixed" at all, despite appearances. It's hard to mix a viscous liquid.

The preceding scenario sounds good for corn syrup. But it doesn't address one key question: Why *doesn't* water behave this way? When you stir cream into your coffee, it immediately swirls into a complex, **turbulent** pattern. Nor does the fluid motion stop when you stop stirring; the coffee's momentum continues to carry it along. In just a few seconds, an initial blob of cream gets stretched to a thin ribbon only a few molecules thick; diffusion can then quickly and irreversibly obliterate the ribbon. Stirring in the opposite direction won't reassemble the blob. It's easy to mix a nonviscous liquid.

5.2 LOW REYNOLDS NUMBER

To summarize, the last two paragraphs of Section 5.1.3 served to refocus our attention, away from the striking observed distinction between mixing and nonmixing flows and onto a more subtle underlying distinction, between turbulent and laminar flows. To make progress, we need some physical criterion that explains why corn syrup (and other fluids like glycerine and crude oil) will undergo laminar flow, whereas water (and other fluids like air and alcohol) commonly exhibit turbulent flow. The surprise will be that the criterion depends not only on the nature of the fluid but also on the *scale* of the process under consideration. In the nanoworld, water will prove to be effectively *much thicker* than the corn syrup in your experiment; thus, essentially all flows in the nanoworld are laminar.

5.2.1 A critical force demarcates the physical regime dominated by friction

Because viscosity certainly has something to do with the distinction between mixing and nonmixing flows, let's look a bit more closely at what it means. The planar geometry sketched in Figure 5.2b is simpler than that of a spherical ball, so we use it for our formal definition of viscosity. Imagine two flat parallel plates separated by a layer of fluid of thickness d. We hold one plate fixed while sliding the other sideways (the z direction in Figure 5.2b) at speed v_0. This motion is called **shear**. Then the dragged plate feels a resisting viscous force directed against v_0; the stationary plate feels an equal and opposite force (called an entraining force) parallel to v_0.

The viscous force f will be proportional to the area A of each plate. It will increase with increasing speed v_0 but decrease as we increase the plate separation. Empirically, for small enough v_0, many fluids indeed show the simplest possible force rule consistent with these expectations:

$$f = -\eta v_0 A/d. \qquad \text{viscous force in a Newtonian fluid, planar geometry}$$

(5.4)

The constant of proportionality η is the fluid's **viscosity**. Equation 5.4 separates out all the situation-dependent factors (area, gap, speed), thereby exposing η as the one factor intrinsic to the type of fluid. The minus sign reminds us that the drag force opposes the imposed motion.

Your Turn 5D

Verify that the units work out in Equation 5.4, by using your result in Your Turn 5C(a).

Any fluid obeying Equation 5.4 is called a **Newtonian fluid** after the ubiquitous Isaac Newton. Most Newtonian fluids are, in addition, **isotropic** (the same in every direction; anisotropic fluids will not be discussed in this book). Such a fluid is completely characterized by its viscosity and its mass density ρ_m.

We are pursuing the suggestion that simple, laminar flow ensues when η is "large," whereas we get complex, turbulent flow when it's "small." But the question immediately arises, "Large relative to what?" The viscosity is not dimensionless, so there's no absolute meaning to saying that it's large (see Section 1.4.1 on page 18): No fluid can be deemed viscous in an absolute sense. Nor can we form any dimensionless quantity by combining viscosity (dimensions $\mathbb{M}\mathbb{L}^{-1}\mathbb{T}^{-1}$) with mass density

Table 5.1: Density, viscosity, and viscous critical force for common fluids at 25°C.

fluid	ρ_m, kg m^{-3}	η, Pa s	f_{crit}, N
air	1	$2 \cdot 10^{-5}$	$4 \cdot 10^{-10}$
water	1000	0.0009	$8 \cdot 10^{-10}$
olive oil	900	0.08	$7 \cdot 10^{-6}$
glycerine	1300	1	0.0008
corn syrup	1000	5	0.03

(dimensions $\mathbb{ML}^{-3}$). But we *can* form a characteristic quantity with the dimensions of *force:*

$$\boxed{f_{crit} = \eta^2/\rho_m. \qquad \textbf{viscous critical force}} \qquad (5.5)$$

The motion of any fluid will have two physically distinct regimes, depending on whether we apply forces bigger or smaller than that fluid's critical force. Equivalently, we can say that

    a. *There's no dimensionless measure of viscosity and, hence, no **intrinsic** distinction between "thick" and "thin" fluids, but ...*

    b. *Nevertheless, there is a **situation-dependent** characterization*   (5.6) *of when a fluid's motion will be viscous, namely, when the dimensionless ratio f/f_{crit} is small.*

For a given applied force f, we can get a large ratio f/f_{crit} by choosing a fluid with a large mass density or small viscosity. Then inertial effects (proportional to mass) will dominate over frictional effects (proportional to viscosity), and we expect turbulent flow (the fluid keeps moving after we stop applying force). In the opposite case, friction will quickly damp out inertial effects and we expect laminar flow.

Summarizing the discussion so far, Section 5.1.3 began with the distinction between mixing and nonmixing flows. This section first rephrased the issue as the distinction between turbulent and laminar flow, then finally as a distinction between flows dominated by inertia or viscous friction, respectively. We found a criterion for making this distinction in a given situation by using dimensional analysis.

Let's examine some rough numbers for familiar fluids. Table 5.1 shows that, if we pull a marble through corn syrup with a force much less than 0.03 N, then we may expect the motion to be dominated by friction. Inertial effects will be negligible; and, indeed, in the corn-syrup experiment, there's no swirling after we stop pushing the stirring rod. In water, on the other hand, even a millinewton push puts us well into the regime dominated by inertia, not friction; turbulent motion then ensues.

What's striking about the table is that it predicts that water will appear just as viscous to a tiny creature exerting forces less than a nanonewton as glycerine does to us! Indeed, we'll see in Chapter 10 that the typical scale of forces inside cells is more like a thousand times smaller than f_{crit} (the *pico*newton range). Friction rules the world of the cell.

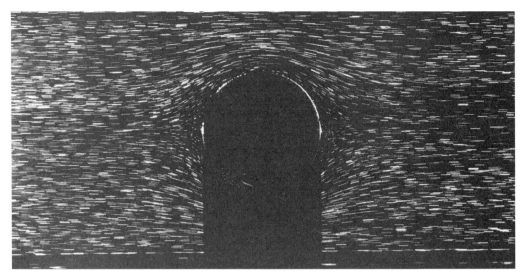

Figure 5.3: (Photograph.) Low Reynolds-number fluid flow past a sphere. The fluid flows from left to right at $\mathcal{R} = 0.1$. The flow lines have been visualized by illuminating tiny suspended metal flakes with a sheet of light coming from the top. (The black area below the sphere is just its shadow.) Note that the figure is symmetrical; the time-reversed flow from right to left would look exactly the same. Note also the orderly, laminar character of the flow. If the sphere were a single-cell organism, a food particle located in its path would get carried around it without ever encountering the cell at all. [From Coutanceau, 1968.]

It's not size per se that counts, but force. To understand why, recall that the flows of a Newtonian fluid are completely determined by its mass density and viscosity, and convince yourself that there is no combination of these two quantities with the dimensions of length. We say that a Newtonian fluid "has no intrinsic length scale," or is "scale invariant." Thus, even though we haven't worked out the full equations of fluid motion, we already know that they won't give qualitatively different physics on scales larger and smaller than some critical length scale, because dimensional analysis has just told us that there can be no such scale! A large object—even a battleship—will move in the friction-dominated regime, if we push on it with less than a nanonewton of force. Similarly, macroscopic experiments, like the one shown in Figure 5.3, can tell us something relevant to a *micro*scopic organism.

$\boxed{T_2}$ *Section 5.2.1′ on page 187 sharpens the idea of friction as dissipation, by reinterpreting viscosity as a form of diffusion.*

5.2.2 The Reynolds number quantifies the relative importance of friction and inertia

Dimensional analysis is powerful, but it can move in mysterious ways. Section 5.2.1 proposed the logic that (*i*) two numbers, ρ_m and η, characterize a simple (that is, isotropic Newtonian) fluid; (*ii*) from these quantities, we can form another, f_crit, with dimensions of force; (*iii*) something interesting must happen at around this range

Figure 5.4: (Schematic.) Motion of a small fluid element, of size ℓ, as it impinges on an obstruction of radius R (see Figure 5.3).

of externally applied force. Such arguments generally strike students as dangerously sloppy. Indeed, when faced with an unfamiliar situation, a physical scientist *begins* with dimensional arguments to raise certain expectations but then proceeds to justify those expectations with more detailed analysis. This section will begin this process, deriving a more precise criterion for laminar flow. Even here, however, we will not bother with numerical factors like 2π and so on; all we want is a rough guide to the physics.

Let's begin with an experiment. Figure 5.3 shows a beautiful example of laminar flow past an obstruction, a sphere of radius R. Far away, each fluid element is in uniform motion at some velocity $\mathbf{v}$. We'd like to know whether the motion of the fluid elements is mainly dominated by inertial effects or by friction.

Consider a small lump of fluid of size ℓ, which is carried by a flow on a collision course with the sphere (Figure 5.4). To sidestep the sphere, the fluid element must accelerate: The velocity must change *direction* during the encounter time $\Delta t \approx R/v$. The magnitude of the change in $\mathbf{v}$ is comparable to that of $\mathbf{v}$ itself, so the rate of change of velocity (that is, the acceleration $d\mathbf{v}/dt$) has magnitude $\approx v/(R/v) = v^2/R$. The mass m of the fluid element is the density ρ_m times the volume.

Newton's Law of motion says that our fluid element obeys

$$f_{ext} + f_{frict} \equiv f_{tot} = \text{mass} \times \text{acceleration}. \tag{5.7}$$

Here f_{ext} denotes the external force from the surrounding fluid's pressure and f_{frict} is the net force on the fluid element from viscous friction. In terms of the quantities defined in the previous paragraph, the right-hand side of Newton's Law (the "inertial term") is

$$\text{inertial term} = \text{mass} \times \text{acceleration} \approx (\ell^3 \rho_m)v^2/R. \tag{5.8}$$

We wish to compare the magnitude of this inertial term with that of f_{frict}. If one of these terms is much larger than the other, then we can drop the smaller term in Newton's Law.

To estimate the frictional force, we first generalize Equation 5.4 to the case where the velocity of the fluid is not a uniform gradient (as it was in Figure 5.2b). To do so, replace the finite velocity difference v_0/d by the derivative, dv/dx. When a fluid element slides past its neighbor, then, they exert forces per unit area on each other equal to[2]

$$\frac{f}{A} = -\eta \frac{\mathrm{d}v}{\mathrm{d}x}. \tag{5.9}$$

In the situation sketched in Figure 5.4, the surface area A of one face of the fluid element is $\approx \ell^2$. The *net* frictional force f_{frict} on the fluid element is the force exerted on it by the one above it, minus the force it exerts on the one below it. We can estimate this difference as ℓ times the derivative $\mathrm{d}f/\mathrm{d}x$, or $f_{\text{frict}} \approx \eta \ell^3 (\mathrm{d}^2 v/\mathrm{d}x^2)$. To estimate the derivative, again note that v changes appreciably over distances comparable to the obstruction's size R; accordingly, we estimate $\mathrm{d}^2 v/\mathrm{d}x^2 \approx v/R^2$. Putting everything together gives

$$\text{friction term} = f_{\text{frict}} \approx \eta \ell^3 v/R^2. \tag{5.10}$$

We are ready to compare Equations 5.8 and 5.10. Dividing these two expressions yields a characteristic dimensionless quantity:[3]

$$\boxed{\mathcal{R} = vR\rho_{\mathrm{m}}/\eta. \qquad \text{the \textbf{Reynolds number}}} \tag{5.11}$$

When $\mathcal{R}$ is small, friction dominates. Stirring produces the least possible response, namely, laminar flow; and the flow stops immediately after the external force f_{ext} stops. (Engineers often use the synonym "creeping flow" for low Reynolds-number flow.) When $\mathcal{R}$ is big, inertial effects dominate, the coffee keeps swirling after you stop stirring, and the flow is turbulent.

We obtained the Reynolds number criterion by considering flow impinging on a sphere, but it is more generally applicable to any situation where the geometry is characterized by some length scale R. Consider, for example, the flow of fluid down a pipe of radius R. In a series of careful experiments in the 1880s, O. Reynolds found that generally the transition to turbulent flow occurs around $\mathcal{R} \approx 1000$. Reynolds varied all the parameters describing the situation (pipe size, flow rate, fluid mass density, and viscosity) and found that the onset of turbulence always depended on just one combination of the parameters, namely, the one given in Equation 5.11.

Let's connect Reynolds's result to the concept of critical force discussed in Section 5.2.1:

Example: Suppose that the Reynolds number is small, $\mathcal{R} \ll 1$. Compare the external force needed to anchor the obstruction in place with the viscous critical force.

Solution: At low Reynolds number, the inertial term is negligible, so f_{ext} is essentially equal to the frictional force (Equation 5.10). To estimate this force, take the fluid

---

[2] $\boxed{T_2}$ Equation 5.9 is valid only in planar geometry (see Problem 5.9). Nevertheless, it gives an adequate estimate of the viscous force for our purposes here.
[3] Notice that the arbitrary size ℓ of our fluid element dropped out of this expression, as it should.

element size ℓ to be that of the obstruction itself; then

$$\frac{f_{\text{frict}}}{f_{\text{crit}}} \approx \frac{\eta R^3 v}{R^2} \frac{1}{\eta^2/\rho_m} = \frac{vR\rho_m}{\eta} = \mathcal{R}.$$

So, indeed, the force applied to the fluid is much smaller than f_{crit} when $\mathcal{R}$ is small.

**Your
Turn
5E**

Suppose that the Reynolds number is big, $\mathcal{R} \gg 1$. Compare the external force needed to anchor the obstruction in place with the viscous critical force.

As always, we need to make some estimates. A 30 m whale, swimming in water at $10\,\text{m}\,\text{s}^{-1}$, has $\mathcal{R} \approx 300\,000\,000$. But a $1\,\mu\text{m}$ bacterium, swimming at $30\,\mu\text{m}\,\text{s}^{-1}$, has $\mathcal{R} \approx 0.000\,03$! Indeed, Section 5.3.1 will show that the locomotion of bacteria works quite differently from the way large creatures swim.

$\boxed{T_2}$ *Section 5.2.2′ on page 188 outlines more precisely the sense in which fluids have no characteristic length scale.*

5.2.3 The time-reversal properties of a dynamical law signal its dissipative character

Now that we have a criterion for laminar flow, we can make our resolution of the mixing/unmixing puzzle (Section 5.1.3) a bit more explicit.

Unmixing The full equations of fluid mechanics are rather complicated, but it's not hard to guess the minimal response of a fluid to the shearing force applied in Figure 5.2b. Because everything is uniform in the y, z directions, we can think of the fluid layer as a stack of thin parallel sheets, each of thickness dx, and apply Equation 5.9 to each layer separately. Denoting the relative velocity of two neighboring sheets by dv_z, each sheet pulls its neighbor with a force per area of

$$\frac{f}{A} = -\eta \frac{dv_z}{dx}.$$

In particular, the sheet of fluid immediately next to a solid wall must move with the same speed as the wall (the **no-slip boundary condition**), because otherwise v would have an infinite derivative at that point, and the required viscous force would be infinite, too.

Because every sheet of fluid moves uniformly (does not accelerate), Newton's Law of motion says the forces on each slab must balance. Thus each must exert on its neighbor above the same force as that exerted on it by its neighbor below, or

$$\frac{dv_z}{dx}$$

must be a constant, independent of x. A function with constant derivative must be a *linear* function. Because v must go from v_0 on the top plate to zero on the bottom plate, we find $v_z(x) = (x/d)v_0$.

Thus a volume element of water initially at (x_0, z_0) moves in time t to $(x_0, z_0 + (x_0/d)v_0 t)$. It's this motion that stretches out an initially spherical blob of ink (Figure 5.2b). If we reverse the force pulling the top plate for an equal time t, we find that every fluid element returns to exactly its original starting point. The blob reassembles; if it had originally been stretched so far as to appear mixed, it now appears to "unmix" (Figure 5.1).

Now suppose that we don't insist on steady motion and instead apply a time-dependent force $f(t)$ to the top plate. This time, the forces on each slab won't quite balance; instead, the net force equals the mass of fluid in the slab times its acceleration, by Newton's Law of motion. As long as the force is well below the viscous critical force, however, this correction will be negligible and all the same conclusions as before apply: Once the top plate has returned to its initial position, each fluid element has also returned. It's a bit like laying a deck of cards on the table and pushing the top card sideways, then back. Regardless of whether the return stroke is hard and short, or gentle and long, as soon as the top plate returns to its original position, so have all the fluid elements (apart from a small amount of true, diffusive mixing).

Time reversal The "unmixing" phenomenon points up a key qualitative feature of low Reynolds-number fluid flow. To understand this feature, let's contrast such flows with the more familiar world of Newtonian mechanics.

If we throw a rock up in the air, it goes up and then down in the familiar way: $z(t) = v_0 t - \frac{1}{2}gt^2$. Now imagine a related process, in which the position $z_r(t)$ is related to the original one by **time reversal**; that is, $z_r(t) \equiv z(-t) = -v_0 t - \frac{1}{2}gt^2$. The time-reversed process is also a legitimate solution of Newton's laws, albeit with a different initial velocity from the original process. Indeed, we can see directly that Newton's Law has this property, just by inspecting it: Writing the force as the derivative of a potential energy gives

$$-\frac{dU}{dx} = m\frac{d^2x}{dt^2}.$$

Because this equation contains two time derivatives, it is unchanged under the substitution $t \to -t$. Ballistic motion is time-reversal invariant.

A second example may reinforce the point. Suppose you're stopped at a traffic light when someone rear-ends you. Starting at time $t = 0$, the position $x(t)$ of your head suddenly accelerates forward. The force needed to make this happen comes from your headrest; it's also directed forward, according to

$$f = m\frac{d^2x}{dt^2}.$$

Now imagine another process, in which your head moves along the time-reversed trajectory $x_r(t) \equiv x(-t)$. Physically, x_r describes a process where your car is initially rolling backward, then hits a wall behind you and stops. Once again your head's acceleration points *forward*, as its velocity jumps from negative to zero. Once again your headrest pushes *forward* on your head. In other words,

> *In Newtonian physics, the time-reversed process is a solution to the equations of motion with the same sign of force as the original motion.* (5.12)

In contrast, the viscous friction rule is *not* time-reversal invariant: The time-reversed trajectory doesn't solve the equation of motion with the same sign of the force. Certainly a pebble in molasses never falls *upward*, regardless what starting velocity we choose! Instead, to get the time-reversed motion we must apply a force that is time reversed and *opposite in direction* to the original. To see this in the mathematics, let's reconsider the equation of motion we found for diffusion with drift, $v_{drift} = f/\zeta$ (Equation 4.12), and rephrase it using $\bar{x}(t)$, the position of the particle at time t averaged over many collision times. ($\bar{x}(t)$ shows us the net drift but not the much faster thermal jiggling motion.) In this language, our equation of motion reads

$$\frac{d\bar{x}}{dt} = \frac{f(t)}{\zeta}. \qquad (5.13)$$

The solution $\bar{x}(t)$ to Equation 5.13 could be uniform motion (if the force $f(t)$ is constant) or accelerated motion (otherwise). But think about the time-reversed motion, $\bar{x}_r(t) \equiv \bar{x}(-t)$. We can find its time derivative by using the chain rule from calculus; it won't be a solution of Equation 5.13 unless we replace $f(t)$ by $-f(-t)$.

The failure of time-reversal invariance is a signal that something *irreversible* is happening in frictional motion. Phrased this way, the conclusion is not surprising: We already knew that friction is the one-way dissipation, or degradation, of ordered motion into disordered motion. Our simple model for friction in Section 4.1.4 explicitly introduced this idea, via the assumption of randomizing collisions.

Here is another example of the same analysis. Section 4.6 gave some solutions to the diffusion equation (Equation 4.20 on page 131). Taking any solution $c_1(x, t)$, we can consider its time-reversed version $c_2(x, t) \equiv c_1(x, -t)$, and its space-reflected version $c_3(x, t) \equiv c_1(-x, t)$. Take a moment to visualize c_2 and c_3 for the example shown in Figure 4.12a on page 134.

> **Your Turn 5F**
>
> Substitute c_2 and c_3 into the diffusion equation and see whether they also are solutions. [*Hint:* Use the chain rule to express derivatives of c_2 or c_3 in terms of those of c_1.] Then explain in words why the answer you got is right.

The distinction between fluids and solids also hinges upon their time-reversal behavior. Suppose we put an elastic solid, like rubber, between the plates in Figure 5.2b. The plates have area A and are separated by a distance d. If we slide the plates a distance Δz, the rubber resists with a force given by a **Hooke relation:** $f = -k(\Delta z)$. The spring constant k in this relation depends on the geometry of the sample; for simple materials, it takes the form $k = \mathcal{G}A/d$, where the **shear modulus** $\mathcal{G}$ is a property of the material. Thus

$$\frac{f}{A} = -\left(\frac{\Delta z}{d}\right)\mathcal{G}. \tag{5.14}$$

The quantity f/A is called the **shear stress**; $(\Delta z)/d$ is the **shear strain**. A fluid, in contrast, has $f/A = -\eta v/d$ (Equation 5.4).

In short, for solids, the stress is proportional to the strain $(\Delta z)/d$, whereas for fluids, it's proportional to the strain *rate*, v/d. A simple elastic solid doesn't care about the rate; you can shift the plates and then hold them stationary, and an elastic solid will continue resisting forever. Fluids, in contrast, have no memory of their initial configuration; they only notice how fast you're *changing* their boundaries.

The difference is one of symmetry: In each case, if we reverse the applied distortion spatially, the opposing force also reverses. But for fluids, if we *time-reverse* the distortion $\Delta z(t)$, then the force reverses direction; whereas, for solids, it doesn't. The equation of motion for distortion of an elastic solid is time-reversal invariant, a signal that there's no dissipation.

$\boxed{T_2}$ *Section 5.2.3' on page 188 describes an extension of the ideas just discussed to materials with both viscous and elastic behavior.*

5.3 BIOLOGICAL APPLICATIONS

Section 5.2.3 brought us close to the idea of entropy, promised in Chapter 1. Entropy will measure precisely *what* is increasing irreversibly in a dissipative process like diffusion. Before we finally define it in Chapter 6, the next section will give some immediate consequences of these ideas, in the world of swimming bacteria.

5.3.1 Swimming and pumping

Section 5.2.1 discussed how, in the low Reynolds-number world, applying a force to fluid generates a motion that can be canceled completely by applying minus the time-reversed force. These results may be amusing to us, but they are matters of life and death to microorganisms.

An organism suspended in water may find it advantageous to swim about. It can only do so by changing the shape of its body in some periodic way. It's not as simple as it may seem. Suppose you flap a paddle, then bring it back to its original position by the same path (Figure 5.5a). You then look around and discover that you have made no net progress, just as every fluid element returned to its original position in the unmixing experiment (Figures 5.1 and 5.2). A more detailed example can help make this clearer.

Consider an imaginary microorganism, trying to swim by pushing a part of its body ("paddles") relative to the rest ("body") (see Figure 5.6). To simplify the math, we'll suppose that the creature can move only in one direction, and the relative motion of paddles and body also lies in the same direction. The surrounding fluid is at rest. We know that in low Reynolds-number motion, moving the body through

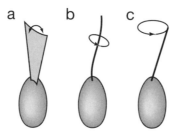

Figure 5.5: (Schematic.) Three swimmers. (a) The flapper makes reciprocal motion. (b) The twirler cranks a stiff helical rod. (c) The spinner swings a stiff, straight rod.

the fluid requires a force determined by a viscous friction coefficient ζ_0. Moving the paddles through the fluid requires a force determined by a different constant ζ_1.

Initially, the body is located at $x = 0$. Then it pushes its paddles backward (toward negative x) relative to its body at a relative speed v for a time t. Next it pushes the paddles forward at a different relative speed v' to return them to their original location. The cycle repeats. Your friend suggests that, by making the "recovery" stroke slower than the "power" stroke (that is, taking $v' < v$), the creature can make net progress.

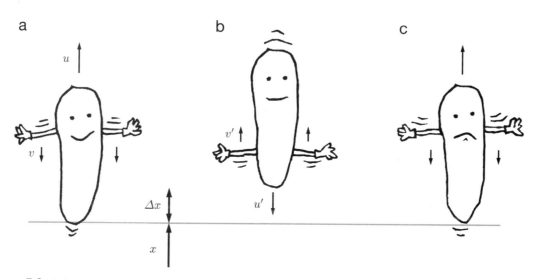

Figure 5.6: (Schematic.) A microscopic swimmer trying to make progress by cycling between forward and backward strokes of its paddles. (a) On the first stroke, the paddles move backward relative to the body at relative speed v, propelling the body through the fluid at speed u. (b) On the second stroke, the paddles move forward at relative speed v', propelling the body backward at speed u'. (c) Then the cycle repeats. The progress made on the first stroke is all lost on the second stroke; reciprocal motion like this cannot give net progress in low Reynolds-number fluid mechanics. [Cartoon by Jun Zhang.]

Example:

a. The actual speed at which the paddles move through the water depends both on the given v and on the speed u of the body, which you don't know yet. Find u for the first half of the cycle.

b. How far and in what direction does the body move in the first stroke?

c. Repeat (a,b) for the second (return) stroke.

d. Your friend proposes to choose v and v' to optimize this process. How do you advise him?

Solution:

a. The velocity of the paddles relative to the surrounding fluid is the relative velocity, $-v$, plus u. Balancing the resulting drag force on the paddles against the drag force on the body gives $u = \zeta_1 v/(\zeta_0 + \zeta_1)$.

b. $\Delta x = tu$, forward, where u is the quantity found in (a).

c. $u' = \zeta_1 v'/(\zeta_0 + \zeta_1)$, $\Delta x' = -t'u'$. We must take $t'v' = tv$ if we want the paddles to return to their original positions on the body. Thus

$$\Delta x' = -t'v'\frac{\zeta_1}{\zeta_0 + \zeta_1} = -tv\frac{\zeta_1}{\zeta_0 + \zeta_1} = -tu.$$

d. It won't work. The answers to (b) and (c) always cancel, regardless of what we take for v and v'. For example, if the "recovery" stroke is half as fast as the "power" stroke, the corresponding net motion is also half as fast. But such a recovery stroke must last twice as long as the power stroke in order to prepare the creature for another cycle!

So a strictly reciprocating motion won't work for swimming in the low Reynolds-number world. What other options does a microorganism have? The required motion must be periodic, so that it can be repeated. It can't be of the reciprocal (out-and-back) type described in the Example. Here are two examples.

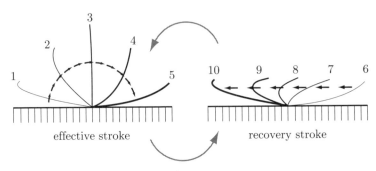

effective stroke   recovery stroke

Figure 5.7: (Schematic.) The ciliary cycle. The effective stroke (*left*) alternates with the recovery stroke (*right*). The motion is not reciprocal, so the cilium can make net progress in sweeping fluid past the surface.

Ciliary propulsion Many cells use cilia, which are whiplike appendages 5–10 μm long and 200 nm in diameter, to generate net thrust. Motile cells (such as *Paramecium*) use cilia to move. Stationary cells (such as the ones lining our air passages) use them to pump fluid or sweep food to themselves (see Figure 2.10 on page 45).

Each cilium contains internal filaments and molecular motors that can slide the filaments across one another, thereby creating an overall bend in the cilium. The motion in Figure 5.7 is typical; it is periodic but not reciprocal. To see how it generates propulsion, we need one intuitive result from low Reynolds-number fluid mechanics (whose mathematical proof is beyond the scope of this book):

> *A rod dragged along its axis at velocity* **v** *feels a resisting force proportional to* $-$**v** *(that is, also directed along the axis). Similarly, a rod dragged perpendicular to its axis feels a resisting force also proportional to* $-$**v** *(that is, also directed perpendicular to the axis). However, the viscous friction coefficient* $\zeta_\parallel$ *for motion parallel to the axis is smaller than the one* $\zeta_\perp$ *for perpendicular motion.*   (5.15)

The ratio between the two friction coefficients depends on the length of the rod; we will use the illustrative value $\frac{2}{3}$.

Figure 5.7 shows a cilium initially lying parallel to the cell surface, pointing to the left. During the effective stroke (left panel), the entire cilium moves perpendicular to its axis, whereas during the recovery stroke (right panel) most of it is moving nearly parallel to its axis. Thus the motion of the fluid created by the power stroke gets only partly undone by the backflow created by the recovery stroke. The difference between these flows is the net pumping of one cycle.

Bacterial flagella What if the speed **v** is neither parallel nor perpendicular to the axis? In this case, Figure 5.8 shows that the resulting drag force will also be somewhere in between the parallel and perpendicular directions, but *not along* **v**. Instead, the

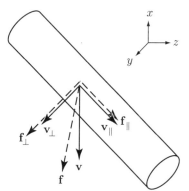

Figure 5.8: (Schematic.) A thin rod is dragged at low Reynolds number with velocity **v**. The force **f** needed to drag the rod is the resultant of two forces $\mathbf{f}_\parallel$ and $\mathbf{f}_\perp$ coming from the components of **v** parallel to and perpendicular to the rod's axis. Even if $\mathbf{v}_\parallel$ and $\mathbf{v}_\perp$ are the same length, as shown, the resulting components of **f** will not be equal; thus **f** will not point parallel to **v**.

force points closer to the perpendicular direction than does the velocity; the larger $\zeta_\perp$ "wins" over the smaller $\zeta_\parallel$. *E. coli* bases its propulsion on this fact.

Unlike cilia, *E. coli*'s flagella do not flex; they are rigid, helical objects, like twisted coathangers, so they cannot solve the propulsion problem by the means shown in Figure 5.7. Because they are only 20 nm thick, it's not easy to visualize their three-dimensional motion under the microscope. Initially, some people claimed that the bacterium waves them back and forth, but we know this can't work: It's a reciprocal motion. Others proposed that a wave of bending travels down the flagellum, but there hardly seemed to be room for any of the required machinery inside such a thin object. In 1973, H. Berg and R. Anderson argued that instead the bacterium *cranked* the flagellum at its base in a rigid *rotary* motion (like the twirler in Figure 5.5b). This was a heretical idea. At that time, no true rotary engine had ever been seen in any living creature (we will, however, meet another example in Chapter 11). Nor was it easy to imagine how to prove such a theory—it's hard to judge the three-dimensional character of a motion seen under the microscope.

M. Silverman and M. Simon found an elegant solution to the experimental problem. They used a mutant *E. coli* strain that lacks most of its flagellum, having instead only a stump (called the "hook"). They anchored the cells to a glass coverslip by their hooks. The flagellar motor, unable to spin the anchored flagellar hook, instead spun the whole bodies of the bacteria, a process easily visible in the microscope! Today we know that the flagellar motor is a marvel of nanotechnology, a rotary engine just 45 nm wide (Figure 5.9).

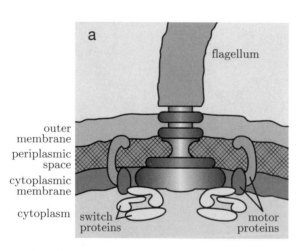
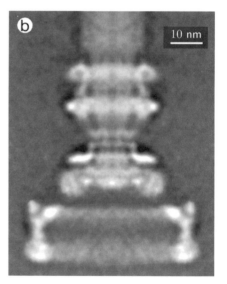

Figure 5.9: (Schematic; reconstruction from electron microscopy.) (a) The bacterial flagellar motor, with elements analogous to those of a macroscopic rotary motor. The inner part of the motor assembly develops a torque relative to the outer part, which is anchored to the polymer network (the peptidoglycan layer), thereby turning the flagellum. The peptidoglycan layer provides the rigid framework of the cell wall; it is located in the periplasmic space between the cell's two membranes. (b) Composite electron micrograph of the actual structure of the motor assembly. [Digital image kindly supplied by D. Derosier; see Derosier, 1998.]

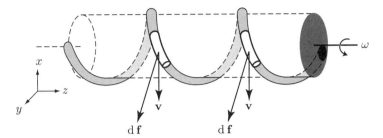

Figure 5.10: (Schematic.) Principle of flagellar propulsion in bacteria. A thin, rigid, helical rod is cranked about its helix axis at angular speed ω. For better visualization, a phantom cylinder has been sketched, with the rod lying on its surface. Two short segments of the rod have been singled out for study, both lying on the near side of the helix and separated by one turn. The rod is attached (*black circle*) to a disk and the disk is rotated, cranking the helix about its axis. The two short segments then move downward, in the plane of the page. Thus, d**f** lies in the plane of the page, but tipped slightly to the left as shown (see Figure 5.8). A net force with a negative z-component is required to keep the helix spinning in place.

Rotary motion certainly meets our criterion of being periodic but not reciprocal. And we are familiar with other spinning helical objects that develop thrust along their axis, namely, submarine and boat propellers. But the details are quite different in the low Reynolds-number case. Figure 5.10 shows a schematic of the situation. A rigid helical object (representing the flagellum) is cranked about its axis (by the flagellar motor). Two short segments of the helix have been singled out for study. The net force d**f** exerted on one short segment by its two neighbors must balance the viscous drag force on that segment. Thus for the helix to undergo the desired rotational motion, d**f** must be the vector shown in Figure 5.8. Adding up all the contributions from every rod segment, we see that the components in the xy plane all cancel (think about the corresponding segments on the far side of the helix, whose velocity vectors point upward). But d**f** also has a small component directed along the $-\hat{z}$ direction, and the df_z's do *not* cancel. Rather, a net leftward force must be supplied to spin the flagellum in place (in addition to a torque about the axis).

Suppose the flagellum is not anchored but, instead, is attached to a bacterium at its rightmost end. Then there is nothing to supply a net leftward force; cranking the flagellum will therefore pull the bacterium to the *right*. This is the propulsion mechanism we sought. Interestingly, mutant bacteria have been found with straight flagella. They spin and spin, but never go anywhere.

$\boxed{T_2}$ *Section 5.3.1' on page 189 discusses the ratio of parallel and perpendicular friction constants in greater detail.*

5.3.2 To stir or not to stir?

It's surprisingly difficult to get anything to eat when you're tiny. We get a hint of why when we examine the experimental photograph, Figure 5.3 on page 166. At low Reynolds number, the flow lines just part majestically as they come to the surface of the sphere; any food molecules carried in the fluid follow the flow lines and never arrive at the surface.

Things are not as bad as they seem. The macroscopic experiment shown in Figure 5.3 doesn't show the effects of diffusion, which *can* carry molecules to receptors on a cell's surface. Diffusion will bring food even to a lazy, motionless cell! Similarly, diffusion will carry waste *away*, even if the cell is too lazy to move away from its waste. So why bother swimming?

Similar remarks apply to *stirring*. It was once believed that a major job of cilia was to sweep fresh fluid to the cell, thereby enhancing its intake relative to passively waiting. To evaluate such arguments, imagine the cilium as moving at some characteristic speed v and swinging through a length d. These parameters determine a time scale $t = d/v$, the time in which the cilium can replace its surrounding fluid with fresh, outside fluid. On the other hand, movement of molecules a distance d will occur just by diffusion in a characteristic time d^2/D, according to the diffusion law (Idea 4.5a on page 115). So stirring will only be worthwhile (more effective than diffusion) if $d/v < d^2/D$, or

$$v > \frac{D}{d}. \qquad (5.16)$$

(Some authors call the dimensionless ratio vd/D the Peclet number.) Taking a cilium to be about $d = 1\,\mu\text{m}$ long, the criterion for stirring to be worthwhile is then that $v > 1000\,\mu\text{m s}^{-1}$. This is also the criterion for swimming to enhance food intake significantly.

But bacteria do not swim anywhere near this fast. *Stirring and swimming don't help enhance food intake for bacteria.* (The story is different for larger creatures, even protozoa, for which the Reynolds number is still small but d and v are both bigger.) There is experimental support for this conclusion. Mutant bacteria with defective flagellar systems manage about as well as their wild-type cousins when food is plentiful.

5.3.3 Foraging, attack, and escape

Foraging Section 5.3.2 may have left you wondering why wild-type bacteria *do* swim. The answer is that life in the mean, real world can be more challenging than life in a nice warm flask of broth. Although bacteria don't need to swim around systematically scooping up available food, still it may be necessary for a cell to find a food supply. The word *find* implies a degree of volition; and mind-boggling as it may be, supposedly primitive organisms like *E. coli* can indeed perform the computations needed to hunt for food.

The strategy is elegant. *E. coli* swims in a burst of more or less straight-line motion, pauses, and then takes off in a new, randomly chosen direction. While swimming, the cell continuously samples its environment. If the concentration of food is increasing, the bacterium extends its run. If the food concentration is decreasing, the cell terminates the run and starts off in a new direction sooner than it would have in an improving environment. Thus the cell executes a form of biased random walk, with a net drift toward higher food concentrations.

But there's no point in making a run so short that the environment won't be appreciably different at the end. Because diffusion constantly tries to equalize the concentration of food (and everything else), then, it's necessary for the bacterium to *outrun diffusion* if swimming is to be of any use in navigating food gradients. We have already found the criterion, Equation 5.16. Now, however, we take $v \approx 30\,\mu\text{m s}^{-1}$ to be the known swimming speed and d to be the length of the run, not the length of the cell. Then we find that, to navigate up food gradients, a bacterium must swim at least $30\,\mu$m, or 30 body lengths, before changing direction. And ... that's what they really do.

Attack and escape Take another look at Figure 5.3 on page 166. Clearly a solid object, gliding through a liquid at low Reynolds number, *disturbs the fluid* out to a distance comparable to its own diameter. This fact can be a liability if your livelihood depends on stealth, for example, if you need to grab your dinner before it escapes. Moreover, swimming up to a tasty morsel will actually tend to push it away, just like your colored blob in the experiment described in Section 5.1.3 on page 161. That's why many medium-small creatures, not so deeply into the low Reynolds-regime as bacteria, put on a burst of speed to push themselves momentarily up to high Reynolds number for the kill. For example, the tiny crustacean *Cyclops* makes its strike by accelerating at up to $12\,\text{m s}^{-2}$, briefly hitting Reynolds numbers as high as 500.

In the same spirit, *escaping* from an attacker will just tend to drag it along with you at low Reynolds number! Here again, a burst of speed can make all the difference. The sessile protozoan *Vorticella*, when threatened, contracts its stalk from 0.2–0.33 mm down to less than half that length at speeds up to $80\,\text{mm s}^{-1}$, the most rapid shortening of any contractile element in any animal. This impressive performance garners the name "spasmoneme" for the stalk.

5.3.4 Vascular networks

Bacteria can rely on diffusion to feed them, but large organisms need an elaborate infrastructure of delivery and waste-disposal systems. Virtually every macroscopic creature thus has one or more **vascular networks** carrying blood, sap, air, lymph, and so on. Typically these networks have a hierarchical, branching structure: The human aorta splits into the iliac arteries, and so on, down to the capillary beds that actually nourish tissue. To get a feeling for some of the physical constraints governing such networks, let's take a moment to work out one of the simplest fluid-flow problems: the steady, laminar flow of a simple Newtonian fluid through a straight, cylindrical pipe of radius R (Figure 5.11a). In this situation, the fluid does not accelerate at all, so we can neglect the inertial term in Newton's Law even if the Reynolds number is not very small.

We must push a fluid to make it travel down a pipe, in order to overcome viscous friction. The frictional loss occurs throughout the pipe, not just at the walls. Just as in Figure 5.2 on page 163, where each layer of fluid slips on its neighbor, in the cylindrical geometry, the shear will distribute itself across the whole cross section of the pipe. Imagine the fluid as a nested set of cylindrical shells. The shell at distance r from the center moves forward at a speed $v(r)$, which we must find. The unknown

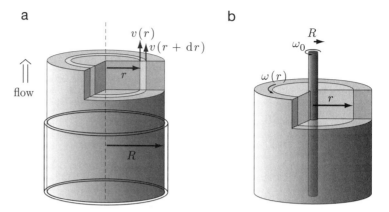

Figure 5.11: (Sketches.) (a) In laminar pipe flow, the inner fluid moves faster than the outer fluid, which must be motionless at the pipe wall (the no-slip boundary condition). We imagine concentric cylindrical layers of fluid sliding over one another. (b) The torsional drag on a spinning rod in viscous fluid. This time the inner fluid *rotates* faster than the outer fluid, which must be at rest far away from the rod. Again we imagine concentric cylindrical layers of fluid sliding over one another; the angular velocity $\omega(r)$ is not constant but decreases with r.

function $v(r)$ interpolates between the stationary walls (with $v(R) = 0$) and the center (with unknown fluid velocity $v(0)$).

To find $v(r)$, we balance the forces acting on the shell lying between r and $r + dr$. The cross-sectional area of this shell is $2\pi r\, dr$. Hence the pressure drop between the ends of the pipe, p, contributes a force $df_1 = 2\pi r p\, dr$ directed along the pipe axis. A viscous force df_2 from the slower-moving fluid at larger r pulls backward on the shell, whereas the faster-moving fluid at smaller r drags it forward with a third force, df_3. For a pipe of length L, the viscous force rule (Equation 5.9 on page 168) gives

$$df_3 = -\eta(2\pi r L)\frac{dv(r)}{dr} \quad \text{and} \quad df_2 = \eta(2\pi(r+dr)L)\frac{dv(r')}{dr'}\bigg|_{r'=r+dr}.$$

Notice that v decreases with r, so f_2 is a negative quantity, whereas f_3 is positive. Force balance is then the statement that $df_1 + df_2 + df_3 = 0$.

Because dr is very small, we can evaluate dv/dr at the point $(r + dr)$ by using a series expansion, dropping terms with more than one power of dr:

$$\frac{dv(r')}{dr'}\bigg|_{r'=r+dr} = \frac{dv(r)}{dr} + dr \times \frac{d^2v}{dr^2} + \cdots.$$

Thus adding df_2 to df_3 gives

$$2\pi \eta L\, dr \times \left(\frac{dv}{dr} + r\frac{d^2v}{dr^2}\right).$$

Adding df_1 and requiring the sum to be zero gives

$$\frac{rp}{L\eta} + \frac{dv}{dr} + r\frac{d^2v}{d^2r} = 0.$$

This is a differential equation for the unknown function $v(r)$. You can check that its general solution is $v(r) = A + B\ln r - r^2 p/(4L\eta)$, where A and B are constants. We had better choose $B = 0$, because the velocity cannot be infinite at the center of the pipe. And we need to take $A = R^2 p/(4L\eta)$ to get the fluid to be stationary at the stationary walls. These conditions fix our solution, the flow profile for laminar flow in a cylindrical pipe:

$$v(r) = \frac{(R^2 - r^2)p}{4L\eta}. \tag{5.17}$$

> **Your Turn 5G**
>
> After going through the math to check the solution (Equation 5.17), explain in words why every factor (except the 4) "had" to be there.

Now we can see how well the pipe transports fluid. The velocity v can be thought of as the flux of volume j_v, or the volume per area per time transported by the pipe. The total flow rate Q, with the dimensions of volume per time, is then the volume flux $j_v = v$ from Equation 5.17, integrated over the cross-sectional area of the pipe:

$$Q = \int_0^R 2\pi r\,dr\,v(r) = \frac{\pi R^4}{8L\eta}p. \tag{5.18}$$

Equation 5.18 is the **Hagen–Poiseuille relation** for laminar pipe flow. Its applicability extends somewhat beyond the low Reynolds-number regime studied in most of this chapter: All we really assumed was laminar flow. This regime includes all but the largest veins and arteries in the human body (or the entire circulatory system of a mouse).

The general form of Equation 5.18 can be expressed as $Q = p/Z$, where the **hydrodynamic resistance** $Z = 8\eta L/(\pi R^4)$. The choice of the word *resistance* is no accident. The Hagen–Poiseuille relation says that the rate of transport of some conserved quantity (volume) is proportional to a driving force (the pressure drop p), just as Ohm's law says that the rate of transport of charge is proportional to a driving force (potential drop). In each case, the constant of proportionality is called resistance. In the context of low Reynolds-number fluid flow, transport rules of the form $Q = p/Z$ are quite common and are collectively called **Darcy's law**. (At high Reynolds number, turbulence complicates matters; and no such simple rule holds.) Another example is the passage of fluid across a membrane (see Problem 4.10). In this context, we write $Z = 1/(AL_p)$ for the resistance, where A is the membrane area and L_p is called the filtration coefficient (some authors use the synonym hydraulic permeability).

A surprising feature of the Hagen–Poiseuille relation is the very rapid decrease of resistance as the pipe radius R increases. Two pipes in parallel will transport twice as much fluid at a given pressure as will one. But a single pipe with twice the area will transport *four* times as much, because $\pi R^4 = (1/\pi)(\pi R^2)^2$, and πR^2 has doubled. This exquisite sensitivity allows our blood vessels to regulate flow with only small dilations or contractions:

Example: Find the change in radius needed to increase the hydrodynamic resistance of a blood vessel by 30%, other things being equal. (Idealize the situation as laminar flow of a Newtonian fluid.)

Solution: We want p/Q to increase to 1.3 times its previous value. Equation 5.18 says that this happens when $(R')^{-4}/R^{-4} = 1.3$, or $R'/R = (1.3)^{-1/4} \approx 0.94$. Thus the vessel need only change its radius by about 6%.

5.3.5 Viscous drag at the DNA replication fork

To finish the chapter, let's descend from physiology to the realm of molecular biology, which will occupy much of the rest of this book.

A major theme of the chapters to come will be that DNA is not just a database of disembodied information but a *physical object* immersed in the riotous thermal environment of the nanoworld. This is not a new observation. As soon as the double-helix model of DNA structure was announced, people asked: How do the two strands separate for replication, when they're wound around each other? One solution is shown in Figure 5.12. The figure shows a Y-shaped junction where the original strand (top) is being disassembled into two single strands. Because the two single strands cannot pass through each other, the original must continually rotate (arrow).

The problem with the mechanism sketched in the figure is that the upper strand extends for a great distance (DNA is long). If one end of this strand rotates, then it would seem that the whole thing must also rotate. Some people worried that the frictional drag resisting this rotation would be enormous. Following C. Levinthal and H. Crane we can estimate this drag and show that, on the contrary, it's negligible.

Consider cranking a long, thin, straight rod in water (Figure 5.11b). This model is not as drastic an oversimplification as it may at first seem. DNA in solution is not really straight, but, when cranked, it can rotate in place, like a tool for unclogging drains. Our estimate will be roughly applicable for such motions. Also, the cell's cytoplasm is not just water; but for small objects (like the 2 nm thick DNA double helix), it's not a bad estimate to use water's viscosity (see Appendix B).

The resistance to rotary motion should be expressed as a *torque*. The drag torque τ will be proportional to the viscosity and to the cranking rate, just as it is in Equation 5.4 on page 164. It will also be proportional to the rod's length L, because there will be a uniform drag on each segment. The cranking rate is expressed as an angular velocity ω, with dimensions $\mathbb{T}^{-1}$. (We know ω once we've measured the rate of replication, because every helical turn contains about 10.5 basepairs.) In short, we must have $\tau \propto \omega \eta L$. Before we can evaluate this expression, however, we need an estimate for the constant of proportionality.

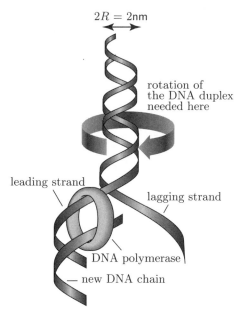

$2R = 2\text{nm}$

rotation of
the DNA duplex
needed here

leading strand

lagging strand

DNA polymerase

— new DNA chain

Figure 5.12: (Schematic.) Replication of DNA requires that the original double helix (*top*) be unwound into its two strands. Molecular machines called DNA polymerase sit on the single strands synthesizing new, complementary strands. The process requires the original strand to spin about its axis, as shown. Another molecular machine called DNA helicase (*not shown*) sits at the opening point and walks along the DNA, unwinding the helix as it goes along. [Adapted from Alberts et al., 2002.]

Certainly the drag will also depend on the rod's radius, R. From the first-year physics formula $\tau = \mathbf{r} \times \mathbf{f}$, we find that torque has the same dimensions as energy. Dimensional analysis then shows that the constant of proportionality we need has dimensions $\mathbb{L}^2$. We have already taken into account the dependence on L. The only other parameter in the problem with the dimensions of length is R (recall that water itself has no intrinsic length scale, Section 5.2.1). Thus the constant of proportionality we seek must be R^2 times some dimensionless number C, or

$$\tau = -C \times \omega\eta R^2 L. \tag{5.19}$$

Problem 5.9 shows that this result is indeed correct and that $C = 4\pi$; but we don't need the precise value for what follows.

The rate at which we must do work to crank the rod is the product of the applied torque times the rotation rate: $-\tau\omega = C\omega^2\eta R^2 L$. Because the rod rotates through 2π radians for each helical turn, we can instead quote the mechanical work needed per helical turn, as

$$W_{\text{frict}} = -2\pi\tau = 2\pi C \times \omega\eta R^2 L. \tag{5.20}$$

An enzyme called DNA polymerase synthesizes new DNA in *E. coli* at a rate of about 1000 basepairs (abbreviated bp) per second, or

$$\omega = 2\pi \, \frac{\text{radian}}{\text{revolution}} \times \frac{1000 \, \text{bp s}^{-1}}{10.5 \, \text{bp/revolution}} \approx 600 \, \text{s}^{-1}.$$

Equation 5.20 then gives $W_{\text{frict}} \approx (2\pi)(4\pi)(600 \, \text{s}^{-1})(10^{-3} \, \text{Pa s})(1 \, \text{nm})^2 L \approx (4.7 \cdot 10^{-17} \, \text{J m}^{-1})L$.

A second enzyme, called **DNA helicase**, does the actual cranking. Helicase walks along the DNA in front of the polymerase, unzipping the double helix as it goes along. The energy required to do this comes from the universal energy-supply molecule ATP. Appendix B lists the useful energy in a single molecule of ATP as $\approx 20 k_B T_r = 8.2 \cdot 10^{-20}$ J. Let's suppose that one ATP suffices to crank the DNA by one full turn. Then the energy lost to viscous friction will be negligible as long as L is much smaller than $(8.2 \cdot 10^{-20} \, \text{J})/(4.7 \cdot 10^{-17} \, \text{J m}^{-1})$, or about 2 mm, a very long distance in the nanoworld. Levinthal and Crane correctly concluded that rotational drag is not an obstacle to replication.

Today we know that another class of enzymes, the topoisomerases, remove the excess twisting generated by the helicase in the course of replication. The preceding estimate should thus be applied only to the region from the replication fork to the first topoisomerase, and hence viscous rotary drag is even less significant than the previous paragraphs makes it seem. In any case, a physical argument let Levinthal and Crane dismiss an objection to the double-helix model for DNA, long before any of the details of the cellular machinery responsible for replication were known.

5.4 EXCURSION: THE CHARACTER OF PHYSICAL LAWS

We are starting to amass a large collection of statements called "laws." (This chapter alone has mentioned Newton's Law of motion, the Second Law of thermodynamics, and Ohm's and Fick's laws.) Generally these terms were born like any other new word—someone noticed a certain degree of generality to the statement, coined the name, a few others followed, and the term stuck. Physicists, however, tend to be a bit less promiscuous in attaching the term *physical Law* to an assertion. Although we cannot just rename terms hallowed by tradition, this book attempts to make the distinction by capitalizing the word *Law* on those statements that seem to meet the physicist's criteria, elegantly summarized by Richard Feynman in 1964.

To summarize Feynman's summary, physical Laws seem to share some common characteristics. Certainly there is an element of subjectivity in the canonization of a Law; but, in the end, there is generally more consensus than dispute on any given case.

- Certainly we must insist on a *very great degree of generality,* an applicability to an extremely broad class of phenomena. Thus, many electrical conductors do not obey "Ohm's law," even approximately, whereas any two objects in the Universe really do seem to attract each other with a gravitational force described (approximately!) by Newton's Law of gravitation.

- Although they are general, physical Laws *need not be, and generally cannot be, exact.* Thus, as people discovered more and deeper layers of physical reality, Newton's Law of motion had to be replaced by a quantum-mechanical version; his Law of gravitation was superseded by Einstein's, and so on. The older, approximate laws remain valid and useful in the very large domain where they were originally discovered, however.

- Physical Laws all seem to be *intrinsically mathematical in their expression.* This characteristic may give them an air of mystery, but it is also the key to their great simplicity. There is very little room in the terse formula $f = ma$ to hide any sleight-of-hand, little room to bend a simple formula to accommodate a new, discrepant experiment. When a physical theory starts to acquire too many complicating features, added to rescue it from various new observations, physicists begin to suspect that the theory was false to begin with.

- Yet, out of the simplicity of a Law, there always emerge subtle, unexpected, and true *conclusions revealed by mathematical analysis.* Word-stories are often invented later to make these conclusions seem natural, but generally the clearest, most direct route to get them in the first place is mathematical.

An appreciation of these ideas may not make you a more productive scientist. But many people have drawn inspiration, even sustenance, from their wonder at the fact that Nature should have any such unifying threads at all.

THE BIG PICTURE

Returning to the Focus Question, we've seen that the key difference between the nanoworld and our everyday life is that viscous dissipation completely dominates inertial effects. A related result is that objects in the nanoworld are essentially unable to store any significant, nonrandom kinetic energy—they don't coast after they stop actively pushing themselves (see Problem 5.4). These results are reminiscent of the observation in Chapter 4 that diffusive transport, another dissipative process, is fast on small length scales; indeed, we saw in Section 5.3.2 that diffusion beats stirring in the submicrometer world.

We saw how to express the distinction between dissipative and nondissipative processes in a very concise form by describing the invariance properties of the appropriate equations of motion: Frictionless Newtonian physics is time-reversal invariant, whereas the friction-dominated world of low Reynolds number is not (Section 5.2.3).

Hiding in the background of all this discussion has been the question of *why* mechanical energy tends to dissipate. Chapter 1 alluded to the answer—the Second Law of thermodynamics. Our task in the next chapter is to make the Second Law, and its cardinal concept of entropy, more precise.

KEY FORMULAS

- *Viscosity:* Suppose a wall is perpendicular to the x direction. The viscous force per area in the $\hat{z}$ direction exerted by a fluid on a wall is $-\eta\, dv_z/dx$ (Equation 5.9).

  $\boxed{T_2}$ The kinematic viscosity is defined as $\nu = \eta/\rho_m$, where ρ_m is the fluid mass density, and has the units of a diffusion constant (see Section 5.2.1′).

- *Reynolds:* The viscous critical force for a fluid is $f_{crit} = \eta^2/\rho_m$, where ρ_m is the mass density of the fluid and η its viscosity (Equation 5.5). The Reynolds number for a fluid flowing at velocity v and negotiating obstacles of size R is $\mathcal{R} = vR\rho_m/\eta$ (Equation 5.11). Laminar flow switches to turbulent flow when $\mathcal{R}$ exceeds about 1000.

- *Rotary drag:* For a macroscopic (many nanometers) cylinder of radius R and length L, spinning on its axis in a fluid at low Reynolds number, the drag torque is $\tau = -4\pi\omega\eta R^2 L$ (Equation 5.19 and Problem 5.9), where η is the fluid viscosity.

- *Hagen–Poiseuille:* The volume flux through a pipe of radius R and length L, in laminar flow, is

$$Q = \frac{\pi R^4}{8L\eta}\, p,$$

where p is the pressure drop (Equation 5.18). The velocity profile is parabolic, that is, $v(r)$ is a constant times $R^2 - r^2$, where r is the distance from the center of the pipe.

FURTHER READING

Semipopular:
Fluid flows: van Dyke, 1982.
The idea of physical Law: Feynman, 1965.

Intermediate:
Much of this chapter was drawn from E. Purcell's classic lecture (Purcell, 1977), and H. Berg's book (Berg, 1993) (particularly Chapter 6); see also Berg, 2000.
Fluids: Feynman et al., 1963b, §§40–41; Vogel, 1994, Chapters 5 and 15.
Flow in blood vessels: Hoppensteadt & Peskin, 2002 .
Microfluidics with biotechnology applications: Austin, 2002.

Technical:
Bacterial flagellar propulsion: Berg & Anderson, 1973; Silverman & Simon, 1974.
Other bacterial strategies: Berg & Purcell, 1977.
Low Reynolds-number fluid mechanics: Happel & Brenner, 1983
Vascular flows: Fung, 1997.

$\boxed{T_2}$ **5.2.1′ Track 2**

1. Section 4.1.4′ on page 147, point (2), pointed out that our simple theory of frictional drag would break down when the force applied to a particle was too great. We have now found a precise criterion: The inertial (memory) effects neglected in Section 4.1.4 will indeed be significant for forces greater than f_{crit}.

2. The phenomenon of viscosity actually reflects yet another diffusion process. When we have small indestructible particles, so that the number of particles is conserved, we found that random thermal motion leads to diffusive transport of particle number via Fick's law, Equation 4.19 on page 130. Section 4.6.4 on page 142 extended this idea, showing that when particles carry electric charge (another conserved quantity), their thermal motion again leads to a diffusive transport of charge (Ohm's law). Finally, because particles carry energy, yet another conserved quantity, Section 4.4.2′ on page 149 argued for a third Fick-type transport rule, called thermal conduction. Each transport rule had its own diffusion constant, giving rise to the electrical and thermal conductivity of materials.

   One more conserved quantity from first-year physics is the momentum **p**. Random thermal motion should also give a Fick-type transport rule for each component of **p**.

   Figure 5.2b on page 163 shows two flat plates, each parallel to the yz-plane, separated by d in the x direction. Let ρ_{p_z} denote the density of the z-component of momentum. If the top plate is dragged at v_z in the $+z$ direction while the bottom is held stationary, we get a nonuniform ρ_{p_z}, namely, $\rho_m \times v_z(x)$, where ρ_m denotes the mass density of fluid. We expect that this nonuniformity should give rise to a *flux* of p_z whose component in the x direction is given by a formula analogous to Fick's law (Equation 4.19 on page 130):

$$(j_{p_z})_x = -\nu \rho_m \frac{dv_z}{dx}. \qquad \text{(planar geometry)} \qquad (5.21)$$

   The constant ν is a new diffusion constant, called the **kinematic viscosity**. (Check its units.)

   But the rate of loss of momentum is just a *force*; similarly, the *flux* of momentum is a force per unit area. The flux of momentum (Equation 5.21) leaving the top plate exerts a resisting drag force opposing the motion; when this momentum arrives at the bottom plate, it exerts an entraining force along v_z. We have thus found the molecular origin of viscous drag. It's appropriate to name ν a kind of viscosity, because it's related in a simple way to η: Comparing Equation 5.4 to Equation 5.21 shows that $\nu = \eta / \rho_m$.

3. We now have two empirical definitions of viscosity, namely, the Stokes formula (Equation 4.14 on page 119) and our parallel-plates formula (Equation 5.4 on page 164). They look similar, but some work is required to prove that they are equivalent. One must write down the equations of motion for a fluid, containing the parameter η, solve them in both the parallel-plate and moving-sphere geometries, and compute the forces in each case. (The math can be found in Landau & Lifshitz, 1987 or Batchelor, 1967, for example.) But the *form* of the Stokes formula just follows from dimensional analysis. Once we know we're in the low-force

regime, we also know that the mass density ρ_m of the fluid cannot enter into the drag force (because inertial effects are insignificant). For an isolated sphere, the only length scale in the problem is its radius R, so the only way to get the proper dimensions for a viscous friction coefficient is to multiply the viscosity by R to the first power. That's what the Stokes formula says, apart from the dimensionless prefactor 6π.

 5.2.2′ Track 2

1. The physical discussion in Section 5.2.2 may have given the impression that the Reynolds-number criterion is not very precise—$\mathcal{R}$ itself looks like the ratio of two rough estimates! A more mathematical treatment begins with the equation of incompressible, viscous fluid motion (the Navier–Stokes equation). This equation is essentially a more general form of Newton's Law than the version used in Equation 5.7.

    Suppose fluid flows through a geometry with a length scale R (for example, the radius of a pipe). Some external agency keeps the fluid moving at overall speed v. Expressing the fluid's velocity field $\mathbf{u}(\mathbf{r})$ in terms of the dimensionless ratio $\bar{\mathbf{u}} \equiv \mathbf{u}/v$, and the position $\mathbf{r}$ in terms of $\bar{\mathbf{r}} \equiv \mathbf{r}/R$, one finds that $\bar{\mathbf{u}}(\bar{\mathbf{r}})$ obeys a set of dimensionless equations and boundary conditions. In these equations the parameters $\rho_m, \eta, v,$ and R enter in only one place, via the dimensionless combination $\mathcal{R}$ (Equation 5.11). Two different flow problems of the same geometrical type, with the same value of $\mathcal{R}$, will therefore be *exactly the same* when expressed in dimensionless form, even if the separate values of the four parameters may differ widely! (See for example Landau & Lifshitz, 1987, §19.) This **hydrodynamic scaling invariance** of fluid mechanics is what lets engineers test submarine designs by building scaled-down models and putting them in bathtubs.

2. Section 5.2.2 quietly shifted from a discussion of flow around an obstruction to Reynolds's results on *pipe* flow. It's important to remember that the critical Reynolds number in any given situation is always *roughly* 1, but this estimate is only accurate to within a couple of orders of magnitude. The actual value in any specified situation depends on the geometry, ranging from about 3 (for exit from a circular hole) to 1000 (for pipe flow, where $\mathcal{R}$ is computed by using the pipe radius).

 5.2.3′ Track 2

1. Section 5.2.3 claimed that the equation of motion for a purely elastic solid has no dissipation. Indeed, a tuning fork vibrates a long time before its energy is gone. Mathematically, if we shake the top plate in Figure 5.2b back and forth, $\Delta z(t) = L \cos(\omega t)$, then Equation 5.14 on page 172 says that for an elastic solid the rate at which we must do work is $f v = (\mathcal{G}A)(L \cos(\omega t)/d)(\omega L \sin(\omega t))$, which is negative just as often as it's positive: All the work we put in on one half-cycle gets returned to us on the next one. In a fluid, however, multiplying the viscous force

by v gives $f v = (\eta A)(L\omega \sin(\omega t)/d)(\omega L \sin(\omega t))$, which is never negative. We're always doing work, which gets converted irreversibly to thermal energy.

2. There's no reason why a substance can't display *both* elastic and viscous response. For example, when we shear a polymer solution there's a transient period when its individual polymer chains are starting to stretch. If the applied force is released during this period, the stretched chains can partially restore the original shape of a blob. Such a substance is called **viscoelastic**. Its restoring force is generally a complicated function of the frequency ω, not simply a constant (as in a solid) nor linear in ω (as in a Newtonian fluid). The viscoelastic properties of human blood, for example, are important in physiology.

3. It's not necessary to apply the exact time-reversed force in order to return to the starting configuration. That's because the left side of Equation 5.13 is more special than simply changing sign under time reversal: It's *first order* in time derivatives. More generally, the viscous force rule (Equation 5.4 on page 164) also has this property. Applying a time-dependent force to a particle in fluid then gives a total displacement $\Delta \mathbf{x}(t) = \zeta^{-1} \int_0^t \mathbf{f}(t') \, dt'$. Suppose we apply some force $f(t)$, thereby moving the particle and all the surrounding fluid. We could bring the particle, and every other fluid element in the sample, back to their original positions by any force whose integral is equal and opposite to the original one. It doesn't matter whether the return stroke is hard and short, or gentle and long, as long as we stay in the low Reynolds-number regime.

T_2 ### 5.3.1′ Track 2

The ratio of parallel to perpendicular drag is not a universal number; instead, it depends on the length of the rod relative to its diameter (the "aspect ratio"). The illustrative value $\frac{2}{3}$ quoted in Section 5.3.1 is appropriate for a rod 20 times as long as its diameter. In the limit of an infinitely long rod, the ratio falls to $\frac{1}{2}$. (The calculations can be found in Happel & Brenner, 1983, §§5–11.)

PROBLEMS

5.1   *Friction versus dissipation*

Gilbert says: You say that friction and dissipation are two manifestations of the same thing. So high viscosity must be a very dissipative situation. Then why do I get beautifully ordered, laminar motion only in the *high*-viscosity case? Why does my ink blob miraculously reassemble itself only in this case?

Sullivan: Um, uh . . .

Help Sullivan out.

5.2   *Density profile*
Finish the derivation of particle density in an equilibrium colloidal suspension (begun in Section 5.1.1) by finding the constant prefactor in Equation 5.1. That is, find a formula for the equilibrium number density $c(x)$ of particles with net weight $m_{net}g$ as a function of the height x. The total number of particles is N; the height of the test tube is h and its cross sectional area is A.

5.3   *Archibald method*
Sedimentation is a key analytical tool in the lab for the study of big molecules. Consider a particle of mass m and volume V in a fluid of mass density ρ_m and viscosity η.

a. Suppose a test tube is spun in the plane of a wheel, pointing along one of the "spokes." The artificial gravity field in the centrifuge is not uniform; rather, it is stronger at one end of the tube than the other. Hence the sedimentation rate will not be uniform either. Suppose that one end lies a distance r_1 from the center, and the other end is at $r_2 = r_1 + \ell$. The centrifuge is spun at angular frequency ω. Adapt the formula $v_{drift} = gs$ (Equation 5.3 on page 160) to find an analogous formula for the drift speed in terms of s in the centrifuge case.

Eventually, sedimentation will stop and an equilibrium profile will emerge. It may take quite a long time for the whole test tube to reach its equilibrium distribution. In that case, Equation 5.2 on page 160 is not the most convenient way to measure the mass parameter m_{net}. The **Archibald method** uses the fact that the *ends* of the test tube equilibrate rapidly, as follows.

b. There can be no flux of material through the ends of the tube. Thus, the Fick-law flux must cancel the flux you found in (a). Write down two equations expressing this statement at the two ends of the tube.

c. Derive the following expression for the mass parameter in terms of the concentration and its gradient at one end of the tube:

$$m_{net} = (\text{stuff}) \times \left.\frac{dc}{dr}\right|_{r=r_1},$$

and a similar formula for the other end, where (stuff) is some factors that you are to find. The concentration and its gradient can be measured photometrically in the lab, thus allowing a measurement of m_{net} long before the whole test tube has come to equilibrium.

5.4   *Coasting at low Reynolds*

The chapter asserted that tiny objects stop moving at once when we stop pushing them. Let's see.

a. Consider a bacterium, idealized as a sphere of radius $1\,\mu m$, propelling itself at $1\,\mu m\,s^{-1}$. At time zero, the bacterium suddenly stops swimming and coasts to a stop, following Newton's Law of motion with the Stokes drag force. How far does it travel before it stops? Comment.

b. Our discussion of Brownian motion assumed that each random step was independent of the previous one; thus, for example, we neglected the possibility of a residual drift speed left over from the previous step. In the light of (a), would you say that this assumption is justified for a bacterium?

5.5   *Blood flow*

Your heart pumps blood into your aorta. The maximum flow rate into the aorta is about $500\,cm^3\,s^{-1}$. Assume that the aorta has diameter 2.5 cm, that the flow is laminar (not very accurate), and that blood is a Newtonian fluid with viscosity roughly equal to that of water.

a. Find the pressure drop per unit length along the aorta. Express your answer in SI units. Compare the pressure drop along a 10 cm section of aorta with atmospheric pressure ($10^5\,Pa$).

b. How much power does the heart expend just pushing blood along a 10 cm section of aorta? Compare your answer with your basal metabolic rate, about 100 W, and comment.

c. The fluid velocity in laminar pipe flow is zero at the walls of the pipe and maximum at the center. Sketch the velocity as a function of distance r from the center. Find the velocity at the center. [*Hint:* The total volume flow rate, which you are given, equals $\int v(r)2\pi r dr.$]

5.6 $\boxed{T_2}$ *Kinematic viscosity*

a. Although the kinematic viscosity ν has the same dimensions $\mathbb{L}^2/\mathbb{T}$ as any other diffusion constant, its physical meaning is quite different from that of D, and its numerical value for water is quite different from the value of D for self-diffusion of water molecules. Find the value of ν from η and compare with D.

b. Still, these values are related. Show, by combining Einstein's relation and the Stokes formula, that taking the radius R of a water molecule to be about 0.2 nm leads to a satisfactory order-of-magnitude prediction of ν from D, R, and the mass density of water.

5.7 $\boxed{T_2}$ *No going back*

Section 5.2.3 argued that the motion of a gently sheared, flat layer would retrace its history if we reverse the applied force. When the force is large, so that we cannot ignore the inertial term in Newton's Law of motion, where exactly does the argument fail?

5.8 $\boxed{T_2}$ *Intrinsic viscosity of a polymer in solution*

Section 4.3.2 argued that a long polymer chain in solution would be found in a random-walk conformation at any instant of time.[4] This claim is not easy to verify directly, so let's approach the question indirectly, by examining the *viscosity* of a polymer solution.

Figure 5.2b on page 163 shows two parallel plates separated by distance d, with the space filled with water of viscosity η. If one plate slides sideways at speed v, then both plates feel viscous force $\eta v/d$ per unit area. Suppose now that a small fraction ϕ of the volume between plates is filled with *solid objects*, taking up space previously taken by water. Then, at speed v, the shear strain rate in the remaining fluid must be greater than before, and the viscous force will be greater, too.

a. To estimate the shear strain rate, imagine that all the rigid objects are lying in a solid layer of thickness ϕd attached to the bottom plane, effectively reducing the gap between the plates. Then what is the viscous force per area?

b. We can express the result by saying that the suspension has an "effective viscosity" η' bigger than η. (Your result for the speed of milk separation in Your Turn 5C on page 161 was actually a bit too high, in part because of this effect.) Write an expression[5] for the relative change $(\eta' - \eta)/\eta$. Use $\phi \ll 1$ to simplify your answer.

c. We want to explore the proposition that a polymer N segments long behaves like a sphere with radius $\alpha L N^p$ for some power p. (Here L is the segment length and α is a constant of proportionality; we won't need the exact values of these parameters.) What do we expect p to be? What then is the volume fraction ϕ of a suspension of c such spheres per volume? Express your answer in terms of the total mass M of a polymer, the mass m per monomer, the concentration of polymer c, L, and α.

d. Discuss the experimental data in Figure 5.13 in the light of your analysis. Each set of points joined by a line represents measurements taken on a family of polymers with various numbers N of identical monomers; each monomer has the same mass m. The total mass $M = Nm$ of each polymer is on the x-axis. The quantity $[\eta]_\Theta$ on the vertical axis is called the polymer's intrinsic viscosity; it is defined as $(\eta' - \eta)/(\eta \rho_{m,p})$, where $\rho_{m,p}$ is the mass of dissolved polymer per volume of solvent. [*Hint:* Recall $\rho_{m,p}$ is small. Write everything in terms of the fixed segment length L, the fixed monomer mass m, and the variable total mass M.]

e. What combination of L and m could we measure from the data? (Don't actually calculate it.)

5.9 $\boxed{T_2}$ *Friction as diffusion*

Section 5.2.1' on page 187 claimed that viscous friction can be interpreted as the diffusive transport of momentum. The argument was that, in the planar geometry, when the flux of momentum given by Equation 5.21 leaves the top plate, it exerts a resisting drag force. When this momentum arrives at the bottom plate, it exerts an entraining force. So far, the argument is quite correct.

---

[4]This problem concerns a polymer under "theta conditions" (see Section 4.3.1' on page 148).
[5]The expression you'll get is not quite complete, because of some effects we left out, but its scaling is right when ϕ is small. Einstein obtained the full formula in his doctoral dissertation. (Then he fixed a computational error six years later!)

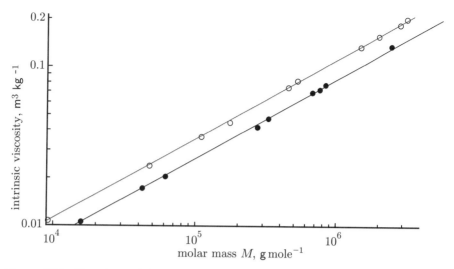

Figure 5.13: (Experimental data.) Log-log plot of the intrinsic viscosity $[\eta]_\Theta$ for polymers with different values for the molar mass M. The two data sets shown represent different combinations of polymer type, solvent type, and temperature, both corresponding to "theta solvent" conditions. *Open circles:* Polyisobutylene in benzene at 24°C. *Solid circles:* Polystyrene in cyclohexane at 34°C. The two *lines* each have logarithmic slope $\frac{1}{2}$. [Data from Flory, 1953.]

Viscous friction is more complicated than ordinary diffusion, however, because momentum is a vector quantity, whereas number density is a scalar. For example, Section 5.2.2 noted that the viscous force law (Equation 5.9 on page 168) needs to be modified for situations other than planar geometry. The required modification really matters if we want to get the correct answer for the spinning-rod problem (Figure 5.11b on page 180).

We consider a long cylinder of radius R with its axis along the $\hat{z}$ direction and centered at $x = y = 0$. Some substance surrounds the cylinder. First suppose that this substance is *solid ice*. When we crank the cylinder, everything rotates as a rigid object with some angular frequency ω. The velocity at position $\mathbf{r}$ is then $\mathbf{v}(\mathbf{r}) = (-\omega y, +\omega x, 0)$. Certainly nothing is rubbing against anything, and there should be no dissipative friction—the frictional transport of momentum had better be zero. And yet, if we examine the point $\mathbf{r}_0 = (r_0, 0, z)$, we find a nonzero gradient

$$\left.\frac{dv_y}{dx}\right|_{\mathbf{r}=\mathbf{r}_0} = \omega.$$

Evidently, our formula for the flux of momentum in planar geometry (Equation 5.21 on page 187) needs some modification for the nonplanar case.

We want a modified form of Equation 5.21 that applies to cylindrically symmetrical flows and vanishes when the flow is rigid rotation. Letting $r \equiv \|\mathbf{r}\| = \sqrt{x^2 + y^2}$, we can write a cylindrically symmetrical flow as

$$\mathbf{v}(\mathbf{r}) = \big(-yg(r), xg(r), 0\big).$$

The case of rigid rotation corresponds to the choice of a constant angular velocity $g(r)$. You are about to find $g(r)$ for a different situation, namely, fluid flow. We can think of this flow as a set of nested cylinders, each with a *different* value of $g(r)$.

Near any point, say, $\mathbf{r}_0$, let $\mathbf{u}(\mathbf{r}) = (-yg(r_0), xg(r_0))$ be the rigidly rotating vector field that agrees with $\mathbf{v}(\mathbf{r})$ at $\mathbf{r}_0$. We then replace Equation 5.21 by

$$(j_{p_y})_x(\mathbf{r}_0) = -\eta \left(\left. \frac{dv_y}{dx} \right|_{r=r_0} - \left. \frac{du_y}{dx} \right|_{r=r_0} \right). \qquad \text{(cylindrical geometry)} \qquad (5.22)$$

In this formula, $\eta \equiv \nu\rho_m$, the ordinary viscosity. Equation 5.22 is the proposed modification of the momentum-transport rule. It says that we compute dv_y/dx and *subtract off* the corresponding quantity with $\mathbf{u}$, to ensure that rigid rotation incurs no frictional resistance.

a. Each cylindrical shell of fluid exerts a torque on the next one and feels a torque from the previous one. These torques must balance. Show that, as a result, the tangential force per area across the surface at fixed r is $(\tau/L)/(2\pi r^2)$, where τ is the external torque on the central cylinder and L is the cylinder's length.

b. Set your result from (a) equal to Equation 5.22 and solve for the function $g(r)$.

c. Find τ/L as a constant times ω. Hence, find the constant C in Equation 5.19 on page 183.

5.10 $\boxed{T_2}$ *Pause and tumble*

In between straight-line runs, *E. coli* pauses. If it just turned off its flagellar motors during the pauses, eventually the bacterium would find itself pointing in a new, randomly chosen direction, as a result of rotational Brownian motion.

If you haven't done Problem 4.9, do it now and compare your answer to part (d) with the measured pause time of 0.14 s. Do you think the bacterium just shuts down its flagellar motors and waits during the pauses? Explain your reasoning.

CHAPTER 6

Entropy, Temperature, and Free Energy

> *The method of "postulating" what we want has many advantages; they are the same as the advantages of theft over honest toil.*
> — Bertrand Russell, 1919

It's time to come to grips with the still rather woolly ideas proposed in Chapter 1 and turn them into precise equations. We can do it, starting from the statistical ideas developed in our study of the ideal gas law and Brownian motion.

Chapter 4 argued that friction in a fluid is the loss of memory of an object's initial, ordered motion. The object's organized kinetic energy passes into the disorganized kinetic energy of the surrounding fluid. The world loses some order as the object merges into the surrounding distribution of velocities. The object doesn't stop moving, nor does its velocity stop changing (it changes with every molecular collision). What stops changing is the *probability distribution* of the particle's many velocities over time.

Actually, friction is just one of several dissipative processes relevant to living cells that we've encountered: All obey similar Fick-type laws and all tend to *erase order.* We need to bring them all into a common framework, the Second Law of thermodynamics introduced in Section 1.2.1. As the name implies, the Second Law has a universality that goes far beyond the concrete situations we've studied so far; it's a powerful way of organizing our understanding of many different things.

To make the formulas as simple as possible, we'll continue to study ideal gases for a while. This may seem like a detour, but the lessons we draw will be applicable to all sorts of systems. For example, the Mass Action rule governing many chemical reactions will turn out to be based on the same physics underlying the ideal gas (Chapter 8). Moreover, Chapter 7 will show that the ideal gas law itself is literally applicable to a situation of direct biological significance, namely, osmotic pressure.

The goal of this chapter is to state the Second Law and, with it, the crucial concept of free energy. The discussion here is far from the whole story. Even so, this chapter will be bristling with formulas. So it's especially important to *work through* this chapter instead of just reading it.

The Focus Question for this chapter is

Biological question: If energy is always conserved, how can some devices be more efficient than others?

Physical idea: Order controls when energy can do useful work, and it's *not* conserved.

6.1 HOW TO MEASURE DISORDER

Chapter 1 was a little vague about the precise meaning of disorder. We need to refine our ideas before they become sharp tools.

Flip a coin a thousand times. You get a random sequence HTTTHTTTHTHHHHTHH . . . . We will say that this sequence contains lots of **disorder**, in the following sense: It's impossible to summarize a random sequence. If you want to store it on your computer, you need 1000 bits of hard disk space. You can't compress it; every bit is independent of every other.

Now let's consider the weather, rain/shine. You can take a thousand days of weather and write it as a bit stream RSSSRSSSSRRRSRR . . . . But this stream is *less disordered* than the coin-flip sequence, because today's weather is more likely to be like yesterday's than different. We could change our coding and let 0 = same as yesterday, 1 = different from yesterday. Then our bit stream is 10011000100110 . . . , and it's not perfectly unpredictable: It has more 0's than 1's. We could compress it by reporting instead the length of each run of similar weather.

Here is another point of view: You could make money betting even odds on the weather every day, because you have some a priori knowledge about this sequence. You won't make money betting even odds on a coin flip, because you have no such prior knowledge. The extra knowledge you have about the weather means that any actual string of weather reports is less disordered than a corresponding string of coin flips. Again, *the disorder in a sequence reflects its predictability.* High predictability is low disorder.

We still need to propose a quantitative measure of disorder. In particular, we'd like our measure to have the property that the total amount of disorder in two uncorrelated streams is just the sum of that in each stream separately. It's crucial to have the word *uncorrelated* in the preceding sentence. If you flip a penny a thousand times, and flip a dime a thousand times, those are two uncorrelated streams. If you watch the news and read the newspaper, those are two correlated streams; one can be used to predict the other, so the total disorder is less than the sum of those for the two streams.

Suppose that we have a very long stream of events (for example, coin flips) and that each event is drawn randomly, independently, and with equal probability from a list of M possibilities (for example, $M = 2$ for a coin; or $M = 6$ for rolling a die). We divide our long stream into "messages" consisting of N events. We are going to explore the proposal that a good measure for the amount of disorder per message is $I \equiv N \log_2 M$, or equivalently $KN \ln M$, where $K = 1/\ln 2$.

It's tempting to glaze over at the sight of that logarithm, regarding it as just a button on your calculator. But there's a simple and much better way to see what the formula means: Taking the case $M = 2$ (coin flip) shows that, in this special case, I is just the number of tosses. More generally, we can regard I as the number of binary digits, or **bits**, needed to transmit the message. That is, I is the number of digits needed to express the message as a big binary number.

Our proposal has the trivial property that $2N$ coin tosses give a message with twice as much disorder as N tosses. What's more, suppose that we toss a coin *and* roll a die N times. Then $M = 2 \times 6 = 12$ and $I = KN \ln 12 = KN(\ln 2 + \ln 6)$, by the property of logarithms. That makes sense: We could have reorganized each message as N coin flips followed by N rolls, and we wanted our measure of disorder to be additive. This is why the logarithm function enters our formula. Letting $\Omega = M^N$ be the total number of all possible N-event messages and again $K = 1/\ln 2$, we can rewrite the proposed formula as

$$I = K \ln \Omega. \tag{6.1}$$

We also want to measure disorder in other kinds of event streams. Suppose we have a message N letters long in an alphabet with M letters (let's say, $M = 31$, Russian), and we know in advance that the letter frequency *isn't* uniform: There are N_1 letters "А," N_2 letters "Б," and so on. That is, the *composition* of our stream of symbols is specified, although its *sequence* is not. The probability of getting each letter is then $P_i = N_i/N$, and the P_i aren't necessarily all equal to $\frac{1}{M}$.

The total number of all possible messages is then

$$\Omega = \frac{N!}{N_1! N_2! \cdots N_M!}. \tag{6.2}$$

To justify this formula we extend the logic of the random walk Example (page 112). There are N factorial (written $N!$) ways to take N objects and arrange them into a sequence. But swapping any of the A's among themselves doesn't change the message, so $N!$ overcounts the possible messages: We need to divide by $N_1!$ to eliminate this redundancy. Arguing similarly for the other letters in the message gives Equation 6.2. (It's always best to test theory with experiment, so try it with two apples, a peach, and a pear ($M = 3$, $N = 4$).)

If all we know about the message are the letter frequencies, then any of the Ω possible messages is equally likely. Let's apply the proposed disorder formula (Equation 6.1) to the entire message:

$$I = K \left[\ln N! - \sum_{j=1}^{M} \ln N_j! \right].$$

If the message is very long, we can simplify the preceding expression using Stirling's formula (Equation 4.2 on page 113). For very large N, we only need to keep the terms in Stirling's formula that are proportional to N, namely, $\ln N! \approx N \ln N - N$. Thus the amount of disorder per letter is $\frac{I}{N} = -K \sum_j \frac{N_j}{N} \ln \frac{N_j}{N}$, or

$$\boxed{\frac{I}{N} = -K \sum_{j=1}^{M} P_j \ln P_j. \qquad \text{Shannon's formula}} \tag{6.3}$$

Actually, not every string of letters makes sense in real Russian, even if it has the correct letter frequencies. If we have the extra knowledge that the string consists of real text, then we could take N to be the number of *words* in the message, M to be the number of words listed in the dictionary, P_j the frequencies of usage of each word, and again use Equation 6.3 to get a revised (and smaller) estimate of the amount of disorder of a message in this more restricted class. That is, real text is more predictable, so it carries even less disorder per letter than do random strings with the letter frequencies of real text.

Shannon's formula has some sensible features. First, notice that I is always positive because the logarithm of a number smaller than 1 is always negative. If every letter is equally probable, $P_j = 1/M$, then Equation 6.3 just reproduces our original proposal, $I = KN \ln M$. If, on the other hand, we know that every letter is an "A," then $P_1 = 1$, all the other $P_j = 0$ and we find $I = 0$: A string of all "A's" is perfectly predictable and has zero disorder. Because Equation 6.3 makes sense and came from Equation 6.1, we'll accept the latter as a good measure of disorder.

Shannon's formula also has the reasonable property that the disorder of a random message is maximum when every letter is equally probable. Let's prove this important fact. We maximize I over the P_j, subject to the constraint that they must all add up to 1 (the normalization condition, Equation 3.2 on page 71). To implement this constraint, we replace P_1 by $1 - \sum_{j=2}^{M} P_j$ and maximize over all the remaining P_j's:

$$-\frac{I}{NK} = \left[P_1 \ln P_1 \right] + \left[\sum_{j=2}^{M} P_j \ln P_j \right]$$

$$= \left[\left(1 - \sum_{j=2}^{M} P_j \right) \ln \left(1 - \sum_{j=2}^{M} P_j \right) \right] + \left[\sum_{j=2}^{M} P_j \ln P_j \right].$$

Let's focus on one particular letter, j_0, and set the derivative with respect to P_{j_0} equal to zero. Using the fact that $\frac{d}{dx}(x \ln x) = (\ln x) + 1$ gives

$$0 = \frac{d}{dP_{j_0}} \left(-\frac{I}{NK} \right) = \left[-\ln \left(1 - \sum_{j=2}^{M} P_j \right) - 1 \right] + \left[\ln P_{j_0} + 1 \right].$$

Exponentiating this formula gives

$$P_{j_0} = 1 - \sum_{j=2}^{M} P_j.$$

The right-hand side is always equal to P_1, so all the P_j's are equal. Thus the disorder is maximal when every letter is equally probable—and then it's given by $NK \ln M$.

$\boxed{T_2}$ *Section 6.1' on page 232 shows how to obtain the last result by using the method of Lagrange multipliers.*

6.2 ENTROPY

6.2.1 The Statistical Postulate

What has any of this got to do with physics *or* biology? It's time to start thinking, not of abstract strings of data, but of the string formed by repeatedly examining the detailed state (or **microstate**) of a physical system. For example, in an ideal gas, the microstate consists of the position and speed of every molecule in the system. Such a measurement is impossible in practice. But imagining what we'd get if we *could* do it will lead us to the entropy, which *can* be defined experimentally.

We'll define the disorder of the physical system as the disorder per observation of the stream of successively measured microstates.

Suppose that we have a box of volume V, about which we know absolutely nothing except that it is isolated and contains N ideal gas molecules with total energy E. *Isolated* means that the box is thermally insulated and closed; no heat, light, or particles enter or leave it, and it does no work on its surroundings. Thus the box will *always* have N molecules and energy E. What can we say about the precise states of the molecules in the box, for example, their individual velocities? Of course, the answer is, "Not much": The microstate changes at a dizzying rate (with every molecular collision). We can't say a priori that any one microstate is more probable than any other.

Accordingly, this chapter will begin to explore the idea that after an isolated system has had a chance to come to equilibrium, the actual sequence of microstates we'd measure (if we could measure microstates) would be effectively a random sequence, with each allowed microstate being equally probable. Restating this in the language of Section 6.1 gives the **Statistical Postulate**:

> When an isolated system is left alone long enough, it evolves to thermal equilibrium. Equilibrium is not one particular microstate. Rather, it's that probability distribution of microstates having the greatest possible disorder allowed by the physical constraints on the system. (6.4)

The constraints just mentioned include the facts that the total energy is fixed and the system is confined to a box of fixed size.

To say it a third time, equilibrium corresponds to the probability distribution expressing greatest *ignorance* of the microstate, given the constraints. Even if initially we had some additional knowledge that the system was in a special class of states (for example, that all molecules were in the left half of a box of gas), eventually the complex molecular motions wipe out this knowledge (the gas expands to fill the box). We then know nothing about the system except what is enforced by the physical constraints.

In some very special systems, it's possible to prove Idea 6.4 mathematically instead of taking it as a postulate. We won't attempt this. Indeed, it's not even always true. For example, the Moon in its orbit around the Earth is constantly changing its velocity, but in a predictable way. There's no need for any probability distribution, and no disorder. Nevertheless, the Postulate is a reasonable proposal for a large, complex system, and it does have experimentally testable consequences; we will find that it applies to a wide range of phenomena relevant for life processes. The key to its success is that even when we want to study a single molecule (a small system with relatively few moving parts), in a cell that molecule will inevitably be surrounded by a thermal *environment* consisting of a huge number of other molecules in thermal motion.

The Great Pyramid at Giza is not in thermal equilibrium: Its gravitational potential energy could be reduced considerably, with a corresponding increase in kinetic energy and hence disorder, if it were to disintegrate into a low pile of gravel. This hasn't happened yet. So the phrase *long enough* in the Statistical Postulate must be treated respectfully. There may even be intermediate time scales where some variables are in thermal equilibrium (for example, the temperature throughout the Pyramid is uniform) while others are not (it hasn't yet flowed into a low pile of sand).

Actually, the Pyramid isn't even at uniform temperature: Every day, the surface heats up, but the core remains at constant temperature. Still, every cubic millimeter *is* of quite uniform temperature. So the question of whether we may apply equilibrium arguments to a system depends both on time and on size *scales*. To find how long a given length scale takes to equilibrate, we use the appropriate diffusion equation—in this case, the law of thermal conduction.

$\boxed{T_2}$ *Section 6.2.1' on page 232 discusses the foundations of the Statistical Postulate in more detail.*

6.2.2 Entropy is a constant times the maximal value of disorder

Let's continue to study an isolated statistical system. (Later, we'll get a more general formulation that can be applied to everyday systems, which are not isolated.) We'll denote the *number* of allowed states of N molecules with energy E by $\Omega(E, N, \dots)$, where the dots represent any other fixed constraints, such as the system's volume. According to the Statistical Postulate, in equilibrium a sequence of observations of the system's microstate will show that each one is equally probable; thus Equation 6.1 gives the system's disorder in equilibrium as $I(E, N, \dots) = K \ln \Omega(E, N, \dots)$ bits. As usual, $K = 1/\ln 2$.

Now certainly Ω is very big for a mole of gas at room temperature. It's huge because molecules are so numerous. We can work with less mind-boggling quantities if we multiply the disorder per observation by a tiny constant, like the thermal energy of a single molecule. More precisely, the traditional choice for the constant is k_B/K, which yields a measure of disorder called the **entropy**, denoted S:

$$S \equiv \frac{k_B}{K} I = k_B \ln \Omega. \tag{6.5}$$

Before saying another word about these abstractions, let's pause to evaluate the entropy explicitly, for a system we know intimately.

Example: Find the entropy for an ideal gas.

Solution: We want to count all states allowed by the conservation of energy. We express the energy in terms of the momentum of each particle:

$$E = \sum_{i=1}^{N} \frac{m}{2} \mathbf{v}_i^2 = \frac{1}{2m} \sum_{i=1}^{N} \mathbf{p}_i^2 = \frac{1}{2m} \sum_{i=1}^{N} \sum_{J=1}^{3} \left(p_{i,J} \right)^2.$$

Here $p_{i,J}$ is the component of particle i's momentum along the J-axis, and m is its mass. This formula resembles the Pythagorean formula: In fact, for $N = 1$, it says precisely that $\sqrt{2mE}$ is the distance of the point $\mathbf{p}$ from the origin; or in other words, that the allowed momentum vectors lie on the surface of a sphere (recall Figure 3.4 on page 78).

When there are lots of molecules, the locus of allowed values of $\{p_{i,J}\}$ is the surface of a sphere of radius $r = \sqrt{2mE}$ in $3N$-dimensional space. The number of states available for N molecules in volume V at fixed E is then proportional to the surface area of this hypersphere. Certainly that area must be proportional to the radius raised to the power $3N - 1$ (think about the case of an ordinary sphere, $N = 1$, whose surface area is $4\pi r^2 = 4\pi r^{3N-1}$). Because N is much larger than 1, we can replace $3N - 1$ by just $3N$.

To specify the microstate, we must give not only the momenta but also the locations of each particle. Because each may be located anywhere in the box, the number of available states must also contain a factor of V^N. So Ω is a constant times $(2mE)^{3N/2} V^N$, and $S = Nk_B \ln[(E)^{3/2}V] + $ const.

The complete version of the last result is called the **Sakur–Tetrode formula**:

$$S = k_B \ln \left[\left(\frac{2\pi^{3N/2}}{(3N/2 - 1)!} \right) (2mE)^{3N/2} V^N \frac{1}{N!} (2\pi\hbar)^{-3N} \frac{1}{2} \right]. \tag{6.6}$$

This is a complex formula, but we can understand it by considering each of the factors in turn. The first factor in round parentheses is the area of a sphere of radius 1, in $3N$ dimensions. It can be regarded as a fact from geometry, to be looked up in a book (or see Section 6.2.2′ on page 233). Certainly it equals 2π when $3N = 2$, and that's the right answer: The circumference of a unit circle in the plane really is 2π. The next two factors are what we just found in the Example. The factor of $(N!)^{-1}$ reflects the fact that gas molecules are indistinguishable; if we exchange $\mathbf{r}_1, \mathbf{p}_1$ with $\mathbf{r}_2, \mathbf{p}_2$, we get a different list of $\mathbf{r}_i$'s and $\mathbf{p}_i$'s, but not a physically different state of the system. $\hbar$ is the Planck constant, a constant of Nature with the dimensions $\mathbb{M}\mathbb{L}^2\mathbb{T}^{-1}$. Its origin lies in quantum mechanics; but for our purposes, it's enough to note that *some* constant with these dimensions is needed to make the dimensions in Equation 6.6 work out properly. The actual value of $\hbar$ won't enter any of our physical predictions.

Equation 6.6 looks scary, but many of the factors enter in nonessential ways. For instance, the first 2 in the numerator gets overwhelmed by the other factors when N is big, so we can drop it, or equivalently put in an extra factor of $\frac{1}{2}$, as was done at the end of Equation 6.6. Other factors like $(m/(2\pi^2\hbar^2))^{3N/2}$ just add a constant to the entropy per molecule and won't affect derivatives like dS/dE. (Later on, however, when studying the chemical potential in Chapter 8, we will need to look at these factors again.)

$\boxed{T_2}$ *Section 6.2.2′ on page 233 makes several more comments about the Sakur–Tetrode formula and derives the formula for the area of a higher-dimensional sphere.*

6.3 TEMPERATURE

6.3.1 Heat flows to maximize disorder

Having constructed the rather abstract notion of entropy, it's time to see some concrete consequences of the Statistical Postulate. We begin with the humblest of everyday phenomena, the flow of thermal energy from a hot object to a cool one.

Thus, instead of studying one isolated box of gas, we now imagine connecting *two* such boxes, called A and B, in such a way that they're still isolated from the world but can slowly exchange energy with each other (Figure 6.1). We could put two insulated boxes in contact and make a small hole in the insulation between them, leaving a wall that transmits energy but does not allow particles to cross. (You can imagine this wall as a drumhead, which can vibrate when a molecule hits it.) The two sides contain N_A and N_B molecules, respectively, and the total energy E_{tot} is fixed. Let's explore how the boxes share this energy.

To specify the total state of the combined system, choose any state of A and any state of B, with energies obeying $E_{tot} = E_A + E_B$. The interaction between the systems is assumed to be so weak that the presence of B doesn't significantly affect the allowed states of A, and vice versa.

E_A can go up, as long as E_B goes down to compensate. So E_B isn't free. After the boxes have been in contact a long time, we then shut the thermal door between them, thereby isolating them, and let each come separately to equilibrium. According to Equation 6.5, the total entropy of the combined system is then the sum of the two subsystems' entropies: $S_{tot}(E_A) = S_A(E_A) + S_B(E_{tot} - E_A)$. We can make this formula more explicit, because we have a formula for the entropy of an ideal gas, and we know

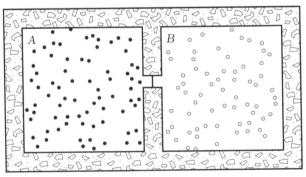

Figure 6.1: (Schematic.) Two systems thermally insulated from the world but only partially insulated from each other. The *hatching* denotes thermal insulation, with one small break on the common wall. The boxes don't exchange particles, only energy. The two subsystems may contain different kinds of molecules.

that energy is conserved. The Sakur–Tetrode formula (Equation 6.6) gives

$$S_{tot}(E_A) = k_B \left[N_A \left(\tfrac{3}{2} \ln E_A + \ln V_A \right) + N_B \left(\tfrac{3}{2} \ln(E_{tot} - E_A) + \ln V_B \right) \right] + \text{const.} \quad (6.7)$$

Equation 6.7 appears to involve logarithms of dimensional quantities, which aren't defined (see Section 1.4.1). Actually, formulas like this one are abbreviations: The term $\ln E_A$ can be thought of as short for $\ln(E_A/(1 \text{ J}))$. The choice of unit is immaterial; different choices just change the value of the constant in Equation 6.7, which wasn't specified anyway.

We can now ask, "What is the most likely value of E_A?" At first it may seem that the Statistical Postulate says that all values are equally probable. But wait. The Postulate says that just before we shut the door between the subsystems, all microstates of the joint system are equally probable. But there are many microstates of the joint system with any given value of E_A, and *the number depends on E_A itself*. In fact, exponentiating the entropy gives this number (see Equation 6.5). So, drawing a microstate of the joint system at random, we are most likely to come up with one whose E_A corresponds to the maximum of the total entropy. To find this maximum, set the derivative of Equation 6.7 to zero:

$$0 = \frac{dS_{tot}}{dE_A} = \frac{3}{2} k_B \left(\frac{N_A}{E_A} - \frac{N_B}{E_B} \right). \quad (6.8)$$

In other words, the systems are most likely to divide their thermal energy in such a way that each has the same average energy per molecule: $E_A/N_A = E_B/N_B$.

This is a very familiar conclusion. Section 3.2.1 argued that in an ideal gas, the average energy per molecule is $\tfrac{3}{2} k_B T$ (Idea 3.21 on page 80). So we have just concluded that *two boxes of gas in thermal equilibrium are most likely to divide their energy in a way that equalizes their temperature.* Successfully recovering this well-known fact of everyday life gives us some confidence that the Statistical Postulate is on the right track.

How likely is "most likely"? To simplify the math, suppose that $N_A = N_B$, so that equal temperature corresponds to $E_A = E_B = \tfrac{1}{2} E_{tot}$. Figure 6.2 shows the entropy maximum and the probability distribution $P(E_A)$ to find A with a given energy after we shut the door. The graph makes it clear that even for just a few thousand molecules on each side, the system is quite likely to be found very close to its equal-temperature point, because the peak in the probability distribution function is very narrow. That is, the observed statistical fluctuations about the most probable energy distribution will be small (see Section 4.4.3 on page 131). For a macroscopic system, where $N_A \approx N_B \approx 10^{23}$, the two subsystems will be *overwhelmingly* likely to share their energy in a way corresponding to *nearly exactly* equal temperatures.

6.3.2 Temperature is a statistical property of a system in equilibrium

The fact that two systems in thermal contact come to the same temperature is not limited to ideal gases! Indeed, the early thermodynamicists found this property of heat to be so significant that they named it the **Zeroth Law** of thermodynamics. Suppose that we put *any* two macroscopic objects into thermal contact. Their entropy

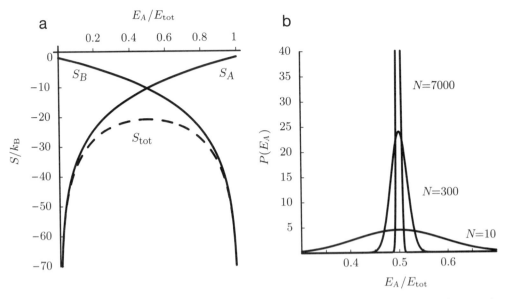

Figure 6.2: (Mathematical functions.) The disorder (entropy) of the joint system is maximal when the two subsystems share the total energy according to Equation 6.8. (a) Entropy of subsystems A (*rising curve*) and B (*descending curve*), as functions of E_A/E_{tot}. Each chamber has $N = 10$ molecules. A constant has been added and the entropy is expressed in units of k_B; thus the actual functions plotted are a constant plus $\ln(E_A)^{3N/2}$ and $\ln(E_{tot} - E_A)^{3N/2}$, respectively. The *dashed line* shows the sum of these curves (total system entropy plus a constant); it's maximal when the subsystems share the energy equally. (b) Probability distribution corresponding to the dashed curve in (a) (*low, wide curve*), and similar distributions with $N = 300$ and 7000 molecules on each side. Compare with the related behavior seen in Figure 4.3 on page 113.

functions won't be the simple one we found for an ideal gas. We *do* know, however, that the total entropy S_{tot} will have a big, sharp, maximum at one value of E_A, because it's the sum of a very rapidly increasing function of E_A (namely, $S_A(E_A)$) plus a very rapidly decreasing function[1] (namely, $S_B(E_{tot} - E_A)$), as shown in Figure 6.2. The maximum occurs when $dS_{tot}/dE_A = 0$.

The previous paragraph suggests that we *define* temperature abstractly as the quantity that comes to equal values when two subsystems exchanging energy come to equilibrium. To implement this idea, let the quantity T for any system be defined by

$$T = \left(\frac{dS}{dE}\right)^{-1}. \qquad \text{fundamental definition of temperature} \qquad (6.9)$$

---

[1]This argument also assumes that both of these functions are concave-down, or $d^2S/dE^2 < 0$. This condition certainly holds in our ideal gas example; and according to Equation 6.9, it expresses the fact that putting more energy into a (normal) system raises its temperature.

**Your
Turn
6A**

a. Verify that the dimensions work in Equation 6.9.

b. For the special case of an ideal gas, use the Sakur–Tetrode formula to verify that the average kinetic energy really is $(3k_B/2)$ times the temperature, as required by Idea 3.21.

c. Show that, quite generally, the condition for maximum entropy in the situation sketched in Figure 6.1 is

$$T_A = T_B. \qquad (6.10)$$

**Your
Turn
6B**

Suppose that we duplicate a system (consider two disconnected, isolated boxes, each with N molecules and each with total energy E). Show that then the entropy doubles but T, defined by applying Equation 6.9 to the combined system, stays the same. That is, find the change in S_{tot} when we add a small amount dE of additional energy to the combined system. It may seem that one needs to know how dE got divided between the two boxes; show that, on the contrary, it doesn't matter. [*Hint:* Use a Taylor series expansion to express $S(E + dE)$ in terms of $S(E)$ plus a correction.]

We say that S is an **extensive** quantity (it doubles when the system is doubled), whereas T is **intensive** (it's unchanged when the system is doubled).

More precisely, the temperature of an isolated macroscopic system can be defined once it has come to equilibrium; it is then a function of how much energy the system has, namely, Equation 6.9. When two isolated macroscopic systems are brought into thermal contact, energy will flow until a new equilibrium is reached. In the new equilibrium, there is no net flux of energy, and each subsystem has the same value of T, at least up to small fluctuations. (They won't have the same *energy*—a coin held up against the Eiffel Tower has a lot less thermal energy than the tower, even though both come to the same temperature.) As mentioned at the end of Section 6.3.1, the fluctuations will be negligible for macroscopic systems. Our result bears a striking resemblance to something we learned long ago: Section 4.4.2 on page 128 showed how a difference in *particle densities* can drive a flux of *particles*, via Fick's law (Equation 4.19).

Subdividing the freezing and boiling points of water into 100 steps and agreeing to call freezing "zero" gives the **Celsius scale**. Using the same step size, but starting at absolute zero, gives the **Kelvin** (or absolute) scale. The freezing point of water lies 273 degrees above absolute zero, which we write as 273 K. We will often evaluate our results at the illustrative value $T_r = 295$ K, which we will call "room temperature."

Temperature is a subtle, new idea, not directly derived from anything you learned in classical mechanics. In fact, a sufficiently simple system, like the Moon orbiting Earth, has no useful concept of T; at any moment, it's in one particular state, so we don't need a statistical description. In a complex system, in contrast, the entropy

S, and hence T, involve *all* allowed microstates. Temperature is a qualitatively new property of a complex system not obviously contained in the microscopic laws of collisions. Such properties are called emergent (see Section 1.2.3).

$\boxed{T_2}$ *Section 6.3.2′ on page 235 gives some more details about temperature and entropy.*

6.4 THE SECOND LAW

6.4.1 Entropy increases spontaneously when a constraint is removed

We can interpret the Zeroth Law as saying that a system with order initially (entropy not maximal; energy separated in such a way that $T_A \neq T_B$) will lose that order (increase in entropy until the temperatures match).

Actually, even before people knew how big molecules were, before people were even quite sure that molecules were real at all, they could still measure temperature and energy. By the mid-nineteenth century, Clausius and Kelvin had concluded that a system in thermal equilibrium had a fundamental property S implicitly defined by Equation 6.9 and obeying a general law, now known as the **Second Law** of thermodynamics:[2]

> *Whenever we release an internal constraint on an isolated macroscopic system in equilibrium, eventually the system comes to a new equilibrium whose entropy is at least as great as before.*   (6.11)

It all sounds very mysterious when you present it from the historical point of view; people were confused for a long time about the meaning of the quantity S, until Ludwig Boltzmann explained that it reflects the *disorder* of a macroscopic system in equilibrium, when all we have is limited, aggregate, knowledge of the state. The Second Law states that, after enough time has passed to reestablish equilibrium, the system will be spending as much time in the newly available states as in the old ones: Disorder will have increased. Entropy is *not* conserved.

Notice that *isolated* means, in particular, that the system's surroundings don't do any mechanical work on it, nor does it do any work on them. Here's an example.

Example: Suppose that we have an insulated tank of gas with a partition down the middle, N molecules on the left side, and none on the right (Figure 6.3). Each side has volume V. At some time, a clockwork mechanism suddenly opens the partition and the gas rearranges. What happens to the entropy?

Solution: Because the gas doesn't push on any moving part, the gas does no work; because the tank is insulated, no thermal energy enters or leaves either. Hence, the gas molecules lose no kinetic energy. So, in Equation 6.6, nothing changes *except* the

---

[2]The *First* Law was just the conservation of energy, including the thermal part (Section 1.1.2).

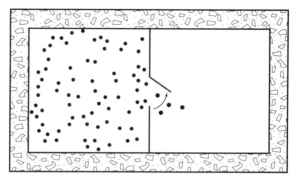

Figure 6.3: (Schematic.) Expansion of gas into vacuum.

factor V^N, and the change of entropy is

$$\Delta S = k_B \left[\ln(2V)^N - \ln(V)^N \right] = N k_B \ln 2, \tag{6.12}$$

which is always positive.

The corresponding increase in disorder after the gas expands, ΔI, is $(K/k_B)\Delta S$, where $K = 1/(\ln 2)$. Substitution gives $\Delta I = N$ bits. That makes sense: Before the change, we knew which side each molecule was on, whereas afterward, we have lost that knowledge. To specify the state to the previous degree of accuracy, we'd need to specify an additional N binary digits. Chapter 1 already made a similar argument, in the discussion leading up to the maximum osmotic work (Equation 1.7 on page 15).

Would this change ever spontaneously happen in reverse? Would we ever look again and find all N molecules on the left side? Well, in principle, yes; but, in practice, no: We would have to wait an impossibly long time for such an unlikely accident.[3] *Entropy increased spontaneously when we suddenly released a constraint,* arriving at a new equilibrium state. We forfeited some *order* when we allowed an uncontrolled expansion; and in practice, it won't ever come back on its own. To get it back, we'd have to *compress* the gas with a piston. Compression requires us to do mechanical work on the system, thereby heating it up. To return the gas to its original state, we'd then have to cool it (remove some thermal energy). In other words,

> *The cost of recreating order is that we must degrade some organized energy into thermal form,* (6.13)

another conclusion foreshadowed in Chapter 1 (see Section 1.2.2 on page 12).

---

[3]How unlikely is it? A mole occupies about 24 L at atmospheric pressure and room temperature. If $V = 1\,L$, then the chance that any observation will see all the gas on one side is $\left(\frac{1}{2}V/V\right)^{N_{mole}V/24\,L} \approx 10^{-7\,470\,000\,000\,000\,000\,000\,000}$.

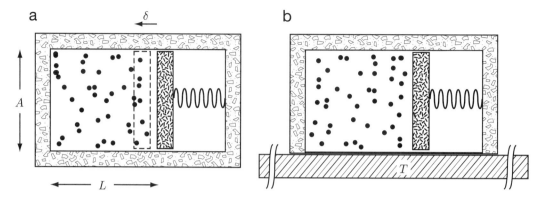

Figure 6.4: (Schematics.) Compression of gas by a spring. (a) Thermally isolated system. The direction of increasing δ is to the left. (b) Subsystem in contact with a heat reservoir, at temperature T. The slab on the bottom of (b) conducts heat, whereas the other walls around the box are thermally insulating. In each case, the chamber on the right (with the spring) contains no gas; only the spring opposes gas pressure from the left side.

Thus the entropy goes up as a system comes to equilibrium. If we fail to harness the escaping gas (as in Figure 6.3), its initial order is lost, as in the parable of the rock falling into mud (Section 1.1.1): We just forfeited knowledge about the system. But now suppose that we *do* harness an expanding gas. We therefore modify the situation in Figure 6.3, this time forcing the gas to do work as it expands.

Again we consider an isolated system, but this time with a sliding piston (Figure 6.4a). The left side of the cylinder contains N gas molecules, initially at temperature T. The right side is empty except for a steel spring. When the piston is at L, the spring exerts a force f directed to the left. Suppose that initially we clamp the piston to a certain position $x = L$ and let the gas come to equilibrium.

Example:

a. Now we unclamp the piston, let it slide freely to a nearby position $L - \delta$, clamp it there, and again let the system come to equilibrium. Here δ is much smaller than L. Find the difference between the entropy of the new state and that of the old one.

b. Suppose we unclamp the piston and let it go where it likes. Its position will then wander thermally, but it's most likely to be found in a certain position L_{eq}. Find this position.

Solution:

a. Suppose, for concreteness, that δ is small and positive, as drawn in Figure 6.4a. The gas molecules initially have total kinetic energy $E_{kin} = \frac{3}{2}k_B T$. The total system energy E_{tot} equals the sum of E_{kin} and the potential energy stored in the spring, E_{spring}. (By definition, in an ideal gas, the molecules' potential energy can be neglected; see Section 3.2.1 on page 78.) The system is isolated, so E_{tot}

doesn't change. Thus the potential energy $f\delta$ lost by the spring increases the kinetic energy E_{kin} of the gas molecules, thereby increasing the temperature and entropy of the gas slightly. At the same time, the loss of volume $\Delta V = -A\delta$ decreases the entropy.

We wish to compute the change in the gas's entropy, using the Sakur–Tetrode formula (Equation 6.6). Note that $\Delta \ln V = (\Delta V)/V$ and similarly, $\Delta \ln E_{kin} = (\Delta E_{kin})/E_{kin}$. These identities give the net entropy change as

$$\Delta S / k_B = \Delta\left(\ln E_{kin}^{3N/2} + \ln V^N\right) = \frac{3}{2}\frac{N}{E_{kin}}\Delta E_{kin} + \frac{N}{V}\Delta V.$$

Replace E_{kin}/N in the first term by $\frac{3}{2}k_B T$. Next use $\Delta E_{kin} = f\delta$ and $\Delta V = -(\delta/L)V$ to find $\Delta S / k_B = ((3/2)f(3k_B T/2)^{-1} - N/L)\delta$, or

$$\Delta S = \frac{1}{T}\left(f - \frac{Nk_B T}{L}\right)\delta.$$

b. The Statistical Postulate says that every microstate is equally probable. Just as in Section 6.3.1, however, there will be far more microstates with L close to the value L_{eq} maximizing the entropy than for any other value of L (recall Figure 6.2). To find L_{eq}, set $\Delta S = 0$ in the preceding formula, which yields $f = Nk_B T/L_{eq}$, or $L_{eq} = Nk_B T/f$.

Dividing the force by the area of the piston yields the pressure in equilibrium, $p = f/A = Nk_B T/(AL) = Nk_B T/V$. We have just recovered the ideal gas law, this time as a consequence of the Second Law: If N is large, then our isolated system will be overwhelmingly likely to have its piston in the location maximizing the entropy. We can characterize this state as the one in which the spring is compressed to the point where it exerts a mechanical force just balancing the ideal-gas pressure.

6.4.2 Three remarks

Some remarks and caveats about the Statistical Postulate (Idea 6.4) are in order before we proceed:

1. The one-way increase in entropy implies a fundamental *irreversibility* to physical processes. Where did the irreversibility come from? Each molecular collision could equally well have happened in reverse. The origin of the irreversibility is not in the microscopic equations of collisions, but *in the choice of a highly specialized initial state*. The instant after the partition is opened, suddenly a huge number of new allowed states open up, and the previously allowed states are suddenly a tiny minority of those now allowed. There is no analogous work-free way to suddenly *forbid* those new states. For example, in Figure 6.3, we'd have to push the gas molecules to get them back into the left side once we let them out. (In principle, we could just wait for them all to be there by a spontaneous statistical fluctuation, but we have already seen that this would be a very long wait.)

Maxwell himself tried to imagine a tiny "demon" who could open the door when he saw a molecule coming from the right but shut it when he saw one coming from the left. It doesn't work; upon closer inspection, one always finds that any physically realizable demon of this type requires an external energy supply (and a heat sink) after all.

2. The formula for the entropy of an ideal gas, Equation 6.6, applies equally to a dilute *solution* of N molecules of solute in some other solvent. Thus, for instance, Equation 6.12 gives the **entropy of mixing** when equal volumes of pure water and dilute sugar solution mix. Chapter 7 will pick up this theme again and apply it to osmotic flow.

3. The Statistical Postulate claims that the entropy of an *isolated, macroscopic* system must not decrease. Nothing of the sort can be said about individual molecules, which are neither isolated (they exchange energy with their neighbors) nor macroscopic. Indeed, individual molecules *can* certainly fluctuate into special states. For example, we already know that any given air molecule in the room will often have energy three times its mean value, because the exponential factor in the velocity distribution (Equation 3.25 on page 84) is not very small when $E = 3 \times \frac{3}{2}k_B T_r$.

$\boxed{T_2}$ *Section 6.4.2′ on page 236 touches on the question of why entropy should increase.*

6.5 OPEN SYSTEMS

Point (3) in Section 6.4.2 will prove very important for us. At first, it may seem like a discouraging remark: If individual molecules don't necessarily tend always toward greater disorder, and if we want to study individual molecules, then what was the point of formulating the Second Law? This section will begin to answer this question by finding a form of the Second Law that is useful when dealing with a small system, which we'll call a, in thermal contact with a big one, called B. We'll call system a **open** to emphasize the distinction from **closed** (that is, isolated) systems. For the moment, we continue to suppose that it is macroscopic. Section 6.6 will then generalize our result to handle the case of microscopic, even single-molecule, subsystems. (Chapter 8 will generalize still further, to consider systems free to exchange molecules, as well as energy, with each other.)

6.5.1 The free energy of a subsystem reflects the competition between entropy and energy

Fixed-volume case Let's return to our gas+piston system (see Figure 6.4b). We'll refer to the subsystem including the gas and the spring as a. As shown in the figure, a can undergo internal motions, but its total volume as seen from outside does not change. In contrast to the gas expansion Example (page 208), this time we'll assume that a is not thermally isolated, but rather rests on a huge block of steel at temperature T (Figure 6.4b). The block of steel (system B) is so big that its temperature is practically unaffected by whatever happens in our small system: We say it's a **thermal reservoir**. The combined system, $a + B$, is still isolated from the rest of the world.

Thus, after we release the piston and let the system come back to equilibrium, the temperature of the gas in the cylinder will not rise, as it did in the gas expansion Example, but will instead stay fixed at T, by the Zeroth Law. Even though all the potential energy lost by the spring went to raise the kinetic energy of the gas molecules temporarily, in the end, this energy was lost to the reservoir. Thus, E_{kin} remains fixed at $\frac{3}{2}Nk_BT$, whereas the total energy $E_a = E_{kin} + E_{spring}$ goes *down* when the spring expands.

Reviewing the algebra shows that this time, the change in entropy for system a is just $\Delta S_a = -\frac{Nk_B}{L}\delta$. Requiring that this expression be positive would imply that the piston always moves to the right—but that's absurd. If the spring exerts more force per area than the gas pressure, the piston will surely move *left*, thereby reducing the entropy of subsystem A. Something seems to be wrong.

Actually, we have already met a similar problem in Section 1.2.2 on page 12, in the context of reverse osmosis. The point is that we have so far looked only at the *subsystem's* entropy, whereas the quantity that must increase is the *whole world's* entropy. We can get the entropy change of system B from its temperature and Equation 6.9: $T(\Delta S_B) = \Delta E_B = -\Delta E_a$. Thus the quantity that must be positive in any spontaneous change of state is not $T(\Delta S_a)$ but $T(\Delta S_{tot}) = -\Delta E_a + T(\Delta S_a)$. Rephrasing this result, we find that the Second Law has a simple generalization to deal with systems that are not isolated:

> *If we bring a small system a into thermal contact with a big system B in equilibrium at temperature T, then B will stay in equilibrium at the same temperature (a is too small to affect it), but a will come to a new equilibrium, which minimizes the quantity $F_a \equiv E_a - TS_a$.* (6.14)

Thus the piston finds its equilibrium position, and is therefore no longer in a position to do mechanical work for us, once its free energy is minimum. The minimum is the point where F_a is stationary under small changes of L; or, in other words, when $\Delta F_a = 0$.

The quantity F_a appearing in Idea 6.14 is called the **Helmholtz free energy** of subsystem a. Idea 6.14 explains the name "free" energy: When F_a is minimum, then a is in equilibrium and won't change any more. Even though the mean energy $\langle E_a \rangle$ isn't zero, nevertheless a won't do any more useful work for us. At this point, the system isn't driving to any state of lower F_a and can't be harnessed to do anything useful along the way.

A system whose free energy is *not* at its minimum is poised to do mechanical or other useful work. This compact principle is just a precise form of what was anticipated in Chapter 1, which argued that the useful energy is the total energy reduced by some measure of disorder. Indeed, Idea 6.14 establishes Equation 1.4 and Idea 1.5 on page 9.

**Your
Turn
6C**
Apply Idea 6.14 to the system in Figure 6.4b, find the equilibrium location of
the piston, and explain why that's the right answer.

The virtue of the free energy is that it focuses all our attention on the subsystem of
interest to us. The surrounding system B enters only in a generic, anonymous way,
through *one number*, its temperature T.

Fixed-pressure case   Another way in which a can interact with its surroundings is by
expanding its volume at the expense of B. We can incorporate this possibility while
still formulating the Second Law solely in terms of a.

Imagine that the two subsystems have volumes V_a and V_B, constrained by a fixed
total volume: $V_a + V_B = V_{tot}$. First, we again define temperature by Equation 6.9,
specifying now that the derivative is to be taken at fixed volume. Next we define pressure in analogy with Equation 6.9: A closed system has

$$ p = T \left. \frac{dS}{dV} \right|_E , \qquad (6.15) $$

where the notation means that the derivative is to be taken holding E fixed. (Throughout this chapter, N is also fixed.)

**Your
Turn
6D**
The factor of T may look peculiar, but it makes sense: Show that the dimensions
of Equation 6.15 work. Then use the Sakur–Tetrode formula (Equation 6.6) to
show that Equation 6.15 does give the pressure of an ideal gas.

Suppose that system B has pressure p, which doesn't change much as a grows or
shrinks because B is so much bigger. Then, by an argument like the one leading to
Idea 6.14, we can rephrase the Second Law to read: *If we bring a small system a into
thermal and mechanical contact with a big system B, then B will stay in equilibrium at
its original temperature T and pressure p, but a will come to a new equilibrium, which
minimizes*

$$ G_a \equiv E_a + pV_a - TS_a. \qquad \textbf{Gibbs free energy} \qquad (6.16) $$

Just as T measures the availability of energy from B, so we can think of p as measuring
the *unwillingness* of B to give up some *volume* to a.

The quantity $H_a \equiv E_a + pV_a$ is called the **enthalpy** of a. We can readily interpret
its second term. If a change of a causes it to grow at the expense of B, then a must do
some mechanical work, $p(\Delta V_a)$, to push aside the bigger system and make room for
the change. This work will partially offset any favorable (negative) ΔE_a.

Chemists often study reactions in which one reactant enters or leaves a gas phase, with a big ΔV, so the distinction between F and G is a significant one. In the chapters to follow, however, we won't worry about this distinction; we'll use the abbreviated term "free energy" without specifying which one is meant. (Similarly, we won't distinguish carefully between energy and enthalpy.) The reactions of interest to us occur in water solution, where the volume does not change much when the reaction takes one step (see Problem 8.4), and hence the difference between F and G is practically a constant, which drops out of before-and-after comparisons. Chapter 8 will use the traditional symbol ΔG to denote the change in free energy when a chemical reaction takes one step.

6.5.2 Entropic forces can be expressed as derivatives of the free energy

A system can also be open if an external mechanical force acts on it. For example, suppose that we eliminate the spring from Figure 6.4b, replacing it by a rod that sticks out of the thermal insulation and that we push with force f_{ext}. Then the free energy of subsystem a is a constant (including E_{kin}) minus TS_a. Dropping the constant gives $F_a = -Nk_BT \ln V = -Nk_BT \ln(LA)$.

The condition for equilibrium cannot be simply $dF_a/dL = 0$, because this condition holds only at $L = \infty$. But it's easy to find the right condition, just by rearranging your result in Your Turn 6C. Our system will have the same equilibrium as the one in Figure 6.4b; it doesn't matter whether the applied force is internal or external. Thus we find that in equilibrium,

$$f_a = -\frac{dF_a}{dL}. \qquad \text{entropic force as a derivative of } F \qquad (6.17)$$

In this formula $f_a = -f_{ext}$ is the force exerted *by* subsystem a on the external world, in the direction of increasing L. We already knew that the subsystem tends to lower its free energy; Equation 6.17 makes precise how hard it's willing to push. Equation 6.17 has intentionally been written to emphasize its similarity to the corresponding formula from ordinary mechanics, $f = -dU/dL$.

Now we can also find the *work* that our subsystem can do against a load. We see that the subsystem will spontaneously expand, even if expansion requires opposing an external load, as long as the opposing force is less than the value in Equation 6.17. To get the maximum possible work, we should continuously adjust the load force to be always just slightly less than the maximum force the system can exert. Integrating Equation 6.17 over L gives a sharpened form of Idea 6.14:

If a subsystem is in a state of greater than minimum free energy, it can do work on an external load. The maximum possible work we can extract is $F_a - F_{a,\min}$.   (6.18)

Gilbert says: By the way, where did the work come from? The internal energy of the gas molecules didn't change, because T didn't change.

Sullivan: That's right: The cylinder has drawn thermal energy from the reservoir (system B) and converted it into mechanical work.

Gilbert: Doesn't that violate the Second Law?

Sullivan: No, our system sacrificed some *order* by letting the gas expand. After the expansion, we don't know as precisely as before where the gas molecules are located. Something—order—did get used up, just as foreshadowed in Chapter 1 (see Section 1.2.2). The concept of free energy makes this intuition precise.

In a nutshell,

> *The cost of upgrading energy from thermal to mechanical form is that we must give up order.*   (6.19)

This statement is just the obverse of the slogan given in Idea 6.13.

6.5.3 Free energy transduction is most efficient when it proceeds in small, controlled steps

Idea 6.18 tells us about the maximum work we can get from a small subsystem in contact with a thermal reservoir. To extract this maximum work, we must continuously adjust the load force to be just slightly less than the force the system can exert. This procedure is generally not practical. We should explore what will happen if we maintain a load that is somewhere between zero (free expansion, Figure 6.3) and the maximum. Also, most familiar engines repeat a *cycle* over and over. Let's construct such an engine in our minds and see how it fits into the framework of our ideas.

Let subsystem a be a cylinder of gas with a piston of area A at one end, held down by weights w_1 and w_2. The initial, equilibrium pressure in the cylinder is the force per unit area: $p_i = (w_1 + w_2)/A$ (Figure 6.5). (For simplicity, suppose that there's no air outside; also, take the weight of the piston itself to be zero.) The cylinder is in thermal contact with a reservoir B at fixed temperature T. Suddenly remove weight w_2 from the piston (slide it off sideways so that this action doesn't require any work). The piston pops up from its initial height L_i with its final height L_f, measured from the bottom, and the pressure goes down to $p_f = w_1/A$.

Example: Find the change in the free energy of the gas, ΔF_a, and compare it with the mechanical work done in lifting weight w_1.

Solution: The final temperature is the same as the initial, so the total kinetic energy $E_{kin} = \frac{3}{2}Nk_BT$ of the gas doesn't change during this process. The pressure in the outside world was assumed to be zero. So all that changes in the free energy is $-TS_a$. The Sakur–Tetrode formula gives the change as

$$\Delta F_a = -Nk_BT \ln \frac{L_f}{L_i}.$$

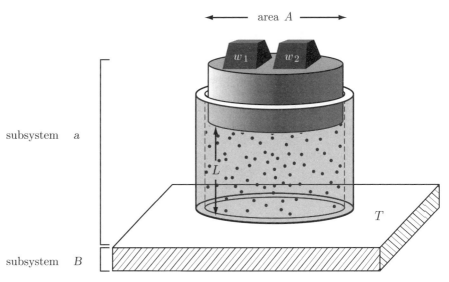

Figure 6.5: (Schematic.) Extracting mechanical work by lifting a weight. After we remove weight w_2, the cylinder rises, lifting weight w_1. A large thermal reservoir (subsystem B) maintains the cylinder at a fixed temperature T.

The ideal gas law now gives the final-state pressure as $p_f L_f = N k_B T / A$, where $p_f = w_1 / A$. Let $X = (L_f - L_i)/L_f$, so X lies between 0 and 1. Then the mechanical work done in raising weight w_1 is $w_1 (L_f - L_i) = N k_B T X$, whereas the free energy changes by $|\Delta F| = -N k_B T \ln(1 - X)$. But $X < -\ln(1 - X)$ when X lies between 0 and 1, so the work is less than the free energy change.

One could do something useful with the mechanical work done lifting w_1 by sliding it off the piston, letting it fall back to its original height L_i, and harnessing the released mechanical energy to grind coffee or whatever. As predicted by Idea 6.18, we can never get more useful mechanical work out of the subsystem than $|\Delta F_a|$.

What could we do to optimize the efficiency of the process, that is, to get out *all* the excess free energy as work? The ratio of work done to $|\Delta F_a|$ equals $-X/(\ln(1 - X))$. This expression is maximum for very small X, that is, for small w_2. In other words, *we get the best efficiency when we release the constraint in tiny, controlled increments*—a **quasistatic process**.

We could get back to the original state by moving our gas cylinder into contact with a different thermal reservoir at a lower temperature T'. The gas cools and shrinks until the piston is back at position L_i. Now we slide the weights back onto it, switch it back to the original, hotter, reservoir (at T), and we're ready to repeat the whole process ad infinitum.

We have just invented a cyclic **heat engine**. Every cycle converts some thermal energy into mechanical form. Every cycle also saps some of the world's order, transferring thermal energy from the hotter reservoir to the colder one, and tending ultimately to equalize them. Figure 6.6 summarizes these words.

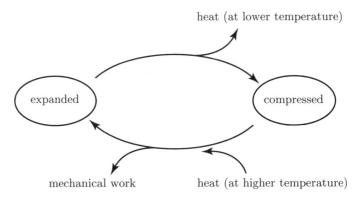

Figure 6.6: (Diagram.) Operating cycle of a heat engine.

That's amusing, but ... biological motors are not cylinders of ideal gas. Nor are they driven by temperature gradients. Your body doesn't have a firebox at one end and a cooling tower at the other, like the electric company's generating plant. So Chapter 10 will turn away from heat engines to motors run by *chemical* energy. Our effort has not been wasted, though. The valuable lesson we learned in this section is based on the Second Law, so it is quite general:

> *Free energy transduction is least efficient when it proceeds by the un-controlled release of a big constraint. It's most efficient when it pro-ceeds by the incremental, controlled release of many small constraints.*   (6.20)

| **Your Turn 6E** | Why do you suppose your body is full of *molecular*-size motors, taking *tiny* steps? Why do you think the electric company only succeeds in capturing a third of the energy content in coal, wasting the rest as heat? |

6.5.4 The biosphere as a thermal engine

The abstract definition of temperature (Equation 6.9) gives us a way to clarify the "quality of energy" concept alluded to in Chapter 1. Consider again an isolated system with two subsystems (Figure 6.1). Suppose that a large, nearly equilibrium system A transfers energy ΔE to system B, which need not be in equilibrium. Then A lowers its entropy by $\Delta S_A = -\Delta E/T_A$. According to the First Law, this transaction raises the energy of B by ΔE. The Second Law says that it must also raise the entropy of B by at least $|\Delta S_A|$ because $\Delta S_A + \Delta S_B \geq 0$.

To give B the greatest possible energy per unit of entropy increase, we need $\Delta E/\Delta S_B$ to be large. We just argued that this quantity cannot exceed $\Delta E/|\Delta S_A|$. Because the last expression equals T_A, our requirement implies that T_A must be large: High-quality energy comes from a high-temperature body.

More precisely, it's not the temperature but the fractional temperature *difference* straddled by a heat engine that determines its maximum possible efficiency. We can see this point in the context of the heat engine Example (page 214) and the following text. Our strategy for extracting work from this system assumed that the first reservoir was hotter than the second, or $T > T'$. Let's see why this assumption was necessary.

The engine did work W on its power stroke (the left-pointing arrow in Figure 6.6). It doesn't change at all in one complete cycle. But it released a quantity of thermal energy Q into the cooler reservoir during the contraction step, thereby increasing the entropy of the outside world by at least Q/T'. Some of this entropy increase is compensated in the next step, where we raise the temperature back up to T: In this process, an amount of thermal energy equal to $Q + W$ flows out of the hotter reservoir, thereby reducing the entropy of the outside world by $(Q + W)/T$. The *net* change in world entropy is then

$$\Delta S_{tot} = Q \left(\frac{1}{T'} - \frac{1}{T} \right) - \frac{W}{T}. \tag{6.21}$$

Because this quantity must be positive, we see that we can get useful work out (that is, $W > 0$) only if $T' < T$. In other words, *The temperature difference is what drives the motor.*

A *perfect* heat engine would convert *all* the input thermal energy to work, exhausting *no* heat. At first this may seem impossible: Setting $Q = 0$ in Equation 6.21 seems to give a decrease in the world's entropy! A closer look, however, shows us another option. If the second reservoir is close to absolute zero temperature, $T' \approx 0$, then we can get near-perfect efficiency, $Q \approx 0$, without violating the Second Law. More generally, *a big temperature difference, T/T', permits high efficiency.*

We can now apply the intuition gleaned from heat engines to the biosphere. The Sun's surface consists of a lot of hydrogen atoms in near-equilibrium at about 5600 K. It's not perfect equilibrium because the Sun is leaking energy into space, but the rate of leakage, inconceivably large as it is, is still small compared to the total. Thus we may think of the Sun as a nearly closed thermal system, connected to the rest of the Universe by a narrow channel, like system A in Figure 6.1 on page 202.

A single chloroplast in a cell can be regarded as occupying a tiny fraction of the Sun's output channel and joining it to a second system B (the rest of the plant in which it's embedded), which is at room temperature. The discussion above suggests that the chloroplast can be regarded as a machine that can extract useful energy from the incident sunlight using the large difference between 5600 K and 295 K. Instead of doing mechanical work, however, the chloroplast creates the high-energy molecule ATP (Chapter 2) from lower-energy precursor molecules. The details of how the chloroplast captures energy involve quantum mechanics, and so are beyond the scope of this book. But the basic thermodynamic argument does show us the *possibility* of its doing so.

6.6 MICROSCOPIC SYSTEMS

Much of our analysis so far has been in the familiar context of macroscopic systems. Such systems have the comforting property that their statistical character is hidden: Statistical fluctuations are small (see Figure 6.2 on page 204), so their gross behavior appears to be deterministic. We invoked this idea each time we said that a certain configuration was "overwhelmingly more probable" than any other, for example, in the discussion of the Zeroth Law (Section 6.3.2).

But as mentioned earlier, we also want to understand the behavior of single molecules. This task is not as hopeless as it may seem: We are becoming familiar with situations in which individual actors are behaving randomly and yet a clear pattern emerges statistically. We just need to replace the idea of a definite state by the idea of a definite *probability distribution* of states.

6.6.1 The Boltzmann distribution follows from the Statistical Postulate

The Boltzmann distribution The key insight needed to get a simple result is that any single molecule of interest (for example, a molecular motor in a cell) is *in contact with a macroscopic thermal reservoir* (the rest of your body). Thus we want to study the generic situation shown in Figure 6.7. The figure shows a tiny subsystem in contact with a large reservoir at temperature T. Although the statistical fluctuations in the energies of a and B are equal and opposite, they're negligible for B but significant for a. We would like to find the probability distribution for the various allowed states of a.

The number of states available to B depends on its energy via $\Omega_B(E_B) = e^{S_B(E_B)/k_B}$ (Equation 6.5). The energy E_B, in turn, depends on the state of a by energy conservation: $E_B = E_{tot} - E_a$. Thus the number of joint microstates where a is in a particular state and B is in *any* allowed state depends on E_a: It equals $\Omega_B(E_{tot} - E_a)$.

The Statistical Postulate says that all allowed microstates of the joint system have the same probability; call it P_0. The addition rule for probabilities then implies that the probability of a being in a particular state, regardless of what B is doing, equals $\Omega_B(E_{tot} - E_a)P_0$, or in other words, it is a constant times $e^{S_B(E_{tot}-E_a)/k_B}$. We can simplify this result by noting that E_a is much smaller than E_{tot} (because a is small),

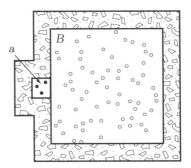

Figure 6.7: (Schematic.) A small subsystem a is in thermal contact with a large system B. Subsystem a may be microscopic, but B is macroscopic. The total system $a+B$ is thermally isolated from the rest of the world.

and by expanding

$$S_B(E_B) = S_B(E_{tot} - E_a) = S_B(E_{tot}) - E_a \frac{dS_B}{dE_B} + \cdots. \qquad (6.22)$$

The dots refer to higher powers of the tiny quantity E_a, which we may neglect. Using the fundamental definition of temperature (Equation 6.9) now gives the probability of observing a in a particular state as $e^{S_B(E_{tot})/k_B} e^{-(E_a/T)/k_B} P_0$, or

> The probability for the small system to be in a state with energy E_a is
> a normalization constant times $e^{-E_a/k_B T}$, where T is the temperature $\qquad (6.23)$
> of the surrounding big system and k_B is the Boltzmann constant.

We have just found the Boltzmann distribution, establishing the proposal made in Chapter 3 (see Section 3.2.3 on page 83).

What makes Idea 6.23 so powerful is that it hardly depends at all on the character of the surrounding big system: The properties of system B enter via only *one number*, its temperature T. We can think of T as the "availability of energy from B": When it's big, system a is more likely to be in one of its higher-energy states, because then $e^{-E_a/k_B T}$ decreases slowly as E_a increases.

Two-state systems Here's an immediate example. Suppose that the small system has only *two* allowed states, S_1 and S_2, and that their energies differ by an amount $\Delta E = E_2 - E_1$. The probabilities of being in these states must obey both $P_1 + P_2 = 1$ and

$$\frac{P_1}{P_2} = \frac{e^{-E_1/k_B T}}{e^{-(E_1 + \Delta E)/k_B T}} = e^{\Delta E/k_B T}. \qquad \text{(simple two-state system)} \qquad (6.24)$$

Solving gives

$$P_1 = \frac{1}{1 + e^{-\Delta E/k_B T}}, \qquad P_2 = \frac{1}{1 + e^{\Delta E/k_B T}}. \qquad (6.25)$$

That makes sense: When the upper state S_2 is far away in energy (ΔE is large), the system is hardly ever excited: It's almost always in the lower state S_1. How far is "far"? It depends on the temperature. At high enough temperature, ΔE will be negligible, and $P_1 \approx P_2 \approx \frac{1}{2}$. As the temperature falls below $\Delta E/k_B$, however, the distribution (Equation 6.25) changes to favor state S_1.

Here is a more involved example:

Your Turn 6F

Suppose a is a small ball tied elastically to some point and free to move in one dimension only. The ball's microstate is described by its position x and velocity v. Its total energy is $E_a(x, v) = \frac{1}{2}(mv^2 + kx^2)$, where k is a spring constant.

a. From the Boltzmann distribution, find the average energy $\langle E_a \rangle$ as a function of temperature. [*Hint:* Use Equation 3.7 on page 73 and Equation 3.14 on page 75.]

b. Now try it for a ball free to move in three dimensions.

Figure 6.8: (Metaphor.) Where the buffalo hop. [Cartoon by Larry Gonick.]

Your result has a name, the **equipartition of energy**. The name reminds us that energy is being equally partitioned among all the places it could go. Both kinetic and potential forms of energy participate.

$\boxed{T_2}$ *Section 6.6.1′ on page 236 makes some connections to quantum mechanics.*

6.6.2 Kinetic interpretation of the Boltzmann distribution

Imagine yourself having just stampeded a herd of buffalo to a cliff. There they are, grunting, jostling, crowded. But there's a small rise before the cliff (Figure 6.8). This barrier corrals the buffalo, although from time to time one falls over the cliff.

   If the cliff is ten meters high, then certainly no buffalo will ever make the trip in the reverse direction. If it's only half a meter high, then they'll occasionally hop back up, although not as easily as they hop down. In the second case, we'll eventually find an equilibrium distribution of buffalo, some up but most down. Let's make these ideas precise.

   Let $\Delta E_{1\to2}$ be the change in gravitational potential energy between the two stable states; abbreviate it as ΔE. The key observation is that thermal equilibrium is not a static state; rather, it's a situation where the backward flow equals—and hence *cancels*—the forward flow.[4] To calculate the flow rates, notice that there's an activation barrier $\Delta E^{\ddagger}$ for falling over the cliff, namely, the gravitational potential energy change needed to move a buffalo up over the barrier. The corresponding energy $\Delta E^{\ddagger} + \Delta E$ in the reverse direction reflects the total height of both the small rise and the cliff itself.

---

[4]Compare our discussion of the Nernst relation (Section 4.6.3 on page 139) or sedimentation equilibrium (Section 5.1 on page 158).

As for buffalo, so for molecules. Here we have the advantage of knowing how rates depend on barriers, from Section 3.2.4 on page 86. We imagine a population of molecules, each of which can spontaneously flip (or **isomerize**) between two configurations. We'll call the states S_1 and S_2 and denote the situation by the shorthand

$$S_2 \underset{k_-}{\overset{k_+}{\rightleftharpoons}} S_1. \tag{6.26}$$

The symbols k_+ and k_- in this formula are called the forward and backward **rate constants**, respectively; we define them as follows.

Initially, there are N_2 molecules in the higher-energy state S_2 and N_1 in the lower-energy state S_1. In a short interval of time dt, the probability that any given molecule in state S_2 will flip to S_1 is proportional to dt; call this probability $k_+ dt$. Section 3.2.4 argued that the probability per time k_+ is a constant C times $e^{-\Delta E^{\ddagger}/k_B T}$, so the average number of conversions per unit time is $N_2 k_+ = CN_2 e^{-\Delta E^{\ddagger}/k_B T}$. The constant C roughly reflects how often each molecule collides with a neighbor.

Similarly, the average rate of conversion in the opposite direction is $N_1 k_- = CN_1 e^{-(\Delta E^{\ddagger}+\Delta E)/k_B T}$. Requiring that the two rates be equal (no net conversion) then gives that the equilibrium populations are related by

$$N_{2,eq}/N_{1,eq} = e^{-\Delta E/k_B T}, \tag{6.27}$$

which is just what we found for P_2/P_1 in Section 6.6.1 (Equation 6.24). Because both isomers live in the same test tube, and so are spread through the same volume, we can also say that the ratio of their number densities, c_2/c_1, is given by the same formula.

We began with an idea about the *kinetics* of molecular hopping between two states, then found in the special case of equilibrium that the relative occupancy of the state is just what the Boltzmann distribution predicts. The argument is like the observation at the end of Section 4.6.3, where we saw that diffusion to equilibrium ends up with the concentration profile expected from the Boltzmann distribution. Our formulas hold together consistently.

We can extract a testable prediction from this kinetic analysis. If we watch a single two-state molecule switching states, we should see it hopping with two rate constants k_+ and k_-. If, moreover, we have some prior knowledge of how ΔE depends on imposed conditions, then the prediction is that $k_+/k_- = e^{\Delta E/k_B T}$. Section 6.7 will describe how such predictions can be confirmed directly in the laboratory.

Notice a key feature of the equilibrium distribution: Equation 6.27 does not contain $\Delta E^{\ddagger}$. All that matters is the energy difference ΔE of the two states. It may take a long time to *arrive* at equilibrium if $\Delta E^{\ddagger}$ is big; but once the system is there, the height of the barrier is immaterial. This result is analogous to the observation at the end of Section 4.6.3 that the value of the diffusion constant D drops out of the Nernst relation, Equation 4.26 on page 141.

Suppose we begin with *non*equilibrium populations of molecules in the two states. Then the number $N_2(t)$ in the upper state *will* change with time, at a net rate given by the difference between the up and down rates just found; similar logic applies to the number $N_1(t)$ in the lower state:

$$\dot{N}_2 \equiv dN_2/dt = -k_+ N_2(t) + k_- N_1(t)$$

$$\dot{N}_1 \equiv dN_1/dt = k_+ N_2(t) - k_- N_1(t). \tag{6.28}$$

The steps leading from Reaction 6.26 to Equations 6.28 were simple but important, so we should pause to summarize them:

To get rate equations from a reaction scheme, we

- *Examine the reaction diagram. Each node (state) of this diagram leads to a differential equation for the number of molecules in the corresponding state.*
- *For each node, find all the links (arrows) impinging on it. The time derivative of the number of this state, $\dot{N}$, has a positive term for each arrow pointing toward its node and a negative term for each arrow pointing away from it.*

$$(6.29)$$

Returning to Equations 6.28, note that the total number of molecules is fixed: $N_1 + N_2 = N_{tot}$. So we can eliminate $N_2(t)$ everywhere, replacing it by $N_{tot} - N_1(t)$. When we do this, we see that one of the two equations is redundant; either one gives

$$\dot{N}_1 = k_+(N_{tot} - N_1) - k_- N_1.$$

This equation is already familiar to us from the concentration decay Example (page 136). Let $N_{1,eq}$ be the population of the lower state in equilibrium. Because equilibrium implies $\dot{N}_1 = 0$, we get $N_{1,eq} = k_+ N_{tot}/(k_+ + k_-)$. Let $x(t) \equiv N_1(t) - N_{1,eq}$ be the deviation from equilibrium. Then

$$\dot{x} = k_+ N_{tot} - (k_+ + k_-)\left(x + \frac{k_+}{k_+ + k_-}N_{tot}\right) = -(k_+ + k_-)x.$$

So, $x(t) = x(0)e^{-(k_+ + k_-)t}$, or

$$\boxed{N_1(t) - N_{1,eq} = (N_1(0) - N_{1,eq})e^{-(k_+ + k_-)t}.} \quad \begin{array}{l}\text{relaxation to}\\\text{chemical equilibrium}\end{array}$$

$$(6.30)$$

In other words, $N_1(t)$ approaches its equilibrium value exponentially, with the decay constant $\tau = (k_+ + k_-)^{-1}$. We again say that a nonequilibrium initial population "relaxes" exponentially to equilibrium. To find how fast it relaxes, differentiate Equation 6.30: The rate at any time t is the deviation from equilibrium at that time, or $(N_1(t) - N_{1,eq})$, times $1/\tau$.

According to the discussion following Equation 6.26,

$$1/\tau = k_+ + k_- = Ce^{-\Delta E^\ddagger/k_B T}(1 + e^{-\Delta E/k_B T}).$$

Thus, unlike the equilibrium populations themselves, a reaction's rate *does* depend on the barrier $\Delta E^{\ddagger}$, a key qualitative fact. Indeed, many important biochemical reactions proceed spontaneously at a negligible rate, because of their high activation barriers. Chapter 10 will discuss how cells use molecular devices—enzymes—to facilitate such reactions when and where they are desired.

Another aspect of exponential relaxation will prove useful later. Suppose that we follow a single molecule, which we observe to be in state S_1 initially. Eventually the molecule will jump to state S_2. Now we ask, how long does it *stay* in S_2 before jumping back? There is no single answer to this question—sometimes this **dwell time** will be short, other times long.[5] But we can say something definite about the *probability distribution* of dwell times, $P_{2\rightarrow 1}(t)$.

Example: Find this distribution.

Solution: First imagine a large collection of N_0 molecules, all starting in state S_2. The number $N(t)$ surviving in this state after an elapsed time t obeys the equation

$$N(t + \mathrm{d}t) = (1 - k_+ \mathrm{d}t)N(t), \text{ with } N(0) = N_0.$$

The solution to this equation is $N(t) = N_0 e^{-k_+ t}$. Now use the multiplication rule (Equation 3.15 on page 76): The probability of a molecule surviving till time t, *and* then hopping in the next interval $\mathrm{d}t$, is the product $(N(t)/N_0) \times (k_+ \mathrm{d}t)$. Calling this probability $P_{2\rightarrow 1}(t)\mathrm{d}t$ gives

$$P_{2\rightarrow 1}(t) = k_+ e^{-k_+ t}. \tag{6.31}$$

Notice that this distribution is properly normalized: $\int_0^{\infty} P_{2\rightarrow 1}(t)\mathrm{d}t = 1$.

Similarly, the distribution of dwell times before hopping in the other direction is $P_{1\rightarrow 2}(t) = k_- e^{-k_- t}$.

6.6.3 The minimum free energy principle also applies to microscopic subsystems

Section 6.6.1 found the Boltzmann distribution by requiring that every microstate of a combined system $a + B$ be equally probable, or in other words, that the entropy of the total system be maximum. Just as in our discussion of macroscopic subsystems, though (Section 6.5), it's better to characterize our result in a way that directs all our attention onto the subsystem of interest and away from the reservoir. For the case of a macroscopic system a, Idea 6.14 on page 211 did this by introducing the free energy $F_a = E_a - TS_a$, which depended on system B only through its temperature. This section will extend the result to a corresponding formula for the case of a microscopic subsystem a.

In the microscopic case, the energy of a can have large relative fluctuations, so it has no definite energy E_a. We must therefore first replace E_a by its average over all

---

[5] Some authors use the synonym *waiting time* for dwell time.

allowed states of a, or $\langle E_a \rangle = \sum_j P_j E_j$. To define the entropy S_a, note that the Boltzmann distribution assigns different probabilities P_j to different states j of system a. Accordingly, we must define the entropy of a using Shannon's formula (Equation 6.3 on page 197). This substitution gives $S_a = -k_B \sum_j P_j \ln P_j$. A reasonable extension of the macroscopic free energy formula is then

$$F_a = \langle E_a \rangle - TS_a. \qquad \text{free energy of a molecular-scale subsystem} \qquad (6.32)$$

Your Turn 6G

Following the steps in Section 6.1 on page 196, show that the Boltzmann probability distribution is the one that minimizes the quantity F_a defined in Equation 6.32.

Thus, if initially the subsystem has a probability distribution differing from the Boltzmann formula (for example, just after releasing a constraint), it is out of equilibrium and can, in principle, be harnessed to do useful work.

Example: What *is* the minimal value of F_a? Show that it's just $-k_B T \ln Z$, where the **partition function** Z is defined as

$$Z = \sum_j e^{-E_j/k_B T}. \qquad \text{partition function} \qquad (6.33)$$

Solution: Use the probability distribution you found in Your Turn 6G to evaluate the minimum free energy:

$$F_a = \langle E_a \rangle - TS_a$$
$$= \sum_j Z^{-1} e^{-E_j/k_B T} E_j + k_B T \sum_j Z^{-1} e^{-E_j/k_B T} \ln \left(Z^{-1} e^{-E_j/k_B T} \right)$$
$$= \sum_j Z^{-1} e^{-E_j/k_B T} E_j + k_B T \sum_j Z^{-1} e^{-E_j/k_B T} \left(\ln (e^{-E_j/k_B T}) - \ln Z \right).$$

The second term equals $-\left(\sum_j Z^{-1} e^{-E_j/k_B T} E_j \right) - k_B T \ln Z$, so $F_a = -k_B T \ln Z$.

In the preceding formulas, the summation extends over all allowed states; if M allowed states all have the same value of E_j, then they contribute $M e^{-E_j/k_B T}$ to the sum. (We say that M is the **degeneracy** of that energy level.) The trick of evaluating the free energy by finding the partition function will prove useful when we work out entropic forces in Chapters 7 and 9.

$\boxed{T_2}$ *Section 6.6.3' on page 237 makes some additional comments about free energy.*

6.6.4 The free energy determines the populations of complex two-state systems

The discussion of two-state systems in Sections 6.6.1 and 6.6.2 may seem too over-simplified to be useful for any real system. Surely the complex macromolecules of interest to biologists never literally have just two relevant states!

Suppose that subsystem a is itself a complex system with many states but that the states may usefully be divided into two classes (or "ensembles of substates"). For example, the system may be a macromolecule; states $S_1, \ldots, S_{N_I}$ may be conformations with an overall "open" shape, and states $S_{N_I+1}, \ldots, S_{N_I+N_{II}}$ constitute the "closed" shape. We call N_I and N_{II} the multiplicities of the open and closed conformations.

Consider first the special situation in which all the states in each class have the same energy. Then $P_I/P_{II} = (N_I e^{-E_I/k_B T})/(N_{II} e^{-E_{II}/k_B T})$: In this special case, we just weight the Boltzmann probabilities of the two classes by their respective degeneracies.

More generally, the probability to be in class I is $P_I = Z^{-1} \sum_{j=1}^{N_I} e^{-E_j/k_B T}$, and similarly for class II, where Z is the full partition function (Equation 6.33). Then the ratio of probabilities is $P_I/P_{II} = Z_I/Z_{II}$, where Z_I is the part of the partition function from class I and so on.

Although the system is in equilibrium and hence visits all its available states, nevertheless, many systems have the property that they spend a long time in one class of states, then hop to the other class and spend a long time there. In that case, it makes sense to apply the definition of free energy (Equation 6.32) to each of the classes separately. That is, we let $F_{a,I} = \langle E_a \rangle_I - TS_{a,I}$, where the subscripts denote quantities referring only to class I.

Your Turn 6H

Adapt the result in the free energy formula Example (page 224) to find that

$$\frac{P_I}{P_{II}} = e^{-\Delta F/k_B T}, \tag{6.34}$$

where $\Delta F \equiv F_{a,I} - F_{a,II}$. Interpret your result in the special case where all the substates in each class have the same energy.

Your result says that our simple formula for the population of a two-state system (Equation 6.24) also applies to a complex system, once we replace energy by free energy.[6]

Just as in Section 6.6.2, we can rephrase our result on equilibrium populations as a statement about the rates of hopping between the two classes of substates:

$$k_{I \to II}/k_{II \to I} = e^{\Delta F/k_B T}. \qquad \text{complex two-state system} \tag{6.35}$$

---

[6] $\boxed{T_2}$ You can generalize this discussion to the fixed-pressure case; the Gibbs free energy appears in place of F (see Section 6.6.3′ on page 237).

6.7 EXCURSION: "RNA FOLDING AS A TWO-STATE SYSTEM" BY J. LIPHARDT, I. TINOCO, JR., AND C. BUSTAMANTE

Recently, we set out to explore the mechanical properties of RNA, an important biopolymer. In cells, RNA molecules store and transfer information, and catalyze biochemical reactions. We knew that numerous biological processes like cell division and protein synthesis depend on the ability of the cell to unfold RNA (as well as to unfold proteins and DNA) and that such unfolding involves mechanical forces, which one might be able to reproduce by using biophysical techniques. To investigate how RNA might respond to mechanical forces, we needed to find a way to grab the ends of individual molecules of RNA. Then we wanted to pull on them and watch them buckle, twist, and unfold under the effect of the applied external force.

We used an **optical tweezer** apparatus, which allows small objects, like polystyrene beads with a diameter of $\approx 3\,\mu$m, to be manipulated by using light (Figure 6.9). Although the beads are transparent, they do bend incoming light rays, transferring some of the light's momentum to each bead, which accordingly experiences a force. A pair of opposed lasers, aimed at a common focus, can thus be used to hold the beads in prescribed locations. Because the RNA is too small to be trapped by itself, we attached it to molecular "handles" made of DNA, which were chemically modified to stick to specially prepared polystyrene beads (Figure 6.9, inset). As

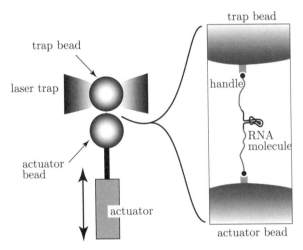

Figure 6.9: (Schematic.) Optical tweezer apparatus. A piezoelectric actuator controls the position of the bottom bead. The top bead is captured in an optical trap formed by two opposing lasers, and the force exerted on the polymer connecting the two beads is measured from the change in momentum of light that exits the optical trap. Molecules are stretched by moving the bottom bead vertically. The end-to-end length of the molecule is obtained as the difference of the position of the bottom bead and the top bead. *Inset:* The RNA molecule of interest is coupled to the two beads via DNA "handles." The handles end in chemical groups that stick to complementary groups on the bead. The drawing is not to scale: Relative to the diameter of the beads (≈ 3000 nm), the RNA is tiny (≈ 20 nm). [Figure kindly supplied by J. Liphardt.]

sketched in the inset, the RNA sequence we studied has the ability to fold back on itself, thereby forming a "hairpin" structure (see Figure 2.16 on page 52).

When we pulled on the RNA via the handles, we saw the force initially increase smoothly with extension (Figure 6.10a, black curve), just as it did when we pulled on the handles alone: The DNA handles behaved much like a spring (a phenomenon to be discussed in Chapter 9). Then, suddenly, at $f = 14.5 \, \mathrm{pN}$, there was a small discontinuity in the force-extension curve (points labeled a and b). The change in length ($\Delta z \approx 20 \, \mathrm{nm}$) of that event was consistent with the known length of the part of the RNA that could form a hairpin. When we reduced the force, the hairpin refolded and the handles contracted. Different samples gave slightly different values for the critical force, but in every case it was sharply defined.

To our surprise, the observed properties of the hairpin were entirely consistent with those of a two-state system. Even though the detailed energetics of RNA folding are known to be rather complex, involving hydration effects, Watson–Crick base-pairing and charge shielding by ions, the overall behavior of the RNA hairpin under external force was that of a system with just two allowed states, folded and unfolded. We stretched and relaxed the RNA hairpin many times and then plotted the fraction of folded hairpins versus force (Figure 6.10b). As the force increased, the fraction folded decreased, and that decrease could be fit to a model used to describe two-state systems (Equation 6.34 and Figure 6.10b, inset). Just as an external magnetic field can be used to change the probability of an atomic magnet to point up or down,[7] the work done by the external force ($f \, \Delta z$) was apparently changing the free energy difference $\Delta F = F_{\mathrm{open}} - F_{\mathrm{closed}}$ between the two states and thus controlling the probability $P(f)$ of the hairpin being folded. But if the ΔF could be so easily manipulated by changing the external force, it might be possible to watch a hairpin "hop" between the two states if we tuned the strength of the external force to the right critical value (such that $P(f) \approx \frac{1}{2}$) and held it there by force-feedback.

Indeed, about one year after starting our RNA unfolding project, we were able to observe this predicted behavior (Figure 6.10c). After showing RNA hopping to everyone who happened to be in the Berkeley physics building that night, we began to investigate this process more closely to see how the application of increasing force tilts the equilibrium of the system toward the longer, unfolded form of the molecule. At forces slightly below the critical force, the molecule stayed mostly in the short folded state except for brief excursions into the longer unfolded state (Figure 6.10c, lower curves). When the force was held at 14.1 pN, the molecule spent roughly equal times in either state ($\approx 1 \, \mathrm{s}$). Finally, at 14.6 pN, the effect was reversed: The hairpin spent more time in the extended, unfolded form and less time in the short, folded form. Thus, it is possible to control the thermodynamics and kinetics of the folding reaction in real time, simply by changing the external force. The only remaining question had to do with the statistics of the hopping reaction. Was RNA hopping a simple process characterized by a constant probability of hopping per unit time at a given force? It appears so: Histograms of the dwell times can be fit to simple exponentials (see Figure 6.10d and Equation 6.31).

---

[7] See Problem 6.5.

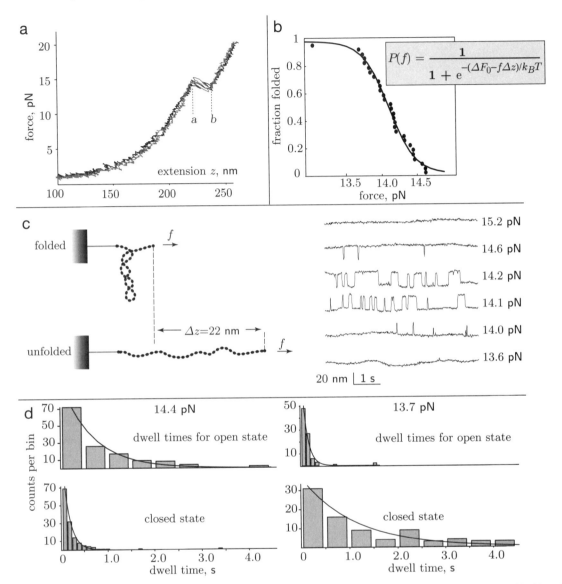

Figure 6.10: (Experimental data) (a) Force-extension curves of an RNA hairpin with handles. Stretching (*black*) and relaxing (*gray*) curves are superimposed. Hairpin unfolding occurs at about 14.5 pN (labeled *a*). (b) Fraction $P(f)$ of hairpins folded versus force. Data (*filled circles*) are from 36 consecutive pulls of a single RNA hairpin. *Solid line*, probability versus force for a two-state system (see Equation 6.34 on page 225). Best-fit values, $\Delta F_0 = 79k_B T_r$, $\Delta z = 22$ nm, consistent with the observed Δz seen in panel (a). (c) Effect of mechanical force on the rate of RNA folding. Length versus time traces of the RNA hairpin at various constant forces. Increasing the external force increases the rate of unfolding and decreases the rate of folding. (d) Histograms of the dwell times in the open and closed states of the RNA hairpin at two different forces ($f = 14.4$ and 13.7 pN). The solid lines are exponential functions fit to the data (see Equation 6.31), giving rate constants for folding and unfolding. At 13.7 pN, the molecule is mostly folded, with $k_{open} = 0.9\,s^{-1}$, and $k_{fold} = 8.5\,s^{-1}$. At 14.4 pN, the unfolded state predominates, with $k_{open} = 7\,s^{-1}$ and $k_{fold} = 1.5\,s^{-1}$. [Figure kindly supplied by J. Liphardt.]

Your Turn 6I	Using the data given in Figure 6.10 and its caption, check how well Equation 6.35 is obeyed by comparing the folding and opening rates at two different forces. That is, find a combination of the rates that does not involve ΔF_0 (which we don't know a priori). Then substitute the experimental numbers and see how well your prediction is obeyed.

The first time one encounters a complicated process, it's natural (and frequently the only thing you can do) to try to strip away as much detail as possible. Then again, such simplification certainly has risks—what is one missing, and are the approximations really that good? In this case, the simple two-state model seems to fit the observations very well, and (so far) we have not detected any behaviors in our RNA hairpin system that would force us to replace this model with something more elaborate.

For more details See Liphardt et al., 2001, and the on-line supplemental material. *Carlos Bustamante is the Howard Hughes Medical Institute Professor of Physics and Molecular and Cell Biology at the University of California, Berkeley. His work involves the development of methods of single molecule manipulation and their application to study complex biochemical processes. Jan Liphardt is currently a Divisional Fellow at Lawrence Berkeley National Lab. He is interested in nonequilibrium statistical mechanics, and in how small systems respond to local application of mechanical forces. Ignacio Tinoco, Jr., is Professor of Chemistry at the University of California, Berkeley. He was chairman of the Department of Energy committee that recommended in 1987 a major initiative to sequence the human genome.*

THE BIG PICTURE

Returning to the Focus Question, we found that a system's useful energy (the portion that can be harnessed to do mechanical or other useful work) is generally less than its total energy content. A machine's efficiency involves how much of this useful energy actually turns into work (with the rest turning into waste heat). We found a precise measure of useful energy, called free energy.

This chapter has been quite abstract, but that's just the obverse of being very generally applicable. Now it's time to look at the fascinating details of how the abstract principles are implemented—to see some concrete realizations of these ideas in living cells. Thus, Chapter 7 will extend the idea of entropic forces to situations relevant in cells, Chapter 8 will look at self-assembly, Chapter 9 will develop the mechanics of macromolecules, and Chapter 10 will examine the operation of molecular motors. All of these biophysical developments will rest on ideas introduced in this chapter.

KEY FORMULAS

- *Entropy:* The disorder of a string of N random, uncorrelated letters drawn from an alphabet of M letters is $I = K \ln \Omega$ bits, where $\Omega = M^N$ and $K = 1/\ln 2$ (Equation 6.1).

  For a very long message whose letter frequencies P_j are known in advance, the disorder is reduced to $-KN \sum_{j=1}^{M} P_j \ln P_j$ (Shannon's formula) (Equation 6.3). For example, if $P_1 = 1$ and all other $P_j = 0$, then the disorder per letter is zero: The message is predictable.

  Suppose that there are $\Omega(E)$ states available to a physical system with energy E. Once the system has come to equilibrium, its entropy is defined as $S(E) = k_B \ln \Omega(E)$ (Equation 6.5).

- *Temperature:* The temperature is defined as $T = \left(\frac{dS}{dE}\right)^{-1}$ (Equation 6.9). If a system is allowed to come to equilibrium in isolation, then later is brought into thermal contact with another system, then T describes the "availability of energy" that the first system could give the second. If two systems have the same T, then there will be no net exchange of energy (the Zeroth Law of thermodynamics).

- *Pressure:* Pressure in a closed subsystem can be defined as $p = T\frac{dS}{dV}\big|_E$ (Equation 6.15). p can be thought of as the "unavailability of volume" from the subsystem, just as T is the "availability of energy."

- *Sakur–Tetrode:* The entropy of a box of ideal gas of volume V, containing N molecules with total energy E, is $S = Nk_B \ln[E^{3/2} V]$ (Equation 6.6), plus terms independent of E and V.

- *Statistical Postulate:* When a big enough, isolated system, subject to some macroscopic constraints, is left alone long enough, it evolves to an equilibrium. Equilibrium is not one particular microstate; rather, it is a probability distribution. The distribution chosen is the one with the greatest disorder (entropy), that is, the one acknowledging the greatest ignorance of the detailed microstate subject to any given constraints (Idea 6.4).

- *Second Law:* Any sudden relaxation of internal constraints (for example, opening an internal door) will lead to a new distribution, one corresponding to the maximum disorder among a bigger class of possibilities. Hence the new equilibrium state will have entropy at least as great as the old one (Idea 6.11).

- *Efficiency:* Free energy transduction is least efficient when it proceeds by the uncontrolled release of a big constraint. It's most efficient when it proceeds by the incremental, controlled release of many small constraints (Idea 6.20).

- *Two-state systems:* Suppose that a subsystem has only two allowed states (isomers), differing in energy by ΔE. Then the probabilities of being in the two states are (Equation 6.25)

$$P_1 = \frac{1}{1 + e^{-\Delta E/k_B T}}, \quad P_2 = \frac{1}{1 + e^{\Delta E/k_B T}}.$$

Suppose that there is an energy barrier $\Delta E^{\ddagger}$ between the two states. The probability per time k_+ that the subsystem will hop to the lower state, if we know it's initially in the upper state, is proportional to $e^{-\Delta E^{\ddagger}/k_B T}$; the probability per time k_- that it will hop to the higher state, if we know it's initially in the lower state, is proportional to $e^{-(\Delta E + \Delta E^{\ddagger})/k_B T}$. For a complex but effectively two-state system, analogous formulas apply with ΔF or ΔG in place of ΔE (Equation 6.35).

If we prepare a collection of molecules in two isomeric forms with populations N_i divided in any way other than the equilibrium distribution $N_{i,\text{eq}}$, then the approach to equilibrium is exponential: $N_i(t) = N_{i,\text{eq}} \pm A e^{-(k_+ + k_-)t}$ (Equation 6.30). Here A is a constant set by the initial conditions.

- *Free energy:* Consider a small system a of fixed volume, sealed so that matter can't go in or out. If we bring system a into thermal contact with a big system B in equilibrium at T, then B will stay in equilibrium at the same temperature (a is too small to affect it), but a will come to a new equilibrium, which minimizes its Helmholtz free energy $F_a = E_a - T S_a$ (Idea 6.14). If a is not macroscopic, we replace E_a by $\langle E_a \rangle$ and use Shannon's formula for S_a (Equation 6.32).

  Suppose, instead, that small system a can also exchange volume with (push on) the larger system and that the larger system has pressure p. Then a will minimize its Gibbs free energy $G_a = E_a - T S_a + p V_a$ (Equation 6.16). When chemical reactions occur in water, V_a is essentially constant, so the difference between F_a and G_a is also essentially a constant.

- *Equipartition:* When the potential energy of a subsystem is a sum of terms of the form kx^2 (that is, an ideal spring), then each of the displacement variables shares average thermal energy $\frac{1}{2} k_B T$ in equilibrium (Your Turn 6F).

- *Partition function:* The partition function for system a in contact with a thermal reservoir B at temperature T is $Z = \sum_j e^{-E_j/k_B T}$ (Equation 6.33). The free energy can be reexpressed as $F_a = -k_B T \ln Z$.

FURTHER READING

Semipopular:

The basic ideas of statistical physics: Feynman, 1965, Chapter 5; Ruelle, 1991, Chapters 17–21; von Baeyer, 1999.

Intermediate:

Many books take a view of this material similar to the one presented here, for example: Schroeder, 2000 and Feynman et al., 1996, Chapter 5; see also Widom, 2002.

Heat engines: Feynman et al., 1963a, §44.

Technical:

Statistical physics: Callen, 1985, Chapters 15–17; Chandler, 1987.

Maxwell's demon: Leff & Rex, 1990.

More on optical tweezers: Bustamante et al., 2000; Mehta et al., 1999; Svoboda & Block, 1994.

The DNA–repressor system as a two-state system: Finzi & Gelles, 1995.

$\boxed{T_2}$ **6.1′ Track 2**

1. Communications engineers are also interested in the compressibility of streams of data. They refer to the quantity I as the "information content" per message. This definition has the unintuitive feature that random messages carry the most information! This book will use the word *disorder* for I; the word *information* will only be used in its everyday sense.

2. Here is another, more elegant, proof that uniform probability gives maximum disorder. We'll repeat the previous derivation, this time using the method of Lagrange multipliers. This trick proves indispensable in more complicated situations. (For more about this method, see for example Shankar, 1995.) We introduce a new parameter α (the Lagrange multiplier) and add a new term to I. The new term is α times the constraint we wish to enforce (that all the P_i add up to 1). Finally, we extremize the modified I over *all* the P_j independently, *and* over α:

$$0 = \frac{d}{dP_{j_0}}\left(-\frac{I}{NK} - \alpha\left(1 - \sum_{j=1}^{M} P_j\right)\right) \text{ and } 0 = \frac{d}{d\alpha}\left(-\frac{I}{NK} - \alpha\left(1 - \sum_{j=1}^{M} P_j\right)\right)$$

$$0 = \frac{d}{dP_{j_0}}\left(\sum_{j=1}^{M} P_j \ln P_j - \alpha\left(1 - \sum_{j=1}^{M} P_j\right)\right) \text{ and } 1 = \sum_{j=1}^{M} P_j.$$

Proceeding as before,

$$0 = \ln P_{j_0} + 1 + \alpha;$$

once again we conclude that all the P_j are equal.

$\boxed{T_2}$ **6.2.1′ Track 2**

1. Why do we need the Statistical Postulate? Most people would agree at first that a single helium atom, miles away from anything else, shielded from external radiation, is not a statistical system. For instance, the isolated atom has definite energy levels. Or does it? If we put the atom into an excited state, it decays at a randomly chosen time. One way to understand this phenomenon is to say that even an isolated atom interacts with ever-present weak, random quantum fluctuations of the vacuum. *No physical system can ever be totally disconnected from the rest of the world.*

   We don't usually think of this effect as making the atom a statistical system simply because the energy levels of the atom are so widely spaced relative to the energies of vacuum fluctuations. Similarly, a billion helium atoms, each separated from its neighbor by a meter, will also have widely spaced energy levels. But if those billion atoms condense into a droplet of liquid helium, then the energy levels get split, typically into sublevels a billion times closer in energy than the original one-atom levels. Suddenly, the system becomes much more susceptible to its environment.

With macroscopic samples, in the range of N_{mole} atoms, this environmental susceptibility becomes even more extreme. If we suspend a gram of liquid helium in a thermally insulating flask, we may well manage to keep it "thermally isolated" in the sense that it won't vaporize for a long time. But we can never isolate it from random environmental influences sufficient to change its substate. Thus, determining the detailed evolution of the microstate from first principles is hopeless. This property is a key difference between bulk matter and single atoms. We therefore need a new principle to get some predictive power for bulk samples of matter. We propose to use the Statistical Postulate for this purpose and see whether it gives experimentally testable results. For more on this viewpoint, see Callen, 1985, §15–1; see also Sklar, 1993.

2. The Statistical Postulate is certainly not graven in stone the way Newton's Laws are. Point (1) has already mentioned that the dividing line between "statistical" and "deterministic" systems is fuzzy. Moreover, even macroscopic systems may not actually explore all their allowed states in any reasonable amount of time, a situation called nonergodic behavior, even though they do make rapid transitions within some subset of their allowed states. For example, an isolated lump of magnetized iron won't spontaneously change the direction of its magnetization. We will ignore the possibility of nonergodic behavior in our discussion, but misfolded proteins, such as the prions thought to be responsible for neurological diseases like scrapie, may provide examples. In addition, single enzyme molecules have been found to enter into long-lived substates with catalytic activity significantly different from those of chemically identical neighboring molecules. Even though the enzymes are constantly bombarded by the thermal motion of the surrounding water molecules, such bombardment seems unable to shake them out of these states of different "personality," or to do so extremely slowly.

T_2 ### 6.2.2' Track 2

1. The Sakur–Tetrode formula (Equation 6.6) is derived in a more careful way in Callen, 1985, §16–10.

2. Equation 6.6 has another key feature: S is **extensive**. This property means that the entropy doubles when we consider a box with twice as many molecules, twice the volume, and twice the total energy.

Your Turn 6J

a. Verify this claim (as usual, suppose that N is large).

b. Also show that the entropy *density* of an ideal gas is

$$S/V = -c k_B[\ln(c/c_*)]. \qquad (6.36)$$

Here the number density $c = N/V$ as usual, and c_* is a constant depending only on the energy per molecule, not on the volume.

As you work through (a), you'll notice that the factor of $N!$ in the denominator is crucial to getting the desired result; before people knew about this factor, they were puzzled by the apparent failure of the entropy to be extensive.

3. Those who question authority can find the area of a higher-dimensional sphere as follows. First, let's return to a deferred promise (from the Gaussian normalization example, page 73) to compute $\int_{-\infty}^{+\infty} dx\, e^{-x^2}$. We'll call this unknown number Y. The trick is to evaluate the expression $\int dx_1 dx_2\, e^{-(x_1{}^2 + x_2{}^2)}$ in two different ways. On one hand, it's just $\int dx_1\, e^{-x_1{}^2} \times \int dx_2\, e^{-x_2{}^2}$, or Y^2. On the other hand, because the integrand depends only on the length of $\mathbf{x}$, this integral is easy to do in polar coordinates. We simply replace $\int dx_1 dx_2$ by $\int r dr d\theta$. We can do the θ integral right away (because nothing depends on θ), so the integrals become $2\pi \int r dr$. Comparing our two expressions for the same thing then gives $Y^2 = 2\pi \int r dr\, e^{-r^2}$. But this new integral is easy. Changing variables to $z = r^2$ shows that it equals $\frac{1}{2}$; so $Y = \sqrt{\pi}$, as claimed in Chapter 3.

To see what that's got to do with spheres, notice that the factor of 2π arising above is the circumference of the unit circle, which we can think of as the analog of the surface area for a "sphere" in two-dimensional space. In the same way, let's now evaluate $\int_{-\infty}^{+\infty} dx_1 \cdots dx_{n+1}\, e^{-x^2}$ in two ways. On one hand, it's just Y^{n+1}; but it's also $\int_0^\infty (A_n r^n dr)\, e^{-r^2}$, where now A_n is the surface area of an n-dimensional sphere, our quarry. Let's call the integral in this expression H_n. Using $Y = \pi^{1/2}$, we conclude that $A_n = \pi^{(n+1)/2}/H_n$.

We already know that $H_1 = \frac{1}{2}$. Next consider that

$$-\frac{d}{d\beta}\bigg|_{\beta=1} \int_0^\infty dr\, r^n e^{-\beta r^2} = \int_0^\infty dr\, r^{n+2} e^{-r^2}$$

(a trick recycled from Equation 3.14). The right side is just H_{n+2}, whereas on the left, we change variables to $r' = \sqrt{\beta} r$, finding $-\frac{d}{d\beta}\big|_{\beta=1}[\beta^{-(n+1)/2} \times H_n]$. So $H_3 = H_1$, $H_5 = 2H_3$; and in general, for any odd number n, we have $H_n = \frac{1}{2} \times \left(\frac{n-1}{2}\right)!$. Substituting into the earlier formula for the sphere area A_n and taking $n = 3N - 1$ gives the first factor quoted in Equation 6.6. (Think for yourself about the case where n is even.)

4. Why did we need the Planck constant $\hbar$ in Equation 6.6? No such constant appears in Equation 6.5. Actually, we might have expected that constants with dimensions would appear when we passed from purely mathematical constructions to physical ones. One way to explain the appearance of $\hbar$ is to note that in classical physics, position and momentum are continuous variables with dimensions. Thus the "number of allowed states" is really a *volume* in the space of positions and momenta; so it has dimensions. But you can't take the logarithm of a number with dimensions! Thus we needed to divide our expression by enough powers of a number with the dimensions $\mathbb{L} \times \mathbb{MLT}^{-1}$. The Planck constant is such a number.

Quantum mechanics gives a deeper answer to this question. In quantum mechanics, the allowed states of a system really are discrete, and we can count them in the naïve way. The number of states corresponding to a volume of position/momentum space involves $\hbar$ via the Uncertainty Principle, so $\hbar$ enters the entropy.

5. We must be careful when formulating the Statistical Postulate, because the form of a probability distribution function will depend on the choice of variables used to specify states. To formulate the Postulate precisely, we must therefore specify that equilibrium corresponds to a probability distribution that is uniform (constant), *when* expressed in terms of a particular choice of variables.

   To find the right choice of variables, recall that equilibrium is supposed to be a situation where the probability distribution doesn't change with time. Next, we use a beautiful result from mechanics (Liouville's theorem): A small region $d^{3N}\mathbf{p}\,d^{3N}\mathbf{r}$ evolves in time to a new region with the same volume in $\mathbf{r}$-$\mathbf{p}$ space. (Other choices of coordinates on this space, like $(\mathbf{v}_i, \mathbf{r}_i)$, do not have this property.) Thus, if a probability distribution is a constant times $d^{3N}\mathbf{p}\,d^{3N}\mathbf{r}$ at one moment, it will have the same form at a later time; such a distribution is suitable for describing equilibrium.

$\boxed{T_2}$ **6.3.2′ Track 2**

1. What's the definition of T (Equation 6.9 on page 204) got to do with older ideas of temperature? One could define temperature by making a mercury thermometer, marking the places where water boils and freezes, and subdividing into 100 equal divisions. That's not very fundamental. If we did the same thing with an alcohol thermometer, the individual markings wouldn't agree: The expansion of liquids is slightly nonlinear. Using the expansion of an ideal gas would be better, but the fact is that each of these standards depends on the properties of some specific material—they're not universal. Equation 6.9 *is* universal, and we've seen that, when we define temperature this way, *any* two big systems (not just ideal gases) come to equilibrium at the same value of T.

2. In Your Turn 6B on page 205, you showed that duplicating an isolated system doubles its entropy. Actually, the extensive property of the entropy is more general than this. Consider two weakly interacting subsystems, for example, our favorite system of two insulated boxes touching each other on small uninsulated patches. The total energy of the state is dominated by the internal degrees of freedom deep in each box, so the counting of states is practically the same as if the boxes were independent, and the entropy is thus additive, giving $S_{\text{tot}} \approx S_A + S_B$. Even if we draw a purely *imaginary* wall down the middle of a big system, the two halves can be regarded as only weakly interacting because the two halves exchange energy (and particles, and momentum ...) only across their boundary. If each subsystem is big enough, the surface-to-volume ratio is small; again, the total energy is dominated by the degrees of freedom interior to each subsystem, and the subsystems are statistically nearly independent except for the constraints of fixed total energy (and volume). Then the total entropy will again be the sum of two independent terms. More generally still, in a macroscopic sample, the entropy will be an en-

tropy *density* times the total volume, as you have already shown in Your Turn 6J for the extreme case of a noninteracting (ideal) gas.

More precisely, we define a macroscopic system as one that can be subdivided into a large number of subsystems, each of which still contains many internal degrees of freedom and interacts weakly with the others. The previous paragraph sketched an argument that the entropy of such a system will be extensive. See Landau & Lifshitz, 1980, §§2, 7 for more on this important point.

$\boxed{T_2}$ **6.4.2′ Track 2**

One can ask *why* the Universe started out in such a highly ordered state, that is, so far from equilibrium. Unfortunately, it's notoriously tricky to apply thermodynamics to the whole Universe. For one thing, the Universe can never come to equilibrium: At $t \to \infty$ it either collapses or at least forms black holes. But a black hole has negative specific heat; hence it can never be in equilibrium with matter!

For our purposes, it's enough to note that the Sun is a hot spot in the sky, and most other directions in the sky are cold. This unevenness of temperature is a form of order. It's what (most) life on Earth ultimately feeds on.

$\boxed{T_2}$ **6.6.1′ Track 2**

1. Our derivation of Idea 6.23 implicitly assumed that only the probabilities of occupying the various allowed states of a depend on temperature; the list of possible states themselves, and their energies, was assumed to be temperature independent. As mentioned in Section 1.5.3 on page 26, the states and their energy levels come (in principle) from quantum mechanics and hence are outside the scope of this book. All we need to know is that there is some list of allowed states.

2. Skeptics may ask why we were allowed to drop the higher-order terms in Equation 6.22. The justification goes back to a remark in Section 6.2.2′: The disorder of a macroscopic system is *extensive*. If you double the system size, the first expansion coefficient I_B in Equation 6.22 doubles, the second one dI_B/dE_B stays the same, and the next one $\frac{1}{2}(d^2I_B/dE_B{}^2)$ drops to half. So each successive term of Equation 6.22 is smaller by an extra factor of a small energy E_a divided by a big energy E_B.

Doesn't this throw out baby with the bathwater? Shouldn't we truncate Equation 6.22 after the *first* term? No: When we exponentiate Equation 6.22, the first term is a constant, which got wiped out when we normalized our probability distribution for system a. So the second term is actually the leading one.

Because this is such an important point, let's make it yet again in a slightly different way. System B is macroscopic, so we can subdivide it equally into $M = 1000$ little systems, each one itself macroscopic and weakly interacting with the others and with a. We know that these little systems are overwhelmingly likely to share E_B equally, because they're all identical and have come to the same temperature.

Hence, each has the same number of allowed states,

$$\Omega_i = e^{S_i/k_B} = \Omega_{0,i} \exp\left(\frac{\delta E}{M}\frac{dS_i}{dE_i} + \cdots\right) \qquad i = 1, \ldots, M.$$

Here $\delta E = E_{\text{tot}} - E_a$, $\Omega_{0,i}$ is the number of allowed states when subsystem i has energy E_{tot}/M, and the dots denote terms with higher powers of $\frac{1}{M}$. The derivative dS_i/dE_i is just $1/T$ and is independent of M, so we can take M large enough to justify truncating the Taylor series expansion after the first term. The total number of states available to system B is then $(\Omega_i)^M$, which is indeed a constant times $e^{\delta E/k_B T}$, and we recover Equation 6.23.

3. The equipartition formula is not valid in quantum statistical physics: Modes whose excitation energy exceeds the thermal energy get "frozen out" of equipartition. Historically, this observation held the key to understanding black-body radiation and, hence, to the creation of quantum theory itself.

$\boxed{T_2}$ **6.6.3′ Track 2**

1. The entropy S_a defined above Equation 6.32 on page 224 can't simply be added to S_B to get the total system entropy, because we can't assume that a is weakly interacting with B (a may not be macroscopic; hence, surface energies needn't be smaller than interior energies).

   But suppose that a *is* itself macroscopic (but still smaller than B). Then the fluctuations in E_a are negligible, so we can omit the averaging symbol in $\langle E_a \rangle$. In addition, all microstates with this energy are equally probable, so $-\sum_{j=1} P_j \ln P_j = (\sum_{j=1} P)(\ln P^{-1}) = 1 \times \ln((\Omega_a)^{-1})^{-1} = \ln \Omega_a$, and S_a reduces to the usual form, Equation 6.5. Thus the formula Equation 6.32 derived in Section 6.6.3 reduces to our prior formula (Idea 6.14) for the special case of a macroscopic subsystem.

2. The Gibbs free energy (Equation 6.16) has a similar generalization to microscopic subsystems, namely,

$$G_a = \langle E_a \rangle + p\langle V_a \rangle - TS_a. \tag{6.37}$$

3. Equation 6.32 gives an important formula for the free energy of an ideal gas of uniform temperature, but nonuniform density $c(\mathbf{r})$. (For example, an external force like gravity may act on the gas.) Suppose a container has N molecules, each moving independently of the others but with specified potential energy $U(\mathbf{r})$. Divide space into many small volume elements Δv. Then the probability for any one molecule to be in the element centered at $\mathbf{r}$ is $P(\mathbf{r}) = c(\mathbf{r})\Delta v/N$. Equation 6.32 on page 224 then gives the free energy per molecule as $F_1 = \sum_{\mathbf{r}}(c(\mathbf{r})\Delta v/N)(U(\mathbf{r}) + k_B T \ln(c(\mathbf{r})\Delta v/N))$. (We can omit the kinetic energy, which just adds a constant to F_1.) Multiplying by N gives the total free energy.

**Your
Turn
6K**

a. Show that

$$F = \int d^3\mathbf{r}\, c(\mathbf{r})\, (U(\mathbf{r}) + k_B T \ln(c(\mathbf{r})/c_*)), \qquad (6.38)$$

for some constant c_*. Why don't we care about the value of this constant?

b. Compare the entropic part of your result with the one for an isolated system, Your Turn 6J on page 233. Which result was easier to derive?

PROBLEMS

6.1   Tall tale
The mythical lumberjack Paul Bunyan usually cut down trees, but on one occasion he attempted to diversify and run his own sawmill. As the historians tell it, "Instead of turning out lumber the mill began to take in piles of sawdust and turn it back into logs. They soon found out the trouble: A technician had connected everything up backwards."

Can we reject this story on the basis of the Second Law?

6.2   Entropy change upon equilibration
Consider two boxes of ideal gas. The boxes are thermally isolated from the world and, initially, from each other as well. Each box holds N molecules in volume V. Box 1 starts with temperature $T_{i,1}$, whereas box 2 starts with $T_{i,2}$. (The subscript "i" means "initial," and "f" will mean "final.") So the initial total energies are $E_{i,1} = N\frac{3}{2}k_B T_{i,1}$ and $E_{i,2} = N\frac{3}{2}k_B T_{i,2}$.

Now we put the boxes into thermal contact with each other but still isolated from the rest of the world. We know they'll eventually come to the same temperature, as argued in Equation 6.10.

a. What is this temperature?

b. Show that the change of total entropy S_{tot} is then

$$k_B \frac{3}{2} N \ln \frac{(T_{i,1} + T_{i,2})^2}{4 T_{i,1} T_{i,2}}.$$

c. Show that this change is always ≥ 0. [*Hint:* Let $X = \frac{T_{i,1}}{T_{i,2}}$ and express the change of entropy in terms of X. Plot the resulting function of X.]

d. Under a special circumstance, the change in S_{tot} will be zero: When? Why?

6.3   Bobble Bird
The Bobble Bird toy dips its beak into a cup of water, rocks back until the water has evaporated, then dips forward and repeats the cycle. All you need to know about the internal mechanism is that after each cycle, it returns to its original state: There is no spring winding down and no internal fuel getting consumed. You could even attach a little ratchet to the toy and extract a little mechanical work from it, maybe lifting a small weight.

a. Where does the energy to do this work come from?

b. Your answer in (a) may at first seem to contradict the Second Law. Explain why it does not. [*Hint:* What system discussed in Chapter 1 does this device resemble?]

6.4   Efficient energy storage
Section 6.5.3 discussed an energy-transduction machine. We can see some similar lessons from a simpler system, an energy-*storage* device. Any such device in the cellular world will inevitably lose energy, as a result of viscous drag, so we imagine pushing a ball through a viscous fluid. We push with constant external force f; as the ball moves, it compresses a spring (Figure 6.11). According to the Hooke relation, the

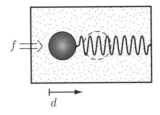

Figure 6.11: A simple energy-storage device. A tank filled with a viscous fluid contains an elastic element (spring) and a bead, whose motion is opposed by viscous drag.

spring resists compression with an elastic force $f = kd$, where k is the spring constant.[8] When this force balances the external force, the ball stops moving, at $d = f/k$.

Throughout the process, the applied force was fixed, so by this point we've done work $fd = f^2/k$. But integrating the Hooke relation shows that our spring has stored only $\int_0^d f(x)\,dx = \frac{1}{2}kd^2$, or $\frac{1}{2}f^2/k$. The rest of the work we did went to generate heat. Indeed, at every position x along the way from 0 to d, some of the applied force compresses the spring while the rest goes to overcome viscous friction.

Nor can we get back all the stored energy, $\frac{1}{2}f^2/k$, because we lose even more to friction as the spring relaxes. Suppose that we suddenly reduce the external force to a value f_1 that is smaller than f.

a. Find how far the ball moves and how much work it does against the external force. We'll call the latter quantity the "useful work" recovered from the storage device.

b. For what constant value of f_1 will the useful work be maximal? Show that even with this optimal choice, the useful work output is only half of what was stored in the spring, or $\frac{1}{4}f^2/k$.

c. How could we make this process more efficient? [*Hint:* Keep in mind Idea 6.20.]

6.5   *Atomic polarization*

Suppose that we have a lot of noninteracting atoms (a gas) in an external magnetic field. You may take as given the fact that each atom can be in one of two states, whose energies differ by an amount $\Delta E = 2\mu B$, depending on the strength of the magnetic field B. Here μ is some positive constant, and B is also positive. Each atom's magnetization is taken to be $+1$ if it's in the lower energy state or -1 if it's in the higher state.

a. Find the *average* magnetization of the entire sample as a function of the applied magnetic field B. [*Remark:* Your answer can be expressed in terms of ΔE by using a hyperbolic trigonometric function; if you know these, then write it this way.]

b. Discuss how your solution behaves when $B \to \infty$ and when $B \to 0$, and why your results make sense.

6.6   *Polymer mesh*

D. Discher studied the mechanical character of the red blood cell cytoskeleton, a polymer network attached to its inner membrane. Discher attached a bead of diameter

---

[8]Another Hooke relation appeared in Chapter 5, where the force resisting a shear deformation was proportional to the size of the deformation (Equation 5.14 on page 172).

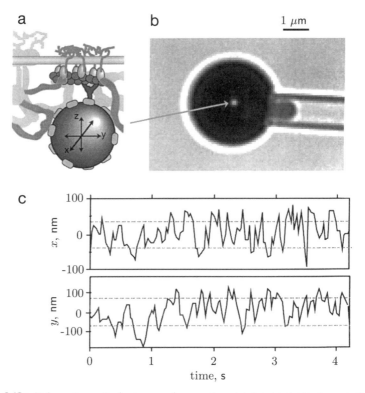

Figure 6.12: (Schematic; optical micrograph; experimental data.) (a) Attachment of a single fluorescent nanoparticle to actin in the red blood cell cortex. (b) The red cell, with attached particle, is immobilized by partially sucking it into a micropipette (*right*) of diameter $1\,\mu$m. (c) Tracking of the thermal motion of the nanoparticle gives information about the elastic properties of the cortex. [Digital image kindly supplied by D. Discher; see Discher, 2000.]

40 nm to this network (Figure 6.12a). The network acts as a spring, constraining the free motion of the bead. He then asked, "What is the stiffness (spring constant) of this spring?"

In the macroworld, we'd answer this question by applying a known force to the bead, measuring the displacement Δx in the x direction, and using $f = k\Delta x$. But it's not easy to apply a known force to such a tiny object. Instead, Discher just passively observed the thermal motion of the bead (Figure 6.12c). He found the bead's root-mean-square deviation from its equilibrium position, at room temperature, to be $\sqrt{\langle(\Delta x)^2\rangle} = 35\,$nm; from this, he computed the spring constant k. What value did he find?

6.7 *Inner ear*

A. J. Hudspeth and coauthors found a surprising phenomenon while studying signal transduction by the inner ear. Figure 6.13a shows a bundle of stiff fibers (called stereo-ocilia) projecting from a sensory cell. The fibers sway when the surrounding inner-ear

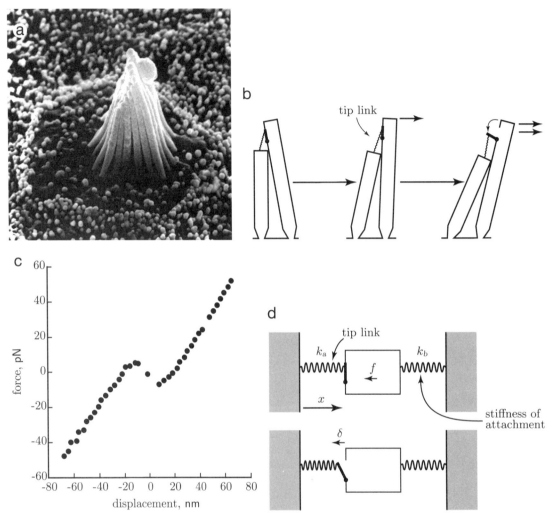

Figure 6.13: (Scanning electron micrograph; diagram; experimental data; diagram) (a) Bundle of stereocilia projecting from an auditory hair cell. (b) Pushing the bundle to the right causes a relative motion between two neighboring stereocilia in the bundle, stretching the tip link, a thin filament joining them. At large enough displacement, the tension in the tip link can open a "trap door." (c) Force exerted by the hair bundle in response to imposed displacements. Positive values of f correspond to forces directed to the left in (b); positive values of x correspond to displacements to the right. (d) Mechanical model for stereocilia. The *left spring* represents the tip link. The spring on the *right* represents the stiffness of the attachment point where the stereocilium joins the main body of the hair cell. The two springs exert a combined force f. The model envisions N of these units in parallel. [(a) Digital image kindly supplied by A. J. Hudspeth; (c) data from Martin et al., 2000.]

fluid moves. Other micrographs (not shown) revealed thin, flexible filaments (called tip links) joining each fiber in the bundle to its neighbor (wiggly line in the sketch, Figure 6.13b).

The experimenters measured the force-displacement relation for the bundle by using a tiny glass fiber to poke it. A feedback circuit maintained a fixed displacement for the bundle's tip and reported back the force needed to maintain this displacement. The surprise is that the experiments gave the complex curve shown in panel (c). A simple spring has a stiffness $k = \frac{df}{dx}$ that is constant (independent of x). The diagram shows that the bundle of stereocilia behaves like a simple spring at large deflections; but in the middle, it has a region of *negative* stiffness!

To explain their observations, the experimenters hypothesized a trap door at one end of the tip link (top right of the wiggly line in Figure 6.13b), and proposed that the trap door was effectively a two-state system.

a. Explain qualitatively how this hypothesis helps us to understand the data.

b. In particular, explain why the bump in the curve is rounded, not sharp.

c. In its actual operation, the hair bundle is not clamped; its displacement can wander at will, subject to applied forces from motion of the surrounding fluid. At zero applied force, the curve shows *three* possible displacements, at about -20, 0, and $+20$ nm. But really, we will never observe one of these three values. Which one? Why?

6.8 $\boxed{T_2}$ *Energy fluctuations*

Figure 6.2 implies that the relative fluctuations of energy between two macroscopic subsystems in thermal contact will be very small in equilibrium. Confirm this statement by calculating the root-mean-square deviation of E_A as a fraction of its mean value. [*Hints:* Suppose the two subsystems are identical, as assumed in the figure. Work out the probability $P(E_A)$ that the joint system will be in a microstate with E_A on one side and $E_{tot} - E_A$ on the other side. Approximate $\ln P(E_A)$ near its peak by a suitable quadratic function, $A - B(E_A - \frac{1}{2}E_{tot})^2$. Use this approximate form to estimate the RMS deviation.]

6.9 $\boxed{T_2}$ *The Langevin function*

Repeat Problem 6.5 for a slightly different situation: Instead of having just two discrete allowed values, our system has a continuous, unit-vector variable $\hat{n}$ that can point in any direction in space. Its energy is a constant plus $-a\hat{n} \cdot \hat{z}$, or $-a\hat{n}_z = -a\cos\theta$. Here a is a positive constant with units of energy and θ is the polar angle of $\hat{n}$.

a. Find the probability distribution $P(\theta, \varphi)\,d\theta\,d\varphi$ for the directions that $\hat{n}$ may point.

b. Compute the partition function $Z(a)$ and the free energy $F(a)$ for this system. Then compute the quantity $\langle \hat{n}_z \rangle$. (Your answer is sometimes called the **Langevin function**.) Find the limiting behavior at high temperature and make sure your answer is reasonable.

6.10 $\boxed{T_2}$ *Gating compliance*

(Continuation of Problem 6.7.) We can model the system in Figure 6.13 quantitatively as follows. We think of the bundle of stereocilia as a collection of N elastic units in parallel. Each element has two springs: One, with spring constant k_a and equilibrium position x_a, represents the elasticity of the tip link filament. The other spring, characterized by k_b and x_b, represents the stiffness of the stereocilium's attachment point (provided by a bundle of actin filaments). See panel (d) of the figure.

The first spring attaches via a hinged element (the "trap door"). When the hinge is in its open state, the attachment point is a distance δ to the left of its closed state relative to the body of the stereocilium. The trap door is itself a two-state system with a free energy change ΔF_0 to jump to its open state.

a. Derive the formula $f_{\text{closed}}(x) = k_a(x - x_a) + k_b(x - x_b)$ for the net force on the stereocilium in the closed state. Rewrite this in the more compact form $f_{\text{closed}} = k(x - x_1)$ and find the effective parameters k and x_1 in terms of the earlier quantities. Then find the analogous formula for the state in which the trap door is open.

b. The total force f_{tot} is the sum of N terms. In $P_{\text{open}}N$ of these terms, the trap door is open; in the remaining $(1 - P_{\text{open}})N$, it is closed. To find the open probability using Equation 6.34 on page 225, we need the free energy difference $\Delta F(x)$ between the system's two states (at fixed x). This difference is a constant, ΔF_0, plus a term involving the energy stored in spring a. Get a formula for $\Delta F(x)$.

c. Assemble the pieces of your answer to get the force $f_{\text{tot}}(x)$ in terms of the unknown parameters N, k_a, k_b, x_a, x_b, δ, and ΔF_1, where $\Delta F_1 \equiv \Delta F_0 + \frac{1}{2} k_a \delta^2$. That's a lot of parameters, but some of them enter only in fixed combination. Show that your answer can be expressed as

$$f_{\text{tot}}(x) = K_{\text{tot}}x + f_0 - \frac{Nz}{1 + e^{-z(x - x_0)/k_B T}},$$

and find the quantities K_{tot}, f_0, z, and x_0 in terms of the earlier parameters.

d. Hudspeth and coauthors fit this model to their data and to other known facts. They found $N = 65$, $K_{\text{tot}} = 1.1 \text{ pN nm}^{-1}$, $x_0 = -2.2 \text{ nm}$, and $f_0 = 25 \text{ pN}$. Graph the formula in (c), using these values. Use various trial values for z, starting from zero and moving upward. What value of z gives a curve resembling the data?

e. The authors also estimated that $k_a = 2 \cdot 10^{-4} \text{ N m}^{-1}$. Use this value and your answer from (d) to find δ. Is this a reasonable value?

CHAPTER 7

Entropic Forces at Work

> *If someone points out to you that your pet theory of the Universe is contradicted by experiment, well, these experimentalists do bungle things sometimes. But if your theory is found to be against the Second Law, I can give you no hope; there is nothing for it but to collapse in deepest humiliation.*
>
> — Sir Arthur Eddington, 1944

Chapter 6 argued that all transactions in a fixed-temperature system are paid for in a single unified currency, the free energy $F = E - TS$. The irreversible processes discussed up to that point emerge as particular cases. For example, a freely falling rock converts its gravitational potential energy to kinetic energy, with no net change in the mechanical energy E. If it lands in mud, however, its organized kinetic energy gets irreversibly converted to thermal form, thereby lowering E and hence F. Similarly, ink diffuses in water to maximize its entropy, thereby raising S and again lowering F. More generally, if *both* energy and entropy change, a macroscopic system in contact with a thermal reservoir will change to lower its free energy, even if

- The change actually increases the energy (but increases TS more), or
- The change actually decreases the entropy (but decreases E/T more).

   In first-year physics, the change in potential energy as some state variable changes is called a mechanical force. More precisely, we write

$$f = -\mathrm{d}U/\mathrm{d}x.$$

Section 6.5.2 extended this identification to statistical systems, starting with the simplest sort of **entropic force**, namely, gas pressure. We found that the force exerted by a gas can be regarded as the derivative of $-F$ with respect to the position of a confining piston. This chapter will elaborate the notion of an entropic force, extending it to cases of greater biological relevance. For example, Chapter 2 claimed that the amazing specificity of enzymes and other molecular devices stems from the precise shapes of their surfaces and from short-range physical interactions between those surfaces and the molecules on which they act. This chapter will explore the origins of some of these entropic forces: the electrostatic, hydrophobic, and depletion effects. As always, we will look for quantitative confirmation of our formal derivations in controlled experiments.

245

The Focus Questions for this chapter are

Biological question: What keeps cells full of fluid? How can a membrane push fluid against a pressure gradient?

Physical idea: Osmotic pressure is a simple example of an entropic force.

7.1 MICROSCOPIC VIEW OF ENTROPIC FORCES

Before proceeding to new ideas, let's take two last looks at the ideal gas law. We already understand the formula for pressure, $p = k_B T N/V$, from a mechanistic point of view (Section 3.2.1 on page 78). Let's now recover this result by using the partition function—a useful warm-up for our study of electrostatic forces later in this chapter and for single-molecule stretching in Chapter 9.

7.1.1 Fixed-volume approach

Suppose that a chamber with gas is in thermal contact with a large body at fixed temperature T. In an ideal gas, N particles of mass m move independently of one another, constrained only by the walls of their vessel, a cube with sides of length L. The total energy is then just the sum of the molecules' kinetic energies, plus a constant for their unchanging internal energies.

   The free energy formula Example (page 224) gives us a convenient way to calculate the free energy of this system, by first calculating the partition function. Indeed, in this situation, the general formula for the partition function (Equation 6.33 on page 224) becomes very simple. To specify a state of the system, we must give the positions $\{\mathbf{r}_i\}$ and momenta $\{\mathbf{p}_i\}$ of every particle. To sum over all possible states, we therefore must sum over every such collection of $\mathbf{r}_1, \ldots, \mathbf{p}_N$. Because position and momentum are continuous variables, we write the sums as integrals:

$$Z(L) = C \int_0^L d^3\mathbf{r}_1 \int_{-\infty}^{\infty} d^3\mathbf{p}_1 \cdots \int_0^L d^3\mathbf{r}_N \int_{-\infty}^{\infty} d^3\mathbf{p}_N \, e^{-(\mathbf{p}_1^2 + \cdots + \mathbf{p}_N^2)/(2mk_B T)}. \quad (7.1)$$

Don't confuse the vectors $\mathbf{p}_i$ (momentum) with the scalar p (pressure)! The limits on the integrals mean that each of the three components of $\mathbf{r}_i$ runs from 0 to L. The factor C includes a factor of $e^{-\epsilon_i/k_B T}$ for each particle, where ϵ_i is the internal energy of molecule i. For an ideal gas, these factors are all fixed, so C is a constant; we won't need its numerical value. (In Chapter 8, we *will* let the internal energies change, to study chemical reactions.) The free energy formula Example then gives $F(L) = -k_B T \ln Z(L)$.

   Equation 7.1 looks awful; but all we really want is the *change* in free energy as we change the volume of the box, because an entropic force is a derivative of the free energy (see Equation 6.17 on page 213). To get the dimensions of pressure (force/area or energy/volume), we need $-dF(L)/d(L^3)$. But *most of the integrals in Equation 7.1 are just constants,* as we see by rearranging them:

$$Z(L) = C \left(\int_0^L d^3\mathbf{r}_1 \cdots \int_0^L d^3\mathbf{r}_N \right) \left(\int_{-\infty}^{\infty} d^3\mathbf{p}_1 e^{-\mathbf{p}_1^2/(2mk_B T)} \cdots \int_{-\infty}^{\infty} d^3\mathbf{p}_N \, e^{-\mathbf{p}_N^2/(2mk_B T)} \right).$$

The only dependence on L comes via the *limits* of the first $3N$ integrals, and each of these integrals just equals L. Thus Z is a constant times L^{3N}, so $F(L)$ is a constant plus $-k_{\mathrm{B}}TN \ln L^3$. This result is reasonable: In an ideal gas, the total potential energy is a constant (the particles don't change, and they don't interact with one another), and so is the total kinetic energy (it's just $\frac{3}{2}k_{\mathrm{B}}TN$). Hence the free energy $F = E - TS$ is a constant minus TS. So we have recovered a fact we already knew, that the entropy is a constant plus $Nk_{\mathrm{B}} \ln L^3$ (see the ideal gas entropy Example, page 200). Differentiating the free energy recovers the ideal gas law:

$$p = -\frac{\mathrm{d}F}{\mathrm{d}(L^3)} = \frac{k_{\mathrm{B}}TN}{V}. \tag{7.2}$$

7.1.2 Fixed-pressure approach

A slight rephrasing of the argument just given will prove useful for our discussion of macromolecular stretching in Chapter 9. Instead of fixing V and finding p, let's fix the *pressure* and find the equilibrium *volume*. That is, instead of a box of fixed volume, we now imagine a cylinder with a sliding piston. The displacement L of the piston is variable; its area A is fixed. A force f pushes the piston inward; thus, the potential energy of the mechanism pushing the piston is fL. The available volume for the gas molecules is now AL. Thus we'd like to compute the average value $\langle L \rangle$, given the externally supplied force. This average is given by a sum over all states of L times the probability to be in the given state.

The Boltzmann distribution gives the probability of having a specified set of positions and momenta:

$$P(\mathbf{r}_1, \ldots, \mathbf{p}_N, L, \mathbf{p}_{\mathrm{piston}}) = C_1 \exp\left[-\left(\frac{\mathbf{p}_1^2 + \cdots + \mathbf{p}_N^2}{2m} + \frac{(\mathbf{p}_{\mathrm{piston}})^2}{2M} + fL\right)/k_{\mathrm{B}}T\right]. \tag{7.3}$$

In this formula, m is the gas particle mass, M is the mass of the piston, and $\mathbf{p}_{\mathrm{piston}}$ is its momentum.

We wish to calculate $\int L \times P(L, \ldots)$. It's convenient to use the fact that P, like any probability distribution, is normalized;[1] thus, its integral equals 1, and we can rewrite our desired quantity as

$$\langle L \rangle = \frac{\int L \times P(L, \mathbf{r}_1, \ldots)\, \mathrm{d}^3\mathbf{r}_1 \cdots \mathrm{d}^3\mathbf{p}_N \mathrm{d}\mathbf{p}_{\mathrm{piston}}\mathrm{d}L}{\int P(L, \mathbf{r}_1, \ldots)\, \mathrm{d}^3\mathbf{r}_1 \cdots \mathrm{d}^3\mathbf{p}_N \mathrm{d}\mathbf{p}_{\mathrm{piston}}\mathrm{d}L}. \tag{7.4}$$

It was convenient to introduce the denominator in Equation 7.4 because now most of the integrals simply cancel between the numerator and denominator, as does the constant C_1, leaving

$$\langle L \rangle = \frac{\int_0^\infty \mathrm{d}L\, e^{-fL/k_{\mathrm{B}}T} L^N \times L}{\int_0^\infty \mathrm{d}L\, e^{-fL/k_{\mathrm{B}}T} L^N}. \tag{7.5}$$

---

[1]See Section 3.1.1 on page 70.

Your Turn 7A

> **Your**
> **Turn**
> **7A**
>
> **a.** Check that Equation 7.5 equals $(N + 1)k_B T/f$. [*Hint:* Integrate by parts to make the numerator of Equation 7.5 look more like the denominator.]
>
> **b.** Show that we have once again derived the ideal gas law. [*Hint:* Remember that N is so big that $N + 1 \approx N$.]

Here is one last reformulation. The trick of differentiating under the integral sign[2] shows that Equation 7.5 can be written compactly as $\langle L \rangle = \mathrm{d}(-k_B T \ln Z(f))/\mathrm{d}f$, where $Z(f)$ is the partition function of the gas + piston system. Replacing f by pA and L by V/A gives

$$\langle V \rangle = \mathrm{d}(-k_B T \ln Z(p))/\mathrm{d}p = \mathrm{d}F(p)/\mathrm{d}p, \qquad (7.6)$$

where p is the pressure.

$\boxed{T_2}$ *Section 7.1.2′ on page 283 introduces the idea of thermodynamically conjugate pairs.*

7.2 OSMOTIC PRESSURE

7.2.1 Equilibrium osmotic pressure follows the ideal gas law

We can turn now to the problem of osmotic pressure (see Figure 1.3 on page 13). A membrane divides a rigid container into two chambers, one with pure water, the other containing a solution of N solute particles in volume V. The solute could be anything from individual molecules (sugar) to colloidal particles. We suppose the membrane to be permeable to water but not to solute. A very literal example would be an ultrafine sieve, with pores too small to pass solute particles. The system will come to an equilibrium with greater hydrostatic pressure on the sugar side, which we measure (Figure 7.1). We'd like a quantitative prediction for this pressure.

One might think that the situation just described would be vastly more complicated than the ideal-gas problem just studied. After all, the solute molecules are constantly in the crowded company of water molecules; hydrodynamics rears its head, and so on. But examining the arguments of Section 7.1.2, we see that they apply equally well to the osmotic problem. It is true that the solute molecules interact strongly with the water, as do the water molecules among themselves. But, in a dilute solution, the solute particles don't interact much with *each other*, so the total energy of a microstate is unaffected by their locations. More precisely, the integral over the positions of the solute molecules is dominated by the large domain where no two are close enough to interact significantly. (This approximation breaks down for concentrated solutions, just as the ideal gas law fails for dense gases.)

Thus for dilute solutions, we can do all the integrals over the solute particle locations $\int \mathrm{d}^3\mathbf{r}_1 \cdots \mathrm{d}^3\mathbf{r}_N$ first: Just as in the derivation of Equation 7.2, we get V^N. This time V is the volume of only that part of the chamber accessible to solute (right-hand

---

[2]See Equation 3.14 on page 75.

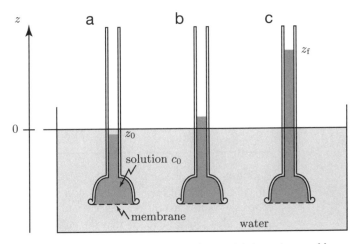

Figure 7.1: (Schematic.) Osmotic pressure experiment. (a) A semipermeable membrane is stretched across a cup-shaped vessel containing sugar solution with concentration c_0. The vessel is then plunged into pure water. Initially, the sugar solution extends to a height z_0 in the neck of the cup. (b) Solution begins to rise in the vessel by osmotic flow, until (c) it reaches an equilibrium height z_f. The pressure in the final equilibrium state is the final height z_f times $\rho_m g$, where ρ_m is the mass density of the solution.

side of Figure 1.3). Because the membrane is essentially invisible to water molecules, *nothing else in the partition function depends on V.* Hence the sum over the positions and momenta of all the water molecules just contributes a constant factor to Z, and such factors cancel from formulas like Equation 7.2.

The equilibrium osmotic pressure p_{equil} in Figure 1.3 is thus given by the ideal gas law:

$$p_{equil} = c k_B T. \qquad \textbf{van 't Hoff relation} \qquad (7.7)$$

Here $c = N/V$ is the number density of solute molecules. p_{equil} is the force per area that we must apply to the solute side of the apparatus to get equilibrium.

The preceding discussion was appropriate to the situation shown in Figure 1.3 on page 13, where we somewhat artificially assumed that there was no air, and hence no atmospheric pressure, outside the apparatus. In the more common situation shown in Figure 7.1, we again get a relation of the form Equation 7.7, but this time for the *difference* in pressure between the two sides of the membrane. Thus $\Delta p = z_f \rho_m g$, where z_f is the final height of the column of fluid, ρ_m is the mass density of solution, and g is the acceleration of gravity. In this case, we conclude that *the equilibrium height of the fluid column is proportional to the solute concentration in the cup.*

The van 't Hoff relation explains a mysterious empirical fact from Chapter 1, the formula for the maximum work that can be done by the osmotic machine (Equation 1.7). Consider again Figure 1.3b on page 13. Suppose that the solvent flows until the volume of the right side (with solute) has doubled. Throughout the flow, the piston has been harnessed to a load. To extract the maximum possible work from the system, we continuously adjust the load to be almost, but not quite, big enough to stop the flow.

> **Your Turn 7B**
>
> Integrate Equation 7.7 to find the maximum total work the piston can do against the load. Compare your answer with Equation 1.7 on page 15 and find the value of the constant of proportionality γ.

Estimates We need some estimates to see whether osmotic pressure is really significant in the world of the cell. Suppose that a cell contains globular proteins, roughly spheres of radius 10 nm, at a concentration such that 30% of the cell's volume is occupied with protein (we say that the **volume fraction** ϕ equals 0.3). This is not an unreasonable picture of red blood cells, which are stuffed with hemoglobin. To find the concentration c in Equation 7.7, we set 0.3 equal to the number of proteins per volume times the volume of one protein:

$$0.3 = c \times \frac{4\pi}{3}(10^{-8}\,\mathrm{m})^3. \tag{7.8}$$

Thus $c \approx 7 \cdot 10^{22}\,\mathrm{m}^{-3}$. To phrase this in more familiar units, remember that one mole per liter corresponds to a concentration of $N_{\mathrm{mole}}/(10^{-3}\,\mathrm{m}^3)$. We'll call a solution of 1 mole/L a one **molar** solution, defining the symbol $\mathrm{M} = \mathrm{mole/L}$. Recalling that in this book the word *mole* is a synonym for Avogadro's number (see Section 1.5.1 on page 23), we find that $c = 1.2 \cdot 10^{-4}\,\mathrm{M}$: It's a 0.12 mM solution.[3]

Thus, if we suspend our cell in pure water, the pressure needed to stop the inward flow of water equals $k_\mathrm{B}T_r c \approx 300\,\mathrm{Pa}$. That's certainly much smaller than atmospheric pressure ($10^5\,\mathrm{Pa}$). But is it big for a cell?

Suppose that the cell has radius $R = 10\,\mu\mathrm{m}$. The excess internal pressure will create tension in the cell's membrane: Every part of the membrane pulls on every other part. We describe tension by imagining a line drawn on the surface; the membrane on the left of the line pulls the membrane on the other side with a certain force per unit length, called the **surface tension** Σ. But force per length has the same units as energy per area; and indeed, to stretch a membrane to greater area, from A to $A+\mathrm{d}A$, we must do work. If we draw two closely spaced, parallel lines of length ℓ, the work to increase their separation from x to $x+\mathrm{d}x$ equals $(\ell\Sigma) \times \mathrm{d}x$. Equivalently, the work equals $\Sigma \times \mathrm{d}A$, where $\mathrm{d}A = \ell\mathrm{d}x$ is the change in *area*. Similarly, to stretch a spherical cell from radius R to $R + \mathrm{d}R$ would increase its area by $\mathrm{d}A = (\mathrm{d}R)\frac{\mathrm{d}A}{\mathrm{d}R} = 8\pi R\mathrm{d}R$ and cost energy equal to $\Sigma \times \mathrm{d}A$.

The cell will stretch until the energy cost of further stretching its membrane balances the free energy reduction from letting the pressurized interior expand. The latter is just $p\mathrm{d}V = p\frac{\mathrm{d}V}{\mathrm{d}R}\mathrm{d}R = p4\pi R^2\mathrm{d}R$. Balancing this gain against $\Sigma \times 8\pi R\mathrm{d}R$

---

[3] In this book, we pretend that dilute-solution formulas are always applicable; so we will not distinguish between molar and molal concentration.

shows that the equilibrium surface tension is

$$\Sigma = Rp/2. \qquad \textbf{Laplace's formula} \qquad\qquad (7.9)$$

Substituting our estimate for p yields $\Sigma = 10^{-5}\,\mathrm{m} \times 300\,\mathrm{Pa}/2 = 1.5 \cdot 10^{-3}\,\mathrm{N\,m^{-1}}$. This tension is roughly enough to rupture a eukaryotic cell membrane, thereby destroying the cell. *Osmotic pressure is significant for cells.*

The situation is even more serious with a small solute like salt. Bilayer membranes are almost impermeable to sodium and chloride ions. And a 1 M salt solution contains about 10^{27} ions per $\mathrm{m^3}$, ten thousand times more than in the protein example just given! Indeed, you cannot dilute red blood cells with pure water; at low concentrations of exterior salt they burst, or **lyse**. Clearly, to escape lysis, living cells must precisely fine-tune their interior concentrations of dissolved solutes, an observation to which we will return in Chapter 11.

7.2.2 Osmotic pressure creates a depletion force between large molecules

Take a look at Figure 7.2. One thing is clear from this picture: It's crowded inside a cell. Not only that, but there is a *hierarchy* of objects of all different sizes, from the enormous ribosomes on down to sugars and tiny single ions (see Figure 2.4 on page 38). This hierarchy can lead to a surprising entropic effect, called the **depletion interaction** or **molecular crowding**.

Consider two large solid objects ("sheep") in a bath containing a suspension of many smaller objects ("sheepdogs") with number density c. (Admittedly, it's an unusual farm where the sheepdogs outnumber the sheep.) We will see that the sheepdogs give rise to an effect tending to herd the sheep together, a purely entropic force having nothing to do with any direct attraction between the large objects.

The key observation, made by S. Asakura and F. Oosawa in 1954, is that each of the large objects is surrounded by a **depletion zone** of thickness equal to the radius R of the small particles; the centers of the small particles cannot enter this zone. Figure 7.3 sketches the idea. Two surfaces of area A approach each other in the presence of smaller particles. The depletion zone reduces the volume available to the small particles; conversely, eliminating it would increase their entropy and hence lower their free energy.

Now let the two surfaces come together. If their shapes match, then as they approach, their depletion zones merge and finally disappear (Figure 7.3b). The corresponding reduction in free energy gives an entropic force driving the surfaces into contact. The effect does not begin until the two surfaces approach each other to within the diameter $2R$ of the small particles: *The depletion interaction is of short range.* Even if the two surfaces' shapes do not match precisely, there will still be a depletion interaction as long as their shapes are similar on the length scale of the small particles. For example, when two big spheres meet (or when a big sphere meets a flat wall), their depletion zones will shrink as long as their radii are much bigger than R, because they look flat to the small spheres.

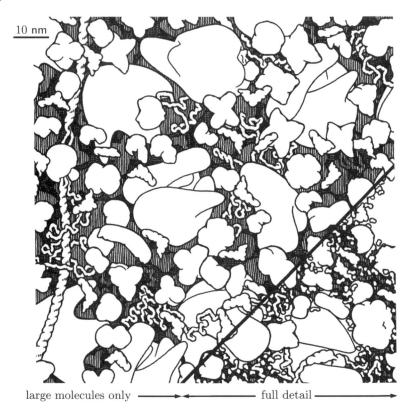

10 nm

large molecules only ——————→ ←—————— full detail —————→

Figure 7.2: (Drawing, based on structural data.) It's crowded inside *E. coli*. For clarity, the main part of the figure shows only the macromolecules; the lower right inset includes smaller molecules (water molecules, however, are still omitted). *Left side*, a strand of DNA (*far left*) is being transcribed to messenger RNA, which is immediately translated into new proteins by ribosomes (*largest objects shown*). Between the ribosomes, proteins of many shapes and sizes are breaking down small molecules for energy and synthesizing new molecules for growth and maintenance. [From Goodsell, 1993.]

We can also interpret the depletion interaction in the language of pressure. Figure 7.3b shows a small particle that attempts to enter the gap but bounces away instead. It's as though there were a semipermeable membrane at the entrance to the gap, admitting water but not particles. The osmotic pressure across this virtual membrane sucks water out of the gap, thereby forcing the two large particles into contact. The pressure is the change of free energy per change of volume (Equation 7.2). As we bring the surfaces into contact, the volume of the depletion zone between them shrinks from $2RA$ to zero. Multiplying this change by the pressure drop ck_BT in the zone gives

$$(\Delta F)/A = ck_BT \times 2R. \tag{7.10}$$

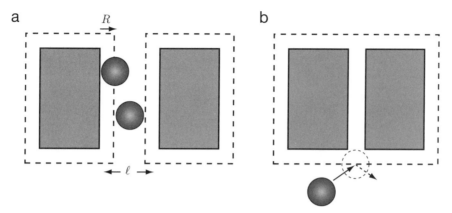

Figure 7.3: (Schematic.) Origin of the depletion interaction. (a) Two surfaces of area A with matching shapes are initially separated by a distance ℓ that is more than twice the radius R of some suspended particles. Each surface is surrounded by a depletion zone of thickness R (*dashed lines*). (b) When the surfaces get closer than $2R$, the depletion zones merge and their combined volume decreases.

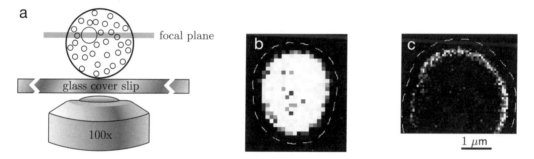

Figure 7.4: (Schematic; experimental data.) An experiment measuring depletion interactions. (a) Experimental setup. A microscope looks at the central plane of a rigid vesicle containing a polystyrene sphere (the "sheep") of radius 0.24 μm. (b) Histogram of the measured location of the large sphere's center over 2000 observations. The solvent in this case contained no smaller objects. Instead of displaying frequencies by the height of bars, the figure shows how often the sphere was found in each location by the shade of the spot at that position; lighter shades denote places where the sphere was more often found. The *dashed line* represents the actual edge of the vesicle; the sphere's center can come no closer than its radius. (c) Conditions similar to (b), except that the vesicle contained a suspension of smaller, 0.04 μm spheres ("sheepdogs") with volume fraction about 30%. Although the "sheepdogs" are not optically visible, they cause the "sheep" to spend most of its time clinging to the wall of the chamber. [Digital images kindly supplied by A. Dinsmore; see Dinsmore et al., 1998.]

The rearrangement of a thin layer around a huge particle may seem unimportant, but the total effect of the depletion interaction can be considerable (see Problem 7.5).

   A. Dinsmore and coauthors gave a clear experimental demonstration of the depletion interaction (Figure 7.4). They prepared a vesicle containing one large particle, about a quarter of a micrometer in radius, and a solution. In one trial, the solution contained a suspension of smaller particles, of radius 0.04 μm; in another trial, these particles were absent, with everything else the same. After carefully arranging conditions to eliminate all other interactions between the large particles (for example, electrostatic forces), they found a dramatic effect: The mere presence of the small "sheepdog" particles forced the large particle to spend most of its time at the wall of the vesicle. By analyzing what fraction of the time the large particle spent at the wall, the experimenters measured the free energy reduction when the particle was sticking there and quantitatively verified the estimate Equation 7.10 (appropriately modified for a curved surface).

Replacing the images of sheep by large macromolecules, and of sheepdogs by polymer coils or small globular proteins, we see that the presence of small objects can significantly help the large macromolecules to find each others' specific recognition sites. For example, introduction of bovine serum albumin (BSA, a protein) or polyethylene glycol (PEG, a polymer) reduces the solubility of deoxyhemoglobin and other large proteins by helping them to stick together; the magnitude of the effect can be a 10-fold reduction of the solubility. Dextran or PEG can also stabilize complexes against thermal disruption: For instance, adding PEG can increase the melting temperature of DNA by several degrees (see Chapter 9) and enhance the association of protein complexes by an order of magnitude or more. In all these examples, we see the general theme that the entropic part of a reaction's free energy change, $-T\Delta S$, is interchangeable with the energetic term in ΔF. Either of these changes can affect the reaction's equilibrium point (see Section 6.6.4 on page 225).

Crowding can also speed up reactions, as the sheepdogs jockey the sheep into their best contact. The presence of a "crowding agent" like PEG or BSA can increase the rate of self-assembly of actin filaments, or the action of various enzymes, by orders of magnitude. We can interpret this result in terms of free energy: The entropic contribution to F lowers an activation barrier to assembly (see Section 6.6.2). Indeed, some cellular equipment, for example, the DNA replication system of *E. coli*, just doesn't work in vitro without some added crowding agent. As our simple physical model predicts, it doesn't matter too much what exactly we choose as our crowding agent—all that matters are its size relative to the assembly and its number density.

It may seem paradoxical that the drive toward *dis*order can *assemble* things. But we must remember that the sheepdogs are much more numerous than the sheep. If the assembly of a few big macromolecules liberates some space for many smaller molecules to explore, then the *total* disorder of the system can go up, not down. In just the same way, we will see later how another entropic force, the hydrophobic interaction, can help drive the exquisitely organized folding of a protein or the assembly of a bilayer membrane from its subunits.

7.3 BEYOND EQUILIBRIUM: OSMOTIC FLOW

The discussion of Section 7.2.1 illustrates the power, the beauty, and the unsatisfied feeling we get from very general arguments. We found a quantitative prediction, which works in practice (see Problem 7.2). But we are still left wondering *why* there should be a pressure drop. Pressure involves an honest, Isaac Newton-type force. Force is a transfer of momentum. But the argument given in Section 7.2.1 makes no mention of momentum; instead, we just manipulated entropy (or disorder). Where exactly does the force come from? How does a change in *order* transmute into a flow of *momentum*?

We met an analogous situation in the context of the ideal gas law: The result obtained abstractly in Section 7.1 would not have been very convincing had we not already given a more concrete, albeit less general, argument in Chapter 3. We need the abstract viewpoint because it can take us safely into situations so complicated that the concrete view obscures the point. But whenever possible, we should *also* seek concrete pictures, even if they're very simplified. Accordingly, this section will revisit osmotic pressure, developing a simplified dynamical view of the van 't Hoff relation. As a bonus, we will also learn about nonequilibrium flow, which will be useful when we study ion transport in Chapters 11 and 12. More generally, our discussion will lay the groundwork for understanding many kinds of free energy transducers. For example, Chapter 10 will use such ideas to explain force generation in molecular machines.

7.3.1 Osmotic forces arise from the rectification of Brownian motion

Osmotic pressure gives a force pushing the pistons in Figure 1.3 on page 13 relative to the cylinder. Ultimately, this force must come from the *membrane* separating the two chambers, because only the membrane is fixed relative to the cylinder. Experimentally, one sees this membrane bow as it pushes the fluid, which in turn pushes against the piston. So what we really want to understand is how, and why, the membrane exerts force on (transmits momentum to) the fluid.

To make the discussion concrete, we'll need a number of simplifying assumptions. Some are approximations, whereas others can be arranged to be literally true in carefully controlled experiments. For example, we will assume that our membrane is totally impermeable to solute particles. Such a membrane is called **semipermeable**; the *semi* reminds us that water *does* pass through such a membrane. We will also take the fluid to be essentially incompressible, like water. Finally, as usual we will suppose that everything is constant in the x and y directions.

Imagine a fluid with an external force acting directly on it, like that of gravity. For example, the pressure in a swimming pool increases with depth because in equilibrium, each fluid element must push upward to balance the weight of the column of fluid above it:

$$p(z) = p_0 + \rho_m g \times (z_0 - z). \tag{7.11}$$

Here p_0 is atmospheric pressure, $z_0 - z$ is the depth, and $\rho_m g$ is the weight (force) per unit volume (a similar expression appears in Equation 3.22 on page 80). More generally, the force acting on a fluid may not be a constant. Let $\mathcal{F}(\mathbf{r})$ be an external force per volume acting in the $+\hat{\mathbf{z}}$ direction at position $\mathbf{r}$ and consider a small cube of fluid centered at $\mathbf{r} = (x, y, z)$. Balancing the forces on the cube again shows that in equilibrium, the pressure cannot be constant but instead must vary (Figure 7.5):

$$\left[-p(z + \tfrac{1}{2}\mathrm{d}z) + p(z - \tfrac{1}{2}\mathrm{d}z) \right] \mathrm{d}x\mathrm{d}y + \mathcal{F}(z)\,\mathrm{d}x\mathrm{d}y\mathrm{d}z = 0. \tag{7.12}$$

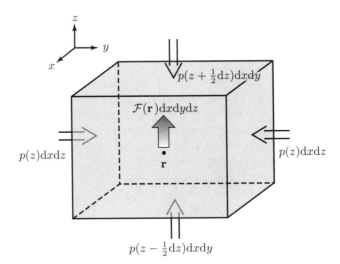

Figure 7.5: (Schematic.) Forces on a small element of fluid. An external force density $\mathcal{F}(\mathbf{r})$ acts on the element's center of mass, at $\mathbf{r} = (x, y, z)$. This force density, and the resulting pressure p, are assumed to be independent of x and y. The fluid's pressure pushes inward on all six sides of the box. The net pressure forces in the x and y directions cancel, but there is a nontrivial requirement for force balance in the $\hat{\mathbf{z}}$ direction.

Taking dz to be small and using the definition of the derivative gives $dp/dz = \mathcal{F}(z)$, the condition for mechanical equilibrium (in this case, called **hydrostatic equilibrium**). Taking the force density $\mathcal{F}$ to be the constant $-\rho_{\mathrm{m}}g$ and solving recovers Equation 7.11 as a special case.

Next imagine a suspension of colloidal particles in a fluid with number density $c(z)$. Suppose that a force $f(z)$ acts along $\hat{\mathbf{z}}$ on *each particle*, depending on the particle's position. (For a literal example of such a situation, imagine two perforated parallel plates in the fluid with a battery connected across them; then a charged particle will feel a force when it's between the plates, but zero force elsewhere.)

In the low Reynolds-number regime, inertial effects are negligibly small (see Chapter 5); so the applied force on each particle is just balanced by a viscous drag from the fluid. The particles, in turn, push back on the fluid, thereby transmitting the applied force to it. Thus, even though the force does not act directly on the fluid, it creates an average force density $\mathcal{F}(z) = c(z)f(z)$ and a corresponding pressure gradient:

$$\frac{dp}{dz} = c(z)f(z). \qquad (7.13)$$

The force on each particle reflects the gradient of that particle's potential energy: $f(z) = -dU/dz$. For example, an impenetrable solid wall creates a zone where the potential energy goes to infinity; the force increases without limit near the wall, pushing any particle away. We'll make the convention that $U \to 0$ far from the membrane (see Figure 7.6b).

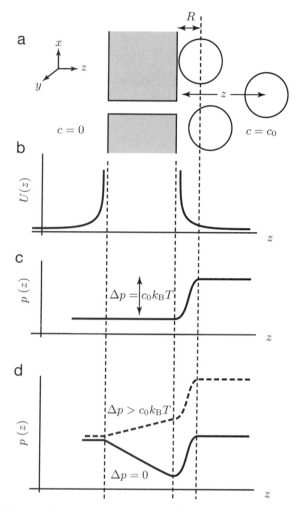

Figure 7.6: (Schematic; sketch graphs.) (a) A literal model of a semipermeable membrane, consisting of a perforated wall with channels too small for suspended particles to pass through. (b) The force along $\hat{z}$ exerted by the membrane on approaching particles is $-dU/dz$, where U is the potential energy of one particle. (c) In equilibrium, the pressure p is constant inside the channel (between the first two *dashed lines*), but p falls in the zone where the particle concentration is decreasing. (d) *Solid curve:* If the pressure on both sides is maintained at the same value, osmotic flow through the channel proceeds at a rate such that the pressure drop across the channel (from viscous drag) cancels the osmotic pressure jump. *Dashed curve:* In reverse osmosis, an external force maintains a pressure gradient even greater than the equilibrium value. The fluid flows in a direction opposite to that seen in ordinary osmotic flow, consistent with the reversed slope of the pressure profile inside the channel.

Equation 7.13 presents us with an apparent roadblock: We have just one equation but two unknown functions, $c(z)$ and $p(z)$. Luckily, we know something else about c: In equilibrium, the Boltzmann distribution gives it as a constant times $e^{-U(z)/k_B T}$, and the constant is just the concentration c_0 in the force-free region, far from the membrane. Then the force density along $\hat{\mathbf{z}}$ is $(c_0 e^{-U/k_B T})(-dU/dz)$, which we rewrite as $c_0 k_B T \frac{d}{dz}[e^{-U(z)/k_B T}]$. According to Equation 7.13, this expression equals dp/dz:

$$\frac{dp}{dz} = k_B T \frac{dc}{dz}.$$

Integrating the equation across the membrane channel and out into the force-free region then shows that $\Delta p = c_0 k_B T$ or more generally, that

> *The equilibrium pressure difference across a semipermeable mem-*
> *brane equals $k_B T$ times the difference in concentration between the*   (7.14)
> *two force-free regions on either side of the membrane.*

We have just recovered the van 't Hoff relation, Equation 7.7 on page 249. Compared with the discussion of Section 7.2.1, however, this time we have a greater level of detail.

Gilbert says: Now I can see the actual mechanism of force generation. When a membrane is impermeable to solute particles, then those particles bounce off the membrane when they approach it. Because of viscous friction, the particles entrain some water as they move, and so water, too, is pulled away from the membrane. But water *can* pass through pores in the membrane, so some is also swept through it. That's osmotic flow; a backward pressure is needed to stop it.

Sullivan: But wait. Even when the particles are *free* (no membrane), their Brownian motion disturbs the surrounding fluid! What's your argument got to do with the osmotic case?

Gilbert: That's true, but the effect you mention is random and averages to zero. In contrast, the membrane exerts only rightward, never leftward, forces on the solute particles. This force does *not* average to zero. So its effect is to **rectify** the Brownian motion of the nearby particles, that is, to create a net motion in one direction.

Sullivan: It still seems like you get something for nothing.

Gilbert: No, the rectification comes at a price: To do useful work, the piston must move, thereby increasing the volume of the side with solute. This change costs *order*, as required by the abstract Idea 6.19 on page 214.

Gilbert has put a finger on where the net momentum flow into the fluid comes from. Particles constantly impinge on the membrane in Figure 7.6a from the right, never from the left. Each time the membrane is obliged to supply a kick to the right. Each kick delivers some momentum; these kicks don't average to zero. Instead, they pull fluid through the channel until equilibrium is reached.

> **Your Turn 7C**
>
> Now suppose that there are particles on *both* sides of the membrane, with concentrations c_1 and c_2. Suppose $c_1 > c_2$. Redraw Figure 7.6 and find the form taken by the van 't Hoff relation in this case.

Our discussion makes clear how misleading it can be to refer to "the osmotic pressure." Suppose we throw a lump of sugar into a beaker. Soon we have a very nonuniform concentration $c(\mathbf{r})$ of sugar. Yet the pressure $p(\mathbf{r})$ is everywhere constant, not equal to $k_B T c(\mathbf{r})$ as we might have expected from a naïve application of the van 't Hoff relation. After all, we know that osmotic pressures can be huge; the fluid would be thrown into violent motion if it suddenly developed such big pressure variations. Instead it sits there quietly, and the concentration spreads by diffusion.

The flaw in the naïve reasoning is the assumption that concentration gradients themselves somehow cause pressure gradients. But pressure can only change if a *force* acts (Equation 7.12). Thus osmotic pressure can only arise if there is a physical object (the semipermeable membrane) present to apply force to the solute particles. In the absence of such an object—for instance, if we just throw a lump of sugar into the water—there is no force and no pressure gradient. Similarly, in the experiment sketched in Figure 1.3 on page 13, initially there will be no osmotic force at all. Only when solute molecules have had a chance to diffuse from the initial lump of sugar to the membrane will the latter begin to rectify their Brownian motion and so transmit force to them, and thence to the fluid.

7.3.2 Osmotic flow is quantitatively related to forced permeation

Section 7.3.1 argued that the membrane repels particles, which in turn drag fluid away from the membrane, thus creating a low-pressure layer there. This layer is the depletion zone; see the solid curve in Figure 7.6c.

Now suppose that we apply *no* force to the pistons in Figure 1.3a. Then there will be no net pressure difference between the sides. After all, pressure is force per area, namely, zero on each side. (More realistically, it's likely to be atmospheric pressure on each side; but still there's no jump.) Doesn't this contradict the van 't Hoff relation? No, the van 't Hoff relation gives, not the actual pressure, but that pressure which *would* be needed to *stop* osmotic flow, that is, the pressure drop *if* the system *were* brought to equilibrium. We can certainly maintain a smaller pressure drop than $c_0 k_B T$; then the osmotic effect will actually pull water through the pores from the $c = 0$ side to the $c = c_0$ side. This process is osmotic flow.

The solid curve in Figure 7.6d summarizes the situation. In equilibrium, the fluid pressure was constant throughout the pore (Figure 7.6c), but now it cannot be. The discussion leading to the Hagen–Poiseuille relation (Equation 5.18 on page 181) then gives the flow rate Q needed to create a uniform pressure drop per unit length of p/L. The system simply chooses that flow rate which gives the pressure drop required by the van 't Hoff relation. These observations apply to reverse osmosis as well (see Section 1.2.2 on page 12): If we push against the natural osmotic flow with a force per area even greater than $c_0 k_B T$, then the flow needed to accommodate the imposed pressure drop goes backward. This situation is shown as the dashed curve in Figure 7.6d.

We can summarize the entire discussion in a single master formula. First we note that, even when we have pure water on both sides of the membrane, there will be flow if we *push* on one piston. Because the pores are generally small and the flow slow, we expect a Darcy-type law for this phenomenon, called hydraulic permeation (see Section 5.3.4 on page 179). If there is a fixed density of pores per unit area, we expect a volume flow (volume per time) proportional to the applied pressure and to the area. The corresponding volume flux is then $j_v = -L_p \Delta p$, where L_p is a constant called the **filtration coefficient** of the membrane (see Problem 4.10 and Section 5.3.4 on page 179). The preceding discussion suggests that there is a generalization of the hydraulic permeation relation to embrace *both* driven and osmotic flow:[4]

$$j_v = -L_p\big(\Delta p - (\Delta c)k_B T\big). \qquad \text{volume flux through a semipermeable membrane} \qquad (7.15)$$

Equation 7.15 establishes a quantitative link between driven permeation and osmotic flow, two seemingly different phenomena. If we apply zero external force, then osmotic flow proceeds at a rate $j_v = L_p k_B T \Delta c$. This is the rate at which the entropic force per area, $(\Delta c)k_B T$, just balances the frictional drag per area, j_v/L_p. As we increase the opposing applied pressure the volume flux slows, drops to zero when $\Delta p = (\Delta c)k_B T$, then *reverses* at still greater Δp, thereby giving reverse osmosis.

Equation 7.15 actually transcends the rather literal model of a membrane as a hard wall pierced with cylindrical channels, introduced earlier for concreteness. It is similar in spirit to the Einstein relation (Equation 4.16 on page 120), as we see from the telltale presence of $k_B T$ linking a mechanically driven transport process to an entropically driven one.

$\boxed{T_2}$ *Section 7.3.1' on page 283 mentions the more general situation of a membrane with some permeability to* both *water and dissolved solute.*

7.4 A REPULSIVE INTERLUDE

Until now, we have studied osmotic forces under the assumption that interactions between solute particles can be neglected. That may be reasonable for sugar, whose molecules are uncharged; but, as we'll see in a moment, electrostatic interactions

---

[4]Some authors introduce the abbreviation $\Pi = c k_B T$ when writing this formula, and call Π the "osmotic pressure." We will avoid this confusing locution and simply call this quantity $c k_B T$.

between the objects contained in a cell can be immense. Accordingly, this section will introduce *mixed* forces, those that are partly entropic and partly energetic.

7.4.1 Electrostatic interactions are crucial for proper cell functioning

Biomembranes and other big objects (such as DNA) are often said to be "electrically charged." The term can cause confusion. Doesn't matter have to be neutral? Let's recall why people said that in first-year physics.

Example: Consider a raindrop of radius $R = 1\,\text{mm}$ suspended in air. How much work would be needed to remove just one electron from just 1% of the water molecules in the drop?

Solution: Removing an electron leaves some water molecules electrically charged. These charged water molecules migrate to the surface of the drop to get away from one another, thereby forming a shell of charge of radius R. Recall from first-year physics that the electrostatic potential energy of such a shell (also called its **Born self-energy**) is $\frac{1}{2}qV(R)$, or $q^2/(8\pi\varepsilon_0 R)$. In this formula, ε_0 is a constant describing the properties of air, the **permittivity**. Appendix B gives $e^2/(4\pi\varepsilon_0) = 2.3 \cdot 10^{-28}\,\text{J m}$. The charge q on the drop equals the number density of water molecules, times the drop volume, times the charge on a proton, times 1%. Squaring gives

$$\left(\frac{q}{e}\right)^2 = \left(\frac{10^3\,\text{kg}}{\text{m}^3}\frac{6 \cdot 10^{23}}{0.018\,\text{kg}} \times \frac{4\pi}{3}(10^{-3}\,\text{m})^3 \times 0.01\right)^2 = 1.9 \cdot 10^{36}.$$

Multiplying by $2.3 \cdot 10^{-28}\,\text{J m}$ and dividing by $2R$ yields about $2 \cdot 10^{11}\,\text{J}$.

Two hundred billion joules is a lot of energy—certainly it's much bigger than $k_\text{B}T_r$! And indeed, macroscopic objects really are electrically neutral (they satisfy the condition of "bulk electroneutrality"). But things look different in the nanoworld.

Your Turn 7D	Repeat the calculation for a droplet of radius $R = 1\,\mu\text{m}$ in water. You'll need to know that the permittivity ε of water is about 80 times bigger than the one for air used in the Example; in other words, the **dielectric constant** $\varepsilon/\varepsilon_0$ of water is about 80. Repeat again for an $R = 1\,\text{nm}$ object in water.

Thus it *is* possible for thermal motion to separate a neutral molecule into charged fragments. For example, when we put an acidic macromolecule such as DNA in water, some of its loosely attached atoms can wander away, leaving some of their electrons behind. In this case, the remaining macromolecule has a net negative charge: DNA becomes a negative **macroion**. This is the sense in which DNA is charged. The lost atoms are positively charged; they are called **counterions**, because their net charge counters (neutralizes) the macroion. Positive ions are also called **cations**, because they'd be attracted to a *cat*hode; similarly, the remaining macroion is called **anionic**.

The counterions diffuse away because they were not bound by chemical (covalent) bonds in the first place and because by diffusing away, they increase their entropy. Chapter 8 will discuss the question of what fraction detach, that is, the problem of partial dissociation. For now, let's study the simple special case of fully dissociated macroions. This is an interesting case, in part because DNA is usually nearly fully dissociated.

The counterions, having left the macroion, now face a dilemma. If they stay too close to home, they won't gain much entropy. But to travel far from home requires lots of energy, to pull away from the opposite charges left behind on the macroion. The counterions thus need to make a compromise between the competing imperatives to minimize energy and maximize entropy. This section will show that for a large flat macroion, the compromise chosen by the counterions is to remain hanging in a cloud near the macroion's surface. After working Your Turn 7D, you won't be surprised to find that the cloud can be a couple of nanometers thick. Viewed from *beyond* the counterion cloud, the macroion appears neutral. Thus, a second approaching macroion won't feel any attraction or repulsion until it gets closer than about twice the cloud's thickness. This behavior is quite different from the behavior of charges in a vacuum: In that case, the electric field doesn't fall off with distance at all![5] In short,

> *Electrostatic interactions are of long range in vacuum. But in solution,*   (7.16)
> *a screening effect reduces this interaction's **effective** range, typically to*
> *a nanometer or so.*

We'd like to understand the formation of the counterion cloud, which is often called the **diffuse charge layer**. Together with the charges left behind in the surface, it forms an **electric double layer** surrounding a charged macroion. The previous paragraph makes it clear that the forces on charged macroions have a mixed character: They are partly electrostatic and partly entropic. Certainly, if we could turn off thermal motion, the diffuse layer would collapse back onto the macroion, thereby leaving it neutral, and there'd be no force at all; we'll see this in the formulas we obtain for the forces.

Before we proceed to calculate properties of the diffuse charge layer, two remarks may help set the biological context.

First, your cells contain a variety of macromolecules. A number of attractive forces are constantly trying to stick the macromolecules together, for example, the depletion force or the more complicated van der Waals force. It wouldn't be nice if they just acquiesced, clumping into a ball of sludge at the bottom of the cell, with the water on top. The same problem bedevils many industrial colloidal suspensions, for example, paint. One way Nature, and we its imitators, avoid this "clumping catastrophe" is to arrange for the colloidal particles to have the same sign of net charge. Indeed, most of the macromolecules in a cell are negatively charged and hence repel one another.

Second, the fact that electrostatic forces are effectively of short range in solution (summarized in Idea 7.16 above) matters crucially for cells, because it means that

---

[5]See Equation 7.20 on page 264.

- Macroions will not feel one another until they're nearby, but
- Once they *are* nearby, the *detailed surface pattern* of positive and negative residues on a protein can be felt by its neighbor, not just the overall charge.

As mentioned in Chapter 2, this observation goes to the heart of how cells organize their myriad internal biochemical reactions. Although thousands of macromolecules may be wandering around any particular location in the cell, typically only those with precisely matching shapes and charge distributions will bind together. We can now see that the root of this amazing specificity is that

> *Even though each individual electrostatic interaction between match-*
> *ing charges is rather weak (relative to $k_B T_r$), still the combined ef-*
> *fect of many such interactions can lead to strong binding of two*   (7.17)
> *molecules—**if** their shapes and orientations match precisely.*

Notice that it's not enough for two matching surfaces to come together; they must also be properly oriented before they can bind. We say that macromolecular binding is **stereospecific**.

Thus, understanding the very fact of molecular recognition, which is crucial for the operation of every cell process, requires that we first understand the counterion cloud around a charged surface.

7.4.2 The Gauss Law

Before tackling statistical systems with mobile charged ions, let's pause to review some ideas about systems of *fixed* charges. We need to recall how a charge distribution gives rise to an electric field $\mathcal{E}$, in the planar geometry shown in Figure 7.7. The figure represents a thin, negatively charged sheet with uniform surface charge density $-\sigma_q$, next to a spread-out layer of positive charge with volume charge density $\rho_q(x)$. Thus σ_q is a positive constant with units $coul\,m^{-2}$, whereas $\rho_q(x)$ is a positive function with units $coul\,m^{-3}$. Everything is constant in the $\hat{y}$ and $\hat{z}$ directions. We'll write $\mathcal{E}$ for the component of the electric field in the $\hat{x}$ direction.

The electric field above the negative sheet is a vector pointing along the $-\hat{x}$ direction, so the function $\mathcal{E}(x)$ is everywhere negative. Just above the sheet, the electric field is proportional to the surface charge density:

$$\mathcal{E}|_{surface} = -\sigma_q/\varepsilon. \qquad \textbf{Gauss Law} \text{ at a flat, charged surface} \qquad (7.18)$$

In this formula, the permittivity ε is the same constant appearing in Your Turn 7D; in water, it's about 80 times the value in air or vacuum.[6] As we move away from

---

[6]Many authors use the notation $\epsilon\epsilon_0$ for the quantity called ε in this book. It's a confusing notation, because then their $\epsilon \approx 80$ is dimensionless while ϵ_0 (which equals our ε_0) *does* have dimensions.

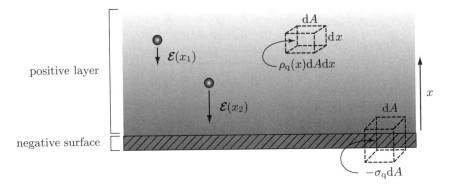

Figure 7.7: (Schematic.) A planar distribution of charges. A thin sheet of negative charge (*hatched, bottom*) lies next to a neutralizing positive layer of free counterions (*shaded, top*). The individual counterions are not shown; the shading represents their average density. The lower box encloses a piece of the surface; so it contains total charge $-\sigma_q dA$, where dA is its cross-sectional area and $-\sigma_q$ is the surface charge density. The upper box encloses charge $\rho_q(x)dAdx$, where $\rho_q(x)$ is the charge density of counterions. The electric field $\mathcal{E}(x)$ at any point equals the electrostatic force on a small test particle at that point, divided by the particle's charge. For all positive x, the field points along the $-\hat{\mathbf{x}}$ direction. The field at x_1 is weaker than that at x_2, because the repelling layer of positive charge between x_1 and $x = 0$ is thicker than that between x_2 and $x = 0$.

the surface, the field gets weaker (less negative): A positively charged particle is still attracted to the negative layer, but the attraction is partially offset by the repulsion of the intervening positive charges. The *difference* in electric fields at two nearby points reflects the reduction of Equation 7.18 by the charge per area in the space between the points. Calling the points $x \pm \frac{1}{2}dx$, we see that this surface charge density equals $\rho_q(x)dx$ (see Figure 7.7). Hence

$$\mathcal{E}(x + \tfrac{1}{2}dx) - \mathcal{E}(x - \tfrac{1}{2}dx) = (dx)\rho_q(x)/\varepsilon. \tag{7.19}$$

In other words,

$$\boxed{\frac{d\mathcal{E}}{dx} = \frac{\rho_q}{\varepsilon}. \qquad \textbf{Gauss Law} \text{ in bulk}} \tag{7.20}$$

Section 7.4.3 will use this relation to find the electric field everywhere outside the surface.

$\boxed{T_2}$ *Section 7.4.2′ on page 284 relates the preceding discussion to the more general form of the Gauss Law.*

7.4.3 Charged surfaces are surrounded by neutralizing ion clouds

The mean field Now we can return to the problem of ions in solution. A typical problem might be to consider a thin, flat, negatively charged surface with surface

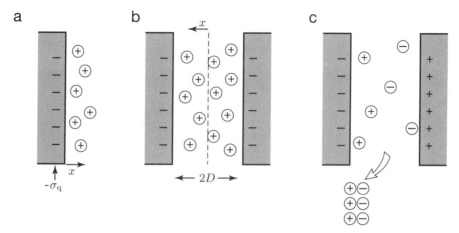

Figure 7.8: (Schematics.) Behavior of counterion near surfaces. (a) Counterion cloud outside a charged surface with surface charge density $-\sigma_q$. (b) When two similarly charged surfaces approach, their counterion clouds begin to get squeezed. (c) When two oppositely charged surfaces approach, their counterion clouds are liberated, and entropy increases.

charge density $-2\sigma_q$ and water on both sides. For example, cell membranes are negatively charged. You might want to coax DNA to enter a cell (say, for gene therapy). Because both DNA and cell membranes are negatively charged, you'd need to know how much they repel.

An equivalent, and slightly simpler, problem is that of a *solid* surface carrying charge density $-\sigma_q$, with water on just one side (Figure 7.8a). Also for simplicity, suppose that the loose positive counterions are **monovalent** (for example, sodium, Na^+). That is, each carries a single charge: $q_+ = e = 1.6 \cdot 10^{-19}$ coul. In a real cell, there will be additional ions of *both* charges from the surrounding salt solution. The negatively charged ones are called **coions** because they have the same charge as the surface. We will neglect the coions for now (see Section 7.4.3′ on page 284).

As soon as we try to find the electric field in the presence of mobile ions, an obstacle arises: We are not given the distribution of the ions, as we were in first-year physics, but instead must *find* it. Moreover, electric forces are of long range. The unknown distribution of ions will thus depend on each ion's interactions not only with its nearest neighbors but also with many other ions! How can we hope to model such a complex system?

Let's try to turn adversity to our advantage. If each ion interacts with many others, perhaps we can approach the problem by thinking of each ion as moving independently of the others' *detailed* locations but under the influence of an electric potential created by the *average* charge density of the others, or $\langle \rho_q \rangle$. We call this approximate electric potential $V(x)$ the **mean field** and this approach the **mean-field approximation**. The approach is reasonable if each ion feels many others; then the relative fluctuations in $V(x)$ about its average will be small (see Figure 4.3 on page 113). To make the notation less cumbersome, we will drop the averaging signs; from now on, ρ_q refers to the average density.

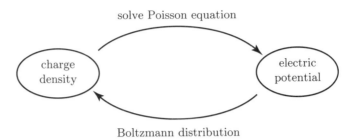

Figure 7.9: (Diagram.) Strategy to find the mean-field solution. Neither the Poisson equation nor the Boltzmann distribution alone can determine the charge distribution, but solving these two equations in two unknowns simultaneously does the job.

The Poisson–Boltzmann equation   We want $c_+(x)$, the concentration of counterions. We are supposing that our surface is immersed in pure water; hence, far away from the surface, $c_+ \to 0$. The electrostatic potential energy of a counterion at x is $eV(x)$. We are treating the ions as moving independently of each other in a fixed potential $V(x)$, so the density of counterions, $c_+(x)$, is given by the Boltzmann distribution. Thus $c_+(x) = c_0 e^{-eV(x)/k_B T}$, where c_0 is a constant. We can add any constant we like to the potential because this change doesn't affect the electric field $\mathcal{E} = -dV/dx$. It's convenient to choose the constant so that $V(0) = 0$. This choice gives $c_+(0) = c_0$; so the unknown constant c_0 is the concentration of counterions at the surface.

Unfortunately, we don't yet know $V(x)$. To find it, apply the second form of the Gauss Law (Equation 7.20), taking ρ_q equal to the density of counterions times e. Remembering that the electric field at x is $\mathcal{E}(x) = -dV/dx$ gives the **Poisson equation**: $d^2V/dx^2 = -\rho_q/\varepsilon$. Given the charge density, we can solve the Poisson equation for the electric potential. The charge density, in turn, is given by the Boltzmann distribution as $ec_+(x) = ec_0 e^{-eV(x)/k_B T}$.

It may seem as though we have a chicken-and-egg problem (Figure 7.9): We need the average charge density ρ_q to get the potential V. But we need V to find ρ_q (from the Boltzmann distribution)! Luckily, a little mathematics can get us out of predicaments like this one. Each of the arrows in Figure 7.9 represents an equation in two unknowns, namely, ρ_q and V. We just need to solve these two equations simultaneously to find the two unknowns. (We encountered the same problem when deriving the van 't Hoff relation, at Equation 7.13 on page 256, and resolved it in the same way.)

Before proceeding, let's take a moment to tidy up our formulas. First, we combine the various constants into a length scale:

$$\ell_B \equiv \frac{e^2}{4\pi \varepsilon k_B T}. \qquad \textbf{Bjerrum length}, \text{ in water} \qquad (7.21)$$

ℓ_B tells us how close together we can push two like-charge ions, if we have energy $k_B T$ available. For monovalent ions in water at room temperature, $\ell_B = 0.71 \text{ nm}$. Next, define the dimensionless rescaled potential $\overline{V}$:

$$\overline{V}(x) \equiv eV(x)/k_B T. \tag{7.22}$$

Now combine the Poisson equation with the Boltzmann distribution to get

$$\frac{d^2 \overline{V}}{dx^2} = -4\pi \ell_B c_0 e^{-\overline{V}}. \qquad \textbf{Poisson–Boltzmann equation} \tag{7.23}$$

The payoff for introducing the abbreviations $\overline{V}$ and ℓ_B is that now Equation 7.23 is less cluttered, and we can verify at a glance that its dimensions work: Both d^2/dx^2 and $\ell_B c_0$ have units m^{-2}.

Like any differential equation, Equation 7.23 has, not one, but a whole *family* of solutions. To get a unique solution, we need to specify additional information, namely, some **boundary conditions** on the unknown function $V(x)$. For example, if you throw a rock upward, Newton's Law says that its height $z(t)$ obeys the equation $d^2z/dt^2 = -g$. But this equation won't tell us how high the rock will go! We also need to specify how hard you threw the rock, or more precisely, its speed and location when it left your hand at time zero. Similarly, we should not expect Equation 7.23 to specify the full solution because it doesn't mention the surface charge density. Instead, the equation has a family of solutions; we must choose the one corresponding to the given value of σ_q.

To see how σ_q enters the problem, we now apply the surface form of the Gauss Law (Equation 7.18), which gives $-\frac{dV}{dx}\big|_{\text{surface}} = -\frac{\sigma_q}{\varepsilon}$, or

$$\frac{d\overline{V}}{dx}\bigg|_{\text{surface}} = 4\pi \ell_B \frac{\sigma_q}{e}. \qquad \text{(when the allowed region is } x > 0\text{)} \tag{7.24}$$

When using this formula, remember that σ_q is a positive number; the surface has charge density $-\sigma_q$.

Example: How does one remember the correct sign in this formula?

Solution: Notice that the electrostatic potential V goes down as we approach a negative object. Thus, approaching counterions feel their potential energy eV decrease as they approach the surface, so they're attracted. If x is the distance from a negatively charged surface, then V will be decreasing as we approach it, or increasing as we leave: $dV/dx > 0$, so the sign is correct in Equation 7.24.

Solution of the Poisson–Boltzmann equation   We have reduced the problem of finding the counterion distribution outside a surface to solving Equation 7.23. This is a differential equation, so we'll need to impose some conditions to determine a unique solution. Furthermore, the equation itself contains an unknown constant c_0, which requires another condition to fix its value. The conditions are

- The boundary condition at the surface (Equation 7.24),
- An analogous condition $d\overline{V}/dx = 0$ at infinity, because no charge is located there, and
- The convention that $\overline{V}(0) = 0$.

It's usually not easy to solve nonlinear differential equations like Equation 7.23. Still, in some special situations, we do get lucky. We need a function whose second derivative equals its exponential. We recall that the logarithm of a power of x has the property that both its derivative and its exponential are powers of x. We don't want $\overline{V}(x) = \ln x$, because that's divergent (equal to infinity) at the surface. Nevertheless, a slight modification gives something promising: $\overline{V}(x) \overset{?}{=} B \ln(1 + (x/x_0))$. This expression has the feature that $\overline{V}(0) = 0$, so we need not add any extra constant to $\overline{V}$.

We now check whether we can choose values for the constants B and x_0 in such a way that the proposed solution solves the Poisson–Boltzmann equation. Substituting $B \ln(1 + (x/x_0))$ into Equation 7.23, we indeed find that it works, provided we take $B = 2$ and $x_0 = 1/\sqrt{2\pi \ell_B c_0}$.

Next we must impose the boundary condition (Equation 7.24). In the present situation, this condition says $2/x_0 = 4\pi \ell_B(\sigma_q/e)$. It may seem as though we have exhausted all our freedom to adjust the trial solution (when we chose values for B and x_0). But the Poisson–Boltzmann equation itself contains an unknown parameter, c_0. You can check that taking this parameter to be $c_0 = 2\pi \ell_B(\sigma_q/e)^2$ ensures that our solution satisfies the boundary condition, and that then

$$V(x) = 2\frac{k_B T}{e} \ln(1 + (x/x_0)), \quad \text{where} \quad x_0 = (2\pi \ell_B \sigma_q/e)^{-1}. \tag{7.25}$$

Your Turn 7F

Find the equilibrium concentration profile $c_+(x)$ away from the surface. Check your answer by calculating the total surface density of counterions, $\int_0^\infty dx\, c_+(x)$, and verifying that the whole system is electrically neutral.

The solution you just found is sometimes called the **Gouy–Chapman layer**; x_0 is called the Gouy–Chapman length. This solution is appropriate in the neighborhood of a flat, charged surface in pure water.[7] Let's extract some physical conclusions from the math.

First, we see from Your Turn 7F that, indeed, a diffuse layer forms, with thickness roughly x_0. As argued physically in Section 7.4.1, the counterions are willing

---

[7] $\boxed{T_2}$ Or more realistically, a highly charged surface in a salt solution whose concentration is low enough; see Section 7.4.3′ on page 284.

to pay some electrostatic potential energy in order to gain entropy. More precisely, the counterions pull some thermal energy from their environment to make this payment. They can do this because doing so lowers the entropic part of their free energy more than it raises the electrostatic part. If we could turn off thermal motion (that is, send $T \to 0$), the energy term would dominate and the layer would collapse. We see this mathematically from the observation that then the Bjerrum length would go to infinity and $x_0 \to 0$.

How much electrostatic energy must the counterions pay to dissociate from the planar surface? We can think of the layer as a planar sheet of charge hovering at a distance x_0 from the surface. When two sheets of charge are separated, we have a parallel-plate capacitor. Such a capacitor, with area A, stores electrostatic energy $E = q_{\text{tot}}^2/(2C)$. Here q_{tot} is the total charge separated; for our case, it's $\sigma_q A$. The capacitance of a parallel-plate capacitor is given by

$$C = \varepsilon A / x_0. \tag{7.26}$$

Combining the preceding formulas gives an estimate for the density of stored electrostatic energy per unit area for an isolated surface in pure water:

$$E/(\text{area}) \approx k_B T (\sigma_q/e). \qquad \text{(electrostatic self-energy, no added salt)} \tag{7.27}$$

That makes sense: The environment is willing to give up about $k_B T$ of energy per counterion. This energy gets stored in forming the diffuse layer.

Is it a lot of energy? A fully dissociating bilayer membrane can have one unit of charge per lipid head group, or roughly $|\sigma_q/e| = 0.7\,\text{nm}^{-2}$. A spherical vesicle of radius $10\,\mu\text{m}$ then carries stored free energy $\approx 4\pi(10\,\mu\text{m})^2 \times (0.7/\text{nm}^2)k_B T_r \approx 10^9 k_B T$. It's a lot! We'll see how to harness this stored energy in Section 7.4.5.

For simplicity, the preceding calculations assumed that a dissociating surface was immersed in pure water. In real cells, however, the cytosol is an **electrolyte**, or salt solution. In this case, the density of counterions at infinity is not zero, and the counterions originally on the surface have less to gain entropically by escaping; so the diffuse charge layer will hug the surface more tightly than it does in Equation 7.25. That is,

> *Increasing salt in the solution shrinks the diffuse layer.* (7.28)

$\boxed{T_2}$ *Section 7.4.3′ on page 284 solves the Poisson–Boltzmann equation for a charged surface in a salt solution, arriving at the concept of the Debye screening length and making Equation 7.28 quantitative.*

7.4.4 The repulsion of like-charged surfaces arises from compression of their ion clouds

Now that we know what it's like near a charged surface, we're ready to go further and compute an entropic *force* between charged surfaces in solution. Figure 7.8b shows the geometry. One might be tempted to say, "Obviously, two negatively charged surfaces will repel." But wait: Each surface, together with its counterion cloud, is an electrically *neutral* object! Indeed, if we could turn off thermal motion, the mobile ions would collapse down to the surfaces, thereby rendering them neutral. Thus the repulsion between like-charged surfaces can only arise as an entropic effect. As the surfaces get closer than about twice their Gouy–Chapman length x_0, their diffuse counterion clouds get squeezed; they then resist with an osmotic pressure. Here are the details.

For simplicity, let's continue to suppose that the surrounding water has no added salt and, hence, no ions other than the counterions dissociated from the surface.[8] This time we'll measure distance from the midplane between *two* surfaces, which are located at $x = \pm D$ (Figure 7.8b). We'll suppose that each surface has surface charge density $-\sigma_q$. We choose the constant in V so that $V(0) = 0$; hence the parameter $c_0 = c_+(0)$ is the unknown concentration of counterions at the midplane. $V(x)$ will then be symmetrical about the midplane, so Equation 7.25 won't work. Keeping the logarithm idea, though, this time we try $\overline{V}(x) = A \ln \cos(\beta x)$, where A and β are unknown constants. Certainly this trial solution is symmetrical and equals zero at the midplane, where $x = 0$.

The rest of the procedure is familiar. Substituting the trial solution into the Poisson–Boltzmann equation (Equation 7.23) gives $A = 2$ and $\beta = \sqrt{2\pi \ell_B c_0}$. The boundary condition at $x = -D$ is again Equation 7.24. Imposing the boundary conditions on our trial solution gives a condition fixing β:

$$4\pi \ell_B (\sigma_q/e) = 2\beta \tan(D\beta). \qquad (7.29)$$

Given the surface charge density $-\sigma_q$, we solve Equation 7.29 for β as a function of the spacing $2D$; then the desired solution is

$$\overline{V}(x) = 2 \ln \cos(\beta x), \quad \text{or} \quad c_+(x) = c_0 (\cos \beta x)^{-2}. \qquad (7.30)$$

As expected, the charge density is greatest near the plates; the potential is maximum in the center.

We want a force. Examining Figure 7.8b, we see that our situation is essentially the opposite of the depletion interaction (Figure 7.3b on page 253): There, particles were forbidden in the gap, whereas now they are *required* to be there, by charge neutrality. In either case, some force acts on individual particles to constrain their Brownian motion; that force gets transmitted to the confining surfaces by the fluid, thereby creating a pressure drop of $k_B T$ times the concentration difference between the force-free regions (see Idea 7.14 on page 258). In our case, the force-free regions are the exterior and the midplane (because $\mathcal{E} = -\frac{dV}{dx} = 0$ there). The corresponding concentrations are 0 and c_0, respectively; so the repulsive force per unit area on the surfaces is just

$$\boxed{f/(\text{area}) = c_0 k_B T. \qquad \text{repulsion of like-charged surfaces, no added salt}}$$

$$(7.31)$$

---

[8] $\boxed{T_2}$ This is not as restrictive as it sounds. Even in the presence of salt, our result will be accurate if the surfaces are highly charged because in this case, the Gouy-Chapman length is less than the Debye screening length (see Section 7.4.3′ on page 284).

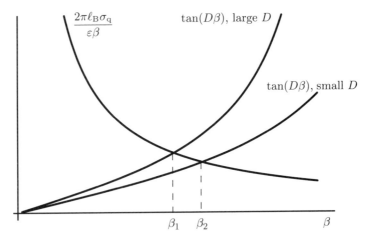

Figure 7.10: (Mathematical functions.) Graphical solution of Equation 7.29. The sketch shows the function $2\pi \ell_B \sigma_q /(\varepsilon \beta)$, as well as $\tan D\beta$ for two values of the plate separation $2D$. The value of β at the intersection of the rising and falling curves gives the desired solution. The figure shows that smaller plate separation gives a larger solution β_2 than does large separation (yielding β_1). Larger β in turn implies a larger ion concentration $c_0 = \beta^2/(2\pi \ell_B)$ at the midplane and larger repulsive pressure.

In this formula, $c_0 = \beta^2/(2\pi \ell_B)$ and $\beta(D, \sigma_q)$ is the solution of Equation 7.29. You can solve Equation 7.31 numerically (see Problem 7.10), but a graphical solution shows qualitatively that β increases as the plate separation decreases (Figure 7.10). Thus the repulsive pressure increases, too, as expected.

Note that the force just found is not simply proportional to the absolute temperature, because β has a complicated temperature dependence. This means that our pressure is not a purely entropic effect (like the depletion interaction, Equation 7.10), but a mixed effect: The counterion layer reflects a *balance* between entropic and energetic imperatives. As remarked at the end of Section 7.4.3, the qualitative effect of adding salt to the solution is to tip this balance away from entropy, thereby shrinking the diffuse layers on the surfaces and *shortening the range* of the interaction.

This theory works (see Figure 7.11). You'll make a detailed comparison with experiment in Problem 7.10, but for now, a simple case is of interest:

Your Turn 7G

Show that at very low surface charge density, $\sigma_q \ll 1/(D\ell_B)$, the density of counterions in the gap is nearly uniform and equals the total charge on the plates divided by the volume of the gap between them, as it must.

Thus, in this case, the counterions act as an ideal solution, and the pressure they exert is that predicted by the van 't Hoff formula.

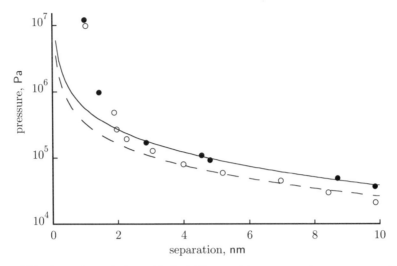

Figure 7.11: (Experimental data with fits.) The repulsive pressure between two positively charged surfaces in water. The surfaces were egg lecithin bilayers containing 5 mole% or 10 mole% phosphatidylglycerol (*open and filled circles*, respectively). The curves show one-parameter fits of these data to the numerical solution of Equations 7.29 and 7.31. The fit parameter is the surface charge density σ_q. The *dashed line* shows the solution with one proton charge per 24 nm^2; the *solid line* corresponds to a higher charge density (see Problem 7.10). At separations below 2 nm, the surfaces begin to touch and other forces besides the electrostatic one appear. Beyond 2 nm, the purely electrostatic theory fits the data well, and the membrane with a larger density of charged lipids is found to have a larger effective charge density, as expected. [Data from Cowley et al., 1978.]

$\boxed{T_2}$ *Section 7.4.4′ on page 286 derives the electrostatic force directly as a derivative of the free energy.*

7.4.5 Oppositely charged surfaces attract by counterion release

Now consider an encounter between surfaces of *opposite* charge (Figure 7.8c on page 265). Without working through the details, we can understand the attraction of such surfaces in solution qualitatively by using the ideas developed earlier. Again, as the surfaces approach from infinity, each presents a net charge density of *zero* to the other; there is no long-range force, unlike the constant attractive force between two such planar surfaces in air. Now, however, as the surfaces approach, they can shed counterion pairs while preserving the system's neutrality. The released counterions leave the gap altogether and hence gain entropy, thereby lowering the free energy and driving the surfaces together. If the charge densities are equal and opposite, the process proceeds until the surfaces are in tight contact, with no counterions left at all. In this case, there is no separation of charge, and no counterions remain in the gap. Thus all the self-energy estimated in Equation 7.27 gets released. We have already estimated that this energy is substantial: Electrostatic binding between surfaces of matching shape can be very strong.

7.5   SPECIAL PROPERTIES OF WATER

Suppose you mix oil and vinegar for your salad, shake it thoroughly, and then the phone rings. When you come back, the mixture has separated. The separation is not caused by gravity; salad dressing also separates (a bit more slowly) on the space shuttle. We might be tempted to panic and declare a violation of the Second Law. But by now we know enough to frame some other hypotheses:

1. Maybe some attractive force pulls the individual molecules of water together (expelling the oil) to lower the total energy (as in the water-condensation example of Section 1.2.1 on page 9). The energy thus liberated would escape as heat, thereby increasing the rest of the world's entropy, perhaps enough to drive the separation.
2. Maybe the decrease of entropy when the small, numerous oil droplets combine is offset by a much larger *in*crease of entropy from some even smaller, even more numerous, objects, as in the depletion interaction (Section 7.2.2 on page 251).

Actually, many pairs of liquids separate spontaneously, essentially for energetic reasons like point (1). What's special about water is that its dislike for oil is unusually strong and has an unusual temperature dependence. Section 7.5.2 will argue that these special properties stem from an additional mechanism, listed as point (2) above. (In fact, some hydrocarbons actually *liberate* energy when mixed with water, so point (1) cannot explain their reluctance to mix.) Before this discussion, however, we first need some facts about water.

7.5.1   Liquid water contains a loose network of hydrogen bonds

The hydrogen bond   The water molecule consists of a large oxygen atom and two smaller hydrogen atoms. The atoms don't share their electrons very fairly: All the electrons spend almost all their time on the oxygen. Molecules that maintain a permanent separation of charge, like water, are called **polar**. A molecule that is everywhere roughly neutral is called **nonpolar**. Common nonpolar molecules include hydrocarbon chains, like the ones making up oils and fats (Section 2.2.1 on page 46).

A second key property of the water molecule is its bent, asymmetrical shape: We can draw a plane slicing through the oxygen atom in such a way that both the hydrogens lie on the same side of the plane. The asymmetry means that an external electric field will tend to *align* water molecules, partly countering the tendency of thermal motion to randomize their orientations. Your microwave oven uses this effect. It applies an oscillating electric field, which shakes the water molecules in your food. Friction then converts the shaking motion into heat. We summarize these comments by saying that the water molecule is a **dipole** and that the ability of these dipoles to align (or "polarize") makes liquid water a highly **polarizable** medium. (Water's polarizability is also the origin of the large value of its permittivity ε; see Section 7.4.1.)

There are many small polar molecules, most of which are dipoles. Among these, water belongs to a special subclass. Note that each hydrogen atom in a water molecule had only one electron to begin with. Once it has lost that electron, each hydrogen

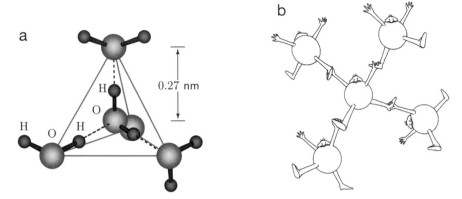

Figure 7.12: (Sketch; metaphor.) (a) Tetrahedral arrangement of water molecules in an ice crystal. The *sticks* depict chemical bonds; the *dashed lines* are hydrogen bonds. The gray outline of the tetrahedron is just to guide the eye. The oxygen atom in the center of the figure has two dashed lines (one is hidden behind the oxygen), coming from the directions most distant from the directions of its own two hydrogen atoms. (b) Crystal structure of ice. [(a) Adapted from Israelachvili, 1991. (b) From Ball, 2000.]

ends up essentially as a naked proton; its physical size is much smaller than that of any neutral atom. The electric field about a point charge grows as $1/r^2$ as the distance r to the charge goes to zero, so the two tiny positive spots on the water molecule are each surrounded by an intense electric field. This effect is specific to hydrogen: Any other kind of atom bonded to oxygen retains its other electrons. Such a partially stripped atom carries about the same charge $+e$ as a proton, but its charge distribution is much larger and hence more diffuse, with milder electric fields than those on a hydrogen.

Each water molecule thus has two sharply positive spots, which are oriented at a definite angle of 104° to each other. That angle is about the same as the angle between rays drawn from the center to two of the corners of a tetrahedron (Figure 7.12a). The molecule will try to orient itself in such a way as to point each of its two positive spots directly at some other molecule's "back side" (the negatively charged region opposite the hydrogens), as far away as possible from the latter's two positive spots. The strong electric fields near the hydrogen atoms make this interaction stronger than the generic tendency for any two electric dipoles to attract, and align with, each other.

The idea that a hydrogen atom in one molecule could interact with an oxygen atom in another molecule, in a characteristic way, was first proposed in 1920 by M. Huggins, an undergraduate student of the chemist G. Lewis. Lewis named this interaction the **hydrogen bond**, or **H-bond**.

As noted earlier, every molecule in a sample of liquid water will simultaneously attempt to point its two hydrogen atoms toward the back sides of other molecules. The best way to arrange this is to place the water molecules at the points of a tetrahedral lattice. Figure 7.12a shows a central water molecule with four nearest neighbors.

Two of the central molecule's hydrogens are pointing directly at the oxygen atoms of neighbors (top and front-right), while its two other neighbors (front-left and back) point *their* hydrogens toward *its* back side. As we lower the temperature, thermal disorder becomes less dominant and the molecules lock into a perfect lattice—an ice crystal. To help yourself imagine this lattice, think of your torso as the oxygen atom, your hands as the hydrogen atoms, and your feet as the docking sites for other hydrogens. Stand with your legs apart at an angle of 104° and your arms at the same angle. Twist 90° at the waist. Now you're a water molecule. Get a few dozen friends to assume the same pose. Now instruct everyone to grab someone's ankle with each hand (this works better in zero gravity). Now you're an ice crystal (Figure 7.12b).

   X-ray crystallography reveals that ice really does have the structure shown in Figure 7.12. Each oxygen is surrounded by four hydrogen atoms. Two are at the distance 0.097 nm appropriate for a covalent bond; the other two are at a distance 0.177 nm. The latter distance is too long to be a covalent bond but shorter than the distance 0.26 nm we'd expect from adding the radii of atomic oxygen and hydrogen. Instead, it reflects the fact that the hydrogen has been stripped of its electron cloud; its size is essentially zero. (One often sees the "length of the H-bond" in water quoted as 0.27 nm. This number actually refers to the distance between the oxygen atoms, that is, the *sum* of the lengths of the sticks and dashed lines in Figure 7.12a.)

   The energy of attraction of two water molecules, oriented to optimize their H-bonding, is intermediate between a true (covalent) chemical bond and the generic attraction of any two molecules; this explains why it merits the separate name "H-bond." More precisely, when two isolated water molecules (in vapor) stick together, the energy change is about $-9k_\mathrm{B}T_\mathrm{r}$. For comparison, the generic (van der Waals) attraction between any two small neutral molecules is typically only 0.6–1.6 $k_\mathrm{B}T_\mathrm{r}$. True chemical bond energies range from 90 to 350 $k_\mathrm{B}T_\mathrm{r}$.

The hydrogen bond network of liquid water   The network of H-bonds shown in Figure 7.12 cannot withstand thermal agitation when the temperature exceeds 273 K: Ice melts. Even liquid water, however, remains partially ordered by H-bonds. It adopts a compromise between the energetic drive to form a lattice and the entropic drive to disorder. Thus, instead of a single tetrahedral network, we can think of water as a collection of many small fragments of such networks. Thermal motion constantly agitates the fragments, moving, breaking, and reconnecting them, but the neighborhood of each water molecule still looks approximately like the figure. In fact, at room temperature, each water molecule maintains most of its H-bonds (averaging about 3.5 of the original 4 at any given time). Because each water molecule still has most of its H-bonds, and these are stronger than the generic attractions between small molecules, we expect that liquid water will be harder to break apart into individual molecules (water vapor) than other liquids of small, but not H-bonding, molecules. And indeed, the boiling point of water is 189 K higher than that of the small hydrocarbon molecule ethane. Methanol, another small molecule capable of making *one* H-bond from its –OH group, boils at an intermediate temperature, 36 K lower than water (with two H-bonds per molecule). In short,

> *The cohesive forces between molecules of water are larger than those between other small molecules that do not form H-bonds.* (7.32)

Hydrogen bonds as interactions within and between macromolecules in solution
Hydrogen bonds will also occur between molecules containing hydrogen covalently
bonded to *any* electronegative atom (specifically oxygen, nitrogen, or fluorine). Thus,
not only water, but also many of the molecules described in Chapter 2 can interact
via H-bonding. We cannot directly apply the estimates just given for H-bond strength
to the water environment, however. Suppose that two parts of a macromolecule are
initially in direct contact, forming an H-bond (for example, the two halves of a DNA
basepair, Figure 2.11 on page 47). When we separate the two parts, their H-bond is
lost. But each of the two will immediately form H-bonds with surrounding water
molecules, partially compensating for the loss! In fact, the *net* free energy cost of
breaking a single H-bond in water is generally only about $1–2k_B T_r$. Other competing
interactions are also smaller in water, however, so the H-bond is still significant. For
example, the dipole interaction, like any electrostatic effect, is diminished by the high
permittivity of the surrounding water (see Your Turn 7D on page 261).

Despite their modest strength, H-bonds in the water environment are nevertheless important in stabilizing macromolecular shapes and assemblies. In fact, the
very weakness and short range of the H-bond are what make it so useful in giving
macromolecular interactions their specificity. Suppose that two objects need several
weak bonds to overcome the tendency of thermal motion to break them apart. The
short range of the H-bond implies that the objects can only make multiple H-bonds
if their shapes and distribution of bonding sites match precisely. Thus, for example,
H-bonds help hold the basepairs of the DNA double helix together, but only if each
base is properly paired with its complementary base (see Figure 2.11 on page 47).
Section 9.5 will also show how, despite their weakness, H-bonds can give rise to large
structural features in macromolecules via cooperativity.

$\boxed{T_2}$ *Section 7.5.1' on page 288 adds more detail to the picture of H-bonding just
sketched.*

7.5.2 The hydrogen-bond network affects the solubility of small molecules in water

Solvation of small nonpolar molecules Section 7.5.1 described liquid water as a
rather complex state, balancing energetic and entropic imperatives. With this picture
in mind, we can now sketch how water responds to—and, in turn, affects—other
molecules immersed in it.

One way to assess water's interaction with another molecule is to measure that
molecule's solubility. Water is quite choosy in its affinities, with some substances mixing freely (for example, hydrogen peroxide, H_2O_2), others dissolving fairly well (for
example, sugars), while yet others hardly dissolve at all (for example, oils). Thus,
when pure water is placed in contact with, say, a lump of sugar, the resulting equilibrium solution will have a higher concentration of sugar than the corresponding
equilibrium with an oil drop in water. We can interpret these observations by saying that the free energy cost for an oil molecule to enter water is larger than that for
sugar (Section 6.6.4 on page 225 relates free energy changes to occupation probabilities).

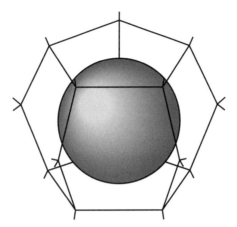

Figure 7.13: (Sketch.) Clathrate cage of H-bonded water molecules, shown as vertices of a polyhedron surrounding a nonpolar object (*gray sphere*). Four lines emerge from each vertex, representing the directions to the four water molecules H-bonded to the one at the vertex. This idealized structure should not be taken as a literal depiction; in liquid water, some of the H-bonds will always be broken. Rather, the figure demonstrates the geometrical possibility of surrounding a small nonpolar inclusion without any loss of H-bonds.

To understand these differences, we first note that hydrogen peroxide, which mixes freely with water, has two hydrogen atoms bonded to oxygens; so the molecule can participate fully in water's H-bond network. Thus, introducing an H_2O_2 molecule into water hardly disturbs the network, and hence incurs no significant free energy cost. In contrast, hydrocarbon chains such as those composing oils are nonpolar (Section 7.5.1), and so offer no sites for H-bonding. We might at first suppose that the layer of water molecules surrounding such a nonpolar intruder would lose some of its energetically favorable H-bonds, thereby creating an energy cost for introducing the oil. Actually, though, water is more clever than this. The surrounding water molecules can form a structure called a **clathrate cage** around the intruder, thereby maintaining their H-bonds with each other with nearly the preferred tetrahedral orientation (Figure 7.13). Hence the average number of H-bonds maintained by each water molecule need not drop very much when a small nonpolar object is introduced.

But energy minimization is not the whole story in the nanoworld. To form the cage structure shown in Figure 7.13, the surrounding water molecules have given up some of their orientational freedom: They cannot point any of their four H-bonding sites toward the nonpolar object and still remain fully H-bonded. Thus the water surrounding a nonpolar molecule must choose between sacrificing H-bonds, with a corresponding increase in electrostatic energy, or retaining them, with a corresponding loss of entropy. Either way, the free energy $F = E - TS$ goes up. This free energy cost is the origin of the poor solubility of nonpolar molecules in water at room temperature, a phenomenon generally called the **hydrophobic effect**.

The change in water structure upon entry of a nonpolar molecule (called hydrophobic solvation) is too complex for an explicit calculation of the sort given in

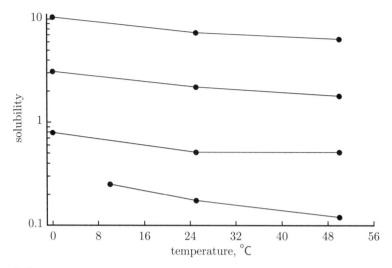

Figure 7.14: (Experimental data.) Semilog plot of the solubilities of small nonpolar molecules in water, as functions of temperature. The vertical axis gives the mass percentage of solute in water, when water reaches equilibrium with the pure liquid. *Top to bottom,* butanol (C_4H_9OH), pentanol ($C_5H_{11}OH$), hexanol ($C_6H_{13}OH$), and heptanol ($C_7H_{15}OH$). Note that the solubilities decrease with increasing chain length. [Data from Lide, 2001.]

Section 7.4.3 for electrostatics. Hence we cannot predict a priori which of the two extremes mentioned earlier (preserving H-bonds or maintaining high entropy) water will choose. At least in some cases, though, we can reason from the fact that certain small nonpolar molecules become less soluble in water as we warm the system starting from room temperature (see Figure 7.14). At first, this observation seems surprising: Shouldn't increasing temperature *favor* mixing? But suppose that for every solute molecule that enters, thereby gaining some entropy with its increased freedom to wander in the water, several surrounding water molecules *lose* some of their orientational freedom, for example, by forming a cagelike structure. In this way, dissolving more solute can incur a net decrease in entropy. Raising the temperature makes this cost more significant, and so makes it harder to keep solute in solution. In short, *solubility trends like the ones shown in Figure 7.14 imply a large entropic component to the free energy cost of hydrophobic solvation.*

   More generally, detailed measurements confirm that, at room temperature, the entropic term $-T\Delta S$ dominates the free energy cost ΔF of dissolving any small nonpolar molecule in water. The *energy* change ΔE may actually be favorable (negative), but, in any case, it is outweighed by the entropic cost. For example, when propane (C_3H_8) dissolves in water, the total free energy change is $+6.4k_B T_r$ per molecule; the entropic contribution is $+9.6k_B T_r$, whereas the energetic part is $-3.2k_B T_r$. (Further evidence for the entropic character of the hydrophobic effect at room temperature comes from computer simulations of water structure, which show that, outside a nonpolar surface, the water's O–H bonds are indeed constrained to lie parallel to the surface.)

The short range of the hydrogen bond suggests that the H-bond network will get disrupted only in the first layer of water molecules surrounding a nonpolar object. The free energy cost of creating an interface should therefore be proportional to its *surface area*; and experimentally, it's roughly true. For example, the solubilities of hydrocarbon chains decrease with increasing chain length (see Figure 7.14). Taking the free energy cost of introducing a single propane molecule into water and dividing by the approximate surface area of one molecule (about $2 \, \mathrm{nm}^2$) gives a free energy cost per surface area of $\approx 3 k_B T_r \, \mathrm{nm}^{-2}$.

Solvation of small polar molecules The preceding discussion contrasted molecules like hydrogen peroxide, which make H-bonds and mix freely with water, with nonpolar molecules like propane. Small *polar* molecules occupy a middle ground between these extremes. Like hydrocarbons, they do not form H-bonds with water; so in many cases, their solvation carries an entropic penalty. Unlike hydrocarbons, however, they do interact electrostatically with water: The surrounding water molecules can point their negative sides toward the molecule's positive parts and away from its negative parts. The resulting reduction in electrostatic energy can compensate the entropic loss, thereby making small polar molecules soluble at room temperature.

Large nonpolar objects The clathrate cage strategy shown in Figure 7.13 only works for sufficiently small included objects. Consider the extreme case of an infinite *planar* surface, for example, the surface of a lake, which is an interface between air and water. Air itself can be regarded as a hydrophobic substance because it too disrupts the H-bond network; the surface tension of the air–water interface is about $0.072 \, \mathrm{J \, m}^{-2}$. Clearly, the water molecules at the surface cannot each maintain four H-bonds directed tetrahedrally! Thus the hydrophobic cost of introducing a large nonpolar object into water carries a significant energy component, reflecting the breaking of H-bonds. Nevertheless, the magnitude of the hydrophobic effect in the large-object case is roughly the same as that of small molecules:

Your Turn 7H	Convert the free energy cost per area given earlier to $\mathrm{J \, m}^{-2}$ and compare it with the measured bulk oil–water surface tension Σ, which equals $\approx$ 0.04–0.05 $\mathrm{J \, m}^{-2}$.

Nonpolar solvents Although this section has mainly been concerned with solvation by water, it is useful to contrast the situation with nonpolar solvents, like oil or the interior of a bilayer membrane. Oils have no network of H-bonds. Instead, the key determinant of solubility is the electrostatic (Born) self-energy of the guest molecule. A polar molecule will prefer to be in water, where its self-energy is reduced by water's high permittivity (see Section 7.4.1 on page 261). Transferring such a molecule into oil thus incurs a large energy cost and is unfavorable. Nonpolar molecules, in contrast, have no such preference and pass more easily into oillike environments. We saw these phenomena at work when studying the permeability of lipid bilayers (Figure 4.13 on page 137): Fatty acids like hexanoic acid, with their hydrocarbon chains, dissolve more readily in the membrane (and hence permeate better) than do polar molecules like urea.

$\boxed{T_2}$ *Section 7.5.2′ on page 289 adds some details to our discussion of the hydrophobic effect.*

7.5.3 Water generates an entropic attraction between nonpolar objects

Section 7.4.5 described a very general interaction mechanism:

1. An isolated object (for example, a charged surface) assumes an equilibrium state (the counterion cloud) that makes the best compromise between entropic and energetic imperatives.
2. Disturbing this equilibrium (by bringing in an oppositely charged surface) can release a constraint (charge neutrality) and hence allow a reduction in the free energy (by counterion release).
3. This change favors the disturbance, thereby creating a force (the surfaces attract).

The depletion force furnishes an even simpler example (see Section 7.2.2 on page 251); here the released constraint is the reduction in the depletion zone's volume as two surfaces come together.

Thinking along these same lines, W. Kauzmann proposed in 1959 that any two nonpolar surfaces in water would tend to *coalesce*, in order to reduce the total nonpolar surface that they present to the water. Because the cost of hydrophobic solvation is largely entropic, so will be the corresponding force, or **hydrophobic interaction**, driving the surfaces together.

It's not easy to derive a quantitative, predictive theory of the hydrophobic interaction, but some simple qualitative predictions emerge from the picture just given. First, the largely entropic character of the hydrophobic effect suggests that the hydrophobic interaction should increase as we warm the system, starting from room temperature. Indeed, in vitro, the assembly of microtubules, driven in part by their monomers' hydrophobic preference to sit next to one another, can be controlled by temperature: Increasing the temperature *enhances* microtubule formation. Like the depletion interaction, the hydrophobic effect can harness entropy to create an apparent *increase* in order (self-assembly) by coupling it to an even greater increase of *dis*order among a class of smaller, more numerous objects (in this case the water molecules). Because the hydrophobic interaction involves mostly just the first layer of water molecules, it is of short range, like the depletion interaction. Thus we add the hydrophobic interaction to the list of weak, short-range interactions that are useful in giving macromolecular interactions their remarkable specificity. Chapter 8 will argue that the hydrophobic interaction is the dominant force driving protein self-assembly.

THE BIG PICTURE

Returning to the Focus Question, we have seen how the concentration of a *solute* can cause a flux of *water* across a membrane, with potentially fatal consequences. Chapter 11 will pick up this thread, showing how eukaryotic cells have dealt with the osmotic threat and even turned it to their advantage. Starting with osmotic pressure, we generalized the approach to include partially entropic forces, like the electrostatic and hydrophobic interactions responsible in part for the crucial specificity of inter-molecular recognition.

Taking a broader view, entropic forces are ubiquitous in the cellular world. To take just one example, each of your red blood cells has a meshwork of polymer strands attached to its plasma membrane. The remarkable ability of red cells to spring back to their disklike shape after squeezing through capillaries many times comes down to the elastic properties of this polymer mesh—and Chapter 9 will show that the elastic resistance of polymers to deformation is another example of an entropic force.

KEY FORMULAS

- *Osmotic:* A semipermeable membrane is a thin, passive partition through which solvent, but not solute, can pass. The pressure jump across a semipermeable membrane needed to *stop* osmotic flow of solvent equals $c k_B T$ for a dilute solution with number density c on one side and zero on the other (Equation 7.7).

    The actual pressure jump Δp may differ from this value. In that case, there is flow in the direction of the net thermodynamic force, $\Delta p - (\Delta c) k_B T$. If that force is small enough, then the volume flux of solvent will be $j_v = -L_p(\Delta p - (\Delta c) k_B T)$, where the *filtration coefficient* L_p is a property of the membrane (Equation 7.15).

- *Depletion interaction:* When large particles are mixed with smaller ones of radius R (for example, globular proteins mixed with small polymers), the smaller ones can push the larger ones together, to maximize their own entropy. If the two surfaces match precisely, the corresponding reduction of free energy per contact area is $\Delta F/A = c k_B T \times 2R$ (Equation 7.10).

- *Gauss:* Suppose that there is a plane of charge density $-\sigma_q$ at $x = 0$ and no electric field at $x < 0$. Then the Gauss Law gives the electric field in the $\hat{x}$ direction, just above the surface: $\mathcal{E}|_{\text{surface}} = -\sigma_q/\varepsilon$ (Equation 7.18).

- *Poisson:* The potential obeys Poisson's equation, $d^2 V/dx^2 = -\rho_q/\varepsilon$, where $\rho_q(\mathbf{r})$ is the charge density at $\mathbf{r}$ and ε is the permittivity of the medium, for example, water or air.

- *Bjerrum length:* $\ell_B = e^2/(4\pi \varepsilon k_B T)$ (Equation 7.21). This length describes how closely two like-charged ions can be brought together with $k_B T$ of energy available. In water at room temperature, $\ell_B = 0.71$ nm.

- *Debye:* $\boxed{T_2}$ The screening length for a monovalent salt solution (for example, NaCl at concentration c_∞), is $\lambda_D = (8\pi \ell_B c_\infty)^{-1/2}$ (Equation 7.35). At room temperature, it's $0.31 \text{ nm}/\sqrt{[\text{NaCl}]}$ (for a 1:1 salt like NaCl), or $0.18 \text{ nm}/\sqrt{[\text{CaCl}_2]}$ (2:1 salt), or $0.15 \text{ nm}/\sqrt{[\text{MgSO}_4]}$ (2:2 salt), where [NaCl] is the concentration measured in moles per liter.

FURTHER READING

Semipopular:
Electrostatics in the cellular context: Gelbart et al., 2000.
Properties of water: Ball, 2000.

Intermediate:
Depletion forces: Ellis, 2001.
Osmotic flow: Benedek & Villars, 2000b, §2.6.
Physical chemistry of water: Tinoco et al., 2001; Dill & Bromberg, 2002; Franks, 2000; van Holde et al., 1998.
Electrostatics in solution: Israelachvili, 1991; Safran, 1994.

Technical:
Depletion forces: Parsegian et al., 2000.
Electrostatic screening: Landau & Lifshitz, 1980, §§78, 92.
Hydrophobic effect: Southall et al., 2002; Israelachvili, 1991; Tanford, 1980.

7.1.2′ Track 2

1. The formal way to explain why we added the term fL to Equation 7.3 on page 247 is to say that we are performing a "Legendre transformation" from the fixed-volume to the fixed-pressure ensemble.

2. There is a symmetry between Equation 7.6 and the corresponding formula from our earlier discussion:

$$p = -\mathrm{d}F(V)/\mathrm{d}V \quad \text{(Equation 7.2)}; \quad \langle V \rangle = \mathrm{d}F(p)/\mathrm{d}p \quad \text{(Equation 7.6)}.$$

Pairs of quantities such as p and V, which appear symmetrically in these two versions, are called thermodynamically conjugate variables.

   Actually, the two formulas just given are not *perfectly* symmetrical because one involves V and the other $\langle V \rangle$. To understand this difference, recall that the first one rested on the entropic force formula, Equation 6.17. The derivation of this formula involved *macroscopic* systems, in effect saying "the piston is overwhelmingly likely to be in the position . . . ." In macroscopic systems, there is no need to distinguish between the expectation value of a variable and the value measured in a particular observation. In contrast, Equation 7.6 is valid even for microscopic systems, so it needs to specify that the expectation value is what is being predicted. The formulation of Equation 7.6 is the one we'll need when we analyze single-molecule stretching experiments in Chapter 9.

 7.3.1′ Track 2

1. The discussion of Section 7.3.1 made an implicit assumption; although it is quite well obeyed in practice, we should spell it out. We assumed that the filtration coefficient L_p was small enough, and hence that the flow was slow enough, to prevent the flow from significantly disturbing the concentrations on each side. So we can continue to use the equilibrium argument of Section 7.2.1 to find Δp. More generally, the osmotic flow rate will be a power series in Δc; we have just computed its leading term.

2. Osmotic effects will occur even if the membrane is not totally impermeable to solute, and indeed real membranes permit *both* solvent and solute to pass. In this case, the roles of pressure and concentration jump are not quite as simple as in Equation 7.15, although they are still related. When both these forces are small, we can expect a linear response combining Darcy's law and Fick's law:

$$\begin{bmatrix} j_{\mathrm{v}} \\ j_{\mathrm{s}} \end{bmatrix} = -\mathsf{P} \begin{bmatrix} \Delta p \\ \Delta c \end{bmatrix}. \tag{7.33}$$

Here P is called the permeability matrix.[9] Thus P_{11} is the filtration coefficient, whereas P_{22} is the solute permeability $\mathcal{P}_{\mathrm{s}}$ (see Equation 4.21 on page 135). The off-diagonal entry P_{12} describes osmotic flow, that is, solvent flow driven by a

---

[9]Section 9.3.1 on page 354 reviews matrix notation.

concentration jump. Finally, P_{21} describes "solvent drag": Mechanically pushing solvent through the membrane pulls along some solute.

Thus, a semipermeable membrane corresponds to the special case with $P_{22} = P_{21} = 0$. If, in addition, the system is in equilibrium, so that both fluxes vanish, then Equation 7.33 reduces to $P_{11}\Delta p = -P_{12}\Delta c$, and the result of Section 7.3.1 becomes, for a semipermeable membrane, $P_{12} = -L_p k_B T$.

More generally, L. Onsager showed in 1931 from basic thermodynamic reasoning that solvent drag is *always* related to solute permeability by $P_{12} = k_B T(c_0^{-1} P_{21} - L_p)$. A kinetic model, similar to the treatment of this chapter, was given in Manning, 1968.

 7.4.2′ Track 2

The formulas in Section 7.4.2 are special cases of the general Gauss law, which states that

$$\int \mathcal{E} \cdot d\mathbf{A} = \frac{q}{\varepsilon}.$$

In this formula, the integral is over any closed surface. The symbol $d\mathbf{A}$ represents a directed area element of the surface; it is defined as $\hat{\mathbf{n}} dA$, where dA is the element's area and $\hat{\mathbf{n}}$ is the outward-pointing vector perpendicular to the element. q is the total charge enclosed by the surface. Applying this formula to the two small boxes shown in Figure 7.7 on page 264 yields Equations 7.20 and 7.18, respectively.

7.4.3′ Track 2

The solution Equation 7.25 has a disturbing feature: The potential goes to infinity far from the surface! It's true that physical quantities like the electric field and concentration profile are well behaved (see Your Turn 7F), but still, this pathology hints that we have missed something. For one thing, no macromolecule is really an infinite plane. But a more important and interesting omission from our analysis is the fact that any real solution has at least some coions; the concentration c_∞ of salt in the surrounding water is never exactly zero.

Rather than introducing the unknown parameter c_0 and then going back to set it, this time we'll choose the constant in $V(x)$ so that $V \to 0$ *far* from the surface; then the Boltzmann distribution reads

$$c_+(x) = c_\infty e^{-eV(x)/k_B T} \quad \text{and} \quad c_-(x) = c_\infty e^{-(-e)V(x)/k_B T}$$

for the counterions and coions, respectively. The corresponding Poisson–Boltzmann equation is

$$\frac{d^2 \overline{V}}{dx^2} = -\frac{1}{2}\lambda_D^{-2}\left[e^{-\overline{V}} - e^{\overline{V}}\right], \tag{7.34}$$

where again $\overline{V} = eV/k_{\mathrm{B}}T$ and λ_{D} is defined as

$$\lambda_{\mathrm{D}} \equiv (8\pi \ell_{\mathrm{B}} c_\infty)^{-1/2}. \qquad \textbf{Debye screening length} \qquad (7.35)$$

In a solution of table salt, with $c = 0.1\,\mathrm{M}$, the screening length is about $1\,\mathrm{nm}$.

The solutions to Equation 7.34 are not elementary functions (they're called elliptic functions), but once again, we get lucky for the case of an isolated surface.

Your Turn 7I

Check that

$$\overline{V}(x) = -2\ln \frac{1 + e^{-(x+x_*)/\lambda_{\mathrm{D}}}}{1 - e^{-(x+x_*)/\lambda_{\mathrm{D}}}} \qquad (7.36)$$

solves the equation. In this formula, x_* is any constant. [*Hint:* It saves some writing to define a new variable, $\zeta = e^{-(x+x_*)/\lambda_{\mathrm{D}}}$, and rephrase the Poisson–Boltzmann equation in terms of ζ, not x.]

Before we can use Equation 7.36, we still need to impose the surface boundary condition. Equation 7.24 fixes x_*, via

$$e^{x_*/\lambda_{\mathrm{D}}} = \frac{e}{2\pi \ell_{\mathrm{B}} \lambda_{\mathrm{D}} \sigma_{\mathrm{q}}} \left(1 + \sqrt{1 + (2\pi \ell_{\mathrm{B}} \lambda_{\mathrm{D}} \sigma_{\mathrm{q}}/e)^2}\right). \qquad (7.37)$$

Your Turn 7J

Suppose that we only want the answer at distances less than some fixed x_{max}. Show that at low enough salt concentration (big enough λ_{D}), the solution Equation 7.36 becomes a constant plus our earlier result, Equation 7.25. How big must λ_{D} be?

We can now look at a more relevant limit for biology: This time, hold the salt concentration fixed and go out to large distances, where our earlier result (Equation 7.25) displayed its pathological behavior. For $x \gg \lambda_{\mathrm{D}}$, Equation 7.36 reduces to

$$\overline{V} \to -(4e^{-x_*/\lambda_{\mathrm{D}}})e^{-x/\lambda_{\mathrm{D}}}. \qquad (7.38)$$

That is,

> *The electric fields far outside a charged surface in an electrolyte are exponentially screened at distances greater than the Debye length λ_{D}.* $\qquad (7.39)$

Idea 7.39 and Equation 7.35 confirm an earlier expectation: Increasing c_∞ decreases the screening length, shrinking the diffuse charge layer and hence shortening the effective range of the electrostatic interaction (Idea 7.28).

In the special case of weakly charged surfaces (σ_q is small), Equation 7.37 gives $e^{-x_*/\lambda_D} = \pi \ell_B \lambda_D \sigma_q / e$; so the potential simplifies to

$$V(x) = -\frac{\sigma_q \lambda_D}{\varepsilon} e^{-x/\lambda_D}. \qquad \text{potential outside a weakly charged surface}$$

$$(7.40)$$

The ratio of the actual prefactor in Equation 7.38 and the form appropriate for weakly charged surfaces is sometimes called **charge renormalization**: Any surface will, at great distances, look the same as a weakly charged surface, but with the "renormalized" charge density $\sigma_{q,R} = (4\varepsilon/\lambda_D)e^{-x_*/\lambda_D}$. The true charge on the surface becomes apparent only when an incoming object penetrates into its strong-field region.

In the presence of added salt, the layer thickness no longer grows without limit as the layer charge gets smaller (as it did in the no-salt case, Equation 7.25); rather, it stops growing when it hits the Debye screening length. For weakly charged surfaces, then, the stored electrostatic energy is roughly that of a capacitor with gap spacing λ_D, not x_0. Repeating the argument at the end of Section 7.4.3, we now find the stored energy per unit area to be

$$E/(\text{area}) \approx k_B T \left(\frac{\sigma_q}{e}\right)^2 2\pi \lambda_D \ell_B. \qquad \begin{array}{l}\text{(electrostatic energy with added} \\ \text{salt, weakly charged surface)}\end{array} \qquad (7.41)$$

$\boxed{T_2}$ **7.4.4′ Track 2**

The crucial last step leading to Equation 7.31 may seem too slick. Can't we work out the force the same way we calculate any entropic force, by taking a derivative of the free energy? Absolutely. Let's compute the free energy of the system of counterions+surfaces, holding fixed the charge density $-\sigma_q$ on each surface but varying the separation $2D$ between the surfaces (see Figure 7.8b on page 265). Then the force between the surfaces will be $pA = -dF/d(2D)$, where A is the surface area, just as in Equation 6.17 on page 213.

First we notice an important property of the Poisson–Boltzmann equation (Equation 7.23 on page 267). Multiplying both sides by $d\overline{V}/dx$, we can rewrite the equation as

$$\frac{d}{dx}\left[\left(\frac{d\overline{V}}{dx}\right)^2\right] = 8\pi \ell_B \frac{dc_+}{dx}.$$

Integrating this equation gives a simpler, first-order equation:

$$\left(\frac{d\overline{V}}{dx}\right)^2 = 8\pi \ell_B (c_+ - c_0). \qquad (7.42)$$

To fix the constant of integration, we noted that the electric field is zero at the midplane, and $c_+(0) = c_0$ there.

Next we need the free energy density per unit area in the gap. You found the free energy density of an inhomogeneous ideal gas (or solution) in Your Turn 6K on page 237. The free energy for our problem is the integral of this quantity, plus the electrostatic energy[10] of the two negatively charged plates at $x = \pm D$:

$$F/(k_B T \times \text{area}) = -\frac{1}{2}\frac{\sigma_q}{e}\left(\overline{V}(D) + \overline{V}(-D)\right) + \int_{-D}^{D} dx \left[c_+ \ln\frac{c_+}{c_*} + \frac{1}{2}c_+\overline{V}\right].$$

In this formula, c_* is a constant whose value will drop out of our final answer (see Your Turn 6K).

We simplify our expression by first noting that $\ln(c_+/c_*) = \ln(c_0/c_*) - \overline{V}$, so the terms in square brackets are $c_+ \ln(c_0/c_*) - \frac{1}{2}c_+\overline{V}$. The first of these terms is a constant times c_+, so its integral is $2(\sigma_q/e)\ln(c_0/c_*)$. To simplify the second term, use the Poisson–Boltzmann equation to write $c_+ = -(4\pi\ell_B)^{-1}(d^2\overline{V}/dx^2)$. Next integrate by parts, obtaining

$$F/(k_B T \times \text{area}) = 2\frac{\sigma_q}{e}\left[\ln\frac{c_0}{c_*} - \frac{1}{2}\overline{V}(D)\right] + \frac{1}{8\pi\ell_B}\frac{d\overline{V}}{dx}\overline{V}\Big|_{-D}^{D} - \frac{1}{8\pi\ell_B}\int_{-D}^{D} dx\left(\frac{d\overline{V}}{dx}\right)^2.$$

We evaluate the boundary terms by using Equation 7.24 on page 267 at $x = -D$ and its analog on the other surface; they equal $-(\sigma_q/e)\overline{V}(D)$.

To do the remaining integral, recall Equation 7.42: it's $-\int_{-D}^{D} dx\,(c_+ - c_0)$, or $2(Dc_0 - (\sigma_q/e))$. Combining these results gives

$$F/(k_B T \times \text{area}) = 2Dc_0 + 2\frac{\sigma_q}{e}\left(\ln\frac{c_0}{c_*} - \overline{V}(D) - 1\right)$$

$$= \text{const} + 2Dc_0 + 2\frac{\sigma_q}{e}\ln\frac{c_+(D)}{c_*}.$$

The concentration at the wall can again be found from Equations 7.42 and 7.24: $c_+(D) = c_0 + (8\pi\ell_B)^{-1}(d\overline{V}/dx)^2 = c_0 + 2\pi\ell_B(\sigma_q/e)^2$.

A few abbreviations will make for shorter formulas. Let $\gamma = 2\pi\ell_B\sigma_q/e$ and $u = \beta D$, where $\beta = \sqrt{2\pi\ell_B c_0}$ as before. Then u and β depend on the gap spacing, whereas γ does not. With these abbreviations,

$$F/(k_B T \times \text{area}) = 2Dc_0 + \frac{\gamma}{\pi\ell_B}\ln\frac{c_0 + \gamma^2/(2\pi\ell_B)}{c_*}.$$

We want to compute the derivative of this expression with respect to the gap spacing, holding σ_q (and hence γ) fixed. We find

$$\frac{p}{k_B T} = -\frac{1}{k_B T}\frac{d\left(F/(k_B T \times \text{area})\right)}{d(2D)} = -c_0 - \left(D + \frac{\gamma}{2\pi\ell_B c_0 + \gamma^2}\right)\frac{dc_0}{dD}.$$

---

[10]Notice that adding any constant to $\overline{V}$ leaves this formula unchanged, because the integral $\int c_+ dx = 2\sigma_q/e$ is a constant, by charge neutrality. To understand the reason for the factor $\frac{1}{2}$ in the first and last terms, think about two point charges q_1 and q_2. Their potential energy at separation r is $q_1q_2/(4\pi\varepsilon r)$ (plus a constant). This is *one half* of the sum $q_1 V_2(r_1) + q_2 V_1(r_2)$. (The same factor of $\frac{1}{2}$ also appeared in the electrostatic self-energy Example on page 261.)

In the last term, we need

$$\frac{dc_0}{dD} = \frac{d}{dD}\left(\frac{u^2}{D^2 2\pi \ell_B}\right) = \frac{u}{\pi \ell_B D^3}\left(D\frac{du}{dD} - u\right).$$

To find du/dD, we write the boundary condition (Equation 7.29 on page 270) as $\gamma D = u\tan u$ and differentiate to find

$$\frac{du}{dD} = \frac{\gamma}{\tan u + u\sec^2 u} = \frac{\gamma u}{D\gamma + u^2 + (D\gamma)^2}.$$

This has gone far enough. In Problem 7.11, you'll finish the calculation to get a direct derivation of Equation 7.31. For a deeper derivation from thermodynamics, see Israelachvili, 1991, §12.7.

 7.5.1′ Track 2

1. The discussion in Section 7.5.1 described the electric field around a water molecule as that due to two positive point charges (the naked protons) offset from a diffuse negative cloud (the oxygen atom). Such a distribution will have a permanent electric dipole moment; and indeed, water is highly polarizable. But we drew a distinction between the H-bond and ordinary dipole interactions. This distinction can be described mathematically by saying that the charge distribution of the water molecule has many higher multipole moments (beyond the dipole term). These higher moments give the field both its great intensity and rapid falloff with distance.

   For comparison, propanone (acetone, or nail-polish remover, CH_3-CO-CH_3) has an oxygen atom bonded to its central carbon. The oxygen grabs more than its share of the carbon's electron cloud, leaving it positive but not naked. Accordingly, propanone has a dipole moment but not the strong short-range fields responsible for H-bonds. And indeed, the boiling point of propanone, although higher than a similar nonpolar molecule, is 44 K lower than that of water.

2. The picture of the H-bond given in Section 7.5.1 was rooted in classical electrostatics, so it is only part of the story. In fact, the H-bond is also partly covalent (quantum-mechanical) in character. Also, the formation of an H-bond between the hydrogen of an –OH group, for example, and another oxygen actually *stretches* the covalent bond in the original –OH group. Finally, the H-bond accepting sites, described rather casually as the "back side" of the water molecule, are in fact more sharply defined than our picture made it seem: The molecule strongly prefers to have all four of its H-bonds directed in the tetrahedral directions shown in Figure 7.12a on page 274.

3. Another feature of the ice crystal structure (Figure 7.12) is important, and general. In every H-bond shown, two oxygens flank a hydrogen, and all lie on a straight line (that is, the H-bond and its corresponding covalent bond are colinear). Quite generally, the H-bond is *directional*: There is a significant loss of binding free energy

if the hydrogen and its partners are not on a line. This additional property of H-bonds makes them even more useful for giving binding specificity to macro-molecules.

4. The books listed at the end of the chapter give many more details about the re-markable properties of liquid water.

7.5.2′ Track 2

1. The term *hydrophobic* can cause confusion, because it seems to imply that oil "fears" water. Actually, oil and water molecules attract each other, by the usual generic (van der Waals) interaction between any molecules; an oil–water mixture has lower energy than equivalent molecules of oil and water floating separately in vacuum. But liquid water attracts *itself* even more than it attracts oil (that is, its undisturbed H-bonding network is quite favorable), so it nevertheless tends to expel nonpolar molecules.

2. In his pioneering work on the hydrophobic effect, W. Kauzmann gave a more precise form of the solubility argument of Section 7.5.2. Figure 7.14 on page 278 shows that at least some nonpolar molecules' solubilities decrease as we raise the temperature beyond room temperature. Le Châtelier's Principle (to be discussed later, in Section 8.2.2′ on page 336) implies that for these substances, solvation releases energy because raising the temperature forces solute out of solution. The translational entropy change for a molecule to enter water is always positive be-cause then the molecule explores a greater volume. If the entropy change of the water itself were also positive, then every term of ΔF would favor solvation, and we could dissolve any amount of solute. That's not the case, so these solutes must induce a negative entropy change in the water upon solvation.

PROBLEMS*

7.1   Through one's pores

a. You are making strawberry shortcake. You cut up the strawberries, then sprinkle on some powdered sugar. A few moments later, the strawberries look juicy. What happened? Where did this water come from?

b. One often hears the phrase "learning by osmosis." Explain what's technically wrong with this phrase, and why "learning by permeation" might describe the desired idea better.

7.2   Pfeffer's experiment

van 't Hoff based his theory on the experimental results of W. Pfeffer. Here are some of Pfeffer's original 1877 data for the pressure needed to stop osmotic flow between pure water and a sucrose solution, across a copper ferrocyanide membrane at $T = 15°C$:

sugar concentration, g/(100 g of water)	pressure, mm of mercury
1	535
2	1016
2.74	1518
4	2082
6	3075

a. Convert these data to our units, m^{-3} and Pa (the molar mass of sucrose is about $342 \, g \, mole^{-1}$) and graph them. Draw some conclusions.

b. Pfeffer also measured the effect of temperature. At a fixed concentration of $(1 \, g \, sucrose)/(100 \, g \, water)$ he found:

temperature, °C	pressure, mm of mercury
7	505
14	525
22	548
32	544
36	567

Again convert to SI units, graph, and draw conclusions.

7.3   Experimental pitfalls

You are trying to make artificial blood cells. You have managed to get pure lipid bilayers to form spherical bags of radius 10 μm, filled with hemoglobin. The first time you did this, you transferred the "cells" into pure water and they promptly burst, spilling

---

*Problem 7.4 is adapted with permission from Benedek & Villars, 2000b.

the contents. Eventually, you found that transferring them to a 1 mM salt solution prevents bursting, leaving the "cells" spherical and full of hemoglobin and water.

a. If 1 mM is good, then would 2 mM be twice as good? What happens when you try this?

b. Later you decide that you don't want salt outside because it makes your solution electrically conducting. How many moles per liter of glucose should you use instead?

7.4 *Osmotic estimate of molecular weight*

Chapter 5 discussed the use of centrifugation to estimate macromolecular weights, but this method is not always the most convenient.

a. The osmotic pressure of blood plasma proteins is usually expressed as about 28 mm of mercury (this unit is defined in Appendix A) at body temperature, 303 K. The quantity of plasma proteins present has been measured to be about 60 g L^{-1}. Use these data to estimate the average molar mass M in g/mole for these plasma proteins, assuming the validity of the dilute limit.

b. The filtration coefficient of capillary membranes is sometimes quoted as $L_p = 7 \cdot 10^{-6}$ cm s^{-1}atm^{-1}. If we put pure water on *both* sides of a membrane with a pressure drop of Δp, the resulting volume flux of water is $L_p \Delta p$. Assume that a normal person has rough osmotic balance across his capillaries but that in a particular individual, the blood plasma proteins have been depleted by 10%, as the result of a nutritional deficiency. What would be the total accumulation of fluid in interstitial space (liters per day), given that the total area of open capillaries is about 250 m^2? Why do you think starving children have swollen bellies?

7.5 *Depletion interaction estimates*

Section 7.2.1 said that a typical globular protein is a sphere of radius 10 nm. Cells have a high concentration of such proteins; for illustration, suppose that they occupy about 30% of the interior volume.

a. Imagine two large, flat objects inside the cell (representing two big macromolecular complexes with complementary surfaces). When they approach each other closer than a certain separation, they'll feel an effective depletion interaction driving them still closer, a force caused by the surrounding suspension of smaller proteins. Draw a picture, assuming that the surfaces are parallel as they approach each other. Estimate the separation at which the force begins.

b. If the contact area is 10 μm^2, estimate the total free energy reduction when the surfaces stick. You may neglect any other possible interactions between the surfaces; and as always, assume that we can still use the van 't Hoff (dilute-suspension) relation for osmotic pressure. Is it significant relative to $k_B T_r$?

7.6 *Effect of hydrogen bonds on water*

According to Section 7.5.1, the average number of H-bonds between a molecule of liquid water and its neighbors is about 3.5. Assume that these bonds are the major interaction holding liquid water together and that each H-bond lowers the energy by about $9k_B T_r$. Using these ideas, find a numerical estimate for the heat of vaporization of water (see Problem 1.6), then compare your prediction with the measured value.

7.7 $\boxed{T_2}$ *Weak-charge limit*

Section 7.4.3 considered an ionizable surface immersed in pure water. Thus, the surface dissociated into a negative plane and a cloud of positive counterions. Real cells, however, are bathed in a solution of salt, among other things; there is an external reservoir of *both* counterions and negative coions. Section 7.4.3′ on page 284 gave a solution for this case, but the math was complicated; here is a simpler, approximate treatment.

Instead of solving Equation 7.34 exactly, consider the case where the surface's charge density is small. Then the potential $V(0)$ at the surface will not be very different from the value at infinity, which we took to be zero. (More precisely, the dimensionless combination $\overline{V}$ is everywhere much smaller than 1.) Approximate the right-hand side of Equation 7.34 by the first two terms of its series expansion in powers of $\overline{V}$. The resulting approximate equation is easy to solve. Solve it, and give an interpretation to the quantity λ_D defined in Equation 7.35.

7.8 $\boxed{T_2}$ *Diffusion increases entropy*

Suppose that we prepare a solution at time $t = 0$ with a nonuniform concentration $c(\mathbf{r})$ of solute. (For example, we could add a drop of ink to a glass of water without mixing it.) This initial state is not a minimum of free energy: Its entropy is not maximal. We know that diffusion will eventually erase the initial order.

Section 7.2.1 argued that for dilute solutions, the dependence of entropy on concentration was the same as that of an ideal gas. Thus the entropy S of our system will be the integral of the entropy density (Equation 6.36 on page 233) over $d^3\mathbf{r}$, plus a constant that we can ignore. Calculate the time derivative of S in a thermally isolated system, using what you know about the time derivative of c. Then comment. [*Hint:* In this problem, you can neglect bulk (convective) flow of water. You can also assume that the concentration is always zero at the boundaries of the chamber; the ink spreads from the center without hitting the walls.]

7.9 $\boxed{T_2}$ *Another mean-field theory*

The aim of this problem is to gain a qualitative understanding of the experimental data in Figure 4.8c on page 125, by using a mean-field approximation pioneered by P. Flory.

Recall that the figure gives the average size of a random coil of DNA attached ("adsorbed") to a two-dimensional surface—a self-avoiding, two-dimensional random walk. To model such a walk, we first review unconstrained (non–self-avoiding) random walks. Notice that Equation 4.28 on page 143 gives the number of N-step paths that start at the origin and end in an area $d^2\mathbf{r}$ around the position $\mathbf{r}$ (in an approximation discussed in Section 4.6.5′ on page 150). Using Idea 4.5b on page 115, this number equals $e^{-r^2/(2NL^2)}d^2\mathbf{r}$ times a normalization constant, where L is the step size. To find the mean-square displacement of an ordinary random walk, we compute the average $\langle \mathbf{r}^2 \rangle$, weighting every allowed path equally. The preceding discussion lets us express the answer as

$$\langle \mathbf{r}^2 \rangle = - \left. \frac{\mathrm{d}}{\mathrm{d}\beta} \right|_{\beta=(2NL^2)^{-1}} \left(\ln \int \mathrm{d}^2 \mathbf{r}\, e^{-\beta r^2} \right) = 2L^2 N. \qquad (7.43)$$

That's a familiar result (see the discussion preceding Equation 4.6).

But we *don't* want to weight every allowed path equally; those that self-intersect should be penalized by a Boltzmann factor. Flory estimated this factor in a simple way. The effect of self-avoidance is to swell the polymer coil to a size larger than what it would be in a pure random walk. This swelling also increases the mean end-to-end length of the coil, so we imagine the coil as a circular blob whose radius is a constant C times its end-to-end distance, $r = |\mathbf{r}|$. The area of such a blob is then $\pi(Cr)^2$. In this approximation, the average surface density of polymer segments in the class of paths with end-to-end distance r is $N/(\pi C^2 r^2)$.

We next idealize the adsorbed coil as having a *uniform* surface density of segments and assume that each of the polymer's segments has a probability of bumping into another that depends on that density.[11] If each segment occupies a surface area a, then the probability of an area element being occupied is $Na/(\pi C^2 r^2)$. The probability of any of the N chain elements landing on a space that is already occupied is given by the same expression, so the number of doubly occupied area elements equals $N^2 a/(\pi C^2 r^2)$. The energy penalty V equals this number times the energy penalty ϵ per crossing. Writing $\bar{\epsilon} = \epsilon a/(\pi C^2 k_B T)$ gives the estimate $V/k_B T = \bar{\epsilon} N^2/r^2$.

Adapt Equation 7.43 by introducing the Boltzmann weighting factor $e^{-V/k_B T}$. Take $L = 1\,\mathrm{nm}$ and $\bar{\epsilon} = 1\,\mathrm{nm}^2$ for concreteness, and work at room temperature. Use some numerical software to evaluate your modified integral, finding $\langle \mathbf{r}^2 \rangle$ as a function of N for fixed segment length L and overlap cost $\bar{\epsilon}$. Make a log-log plot of the answer and show that, for large N, $\langle \mathbf{r}^2 \rangle \to \mathrm{const} \times N^{2\nu}$. Find the exponent ν and compare with the experimental data.

7.10 $\boxed{T_2}$ *Charged surfaces*

Use some numerical software to solve Equation 7.29 for β as a function of plate separation $2D$ for fixed charge density σ_q. For concreteness, take σ_q to equal $e/(20\,\mathrm{nm}^2)$. Now convert your answer into a force by using Equation 7.31 and compare your answer with Figure 7.11. Repeat with other values of σ_q to find (roughly) the one that best fits the upper curve in the figure at separation greater than 2 nm. If this surface were fully dissociated, it would have one electron charge per $7\,\mathrm{nm}^2$. Is it fully dissociated?

7.11 $\boxed{T_2}$ *Direct calculation of a surface force*

Finish the derivation of Section 7.4.4′ on page 286. The goal is to establish Equation 7.31.

---

[11] Substituting this estimate for the actual self-intersection of the conformation amounts to a mean-field approximation, similar in spirit to the one in Section 7.4.3 on page 264.

CHAPTER 8

Chemical Forces and Self-Assembly

The ant has made himself illustrious
Through constant industry industrious.
So What?
Would you be calm and placid
If you were full of formic acid?
—Ogden Nash, 1935

Chapter 7 showed how simple free energy transduction machines, like the osmotic pressure cell (Figure 1.3 on page 13) or the heat engine (Figure 6.5 on page 215), generate mechanical forces from concentration or temperature differences. But even though living creatures do make use of these sources of free energy, their most important energy storage mechanisms involve *chemical* energy. This chapter will establish chemical energy as just another form of free energy, mutually convertible with all the other forms. We will do this by developing further the idea that every molecule carries a definite stored potential energy and by adding that energy into the first term of the fundamental formula for free energy, $F = E - TS$. We will then see how chemical energy drives the self-assembly responsible for the creation of bilayer membranes and cytoskeletal filaments.

The Focus Question for this chapter is

Biological question: How can a molecular machine, sitting in the middle of a well-mixed solution, extract useful work? Doesn't it need to sit at the *boundary* between chambers of different temperature, pressure, or concentration, like a heat engine, turbine, or osmotic cell?

Physical idea: Even a well-mixed solution can contain many different molecular species, at far-from-equilibrium concentrations. The deviation from equilibrium gives rise to a chemical force.

8.1 CHEMICAL POTENTIAL

Cells do not run on temperature gradients. Instead, they eat food and excrete waste. Moreover, the "useful work" done by a molecular machine may be chemical synthesis, not mechanical work.

In short, the machines of interest to us exchange both energy *and molecules* with the outside world. To begin to understand chemical forces, then, we first examine how a small subsystem in contact with a large one chooses to share each kind of molecule, temporarily neglecting the possibility of interconversions among the molecular species.

8.1.1 μ measures the availability of a particle species

We must generalize our formulas from Chapter 6 to handle the case where two systems, A and B, exchange particles as well as energy. As usual, we will begin by thinking about ideal gases. So we imagine an isolated system, for example, an insulated box of fixed volume with N noninteracting gas molecules inside. Let $S(E, N)$ be the entropy of this system. Later, when we want to consider several species of molecules, we'll call their populations $N_1, N_2, \ldots$, or generically N_α, where $\alpha = 1, 2, \ldots$.

The temperature of our system at equilibrium is again defined by Equation 6.9, $\frac{1}{T} = \frac{dS}{dE}$. Now, however, we add the clarification that the derivative is taken holding the N_α's fixed: $T^{-1} = \frac{dS}{dE}\big|_{N_\alpha}$. (Take a moment to review the visual interpretation of this statement in Section 4.5.2 on page 134.) Because we want to consider systems that gain or lose molecules, we'll also need to look at the derivatives with respect to the N_α's: Let

$$\mu_\alpha = -T \left. \frac{dS}{dN_\alpha} \right|_{E, N_\beta, \beta \neq \alpha}. \tag{8.1}$$

The μ_α's are called **chemical potentials**. This time, the notation means that we are to take the derivative with respect to one of the N_α's, holding fixed both the other N's and the total energy of the system. Notice that the number of molecules is dimensionless, so μ has the same dimensions as energy.

You should now be able to show, exactly as in Section 6.3.2, that when two macroscopic subsystems can exchange *both* particles and energy, eventually each is overwhelmingly likely to have energy and particle numbers such that $T_A = T_B$ and the chemical potentials match for each species α:

$$\boxed{\mu_{A,\alpha} = \mu_{B,\alpha}. \qquad \text{matching rule for macroscopic systems in equilibrium}}$$

$$\tag{8.2}$$

When Equation 8.2 is satisfied, we say the system is in **chemical equilibrium**. Just as $T_A - T_B$ gives the entropic force driving energy transfer, so $\mu_{A,\alpha} - \mu_{B,\alpha}$ gives another entropic force driving the net transfer of particles of type α. For instance, this rule is the right tool to study the coexistence of water and ice at $0°$C; in equilibrium, water molecules must have the same μ in each phase.

There is a subtle point hiding in Equation 8.1. Up to now, we have been ignoring the fact that each individual molecule has some internal energy ϵ, for example, the energy stored in chemical bonds (see Section 1.5.3 on page 26). Thus the total energy is the sum $E_{\text{tot}} = E_{\text{kin}} + N_1 \epsilon_1 + \cdots$ of kinetic plus internal energies. In an ideal gas, the particles never change, so the internal energy is locked up: It just gives a constant contribution to the total energy E, which we can ignore. In this chapter, however, we will need to account for the internal energies, which change during a chemical reaction. Thus it's important to note that the derivative in the definition of chemical potential (Equation 8.1) is to be taken while holding fixed the *total* energy, including its internal component.

To appreciate this point, let's work out a formula for the chemical potential in the ideal-gas case and see how ϵ comes in. Our derivation of the entropy of an ideal gas is a useful starting point (see the ideal gas entropy Example on page 200), but we need to remember that E appearing there was only the kinetic energy E_{kin}.

Your Turn 8A

As a first step to evaluating Equation 8.1, calculate the derivative of S with respect to N for an ideal gas, holding fixed the kinetic energy. Take N to be very large and find

$$\frac{dS}{dN}\bigg|_{E_{kin}} = k_B \frac{3}{2} \ln\left(\frac{1}{3\pi} \frac{m}{\hbar^2} \frac{E_{kin}}{N} \left(\frac{V}{N}\right)^{2/3}\right).$$

To finish the derivation of μ, we need to convert the formula you just found to give the derivative at fixed total energy E, not fixed E_{kin}. If we inject a molecule into the system, holding E_{kin} fixed, then *extract* an amount of kinetic energy equal to the internal energy ϵ of that molecule, this combined process has the net effect of holding the total energy E fixed while changing the particle number by $dN = 1$. Thus we need to subtract a correction term from the result in Your Turn 8A.

Example: Carry out the step just described, and show that the chemical potential of an ideal gas can be written as

$$\mu = k_B T \ln(c/c_0) + \mu^0(T). \qquad \text{chemical potential, ideal gas or dilute solution}$$

(8.3)

In this formula, $c = N/V$ is the number density, c_0 is a constant (called the **reference concentration**), and

$$\mu^0(T) = \epsilon - \frac{3}{2} k_B T \ln \frac{m k_B T}{2\pi \hbar^2 c_0^{2/3}}. \qquad \text{(ideal gas)} \qquad (8.4)$$

Solution: Translating the words into math, we need to subtract $\epsilon \frac{dS}{dE_{kin}}\big|_N$ from the result in Your Turn 8A. Combining the resulting formula with Equation 8.1 and using the fact that the average kinetic energy E_{kin}/N equals $\frac{3}{2} k_B T$ then gives

$$\mu = \epsilon + k_B T \ln c - \frac{3}{2} k_B T \ln \left[\frac{4\pi}{3} \frac{m}{(2\pi\hbar)^2} \frac{3}{2} k_B T\right].$$

This formula appears to involve the logarithms of dimensional quantities. To make each term separately well defined, we add and subtract $k_B T \ln c_0$, obtaining Equations 8.3 and 8.4.

We call μ^0 the **standard chemical potential** at temperature T defined with respect to the chosen reference concentration. The choice of the reference value is a convention; the derivation just given makes it clear that its value drops out of the right-hand side of Equation 8.3. Chemists refer to the dimensionless quantity $e^{(\mu - \mu^0)/k_B T}$ as the **activity**. Thus Equation 8.3 states that the activity equals approximately c/c_0 for an ideal gas.

Equation 8.3 also holds for dilute *solutions* as well as for low-density gases. As argued in our discussion of osmotic pressure (Section 7.2 on page 248), the entropic term is the same in either case. For a solute in a liquid, however, the value of $\mu^0(T)$ will no longer be given by Equation 8.4. Instead, $\mu^0(T)$ will now reflect the fact that the solvent (water) molecules themselves are *not* dilute, so the attractions of solvent molecules to one another and to the solute are not negligible. Nevertheless, Equation 8.3 will still hold with some measurable standard chemical potential $\mu^0(T)$ for the chemical species in question at some standard concentration c_0. Usually we don't need to worry about the details of the solvent interactions; we'll regard μ^0 as just a phenomenological quantity to be looked up in tables.

For gases, the standard concentration is taken to be the one obtained at atmospheric pressure and temperature: roughly one mole per 22 L. In this book, however, we will nearly always be concerned with **aqueous solutions** (solutions in water), not with gases. For aqueous solutions, the standard concentrations are all taken to be[1] $c_0 = 1 \, \text{M} \equiv 1 \, \text{mole/L}$, and we introduce the shorthand notation $[X] \equiv c_X/(1 \, \text{M})$ for the concentration of any molecular species X in molar units. A solution with $[X] = 1$ is called a one **molar** solution.

You can generalize Equation 8.3 to situations where in addition to ϵ, each molecule also has an extra potential energy $U(z)$ depending on its position. For example, a particle of mass m in a gravitational field has $U(z) = mgz$, where z is the height. A more important case is that of an electrically charged species, where $U(z) = qV(z)$. In either case, we simply replace μ^0 by $\mu^0 + U(z)$ in Equation 8.3. (In the electric case, some authors call this generalized μ the **electrochemical potential**.) Making this change to Equation 8.3 and applying the matching rule (Equation 8.2) shows that, in equilibrium, every part of an electrolyte solution has the same value of $c(z)e^{qV(z)/k_B T}$. This result is already familiar to us—it's equivalent to the Nernst relation (Equation 4.26 on page 141).

Setting aside these refinements, the key result of this section is that we have found a quantity μ describing the *availability of particles* just as T describes the availability of energy; for dilute systems, it separates into a part with a simple dependence on the concentration, plus a concentration-independent part $\mu^0(T)$ involving the internal energy of the molecule. More generally, we have a fundamental definition of this availability (Equation 8.1) and a result about equilibrium (the matching rule, Equation 8.2) that is applicable to any system, dilute or not. This degree of generality is

---

[1]With some exceptions—see Section 8.2.2.

important because we know that the interior of cells is not at all dilute—it's crowded (see Figure 7.2 on page 252).

The chemical potential goes up when the concentration increases (more molecules are available), but it's also greater for molecules with more internal energy (they're more eager to dump that energy into the world as heat, thereby increasing the world's disorder). In short,

> *A molecular species will be highly available for chemical reactions if*
> *its concentration c is big or its internal energy ϵ is big.*   (8.5)

The chemical potential (Equation 8.3) describes the overall availability.

$\boxed{T_2}$ *Section 8.1.1' on page 335 makes some connection to more advanced treatments and to quantum mechanics.*

8.1.2 The Boltzmann distribution has a simple generalization accounting for particle exchange

From here, it's straightforward to redo the analysis of Section 6.6.1. We temporarily continue to suppose that particles cannot interconvert and that a smaller system a is in equilibrium with a much larger system B. Then the relative fluctuations of N_a (the number of particles in a) can be big because a may not be macroscopic. So we cannot just compute N_a by using Equations 8.1 and 8.2; the best we can do is to give the *probability distribution* P_j of various states j that a may assume. System B, on the other hand, *is* macroscopic; in equilibrium, the relative fluctuations of N_B will therefore be small.

Let state j of subsystem a have energy E_j and particle number N_j. We want the probability P_j for a to be in state j, regardless of what B is doing.

Your Turn 8B

Show that in equilibrium,

$$P_j = \mathcal{Z}^{-1} e^{(-E_j + \mu N_j)/k_B T}, \qquad (8.6)$$

where again the **grand partition function** $\mathcal{Z}$ is the appropriate normalization constant, $\mathcal{Z} = \sum_j e^{(-E_j + \mu N_j)/k_B T}$. [*Hint:* Adapt the discussion in Section 6.6.1 on page 218.]

The probability distribution you just found is sometimes called the **Gibbs**, or **grand canonical**, distribution. It's a generalization of the Boltzmann distribution (Equation 6.23 on page 219). Once again, we see that most of the details about system B don't matter; all that enters are two numbers, the values of its temperature and chemical potential.

Thus, large μ means system a is more likely to contain many particles, justifying the interpretation of μ as the availability of particles from B. It's now straightforward to work out results analogous to Your Turn 6G and the free energy formula Example (page 224), but we won't need these later. (It's also straightforward to include changes in volume as molecules migrate; see Problem 8.8.)

8.2 CHEMICAL REACTIONS

8.2.1 Chemical equilibrium occurs when chemical forces balance

At last we are ready to think about chemical reactions. Let's begin with a very simple situation, in which a molecule has two states (or isomers) $\alpha = 1,2$ differing only in internal energy: $\epsilon_2 > \epsilon_1$. We also suppose that spontaneous transitions between the two states are rare; so we can think of the states as two different molecular species. Thus we can prepare a beaker (system B) with any numbers N_1 and N_2 we like, and these numbers won't change.

But now imagine that, in addition, our system has a "phone booth" (called sub-system a) where, like Superman and his alter ego, molecules of one type can duck in and convert (or isomerize) to the other type. (We can think of this subsystem as a molecular machine, like an enzyme, although we will later argue that the same analysis applies more generally to any chemical reaction.)

Suppose that type 2 walks into the phone booth and type 1 walks out. After this transaction, subsystem a is in the same state as it was to begin with. Because energy is conserved, the big system B also has the same total energy as it had to begin with. But now B has one fewer type 2 and one more type 1 molecule. The difference of internal energies, $\epsilon_2 - \epsilon_1$, gets delivered to the large system B as thermal energy.

No physical law prevents the same reaction from happening in reverse. Type 1 can walk into the phone booth and spontaneously convert to type 2, *drawing* the necessary energy from the thermal surroundings.

Gilbert says: Of course, this would never happen in real life. Energy doesn't spontaneously organize itself from thermal motion to any sort of potential energy. Rocks don't fly out of the mud.

Sullivan: But transformations of individual molecules can go in either direction. If a reaction can go forward, it can also go backward, at least once in a while. Don't forget our buffalo (Figure 6.8 on page 220).

Gilbert: Yes, of course. I meant the *net* number converting to the low-energy state per second must be positive.

Sullivan: But wait! We've seen before how even that isn't necessarily true, as long as somebody pays the disorder bill. Remember our osmotic machine; it can draw thermal energy out of the environment to lift a weight (Figure 1.3a on page 13).

Sullivan has a good point. The preceding discussion, along with the defining Equation 8.1, implies that when the reaction takes one step that converts type 1 to type 2, the world's entropy changes[2] by $(-\mu_2 + \mu_1)/T$. The Second Law says that a net, macroscopic flow in this direction will happen if $\mu_1 > \mu_2$. So it makes sense to refer to the difference of chemical potentials as a "chemical force" driving isomerization. In the situation sketched above, $\epsilon_1 < \epsilon_2$, but ϵ is only a part of the chemical potential (Equation 8.3). If the *concentration* of the low-energy isomer is high (or that of the high-energy isomer is low), then we can have $\mu_1 > \mu_2$, and hence a net flow

---

[2]It's crucial that we defined μ as a derivative holding total energy fixed. Otherwise $(-\mu_2 + \mu_1)/T$ would describe an impossible, energy-nonconserving process.

$1 \to 2$! And, indeed, some spontaneous chemical reactions are endothermic (heat-absorbing): Think of the chemical icepacks used to treat sprains. The ingredients inside the icepack spontaneously put themselves into a higher-energy state, drawing the necessary thermal energy from their surroundings.

What does Sullivan mean by "pays the disorder bill"? Suppose that we prepare a system where initially species 1 far outnumber 2. This is a state with some order. Allowing conversions between the isomers is like connecting two tanks of equal volume but with different numbers of gas molecules. Gas whooshes through the connection to equalize those numbers, thereby erasing that order. It can whoosh through even if it has to turn a turbine along the way and do mechanical work. The energy to do that work came from the thermal energy of the environment, but *the conversion from thermal to mechanical energy was paid for by the increase of disorder as the system equilibrated.* Similarly, in our example, if state 1 outnumbers state 2 there will be an entropic force pushing the conversion reaction in the direction $1 \to 2$, even if this direction is "uphill," that is, even if the reaction raises the stored chemical energy. As the reaction proceeds, the supply of 1 gets depleted (and μ_1 decreases) while that of 2 gets enriched (μ_2 increases), until $\mu_1 = \mu_2$. Then the reaction stalls. In other words,

$$\text{\textit{Chemical equilibrium is the point where the chemical forces balance.}} \qquad (8.7)$$

More generally, if mechanical or electrical forces act on a system, we should expect equilibrium when the net of *all* the driving forces, including chemical ones, is zero.

The preceding paragraph should sound familiar. Section 6.6.2 on page 220 argued that two species with a fixed energy difference ΔE would come to an equilibrium with concentrations related by $c_2/c_1 = e^{-\Delta E/k_B T}$ (Equation 6.24 on page 219). Taking the logarithm of this formula shows that for dilute solutions, it's nothing but the condition that $\mu_2 = \mu_1$. If the two "species" have many internal substates, the discussion in Section 6.6.4 on page 225 applies; we just replace ΔE by the internal *free* energy difference of the two species. The chemical potentials include both the internal entropy and the concentration-dependent part, so the criterion for equilibrium is still $\mu_2 = \mu_1$.

There is another useful interpretation of chemical forces. So far, we have been considering an isolated system and discussing the change in its entropy when a reaction takes one step. We imagined dividing the system into subsystems a (the molecule undergoing isomerization) and B (the surrounding test tube), and required that the entropy of the isolated system $a+B$ increase. But more commonly, $a+B$ is in thermal contact with an even larger world, as, for example, when a reaction takes place in a test tube sitting in our lab. In this case, the entropy change of $a+B$ will *not* be $(-\mu_2 + \mu_1)/T$, because some thermal energy will be exchanged with the world in order to hold the temperature fixed. Nevertheless, the quantity $\mu_2 - \mu_1$ still does control the direction of the reaction:

Your Turn 8C

a. Following Section 6.5.1 on page 210, show that in a sealed test tube held at fixed temperature, the Helmholtz free energy F of $a+B$ changes by $\mu_2 - \mu_1$ when the reaction takes one step.

> **b.** Similarly, for an *open* test tube in contact with the atmosphere at pressure p, show that the Gibbs free energy G changes by $\mu_2 - \mu_1$ when the reaction takes one step.

$\boxed{T_2}$ *Section 8.2.1' on page 335 connects the discussion to the notation used in advanced texts.*

8.2.2 ΔG gives a universal criterion for the direction of a chemical reaction

Section 8.2.1 showed how the condition $\mu_1 = \mu_2$ for the equilibrium of an isomerization reaction recovers some ideas from Chapter 6. Our present viewpoint has a number of advantages over the earlier one, however:

1. The analysis of Section 6.6.2 was concrete, but its applicability was limited to dilute solutions (of buffalo). In contrast, the equilibrium condition $\mu_2 = \mu_1$ is completely general: It's just a restatement of the Second Law. If μ_1 is bigger than μ_2, then the net reaction $1 \rightarrow 2$ increases the world's entropy. Equilibrium is the situation where no such further increase is possible.
2. Interconversions between two isomers are interesting, but there's a lot more to chemistry than that. Our present viewpoint lets us generalize our result.
3. The analysis of Section 6.6.2 gave us a hint of a deep result when we noted that the activation barrier $\Delta E^{\ddagger}$ dropped out of the equilibrium condition. We now see that more generally, *it doesn't matter at all what happens inside the "phone booth"* mentioned at the start of Section 8.2.1. We made no mention of it, apart from the fact that it ends up in the same state in which it started.

Indeed, the "phone booth" may not be present at all: Our result for equilibrium holds even for spontaneous reactions in solution, as long as they are slow enough that we have well-defined initial concentrations c_1 and c_2.

Burning hydrogen Let's follow up on point (2). The burning of hydrogen is a familiar chemical reaction:

$$2H_2 + O_2 \rightharpoonup 2H_2O. \tag{8.8}$$

Let's consider this as a reaction involving three ideal gases. We take an isolated chamber at room temperature containing twice as many moles of hydrogen as oxygen, then set off the reaction with a spark. We're left with a chamber containing water vapor *and* very tiny traces of hydrogen and oxygen. We now ask, how much unreacted hydrogen remains?

Equilibrium is the situation where the world's entropy S_{tot} is a maximum. To be at a maximum, all the derivatives of the entropy must equal zero; in particular, there must be no change in S_{tot} if the reaction takes one step to the left (or right). So, to find the condition for equilibrium, we compute this change and set it equal to zero.

Because atoms aren't being created or destroyed, a step to the right removes one oxygen and two hydrogen molecules from the world and creates two water molecules. Define the symbol ΔG by

$$\Delta G \equiv 2\mu_{H_2O} - 2\mu_{H_2} - \mu_{O_2}. \tag{8.9}$$

With this definition, Equation 8.1 says that the change in the world's entropy for an isolated reaction chamber is $\Delta S_{tot} = -\Delta G/T$. For equilibrium, we require that $\Delta S_{tot} = 0$. Your Turn 8C gave another interpretation of ΔG, namely, as the change of free energy of an open reaction chamber. From this equivalent point of view, setting $\Delta G = 0$ amounts to requiring that the Gibbs free energy be at a minimum.

Oxygen, hydrogen, and water vapor are all nearly ideal gases under ordinary conditions. Thus we can use Equation 8.3 on page 296 to simplify Equation 8.9 and put the equilibrium condition into the form

$$0 = \frac{\Delta G}{k_B T} = \frac{2\mu_{H_2O}^0 - 2\mu_{H_2}^0 - \mu_{O_2}^0}{k_B T} + \ln\left[\left(\frac{c_{H_2O}}{c_0}\right)^2 \left(\frac{c_{H_2}}{c_0}\right)^{-2} \left(\frac{c_{O_2}}{c_0}\right)^{-1}\right].$$

We can lump all the concentration-independent terms of this equation into one package, the **equilibrium constant** of the reaction:

$$K_{eq} \equiv e^{-(2\mu_{H_2O}^0 - 2\mu_{H_2}^0 - \mu_{O_2}^0)/k_B T}. \tag{8.10}$$

With this abbreviation, the condition for equilibrium becomes

$$\frac{(c_{H_2O})^2}{(c_{H_2})^2 c_{O_2}} = K_{eq}/c_0. \qquad \text{(in equilibrium)} \tag{8.11}$$

The left side of this formula is sometimes called the **reaction quotient**. It's also convenient to define a logarithmic measure of the equilibrium constant via

$$pK \equiv -\log_{10} K_{eq}. \tag{8.12}$$

Equation 8.11 is just a restatement of the Second Law. Nevertheless, it tells us something useful: The condition for equilibrium is that a certain combination of the concentrations (the reaction quotient) must equal a concentration-independent constant (the equilibrium constant divided by the reference concentration).

In the situation under discussion (hydrogen gas reacts with oxygen gas to make water vapor), we can make our formulas still more explicit:

Your Turn 8D

a. Show that the results of the gas chemical potential Example (page 296) let us write the equilibrium constant as

$$K_{eq} = \left[e^{(2\epsilon_{H_2} + \epsilon_{O_2} - 2\epsilon_{H_2O})/k_B T}\right] \times \left[c_0 \left(\frac{2\pi\hbar^2}{k_B T} \frac{(m_{H_2O})^2}{(m_{H_2})^2 m_{O_2}}\right)^{3/2}\right]. \tag{8.13}$$

b. Check the dimensions in this formula.

Equation 8.13 shows that the equilibrium constant of a reaction depends on our choice of a reference concentration c_0. (Indeed, it must, because the equilibrium value of the reaction quotient does *not* depend on c_0.)

The equilibrium constant also depends on temperature. Mostly this dependence arises from the exponential factor in Equation 8.13. Hence we get the same behavior as in isomerization (Equation 6.24 on page 219):

- At low temperatures, the first factor becomes extremely large because it is the exponential of a large positive number. Equation 8.11 in turn implies that the equilibrium shifts almost completely in favor of water.
- At very high temperatures, the first factor is close to 1. Equation 8.11 then says that there will be significant amounts of unreacted hydrogen and oxygen. The mechanical interpretation is that thermal collisions are constantly ripping water molecules apart as fast as they form.

Example: Physical chemistry books give the equilibrium constant for Reaction 8.8 at room temperature as $e^{(457\,\mathrm{kJ/mole})/k_B T_r}$. If we begin with 2000 mole of hydrogen and 1000 mole of oxygen in a 22 m^3 room, how much of these reactants will remain after the reaction comes to equilibrium?

Solution: We use Equation 8.11. Let x be the number of moles of unreacted O_2. So we have $2(1000 - x)$ mole of H_2O in the final state and $2x$ moles of unreacted H_2. Recalling that standard free energy changes for gases are computed using the reference concentration 1 mole/22 L, Equation 8.11 says

$$\left(\frac{2(1000 - x)\,\mathrm{mole}}{22\,\mathrm{m}^3} \right)^2 \frac{22\,\mathrm{m}^3}{x\,\mathrm{mole}} \left(\frac{22\,\mathrm{m}^3}{2x\,\mathrm{mole}} \right)^2 = e^{(457\,\mathrm{kJ/mole})/(2.5\,\mathrm{kJ/mole})} \frac{0.022\,\mathrm{m}^3}{\mathrm{mole}}.$$

Almost all the reactants will be used up because the reaction is energetically very favorable. So x will be very small, and we may approximate the numerator on the left, replacing $(1000 - x)$ by 1000. Thus $x = \left(1000^3 e^{-457/2.5} \right)^{1/3}$, or $3.4 \cdot 10^{-24}$. That's just *two molecules* of unreacted oxygen!

General reactions Quite generally, we can consider a reaction among k reactants and $m - k$ products:

$$\nu_1 X_1 + \cdots + \nu_k X_k \rightleftharpoons \nu_{k+1} X_{k+1} + \cdots + \nu_m X_m.$$

The whole numbers ν_k are called the **stoichiometric coefficients** of the reaction. Defining

$$\Delta G \equiv -\nu_1 \mu_1 - \cdots - \nu_k \mu_k + \nu_{k+1} \mu_{k+1} + \cdots + \nu_m \mu_m, \qquad (8.14)$$

we again find that $-\Delta G$ is the free energy change when the reaction takes one forward step, or

A chemical reaction will run forward if the quantity ΔG is a negative number, or backward if it's positive. $\qquad (8.15)$

Idea 8.15 justifies our calling ΔG the net **chemical force** driving the reaction. Equilibrium is the situation where a reaction makes no net progress in either direction, or $\Delta G = 0$. Just as before, we can usefully rephrase this condition by separating ΔG into its concentration-independent part, the **standard free energy change** of the reaction,

$$\Delta G^0 \equiv -\nu_1 \mu_1^0 - \cdots + \nu_m \mu_m^0, \tag{8.16}$$

plus the concentration terms. Defining the μ^0's with standard concentrations $c_0 = 1\,\text{M}$ gives the general form of Equation 8.11:

$$\frac{[X_{k+1}]^{\nu_{k+1}} \cdots [X_m]^{\nu_m}}{[X_1]^{\nu_1} \cdots [X_k]^{\nu_k}} = K_{eq} \text{ in equilibrium, where } K_{eq} \equiv e^{-\Delta G^0/k_B T}.$$

Mass Action rule

$$\tag{8.17}$$

In this expression, $[X]$ denotes $c_X/(1\,\text{M})$.

Even in aqueous solution, where the formula for μ^0 found in the gas chemical potential Example (page 296) won't literally apply, we can still use Equation 8.17 as long as the solutions are dilute. We just find ΔG^0 by looking up the appropriate μ^0's for aqueous solutions.

Your Turn 8E

Actually, chemistry books generally don't tabulate μ_α^0; instead they list values of $\Delta G_{f,\alpha}^0$, the free energy of formation of molecular species α under standard conditions from its elemental constituents. You can use these values in place of μ_α^0 in Equation 8.16. Explain why this works.

Your Turn 8F

Chemistry books sometimes quote the value of ΔG in units of kcal/mole and quote Equation 8.17 in the handy form $K_{eq} = 10^{-\Delta G^0/(??\,\text{kcal/mole})}$. Find the missing number.

Special biochemical conventions Biochemists make some special exceptions to the convention that $c_0 = 1\,\text{M}$:

• In a dilute aqueous solution of any solute, the concentration of water is always about 55 M. Accordingly, we take this value as the reference concentration for water. Then instead of $[H_2O]$, Equation 8.17 has the factor $c_{H_2O}/c_{0,H_2O} = c_{H_2O}/(55\,\text{M}) \approx 1$. With this convention, we can just omit this factor altogether from the Mass Action rule, even if water participates in the reaction.

- Similarly, when a reaction involves hydrogen ions (protons, or H^+), we choose their standard concentration to be 10^{-7} M. Again, this is the same as *omitting* factors of $c_{H^+}/c_{0,H^+}$ when the reaction proceeds at neutral pH (see Section 8.3.2).

In any case, the notation [X] will always refer to $c_X/1$ M.

The choice of standard concentrations for a reaction influences the numerical value of the reaction's standard free energy change and equilibrium constant. When we use the preceding special conventions, we denote the corresponding quantities as $\Delta G'^0$ and K'_{eq} (the standard transformed free energy change and transformed equilibrium constant) to avoid ambiguity.

Beware: Different sources may use additional special conventions for defining standard quantities. Standard conditions also include the specification of temperature (to 25°C) and pressure (to 10^5 Pa, roughly atmospheric pressure).

Actually, it's a bit simplistic to think of ions, like H^+, as isolated objects. We already know from Chapter 7 that any foreign molecule introduced into a solvent like water disturbs the structure of the neighboring solvent molecules, becoming effectively a larger, somewhat blurry object, loosely called a **hydrated ion**. When we speak of an ion like Na^+, we mean this entire complex; the standard potential μ^0 includes the free energy cost to assemble the whole thing. In particular, a proton in water associates especially tightly with *one* of the surrounding water molecules. Even though the proton doesn't bind covalently to its partner molecule, chemists refer to the combined object as a single entity, the **hydronium ion** H_3O^+. We are really referring to this complex when we write H^+.

Disturbing the peace Another famous result is now easy to understand: Suppose that we begin with concentrations *not* obeying Equation 8.17. Perhaps we took an equilibrium and dumped in a little more X_1. Then the chemical reaction will run forward—or in other words, in the direction that *partially undoes* the change we made—in order to reestablish Equation 8.17, and thereby increase the world's entropy. Chemists call this form of the Second Law **Le Châtelier's Principle**.

We have arrived at the promised extension of the matching rule (Equation 8.2) for systems of interconverting molecules. When several species are present in a system at equilibrium, once again each one's chemical potential must be constant throughout the system. But we found that the possibility of interconversions imposes additional conditions for equilibrium:

> *When one or more chemical reactions can occur at rates fast enough to equilibrate on the time scale of the experiment, equilibrium also implies relations **between** the various μ_α, namely, one Mass Action rule (Equation 8.17) for each relevant reaction.* (8.18)

Remarks The discussion of this section has glossed over an important difference between thermal equilibrium and ordinary mechanical equilibrium. Suppose that we gently place a piano on a heavy spring. The piano moves downward, compressing the spring, which stores elastic potential energy. At some point, the gravitational force on the piano equals the elastic force from the spring, and then *everything stops*. But in statistical equilibrium, nothing ever stops. Water continues to permeate the membrane of our osmotic machine at equilibrium; isomer *1* continues to convert to *2* and vice versa. Statistical equilibrium just means that there's no *net* flow of any macroscopic quantity. (We already saw this point in the discussion of buffalo equilibrium, Section 6.6.2 on page 220.)

We are partway to understanding the Focus Question for this chapter. The caveat about reaction rates in Idea 8.18 reminds us that, for example, a mixture of hydrogen and oxygen at room temperature can stay out of equilibrium essentially forever; the activation barrier to the spontaneous oxidation of hydrogen is so great that we instead get an apparent equilibrium, where Equation 8.11 does *not* hold. The deviation from complete equilibrium represents *stored free energy*, waiting to be harnessed to do our bidding. Thus hydrogen can be burned to fuel a car, and so on.

$\boxed{T_2}$ *Section 8.2.2′ on page 336 mentions some finer points about free energy changes and the Mass Action rule.*

8.2.3 Kinetic interpretation of complex equilibria

More complicated reactions have more complex kinetics, but the interpretation of equilibrium is the same. There can be some surprises along the way to this conclusion, however. Consider a hypothetical reaction, in which two diatomic molecules X_2 and Y_2 join and recombine to make two XY molecules: $X_2 + Y_2 \rightarrow 2XY$. It all seems straightforward at first. The rate at which any given X_2 molecule finds and bumps into a Y_2 molecule should be proportional to Y_2's number density, c_{Y_2}. The rate of all such collisions is then this quantity times the total number of X_2 molecules, which in turn is a constant (the volume) times c_{X_2}.

It seems reasonable that at low concentrations a certain fixed fraction of those collisions would overcome an activation barrier. Thus we might conclude that the rate r_+ of the forward reaction (reactions per time) should also be proportional to $c_{X_2} c_{Y_2}$, and likewise for the reverse reaction:

$$r_+ \overset{?}{=} k_+ c_{X_2} c_{Y_2} \quad \text{and} \quad r_- \overset{?}{=} k_- (c_{XY})^2. \tag{8.19}$$

In this formula, k_+ and k_- are the rate constants of the reaction. They are similar to the quantities we defined for the two-state system (Section 6.6.2 on page 220), but with different units: Equation 8.19 shows their units to be $[k_\pm] \sim \text{s}^{-1}\text{M}^{-2}$. We associate rate constants with a reaction by writing them next to the appropriate arrows:

$$X_2 + Y_2 \underset{k_-}{\overset{k_+}{\rightleftharpoons}} 2XY. \tag{8.20}$$

Setting the rates equal, $r_+ = r_-$, gives that at equilibrium, $c_{X_2} c_{Y_2}/(c_{XY})^2 = k_-/k_+$, or

$$\frac{(c_{XY})^2}{c_{X_2} c_{Y_2}} = K_{eq} = \text{const.} \tag{8.21}$$

This seems good—it's the same conclusion we got from Equation 8.17.

Unfortunately, predictions for rates based on the logic leading to Equation 8.19 are often totally wrong. For example, we may find that over a wide range of concentrations, doubling the concentration of Y_2 has almost *no effect* on the forward reaction rate, whereas doubling c_{X_2} *quadruples* the rate! We can summarize such experimental results (for our hypothetical system) by saying that the reaction is of **zeroth order** in Y_2 and **second order** in X_2; this statement means that the forward rate is proportional to $(c_{Y_2})^0(c_{X_2})^2$. Naïvely, we expected it to be first order in both.

What is going on? The problem stems from our assumption that we knew the mechanism of the reaction, that is, that an X_2 smashed into a Y_2 and exchanged one atom, all in one step. Maybe instead, the reaction involves an improbable but necessary first step followed by two very rapid steps:

$$X_2 + X_2 \rightleftharpoons 2X + X_2 \quad \text{(step 1, slow)}$$

$$X + Y_2 \rightleftharpoons XY_2 \quad \text{(step 2, fast)}$$

$$XY_2 + X \rightleftharpoons 2XY \quad \text{(step 3, fast).} \tag{8.22}$$

The slow step of the proposed mechanism is called the **bottleneck**, or **rate-limiting** process. The rate-limiting process controls the overall rate, in this case yielding the pattern of concentration dependences $(c_{Y_2})^0(c_{X_2})^2$. Either reaction mechanism (Reaction 8.20 or 8.22) is logically possible; experimental rate data are needed to rule out the wrong one.

Won't this observation destroy our satisfying kinetic interpretation of the Mass Action rule, Equation 8.19? Luckily, no. The key insight is that in equilibrium, each elementary reaction in Equation 8.22 must *separately* be in equilibrium. Otherwise, there would be a constant pileup of some species, either a net reactant like X_2 or an intermediate like XY_2. Applying the naïve rate analysis to each step separately gives that in equilibrium

$$\frac{(c_X)^2 c_{X_2}}{(c_{X_2})^2} = K_{eq,1} c_0 , \quad \frac{c_{XY_2}}{c_X c_{Y_2}} = \frac{K_{eq,2}}{c_0} , \quad \frac{(c_{XY})^2}{c_{XY_2} c_X} = K_{eq,3}.$$

Multiplying these three equations together reproduces the usual Mass Action rule for the *overall* reaction, with $K_{eq} = K_{eq,1} K_{eq,2} K_{eq,3}$:

> *The details of the intermediate steps in a reaction are immaterial for its overall equilibrium.* (8.23)

This slogan should sound familiar—it's another version of the principle that "equilibrium doesn't care what happens inside the phone booth" (Section 8.2.2).

8.2.4 The primordial soup was not in chemical equilibrium

The early Earth was barren. There was plenty of carbon, hydrogen, nitrogen, oxygen (although not free in the atmosphere as it is today), phosphorus, and sulfur. Could the organic compounds of life have formed spontaneously? Let's look into the equilibrium concentrations of some of the most important biomolecules in a mixture of atoms at atmospheric pressure, with overall proportions C:H:N:O=2:10:1:8 similar to that of our bodies. We optimistically assume a temperature of 500°C, to promote the formation of high-energy molecules. Mostly, we get familiar low-energy, low-complexity molecules H_2O, CO_2, N_2, and CH_4. Then molecular hydrogen comes in at a mole fraction of about 1%, acetic acid at 10^{-10}, and so on. The first really interesting biomolecule on the list is lactic acid, at an equilibrium mole fraction of 10^{-24}! Pyruvic acid is even further down the list, and so on.

Evidently, the exponential relation between free energy and population in Equation 8.17 must be treated with respect. It's a *very* rapidly decreasing function. The concentrations of biomolecules in the biosphere today are nowhere near equilibrium. This is a more refined statement of the puzzle first set out in Chapter 1: *Biomolecules must be produced by the transduction of some abundant source of free energy.* Ultimately, this source is the Sun.[3]

8.3 DISSOCIATION

Before going on, let's survey how our results so far explain some basic chemical phenomena.

8.3.1 Ionic and partially ionic bonds dissociate readily in water

Rock salt (sodium chloride) is "refractory": Heat it in a frypan and it won't vaporize. To understand this fact, we first need to know that chlorine is highly **electronegative**. That is, an isolated chlorine atom, although electrically neutral, will eagerly bind another electron to become a Cl^- ion, because the ion has significantly lower internal energy than the neutral atom. An isolated sodium atom, on the other hand, will *give up* an electron (becoming a sodium ion Na^+) without a very great increase in its internal energy. Thus, when a sodium atom meets a chlorine atom, the joint system can reduce its net internal energy by transferring one electron completely from the sodium to the chlorine. So a crystal of rock salt consists entirely of the ions Na^+ and Cl^-, held together by their electrostatic attraction energy. To estimate that energy, write qV from Equation 1.9 on page 21 as $e^2/(4\pi\varepsilon_0 d)$, where d is a typical ion diameter. Taking $d \approx 0.3\,\text{nm}$ (the atomic spacing in rock salt) gives the energy cost to separate a single NaCl pair as over a hundred times the thermal energy. No wonder rock salt doesn't vaporize until it reaches temperatures of thousands of degrees.

And yet, place that same ionic NaCl crystal in water and it immediately dissociates, even at room temperature. The difference is that, in water, we have an extra factor of $(\varepsilon_0/\varepsilon) \approx 1/80$; thus the energy cost of separating the ions is now comparable to $k_B T_r$. This modest contribution to the free energy cost is overcome by the increase of entropy when an ion pair leaves the solid lump and begins to wander in solution; the overall change in free energy thus favors dissolving.

Ionic salts are not the only substances that dissolve readily in water: Many other molecules dissolve without dissociating at all. For example, sugar and alcohol are

---

[3]As mentioned in Chapter 1, the ecosystems around hot ocean vents are an exception to this general rule.

highly soluble in water. Although their molecules have no net charge, still each has separate positive and negative spots, as does water itself: They are polar molecules. Polar molecules can participate at least partially in the hydrogen-bonding network of water, so there is little hydrophobic penalty when we introduce such intruders into pure water. Moreover, the energy cost of breaking the attraction of their plus ends to their neighbors' minus ends (the **dipole interaction**) is offset by the gain of forming similar conjunctions with water molecules. Because an entropic gain always favors mixing, we expect polar molecules to be highly soluble in water.[4] Indeed, on the basis of this reasoning, we could predict that *any* small molecule with the highly polar **hydroxyl** (or –OH) group found in alcohol should be soluble in water; and in fact, it's generally so. Another example is the amino (or –NH$_2$) group, for example, the one on methylamine. On the other hand, nonpolar molecules, like hydrocarbons, exact such a high hydrophobic penalty that they are poorly soluble in water.

This book won't develop the quantum-mechanical tools to predict a priori whether a molecule will dissociate into polar components. This isn't such a serious limitation, however. Our attitude to all these observations will be simply that there is nothing surprising about the ionic dissociation of a group in water; it's just another simple chemical reaction, to be treated by the usual methods developed in this chapter.

8.3.2 The strengths of acids and bases reflect their dissociation equilibrium constants

Section 7.4.1 discussed the diffuse charge layer that forms near a macromolecule's surface when it dissociates (breaks apart) into a large macroion and many small counterions. The analysis of that section assumed that a constant number of charges per area, σ_q/e, always dissociated, but this is not always a very good assumption. Let's discuss the problem of dissociation in general, starting with small molecules.

Water is a small molecule. Its dissociation reaction is

$$H_2O \rightleftharpoons H^+ + OH^-. \qquad (8.24)$$

Section 8.3.1 argued that reactions of this type need not be prohibitively costly; but still, the dissociation of water does cost more free energy than that of NaCl. Accordingly, the equilibrium constant for Reaction 8.24, although not negligible, is rather small. In fact, pure water has $c_{H^+} = c_{OH^-} = 10^{-7}$ M. (These numbers can be obtained by measuring the electrical conductivity of pure water; see Section 4.6.4 on page 142.) Because the concentration of H_2O is essentially fixed, the Mass Action rule says that water maintains a fixed value of the ion product, defined as

$$K_w \equiv [H^+][OH^-] = (10^{-7})^2. \qquad \textbf{ion product of water} \text{ at room temperature}$$

$$(8.25)$$

---

[4]We still don't expect sugar to *vaporize* readily, the way a small nonpolar molecule like acetone does. Vaporization would break attractive dipole interactions *without* replacing them by anything else.

Suppose that we now disturb this equilibrium, for example, by adding some hydrochloric acid. HCl dissociates much more readily than H_2O, so the disturbance increases the concentration of H^+ from the tiny value for pure water. But Reaction 8.24 is still available, so its Mass Action constraint must still hold in the new equilibrium, regardless of what has been added to the system. Accordingly, the concentration of hydroxyl ions (OH^-) must go *down* to maintain Equation 8.25.

Let's instead add some sodium hydroxide (lye). NaOH also dissociates readily, so the disturbance increases the concentration of OH^-. Accordingly, $[H^+]$ must go down: The added OH^- ions gobble up the tiny number of H^+ ions, making it even tinier. Chemists summarize both situations by defining the **pH** of a solution as

$$pH = -\log_{10}[H^+], \qquad (8.26)$$

a definition analogous to that of pK (Equation 8.12).

We've just seen that

- The pH of pure water equals 7, from Equation 8.25. This value is also called **neutral pH**.
- Adding HCl lowers the pH. A solution with pH less than 7 is called **acidic**. We will call an **acid** any neutral substance that, when dissolved in pure water, creates an acidic solution.
- Adding NaOH raises the pH. A solution with pH greater than 7 is called **basic**. We will call a **base** any neutral substance that, when dissolved in pure water, creates a basic solution.

Many organic molecules behave like HCl, so they are called acids. For example, the carboxyl group –COOH dissociates via

$$-COOH \rightleftharpoons -COO^- + H^+.$$

Familiar examples of this sort of acid are vinegar (acetic acid), lemon juice (citric acid), and DNA (deoxyribonucleic acid). DNA dissociates into many mobile charges plus one big macroion, with two net negative charges per basepair. Unlike hydrochloric acid, however, all these organic acids are only partially dissociating. For example, the pK for dissociation of acetic acid is 4.76; compare this value with the corresponding value of 2.15 for a strong acid like phosphoric acid (H_3PO_4). Dissolving a mole of acetic acid in a liter of water will thus generate a lot of neutral CH_3COOH and only a modest amount of H^+ (see Problem 8.5). We say that acetic acid is a **weak acid**.

Any molecule that gobbles up H^+ will raise the pH. This can happen directly or indirectly. For example, another common motif is the amine group, $-NH_2$, which directly gobbles protons by the equilibrium

$$-NH_2 + H^+ \rightleftharpoons NH_3^+. \qquad (8.27)$$

A special case is ammonia, NH_3, which is simply an amine group attached to a hydrogen atom. We've already seen how other bases (such as lye) work by gobbling protons *indirectly*, liberating hydroxyl ions that push the equilibrium (Reaction 8.24) to the left. Bases can also be strong or weak, depending on the value of their dissociation equilibrium constant (for example, $NaOH \rightleftharpoons Na^+ + OH^-$) or association constant (for example, Reaction 8.27).

Now suppose that we add equal quantities of *both* HCl and NaOH to pure water. In this case, the number of extra H^+ from the acid equals the number of extra OH^- from the base, so we still have $[H^+] = [OH^-]$. The resulting solution of Equation 8.25 again gives $[H^+] = 10^{-7}$, or pH = 7! What happened? The extra H^+ and OH^- gobbled *each other*, combining to become water. The other ions remained, forming a solution of table salt, $Na^+ + Cl^-$. (You could also get a neutral solution by mixing a strong base, NaOH, with a *weak* acid, CH_3COOH, but you'd need a lot more acid than base.)

8.3.3 The charge on a protein varies with its environment

Chapter 2 described proteins as linear chains of monomers, the amino acids. Each amino acid (except proline) contributes an identical group to the protein chain's backbone, $-NH-CH-CO-$, plus a variable group (called a side group) covalently bonded to the central carbon. The resulting polymer is a chain of residues, in a precise sequence specified by the message in the cell's genome coding for that protein. The interactions of the residues with one another and with water determine how the protein folds; the structure of the folded protein determines its function.

In short, proteins are horrendously complicated. How can we say anything simple about such a complex system?

Some amino acids, for example, aspartic or glutamic acid, liberate H^+ from carboxyl groups, like any organic acid. Others, including lysine and arginine, pull H^+ out of solution onto their basic side chains. The corresponding dissociation reactions thus involve the transfer of a proton:

$$\text{acidic side chain:} \quad -COOH \rightleftharpoons -COO^- + H^+$$
$$\text{basic side chain:} \quad -NH_3^+ \rightleftharpoons -NH_2 + H^+. \tag{8.28}$$

The species on the left are the **protonated** forms; those on the right are **deprotonated**.

Each residue of type α has a characteristic equilibrium constant $K_{eq,\alpha}$ for its deprotonation reaction. We find these values tabulated in books. The range of actual values is about $10^{-3.7}$ for the most acidic (aspartic acid) to about $10^{-12.5}$ for the most basic (arginine). The actual probability that a residue of type α will be protonated will then depend on $K_{eq,\alpha}$, and on the pH of the surrounding fluid. Denoting this probability by P_α, we have, for example, $P_\alpha = [-COOH]/([-COOH] + [-COO^-])$. Combining this definition with the equilibrium condition, $[-COO^-][H^+]/[-COOH] = K_{eq,\alpha}$, gives

$$P_\alpha = \frac{1}{1 + K_{eq,\alpha}/[H^+]} = \frac{1}{1 + K_{eq,\alpha}10^{pH}}.$$

It's convenient to rephrase this result, using Equation 8.12, as

$$P_\alpha = (1 + 10^{x_\alpha})^{-1}, \text{ where } x_\alpha = \text{pH} - \text{p}K_\alpha. \qquad \text{probability of protonation}$$

$$(8.29)$$

The average charge on an acidic residue in solution will then be $(-e)(1 - P_\alpha)$. Similarly, the average charge on a basic residue will be eP_α. In both cases, the average charge goes down as pH goes up, as we can see directly from Reactions 8.28.

Actually, in a protein, uncharged and charged residues will affect *each other* instead of all behaving independently. Hence Equation 8.29 says that the degree of dissociation of a residue is a universal function of the pH in its local environment, shifted by the pK of that residue. Equation 8.29 shows that a residue is protonated half the time when the local pH just equals its dissociation pK.

Although we don't know the local pH at a residue, we can guess that it will go up as that of the surrounding solution goes up. For example, we can **titrate** a solution of protein, gradually dripping in a base (starting from a strongly acidic solution). Initially, [H$^+$] is high and most residues are protonated; therefore the acidic ones are neutral, the basic ones are positively charged, and hence the protein is positively charged. Increasing pH decreases [H$^+$], driving each of Reactions 8.28 to the right and decreasing the net charge of the protein. But at first, only the most strongly acidic residues (those with lowest pK) respond. To understand why, note that the universal function $(1+10^x)^{-1}$ is roughly a constant, except near $x = 0$, where it rapidly switches from 1 to 0. Thus only those bases with pK close to the pH respond when pH is changed slightly; the basic residues remain completely protonated as we raise the pH from very acidic to somewhat acidic.

As titration proceeds, each type of residue pops over in turn from protonated to deprotonated, until, under very basic conditions, the last holdouts—the strongly basic residues—finally surrender their protons. By this time, the protein has completely reversed its net charge; now Reactions 8.28 say that the acidic residues are negative and the basic ones neutral. For a big protein, the charge difference between the extremes can be large: For example, titration can change the protonation state of ribonuclease by about 30 protons (Figure 8.1).

8.3.4 Electrophoresis can give a sensitive measure of protein composition

Even though the analysis in Section 8.3.3 was rough, it did explain one key qualitative fact about the experimental data (Figure 8.1): At some critical ambient pH, a protein will be effectively neutral. The value of pH at this point and, indeed, the entire titration curve are fingerprints characteristic of each specific protein.

Section 4.6.4 on page 142 explained how putting an electric field across a salt solution causes the ions in that solution to migrate. Similar remarks apply to a solution of *macroions*, for example, proteins. It is true that the viscous friction coefficient ζ on a large globular protein will be much larger than that on a tiny ion (by the Stokes formula, Equation 4.14 on page 119). But the net driving force on the protein will be

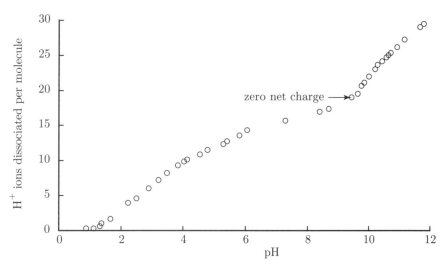

Figure 8.1: (Experimental data.) The protonation state of ribonuclease depends on the pH of the surrounding solution. The *arrow* shows the point of zero net charge. The vertical axis gives the number of H^+ ions dissociated per molecule at 25°C, so the curves show the protein becoming deprotonated as the pH is raised from acidic to basic. [Data from Tanford, 1961.]

huge, too: It's the sum of the forces on each ionized group. The resulting migration of macroions in a field is called **electrophoresis**.

The rule governing the speed of electrophoretic migration is more complicated than the simple $q\mathcal{E}/\zeta$ used in our study of saltwater conductivity. Nevertheless, we can expect that an object with zero net charge has zero electrophoretic speed. Section 8.3.3 argued that any protein has a value of ambient pH at which its net charge is zero (called the protein's isoelectric point). As we titrate through this point, a protein should slow down, stop, and then *reverse* its direction of electrophoretic drift. We can use this property to separate mixtures of proteins.

Not only does every protein have its characteristic isoelectric point; each *variant* of a given protein will, too. A famous example is the defective protein responsible for sickle-cell anemia. In a historic discovery, Linus Pauling and coauthors showed in 1949 that the red blood cells of sickle-cell patients contained a defective form of hemoglobin. Today we know that the defect lies in parts of hemoglobin called the β-globin chains, which differ from normal β-globin by the substitution of a *single amino acid*, from glutamic acid to valine in position six. This tiny change (β-globin has 146 amino acids in all) is enough to create a sticky (hydrophobic) patch on the molecular surface. The mutant molecules clump together to form a solid fiber of fourteen interwound helical strands inside the red cell and give it the sickle shape for which the disease is named. The deformed red cells in turn get stuck in capillaries and then damaged; finally they are destroyed by the body, with the effect of creating anemia.

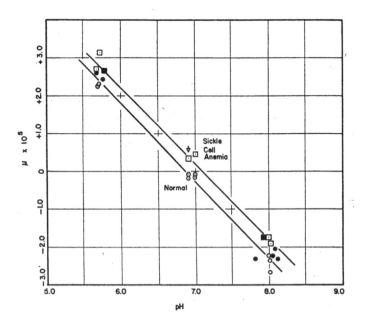

Figure 8.2: (Experimental data.) Pauling and coauthors' original data showing that normal and sickle-cell hemoglobin could be distinguished by their electrophoretic mobility. In this trial, the hemoglobin was bound to carbon monoxide, and its mobility μ (in $(\mathrm{cm\,s^{-1}})/(\mathrm{volt\,cm^{-1}})$) was measured at various values of pH. *Circles:* Normal hemoglobin. *Squares:* Sickle-cell hemoglobin. (Solid black symbols represent trials with dithionite ion present; open symbols are trials without it.) [From Pauling et al., 1949.]

In 1949, the sequence of β-globin was unknown. Nevertheless, Pauling and coauthors pinpointed the source of the disease in a single molecule. They reasoned that a slight chemical modification in hemoglobin could make a correspondingly small change in its titration curve if the differing amino acids had different dissociation constants. Isolating normal and sickle-cell hemoglobin, they indeed found that even though the corresponding titration curves look similar, the two proteins' isoelectric points differ by about a fifth of a pH unit (Figure 8.2). The sign of this difference is just what would be expected for a substitution of valine for glutamic acid: The normal protein is consistently more negatively charged in the range of pH shown than the defective one because

- It has one more acidic (negative) residue, and
- That residue (glutamic acid) has $pK = 4.25$, so it is dissociated throughout the range of pH shown in the graph.

In other physical respects, the two molecules are alike; for example, Pauling and coauthors found that both had the same sedimentation and diffusion constants. Nevertheless, the difference in isoelectric point was enough to distinguish the two versions of the molecule. Most strikingly, Figure 8.2 shows that at pH 6.9, the charges of the normal and defective proteins have opposite signs, and so the two proteins migrate in opposite directions under an electric field. (You'll show in Problem 8.7 that this difference is indeed big enough to separate proteins.)

T_2 Section 8.3.4' on page 336 mentions some more advanced treatments of electrophoresis.

8.4 SELF-ASSEMBLY OF AMPHIPHILES

The pictures in Chapter 2 show a world of complex machinery inside cells, all of which seems to have been constructed by other complex machinery. This arrangement fits with the observation that cells can arise only from other living cells, but it leaves us wondering about the origin of the very first living things. In this light, it's significant that the most fundamental structures in cells—the membranes separating the interior from the world—can actually *self-assemble* from appropriate molecules, just by following chemical forces. This section begins to explore how chemical forces—in particular, the hydrophobic interaction—can drive self-assembly.

Some architectural features of cells blossom quite suddenly at the appropriate moment when they are needed (for example, the microtubules that pull chromosomes apart at mitosis), then just as suddenly melt away. We may well ask, "If self-assembly is automatic, what sort of control mechanism could turn it on and off so suddenly?" Section 8.4.2 will begin to expose an answer to this question.

8.4.1 Emulsions form when amphiphilic molecules reduce the oil–water interface tension

Section 7.5 discussed why salad dressing separates into oil and water, despite the superficial increase in order that such separation entails. Water molecules are attracted to oil molecules, but not as much as they are attracted to one another: The oil–water interface disrupts the network of hydrogen bonds, so droplets of water coalesce to reduce their total surface area. But some people prefer mayonnaise to vinaigrette. Mayonnaise, too, is mostly a mixture of oil and water; yet it does not separate. What's the difference?

One difference is that mayonnaise contains a small quantity of *egg*. An egg is a complicated system, including many large and small molecules. But even very simple, pure substances can stabilize suspensions of tiny oil droplets in water for long periods. Such substances are generically called emulsifiers or surfactants; a suspension stabilized in this way is called an **emulsion**. Particularly important are a class of simple molecules called detergents, and the more elaborate phospholipids found in cell membranes.

The molecular architecture of a surfactant shows us how it works. Figure 8.3a shows the structure of sodium dodecyl sulfate (**SDS**), a strong detergent. One side of this molecule is hydrophobic: It's a hydrocarbon chain. The other side, however, is polar: In fact, it's an ion. This fusion of unlike parts gives the class of molecules with this structure the name **amphiphiles**. These two parts would normally migrate

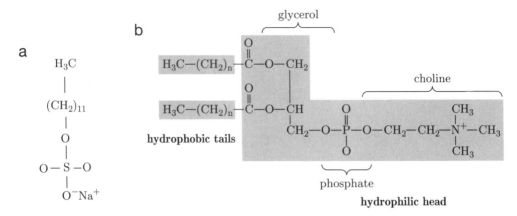

Figure 8.3: (Structural diagrams.) Two classes of amphiphiles. (a) Structure of sodium dodecyl sulfate (SDS), a strong detergent. A nonpolar, hydrophobic, tail (*top*) is chemically linked to a polar, hydrophilic head (*bottom*). In solution, the Na^+ ion dissociates. Molecules from this class form micelles (see Figure 8.5). (b) Structure of a generic phosphatidylcholine, a class of phospholipid molecule. Two hydrophobic tails (*left*) are chemically linked to a hydrophilic head (*right*). Molecules from this class form bilayers (see Figure 8.4).

(or "partition") into the oil phase and water phase, respectively, of an oil–water mixture. But such an amicable separation is not an option—the two parts are handcuffed together by a chemical bond. When added to an oil–water mixture, though, surfactant molecules *can* simultaneously satisfy both of their halves by migrating to the oil–water interface (Figure 8.4). In this way, the polar head can face water while the nonpolar tails face oil. Given enough surfactant to make a layer one molecule thick (that is, a **monolayer**) over the entire interface, the oil and water phases need not make any direct contact at all.

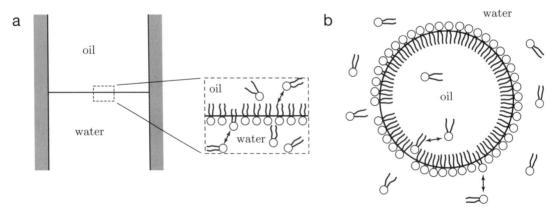

Figure 8.4: (Schematics.) (a) An oil–water interface stabilized by the addition of a small amount of surfactant. Some surfactant molecules are dissolved in the bulk oil or water regions, but most migrate to the boundary as shown in the inset. (b) An oil–water emulsion stabilized by surfactant: The situation is the same as (a), but for a finite droplet of oil.

In mayonnaise, the relevant component of the egg is a phospholipid (lecithin), which migrates to the oil–water interface to minimize its own free energy and at the same time, lower the interfacial energy to the point where rapid coalescence of droplets does not occur. (Other delicacies, for example, sauce béarnaise, also work this way.) Because a monolayer is typically just a couple of nanometers thick, a small quantity of surfactant can stabilize an enormous area of interface. Recall Ben Franklin's teaspoon of oil, which covered a half-acre of pond (Section 1.5.1 on page 23).

Your ***Turn*** ***8G***	You can observe the reduction in surface tension brought about by a tiny amount of dissolved soap in a simple experiment. Carefully float a loop of fine sewing thread on water. Now touch a bar of soap to the part of the water surrounded by the thread. Explain what happens.

You can also see for yourself just how large an area can be covered by one drop of detergent or, equivalently, just how much that drop can be diluted and still change the surface tension over several square centimeters of water. In the same way, a small amount of detergent can clean up a big oily mess by encapsulating oil into stable, hydrophilic droplets small enough to be flushed away by running water.

8.4.2 Micelles self-assemble suddenly at a critical concentration

A mixture of stabilized oil droplets in water may be delicious or useful, but it hardly qualifies as a "self-assembled" structure. The droplets come in many different sizes (that is, they are **polydisperse**) and generally have little structure. Can entropic forces drive the construction of anything more closely resembling what we find in cells?

To answer the preceding question, we begin with another. It may seem from Section 8.4.1 that surfactant molecules in *pure* water would be stymied: With no interface to go to, won't they just have to accept the hydrophobic cost of exposing their tails to the surrounding water?

Figure 8.5 shows that the answer to the second question is "no." Surfactant molecules in solution can assemble into a **micelle**, a sphere consisting of a few dozen molecules. In this way, the molecules can present their nonpolar tails to *one another*, not to the surrounding water. This configuration can be entropically favorable, even though by choosing to associate in this way, each molecule loses some of its freedom to be located anywhere, oriented in any way (see Section 7.5 on page 273).

A remarkable feature of Figure 8.5 is that there is a definite "best" size for the resulting micellar aggregate. If there were too many amphiphilic molecules, then some would be completely in the interior, where their polar heads would be cut off from the surrounding water. But with too few amphiphiles (for example, just one molecule), the tails would not be effectively shielded. Thus amphiphilic molecules can spontaneously self-assemble into objects of fixed, limited, molecular-scale size. The chemical force driving the assembly is not the formation of covalent bonds, but something gentler: the hydrophobic effect, an entropic force.

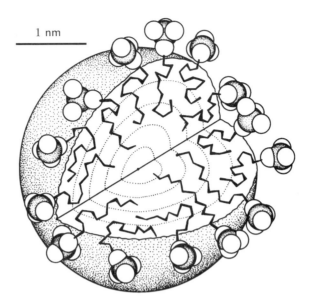

1 nm

Figure 8.5: (Sketch.) A micelle of sodium dodecyl sulfate (SDS). The micelle consists of 60 SDS molecules. The hydrocarbon chains pack in the core at a uniform density roughly the same as that of liquid oil. [From Israelachvili, 1991.]

As early as 1913, J. McBain had deduced the existence of well-defined micelles from his quantitative study of the physical properties of soap solutions. One of McBain's arguments went as follows. We know the total number of molecules in a solution just by measuring how much soap we put in and checking that none of it precipitates out of solution. But we can independently measure how many independently moving objects the solution contains, by measuring its osmotic pressure and using the van 't Hoff relation (Equation 7.7 on page 249). For very dilute solutions, McBain and others found that the osmotic pressure faithfully tracked the total number of amphiphilic ions (solid symbols on the left of Figure 8.6), just as it would for an ordinary salt like potassium chloride (open symbols in Figure 8.6). But the similarity ended at a well-defined point, now called the **critical micelle concentration**, or **CMC**. Beyond this concentration, *the ratio of independently moving objects to all ions dropped sharply* (solid symbols on the right of the graph).

McBain was forced to conclude that beyond the CMC, his molecules didn't stay in an ordinary solution, dispersed throughout the sample. Nor, however, did they aggregate into a separate bulk phase, as oil does in vinaigrette. Instead, they were spontaneously assembling into intermediate-scale objects, bigger than a molecule but still microscopic. Each type of amphiphile, in each type of polar solvent, had its own characteristic value of the CMC. This value typically decreases at higher temperature, thus pointing to the role of the hydrophobic interaction in driving the aggregation (see Section 7.5.2 on page 276). McBain's results were not immediately accepted. But eventually, several physical quantities (for instance, electrical conductivity) were all

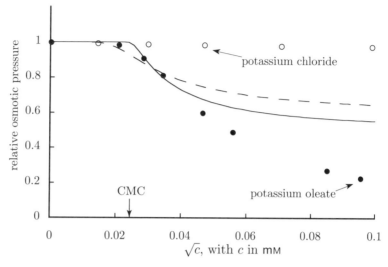

Figure 8.6: (Experimental data, with fits.) Comparison of the osmotic behavior of a micelle-forming substance with that of an ordinary salt. The relative osmotic pressure is defined as the osmotic pressure divided by that of an ideal, fully dissociated solution with the same number of ions. To emphasize the behavior at small concentration, the horizontal axis shows $\sqrt{c}$, where c is the concentration of the solution. *Solid symbols* are experimental data for potassium oleate, a soap; *open symbols* are data for potassium chloride, a fully dissociating salt. The *solid line* shows the result of the model discussed in the text (Equation 8.34 with $N = 30$ and critical micelle concentration 1.4 mM). For comparison, the *dashed line* shows a similar calculation with $N = 5$. The $N = 30$ model accounts for the sharp kink in the relative osmotic activity at the CMC. It fails at higher concentrations, in part because it neglects the fact that the surfactant molecules' head groups are not fully dissociated. [Data from McBain, 1944.]

found to undergo sharp changes at the same critical concentration as that for the osmotic pressure, and the chemical community agreed that he was right.

We can interpret McBain's results with a simplified model. Suppose that the soap he used, potassium oleate, dissociates fully into potassium ions and oleate amphiphiles. The potassium ions contribute to the osmotic pressure by the van 't Hoff relation. But the remaining oleate amphiphiles will instead be assumed to be in thermodynamic equilibrium between individual ions and aggregates of N ions. N is an unknown parameter, which we will choose to fit the data. It will turn out to be just a few dozen, justifying our picture of micelles as objects intermediate in scale between molecules and the macroscopic world.

To work out the details, apply the Mass Action rule (Equation 8.17) to the reaction (N monomers) $\rightleftharpoons$ (one aggregate). Thus the concentration c_1 of free monomers in solution is related to that of micelles, c_N, by

$$c_N/(c_1)^N = \hat{K}_{eq}, \tag{8.30}$$

where $\hat{K}_{eq}$ is a second unknown parameter of the model. ($\hat{K}_{eq}$ equals the dimensionless equilibrium constant for aggregation, K_{eq}, divided by $(c_0)^{N-1}$.) The total concentration of all monomers is then $c_{tot} = c_1 + Nc_N$.

Example: Find the relation between the total number of amphiphilic molecules in solution, c_{tot}, and the number that remain unaggregated, c_1.

Solution:

$$c_{tot} = c_1 + Nc_N = c_1\left(1 + N\hat{K}_{eq}(c_1)^{N-1}\right). \qquad (8.31)$$

We could stop at this point, but it's more meaningful to express the answer, not in terms of $\hat{K}_{eq}$, but in terms of the CMC, c_*. By definition c_* is the value of c_{tot} at which half the monomers are free and half are assembled into micelles. In other words, when $c_{tot} = c_*$ then $c_{1,*} = Nc_{N,*} = \frac{1}{2}c_*$. Substituting into Equation 8.30 gives

$$\left(\frac{1}{2N}c_*\right)\left(\frac{1}{2}c_*\right)^{-N} = \hat{K}_{eq}. \qquad (8.32)$$

We now solve to find $N\hat{K}_{eq} = (2/c_*)^{N-1}$ and substitute into Equation 8.31, finding

$$c_{tot} = c_1(1 + (2c_1/c_*)^{N-1}). \qquad (8.33)$$

Once we have chosen values for the parameters N and c_*, we can solve Equation 8.33 to get c_1 in terms of the total amount of surfactant c_{tot} stirred into the solution. Although this equation has no simple analytical solution, we can understand its limiting behavior. At low concentrations, $c_{tot} \ll c_*$, the first term dominates and we get $c_{tot} \approx c_1$: Essentially all the surfactants are loners. But well above the CMC, the second term dominates and we instead get $c_{tot} \approx Nc_N$; now essentially all the surfactants are accounted for by the micelles.

We can now find the osmotic pressure. The contribution from the potassium ions is $c_{tot}k_BT$, as usual. The contribution from the amphiphiles resembles Equation 8.33, with one key difference: Each micelle counts as just one object, not as N objects.

Your Turn 8H

Show that the total osmotic pressure relative to the value $2c_{tot}k_BT$ in this model is

$$\frac{1}{2}\left(1 + \frac{1 + N^{-1}(2c_1/c_*)^{N-1}}{1 + (2c_1/c_*)^{N-1}}\right). \qquad (8.34)$$

To use this formula, solve Equation 8.33 numerically for c_1 as a function of c_{tot}. Then substitute into Equation 8.34 to get the relative osmotic activity in terms of the total concentration of amphiphiles. Looking at the experimental data in Figure 8.6, we see that we must take c_* to be around 1 mM; the fit shown used $c_* = 1.4$ mM. Two curves are shown: The best fit (solid line) used $N = 30$, whereas the poor fit of the dashed line shows that N is greater than 5.

Certainly more detailed methods are needed to obtain a precise size estimate for the micelles in the experiment shown. But we can extract several lessons from Figure 8.6. First, we have obtained a qualitative explanation of the very sudden onset of micelle formation by the hypothesis that geometrical packing considerations select a narrow distribution of "best" micelle size N. Indeed, the sharpness of the micelle transition could not be explained at all if stable aggregates of two, three, ... monomers could form as intermediates to full micelles. In other words, many monomers must cooperate to create a micelle, and this **cooperativity** *sharpens the transition, mitigating the effects of random thermal motion.* We will revisit this lesson repeatedly in future chapters. Without cooperativity, the curve would fall gradually, not suddenly.

8.5 EXCURSION: ON FITTING MODELS TO DATA

> *If you give me two free parameters, I can describe an elephant.*
> *If you give me three, I can make him wiggle his tail.*
> — Eugene Wigner, 1902–1995

Figure 8.6 shows some experimental data (the solid dots), together with a purely mathematical function (the solid curve). The purpose of graphs like this one is to support an author's claim that some physical model captures an important feature of a real-world system. The reader is supposed to see how the curve passes through the points, then nod approvingly, preferably without thinking too much about the details of either the experiment or the model. But it's important to develop some critical skills to use when assessing (or creating) fits to data.

Clearly the model shown by the solid line in Figure 8.6 is only moderately successful. For one thing, the experimental data show the relative osmotic activity dropping below 50%. Our simplified model can't explain this phenomenon because we assumed that the amphiphiles remain always fully dissociated: Na^+ ions always remain nearly an ideal solution. Actually, however, measurements of the electrical conductivity of micellar solutions show that the degree of dissociation goes down as micelles are formed. We could have made the model look much more like the data simply by assuming that each micelle has an unknown degree of dissociation α, and choosing a value of $\alpha < 1$ that pulled the curve down to meet the data. Why not do this?

Before answering, let's think about the content of Figure 8.6 as drawn. Our model has two unknown parameters, the number N of particles in a micelle and the critical micelle concentration c_*. To make the graph, we adjust their values to match two gross visual features of the data:

- There is a kink in the data at around a millimolar concentration.
- After the kink, the data start dropping with a certain slope.

So the mere fact that the curve resembles the data is perhaps not so impressive as it may seem at first: We dialed two knobs to match two visual features. The real scientific content of the figure comes in two observations we made:

- A simple model, based on cooperativity, explains the qualitative *existence* of a sharp kink, which we don't find in simple two-body association. The osmotic activity of a weak, ordinary acid (for example, acetic acid) as a function of concentration has no such kink: The degree of dissociation, and hence the relative osmotic activity, decreases gradually with concentration.
- The *numerical values* of the fit parameters fit in with the dense web of other facts we know about the world. For example, $N = 30$ implies that the micelles are too small to scatter visible light; and indeed, their solutions are clear, not milky.

Viewed in this light, introducing an ad hoc dissociation parameter to improve the fit in Figure 8.6 would be merely a cosmetic measure: Certainly, a third free parameter would suffice to match a third visual feature in the data, but so what? In short,

> *A fit of a model to data tells us something interesting only insofar as*
> **a.** *One or a few fit parameters reproduce several independent features of the data, or*
> **b.** *The experimental errors on the data points are exceptionally low,* (8.35)
> *and the fit reproduces the data to within those errors, or*
> **c.** *The values of the fit parameters determined by the data mesh with some independently measured facts about the world.*

Here are some examples: (a) Figure 3.7 on page 83 matched the entire distribution of molecular velocities with no fit parameters at all; (b) Figure 9.5 on page 355 in Chapter 9 shows a fit to an exceptionally clean data set; (c) The kink in Figure 8.6 accords with our ideas about the origin of self-assembly.

In case (c), one could fit a third parameter α to the data, try to create an electrostatic theory of the dissociation, then see if it successfully predicted the value of α. But the data shown in Figure 8.6 are too weak to support such a load of interpretation. Elaborate statistical tools exist to determine what conclusions may be drawn from a data set, but most often the judgment is made subjectively. Either way, the maxim is that: *The more elaborate the model, the more data we need to support it.*

8.6 SELF-ASSEMBLY IN CELLS

8.6.1 Bilayers self-assemble from two-tailed amphiphiles

Section 8.4.2 began with a puzzle: How can amphiphilic molecules satisfy their hydrophobic tails in a pure water environment? The answer given there (Figure 8.5) was that they could assemble into a sphere. But this solution may not always be available.

To pack into a sphere, each surfactant molecule must fit into something like a cone shape: Its hydrophilic head must be wider than its tail. More precisely, to form micelles, the volume Nv_{tail} occupied by the tails of N surfactants must be compatible with the surface area Na_{head} occupied by the heads for some N. Although some molecules, like SDS, may be comfortable with this arrangement, it doesn't work for two-tailed molecules like the **phosphatidylcholines** (abbreviated PCs; see Figures 2.14 and 8.3). We have not yet exhausted Nature's cleverness, however. An alternative packing strategy, the bilayer membrane, also presents the hydrophobic tails only to one another. Color Figure 2 shows a slice through a bilayer made of PC. To understand the figure, imagine the double row of molecules shown as extending upward and downward on the page, and out of and into the page, to form a double blanket. Thus the bilayer's midplane is a two-dimensional surface, separating the half-space to the left of the figure from the half-space to the right.

Your Turn 8I

> **a.** Suppose that N amphiphiles pack into a spherical micelle of radius R. Find two relations between a_{head}, v_{tail}, R, and N. Combine these into a single relation between a_{head}, v_{tail}, and R.
>
> **b.** Suppose instead that amphiphiles pack into a planar bilayer of thickness $2d$. Find a relation between a_{head}, v_{tail}, and d.
>
> **c.** In each of the two preceding situations, suppose that the hydrocarbon tails of the amphiphiles cannot stretch beyond a certain length ℓ. Find the resulting geometrical constraints on a_{head} and v_{tail}.
>
> **d.** Why are one-tail amphiphiles likely to form micelles, whereas two-tail amphiphiles are likely to form bilayers?

Two-chain amphiphiles occurring naturally in cells generally belong to a chemical class called phospholipids. We can already understand several reasons why Nature has chosen the phospholipid bilayer membrane as the most ubiquitous architectural component of cells:

- The self-assembly of two-chain phospholipids (like PC) into bilayers is even more avid than that of one-chain surfactants (like SDS) into micelles. The reason is simply that the hydrophobic cost of exposing *two* chains to water is twice as great as that for one chain. This free energy cost ϵ enters the equilibrium constant and hence the CMC, a measure of the chemical drive to self-assembly, via its exponential. There's a big difference between $e^{-\epsilon/k_B T_r}$ and $e^{-2\epsilon/k_B T_r}$, so the CMC for phospholipid formation is tiny. Membranes resist dissolving even in environments with extremely low phospholipid concentration.

- Similarly, phospholipid membranes automatically form closed bags because any edge to the planar structure in Color Figure 2 would expose the hydrocarbon chains to the surrounding water. Such bags, or **bilayer vesicles**, can be almost unlimited in extent; it is straightforward to make "giant" vesicles of radius $10\,\mu m$, the size of eukaryotic cells. This is many thousands of times larger than the thickness of

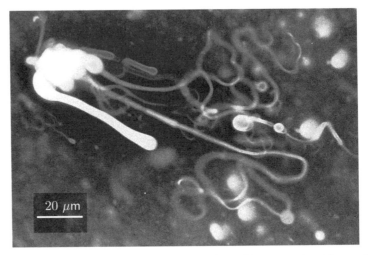

Figure 8.7: (Photomicrograph.) Bilayer structures formed by nonanoic acid, one of several bilayer-forming fatty acids identified in meteorites. The vesicles have been stained with rhodamine, a fluorescent dye. [Digital image kindly supplied by D. Deamer.]

the membrane; giant vesicles are self-assembled structures composed of tens of millions of individual phospholipid molecules.

- Phospholipids are not particularly exotic or complex molecules. They are relatively easy for a cell to synthesize, and phospholipid-like molecules could even have arisen abiotically (from nonliving processes) as a step toward the origin of life. In fact, bilayer membranes are even formed by phospholipid-like molecules that fall to Earth in meteorites (see Figure 8.7)!

- The geometry of phospholipids limits the membrane thickness. This thickness in turn dictates the permeability of bilayer membranes (as we saw in Section 4.6.1 on page 135), their electrical capacitance (using Equation 7.26 on page 269), and even their basic mechanical properties (as we will see in a moment). Choosing the chain length that gives a membrane thickness of a few nanometers turns out to give useful values for all these membrane properties; that's the value Nature has in fact chosen. For example, the permeability to charged solutes (ions) is very low, because the partition coefficient of such molecules in oil is low (see Section 4.6.1 on page 135). Thus bilayer membranes are thin, tough partitions, scarcely permeable to ions.

- Unlike, say, a sandwich wrapper, bilayer membranes are fluid. No specific chemical bond connects any phospholipid molecule to any other, just the generic dislike of water for the hydrophobic tails. Thus the molecules are free to diffuse around one another in the plane of the membrane. This fluidity makes it possible for membrane-bound cells to change their shape, as, for example, when an amœba crawls or a red blood cell squeezes through a capillary.

- Again because of the nonspecific nature of the hydrophobic interaction, membranes readily accept embedded objects; hence they can serve as the doorways to

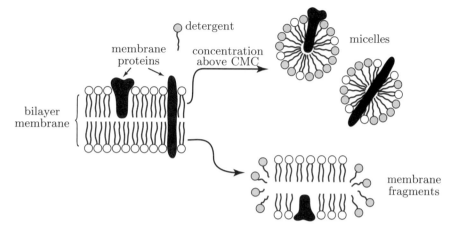

Figure 8.8: (Schematic.) Solubilization of integral membrane proteins (black blobs) by detergent (objects with shaded heads and one tail). *Top right:* At a concentration higher than its critical micelle concentration, a detergent solution can form micelles incorporating both phospholipids (objects with white heads and two tails) and membrane proteins. *Bottom right:* Detergent can also stabilize larger membrane fragments (which would otherwise self-assemble into closed vesicles) by sealing off their edges.

cells (see Figure 2.20 on page 57) and even as the factory floors inside them (see Chapter 11). An object intended to poke through the membrane simply needs to be designed with two hydrophilic ends and a hydrophobic waist; entropic forces then automatically take care of inserting it into a nearby membrane. Understanding this principle also immediately gives us a technological bonus: a technique to isolate membrane-bound proteins (see Figure 8.8).

The physics of bilayer membranes is a vast subject. We will only introduce it, finding an estimate of one key mechanical property of membranes, their bending stiffness.

A bilayer membrane's state of lowest free energy is that of a *flat* (planar) surface. Because the layers are mirror images of each other (see Color Figure 2), there is no tendency to bend to one side or the other. Because each layer is fluid, there is no memory of any previous bent configuration (in contrast to a small patch snipped from a bicycle tire, which remains curved). In short, although it's not impossible to deform a bilayer to a bent shape (indeed, it must so deform in order to close onto itself and form a bag), still bending will entail some free energy cost. We would like to estimate this cost.

Color Figure 2 suggests that the problem with bending is that on one side of the membranes, the polar heads get stretched apart, eventually admitting water into the nonpolar core. In other words, each polar head group normally occupies a particular geometrical area a_{head}; a deviation Δa from this preferred value will incur some free energy cost. To get the mathematical form of this cost for one of the monolayers, we assume that it has a series expansion: $\Delta F = C_0 + C_1 \Delta a + C_2 (\Delta a)^2 + \cdots$. The coeffi-

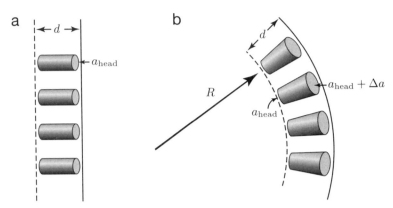

Figure 8.9: (Schematic.) Bilayer membrane bending. d is one-half the total membrane thickness. The cylindrical objects represent individual phospholipid molecules (a second layer on the other side of the dashed line is not shown). (a) The relaxed, flat, membrane conformation gives each head group its equilibrium area a_{head}. (b) Wrapping the membrane about a cylinder of radius R separates the head groups on the outer layer to occupy area $a_{\mathrm{head}} + \Delta a$, where $\Delta a = a_{\mathrm{head}} d / R$.

cient C_0 is a constant and may be dropped. Because the free energy is minimum when $\Delta a = 0$, the contribution from the C_1 term must cancel a corresponding contribution from the other monolayer. For small bends, the higher terms will be negligibly small. Renaming the one remaining coefficient as $\frac{1}{2}k$, then, we expect that the free energy cost will be given by

$$\text{elastic energy per phospholipid molecule} = \frac{1}{2}k(\Delta a)^2. \qquad (8.36)$$

The statement that an elastic energy equals the square of a deformation is a Hooke relation (see Section 5.2.3). The value of the spring constant in this relation, k, is an intrinsic property of the membrane. This relation should hold as long as Δa is much smaller than a_{head} itself. To apply it, suppose that we wrap a small patch of membrane around a cylinder of radius R much bigger than the membrane's thickness d (Figure 8.9). Examining the figure shows that bending the membrane requires that we stretch the outer layer by $\Delta a = a_{\mathrm{head}} d / R$. Thus we expect a bending energy cost per head group of the form $\frac{1}{2}k(a_{\mathrm{head}} d / R)^2$. Because the layer is double, the number of such head groups per area is $2/a_{\mathrm{head}}$. Introducing the new symbol $\kappa \equiv 2kd^2 a_{\mathrm{head}}$ (the **bend stiffness**), we can compactly summarize the discussion by saying:

> *The free energy cost per unit area to bend a bilayer membrane into a cylinder of radius R is of the form $\frac{1}{2}\kappa / R^2$, where κ is an intrinsic parameter describing the membrane. Thus the dimensions of κ are those of energy.* $\qquad (8.37)$

The cost to bend the membrane into a *spherical* patch of radius R is four times as great as in Idea 8.37 because each head group gets stretched in two directions, so

Δa is twice as great. Thus bending the layer into a spherical shape with radius of curvature R costs free energy per unit area $2\kappa/R^2$. The total bending energy to wrap a membrane into a spherical vesicle is then $8\pi\kappa$. This is already an important result: *The total bending energy of a sphere is independent of the sphere's radius.*

To understand the significance of the free energy cost of bending a bilayer (Idea 8.37), we need an estimate of the numerical value of κ. Consider first a single layer at an oil–water interface. Bending the layer into a spherical bulge, with radius of curvature R comparable to the length ℓ_{tail} of the hydrocarbon tails, will spread the heads apart and expose the tails to water. Such a large distortion will incur a hydrophobic free energy cost per unit area, Σ, comparable to that at an oil–water interface. The corresponding cost for a bilayer in water will be roughly twice this value.

We thus have two different expressions for the bending energy of a spherical patch of bilayer, namely, $2\kappa/(\ell_{\text{tail}})^2$ and 2Σ. Equating these expressions lets us estimate κ. Taking typical values $\Sigma \approx 0.05\,\text{J/m}^2$ and $\ell_{\text{tail}} \approx 1.3\,\text{nm}$ gives our estimate: $\kappa \approx 0.8 \cdot 10^{-19}\,\text{J}$. Our estimate is crude, but it's not too far from the measured value $\kappa = 0.6 \cdot 10^{-19}\,\text{J} = 15 k_\text{B} T_\text{r}$ for dimyristoyl phosphatidylcholine (DMPC). The total bending energy $8\pi\kappa$ of a spherical vesicle of DMPC is then around $400 k_\text{B} T_\text{r}$.

We can extract a simple lesson from the measured value of κ. Suppose that we take a flat membrane of area A and impose on it a corrugated (washboard) shape, alternating cylindrical segments of radius R. The free energy cost of this configuration is $\frac{1}{2}\kappa A/R^2$. Taking A to be $1000\,\mu\text{m}^2$, a value corresponding to a typical $10\,\mu\text{m}$ cell, we find that the bending energy cost greatly exceeds $k_\text{B} T_\text{r}$ for any value of R under $10\,\mu\text{m}$. Thus the stiffness of phospholipid bilayer membranes has been engineered to prevent spontaneous corrugation by thermal fluctuations. At the same time, the bending energy needed for gross, overall shape changes (for example, those needed for cell crawling) is only a few hundred times $k_\text{B} T_\text{r}$, so such changes require the expenditure of only a few dozen ATP molecules (see Appendix B). Phospholipid bilayer membrane stiffness is thus in just the right range to be biologically useful.

Not only are cells themselves surrounded by a bilayer plasma membrane. Many of the organelles inside cells are separate compartments, partitioned from the rest by a bilayer. Products synthesized in one part of the cell (the "factory") are also shipped to their destinations in special-purpose transport containers, themselves bilayer vesicles. Incoming complex food molecules awaiting digestion to simpler forms are held in still other vesicles. And Chapter 12 will describe how the activation of one neuron by another across a synapse involves the release of neurotransmitters, which are stored in bilayer vesicles until needed. Self-assembled bilayers are ubiquitous in cells.

$\boxed{T_2}$ *Section 8.6.1' on page 336 mentions some elaborations to these ideas.*

8.6.2 Vista: Macromolecular folding and aggregation

Protein folding   Section 2.2.3 on page 50 sketched a simple-sounding answer to the question of how cells translate the static, one-dimensional data stream in their genome into functioning, three-dimensional proteins. The idea is that the sequence of amino acid residues determined by the genome, together with the pattern of mutual interactions between the residues, determines a unique, properly folded state, called the **native conformation**. Evolution has selected sequences that give rise to useful, functioning native conformations. We can get a glimpse of some of the contributions to the force driving protein folding by using ideas from this chapter and Chapter 7.

Relatively small disturbances in the protein's environment (for example, change of temperature, solvent, or pH) can disrupt the native conformation, or **denature** the protein. Hsien Wu proposed in 1929 that denaturation was in fact precisely the unfolding of the protein from "the regular arrangement of a rigid structure to the irregular, diffuse arrangement of the flexible open chain." In this view, unfolding changes the protein's structure dramatically and destroys its function, without necessarily breaking any chemical bonds. Indeed, restoring physiological conditions returns the balance of driving forces to one favoring folding; for example, M. Anson and A. Mirsky showed that denatured hemoglobin returns to a state physically and functionally identical to its original form when refolded in this way. That is, the folding of a (simple) protein is a *spontaneous* process, driven by the resulting decrease in the free energy of the protein and the surrounding water. Experiments of this sort culminated in the work of C. Anfinsen and coauthors, who showed around 1960 that for many proteins,

- The sequence of a protein completely determines its folded structure, and
- The native conformation is the minimum of the free energy.

The thermodynamic stability of folded proteins under physiological conditions stands in sharp contrast to the random-walk behavior studied in Chapter 4. The discussion there pointed out the immense number of conformations a random chain can assume; protein folding thus carries a correspondingly large entropic penalty. Besides freezing the protein's backbone into a specific conformation, folding also tends to immobilize each amino acid's side chain, with a further cost in entropy. Apparently some even larger free energy gain overcomes these entropic penalties, driving protein folding. It's a delicate balance: At body temperature, the net chemical force driving folding rarely exceeds $20k_B T_r$, the free energy of just a few H-bonds.

What forces drive folding? Section 7.5.1 on page 273 already mentioned the role of hydrogen bonds in stabilizing macromolecules. Walter Kauzmann argued in the 1950s that hydrophobic interactions also supply a major part of the thermodynamic force driving protein folding. Each of the 20 common different amino acids can be assigned a characteristic value of hydrophobicity. Kauzmann argued that a polypeptide chain would spontaneously fold to bury its most hydrophobic residues in its interior, away from the surrounding water, in a manner similar to the formation of a micelle. Indeed, structural data not available at the time has borne out this view: The most hydrophobic residues of proteins tend to be located in the interior of the native (properly folded) conformation.[5] In addition, the study of analogous proteins from different animal species shows that even though they can differ widely in their precise

---

[5]We will see later how the exceptions to this general rule turn out to be important for helping proteins stick to one another.

amino acid sequences, still the hydrophobicities of the core residues hardly differ at all—they are "conserved" under molecular evolution. Similarly, one can create artificial proteins by substituting specific residues in the sequence of some natural protein. Such site-directed mutagenesis experiments show that the resulting protein structure changes most when the substituted residue has a hydrophobicity very different from that of the original residue.

Kauzmann also noted a remarkable thermal feature of protein denaturation. Not only can high temperature (typically $T > 55°C$) unfold a protein, but in many cases, *low* temperature does, too (typically $T < 20°C$). Denaturation by heat fits with an intuitive analogy to melting a solid, but cold denaturation was initially a surprise. Kauzmann pointed out that hydrophobic interactions weaken at lower temperatures (see Section 7.5.3 on page 280), so the phenomenon of cold denaturation points to the role of such interactions in stabilizing protein structure. Kauzmann also noted that proteins can be denatured by transferring them to nonpolar solvents, in which the hydrophobic interaction is absent. Finally, adding even extremely small concentrations of surfactants (for example, 1% SDS) can also unfold proteins. We can interpret this fact by analogy with the solubilization of membranes (Figure 8.8): The surfactants can shield hydrophobic regions of the polypeptide chain, thereby reducing their tendency to associate with one another. For these and other reasons, *hydrophobic interactions are believed to give the dominant force driving protein folding.*

Other interactions can also help to determine a protein's structure. A charged residue, like those studied in Section 8.3.3 on page 311, will have a Born self-energy. Such residues will prefer to sit at the surface of the folded protein, facing the highly polarizable exterior water (see Section 7.5.2 on page 276) rather than being buried in the interior. Positive residues will also seek the company of negatively charged ones, and avoid other positive charges. Although significant, these specific interactions are probably not as important as the hydrophobic effect. For example, if we titrate a protein to zero overall charge, its stability is found not to depend very much on the surrounding salt concentration, even though salt weakens electrostatic effects (see Idea 7.28).

Aggregation Besides supplying *intra*molecular forces driving folding, hydrophobic interactions also give *inter*molecular forces, which can stick neighboring macromolecules together. Section 7.5.3 on page 280 mentioned the example of microtubules, whose tubulin monomers are held together in this way. Section 8.3.4 on page 312 gave another example: Sickle-cell anemia's debilitating effects stem from the unwanted hydrophobic aggregation of defective hemoglobin molecules. Cells can even turn their macromolecules' aggregating tendencies on and off to suit their needs. For example, your blood contains a structural protein called fibrinogen, which normally floats in solution. When a blood vessel gets injured, however, the injury triggers an enzyme that clips off a part of the fibrinogen molecule, exposing a hydrophobic patch. The truncated protein, called fibrin, then polymerizes to form the scaffold on which a blood clot can form.

Hydrophobic aggregation is not limited to the protein–protein case. Chapter 9 will also identify hydrophobic interactions as key to stabilizing the double-helical structure of DNA. Each basepair is shaped like a flat plate; both of its surfaces are nonpolar, so it is driven to stick onto the adjoining basepairs in the DNA chain and form a stack. Hydrophobic interactions also contribute to the adhesion of antibodies to their corresponding antigens.

8.6.3 Another trip to the kitchen

This has been a long, detailed chapter. Let's take another trip to the kitchen.

Besides being a multibillion dollar industry, food science nicely illustrates some of the points made in this chapter. For example, Your Turn 5A on page 159 caricatured milk as a suspension of fat droplets in water. Actually, milk is far more complex than this. In addition to the fat and water, milk contains two classes of proteins, Miss Muffet's curds (the casein complex) and whey (mainly α-lactalbumin and β-lactoglobulin). In fresh milk, the casein complexes self-assemble into micelles with radii around 50 nm. The micelles are kept separate in part by electrostatic repulsion (see Section 7.4.4 on page 269), so the milk is fluid. However, minor environmental changes can induce curdling, which is a coagulation (clumping) of the micelles into a gel (Figure 8.10). In the case of yogurt, the growth of bacteria such as *Lactobacillus bulgaricus* and *Streptococcus thermophilus* creates lactic acid as a waste product (alternatively, you can add acid by hand, for example, lemon juice). The ensuing increase in the concentration of H^+ ions reduces the effective charge on the casein micelles (see Section 8.3.3) and hence also reduces the normal electrostatic repulsion between them. This change tips the balance toward aggregation; milk curdles when its pH is

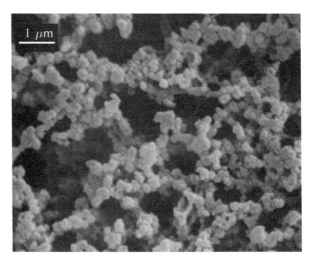

Figure 8.10: (Scanning electron micrograph.) Yogurt. Acid generated by bacteria triggers the aggregation of casein micelles (*spheres* of diameter 0.1 μm in the figure) into a network. The fat globules (*not shown*) are much bigger, with radius 1–3 μm in fresh milk. [Digital image kindly supplied by M. Kalab.]

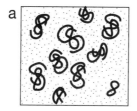 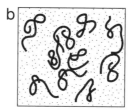 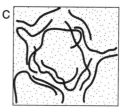

Figure 8.11: (Schematic.) The physics of omelettes. (a) Proteins in their native conformation (b) open up to form random coils upon heating. (c) Neighboring coils then begin to interact with one another to form weak intermolecular bonds. The resulting network can trap water.

lowered from the natural value of 6.5 to below 5.3. The casein network in turn traps the fat globules.[6]

Eggs provide another example of a system of protein complexes. Each protein is a long, chemically bonded chain of amino acids. Most culinary operations do not disrupt the primary structure, or sequence, of this chain because normal cooking temperatures don't supply enough energy to break the peptide bonds. But each protein has been engineered to assume a useful native conformation under the assumption that it will live in an aqueous environment at temperatures below 37°C. When the environment is changed (by introducing air or by cooking), the protein denatures.

Figure 8.11 sketches what can happen. Raising the temperature can convert the precisely folded native structures into random chains. Once the chains open, the various charged and hydrophobic residues on one chain, previously interacting mainly with other residues elsewhere on the same chain, can now find and bind to those on other chains. In this way, a cross-linked network of chains can form. The interstices of this network can hold water, and the result is a solid gel: the cooked egg. As with milk, one may expect that the addition of acid would enhance the coagulation of eggs once the proteins are denatured, and indeed it's so.

Heating is not the only way to denature egg proteins and create a linked network. Merely whipping air into the eggs to create a large surface area of contact with air can totally disrupt the hydrophobic interactions. The ensuing "surface denaturation" of egg proteins like conalbumin is what gives chiffon pie and mousse their structural stability: A network of unfolded proteins arrange themselves with their hydrophobic residues facing the air bubbles and their hydrophilic ones facing the water. This network not only reduces the air–water tension like any amphiphile (see Section 8.4.1), it also stabilizes the arrangement of bubbles because, unlike simple amphiphiles, the proteins are long chains. Other proteins, like ovomucin and globulins, play a supporting role by making the egg so viscous that the initial foam drains slowly, giving the conalbumin time to form its network. Still others, like ovalbumin, support air foams but require heat for their initial denaturation; these proteins are key to supporting the stronger structures of meringue and soufflé. All these attributions of specific roles to

---

[6]The fat globules must themselves be stabilized against coalescing. In fresh milk, they are coated by an amphiphilic membrane and hence form an emulsion (see Figure 8.4).

specific proteins were established by isolating particular proteins and trying them alone or in various combinations.

Eggs also serve as emulsifiers, for example, in the preparation of creamy sauces as mentioned earlier. Preparing such sauces is tricky; a slight deviation from the recipe can turn a nice emulsion into a coagulated mess. Volumes of superstition and folklore have evolved on this subject, claims concerning clockwise versus counterclockwise stirring and so on. Most of these claims have not survived scientific scrutiny. But in a careful study, a group of high school students found that simply adding lecithin, a two-chain phospholipid available at health-food stores, can reliably rescue a failed sauce béarnaise, an unsurprising conclusion in the light of Section 8.4.1.

THE BIG PICTURE

Returning to the Focus Question, we've seen how an activation barrier can lock energy into a molecular species, making the release of that energy negligibly slow. A beaker with a lot of that species may not be doing much, but it is far from equilibrium—and hence, ready to do useful work. We saw how to make such statements quantitative by using the change of free energy when a single molecule enters or leaves a system (the chemical potential μ). We got a formula for μ showing the reciprocal roles of energy and entropy in determining the chemical force driving chemical reactions, and unifying chemical forces with the other sorts of entropic forces studied in this book. Chapters 10 and 11 will extend our understanding from chemical reactions to *mechano*chemical and *electro*chemical reactions, those doing useful work by moving an object against a load force or electrostatic field. Such reactions, and the enzymes that broker them, are central to the functioning of cells.

KEY FORMULAS

- *Chemical potential:* Suppose that there are $\Omega(E, N_1, N_2, \ldots)$ states available to a system with energy E and N_1 particles of type 1, and so on. The chemical potential of species α is then (Equation 8.1)

$$\mu_\alpha = -T \frac{\partial S}{\partial N_\alpha}\bigg|_{E, N_\beta, \beta \neq \alpha}.$$

μ_α describes the "availability of particles" for exchange with another subsystem. If each of the μ_α for one system agrees with the corresponding values for the other, then there will be no net exchange (Equation 8.2).

In an ideal gas or other collection of independent particles (for example, a dilute solution), we have $\mu = k_B T \ln(c/c_0) + \mu^0$ (Equation 8.3). Here c is the number density and c_0 is a conventional reference concentration. μ^0 depends on temperature and on the choice of reference concentration but not on c. For a charged ion in an external electric potential, add $qV(x)$ to μ^0, to get the electrochemical potential.

- *Grand ensemble:* The probability of finding a small subsystem in the microstate i, if it's in contact with a reservoir at temperature T and chemical potentials $\mu_1, \ldots,$ is (Equation 8.6)

$$\mathcal{Z}^{-1} e^{-(E_i - \mu_1 N_{1,i} - \mu_2 N_{2,i} \cdots)/k_B T}.$$

Here $\mathcal{Z}$ is a normalization factor (the partition function) and E_i, $N_{1,i}, \ldots$ are the energy and populations for state i of the subsystem.

- *Mass Action:* Consider a reaction in which ν_1 molecules of species $X_1, \ldots$ react in dilute solution to form ν_{k+1} molecules of species X_{k+1} and so on. Let $\Delta G^0 = -\nu_1 \mu_1^0 - \cdots + \nu_{k+1} \mu_{k+1}^0 + \cdots$, and let ΔG be the similar quantity defined using the μ's. Then the equilibrium concentrations obey (Equation 8.17)

$$\Delta G = 0, \quad \text{or} \quad \frac{[X_{k+1}]^{\nu_{k+1}} \cdots [X_m]^{\nu_m}}{[X_1]^{\nu_1} \cdots [X_k]^{\nu_k}} = K_{eq},$$

where $[X] \equiv c_X/(1\,\text{M})$ and $K_{eq} = e^{-\Delta G^0/k_B T}$. The ratio of concentrations above is called the reaction quotient; if it differs from K_{eq}, the system is not in equilibrium and the reaction proceeds in the net direction needed to move closer to equilibrium.

  Note that ΔG^0 and K_{eq} both depend on the reference concentrations chosen when defining them; Equation 8.17 corresponds to taking the reference concentrations all equal to $1\,\text{M}$. Often it's convenient to define $pK = -\log_{10} K_{eq}$.

- *Chemical force:* If ΔG above is not equal to zero, it amounts to a chemical force, driving the reaction forward if $\Delta G < 0$ and backward otherwise.

- *Acids and bases:* The pH of an aqueous solution is $-\log_{10}[H^+]$. The pH of pure water reflects the degree to which H_2O dissociates spontaneously. It's almost entirely *un*dissociated: $[H^+] = 10^{-7}$, whereas there are 55 mole/L of H_2O molecules.

- *Titration:* Each residue α of a protein has its own pK value for dissociation. The probability of being protonated, P_α, equals $\frac{1}{2}$ when the surrounding solution's pH matches the residue's pK. Otherwise we have (Equation 8.29)

$$P_\alpha = (1 + 10^{x_\alpha})^{-1}, \text{ where } x_\alpha = \text{pH} - pK_\alpha.$$

- *Critical micelle concentration:* In our model, the total concentration of amphiphilic molecules c_{tot} is related to the concentration c_1 of those remaining unaggregated by $c_{tot} = c_1(1 + (2c_1/c_*)^{N-1})$ (Equation 8.33). The critical micelle concentration c_* is the concentration at which half of the amphiphilic molecules are in the form of micelles; its value reflects the equilibrium constant for self-assembly.

FURTHER READING

Semipopular:
On the physics and chemistry of food: McGee, 1984.

Intermediate:
Biophysical chemistry: Atkins, 2001; Dill & Bromberg, 2002; van Holde et al., 1998; Tinoco et al., 2001.
Electrophoresis: Benedek & Villars, 2000c, §3.1.D.
Self-assembly: Evans & Wennerström, 1999; Israelachvili, 1991; Safran, 1994.
Lipidlike molecules from space and the origins of life: Deamer & Fleischaker, 1994.

Technical:
Physical chemistry: Mortimer, 2000.
Physical aspects of membranes: Lipowsky & Sackmann, 1995; Seifert, 1997.
Protein structure: Branden & Tooze, 1999; Dill, 1990.

8.1.1′ Track 2

1. Equation 8.1 on page 295 defined μ as a derivative with respect to the number of molecules N. Chemistry textbooks instead define μ as a derivative with respect to the "amount of substance" n. See the discussion of units in Section 1.5.4′ on page 30.

2. The discussion in the gas chemical potential Example (page 296) amounted to converting a derivative taken with E_{kin} fixed to one taken with E fixed. The formal way to summarize this manipulation is to say that

$$\left.\frac{\partial S}{\partial N}\right|_E = \left.\frac{\partial S}{\partial N}\right|_{E_{kin}} - \epsilon \left.\frac{\partial S}{\partial E_{kin}}\right|_N .$$

3. We have been describing ϵ as if it were a form of potential energy, like a coiled spring inside the molecule. Purists will insist that the energy of a chemical bond is partly potential and partly kinetic, by the Uncertainty Principle. It's true. What lets us lump these energies together, indeed what lets us speak of bond energies at all, is that quantum mechanics tells us that any molecule at rest has a **ground state** with a fixed, definite total energy. Any additional kinetic energy from center-of-mass motion and any potential energy from external fields are given by the usual classical formulas and simply added to the fixed internal energy. That's why we get to use familiar results from classical physics in our analysis.

4. A complicated molecule may have many states of almost equally low energy. Then ϵ will have a temperature-dependent component reflecting in part the likelihood of occupying the various low-energy states. But we won't use ϵ directly; we'll use μ^0, which we already knew was temperature-dependent anyway. This fine point doesn't usually matter because living organisms operate at nearly fixed temperature; once again our attitude is that μ^0 is a phenomenological quantity.

8.2.1′ Track 2

There are other, equivalent definitions of μ besides the one given in Equation 8.1 on page 295. Thus, for example, some advanced textbooks state your results from Your Turn 8C as

$$\mu = \left.\frac{\partial F}{\partial N}\right|_{T,V} = \left.\frac{\partial G}{\partial N}\right|_{T,p} .$$

Two more expressions for the chemical potential are $\left.\frac{\partial E}{\partial N}\right|_{S,V}$ and $\left.\frac{\partial H}{\partial N}\right|_{S,p}$, where H is the enthalpy. The definition in Equation 8.1 was chosen as our starting point because it emphasizes the key role of entropy in determining any reaction's direction.

8.2.2′ Track 2

1. The solutions of interest in cell biology are frequently *not* dilute. In this case, the Second Law still determines a reaction's equilibrium point, but we must use the activity in place of the concentration [X] when writing the Mass Action rule (see Section 8.1.1 on page 295). Dilute-solution formulas are especially problematic in the case of ionic solutions (salts) because our formulas ignore the electrostatic interaction between ions (and indeed all other interactions). Because the electrostatic interaction is of long range, its omission becomes a serious problem sooner as we raise the concentration than that of other interactions. See the discussion in Landau & Lifshitz, 1980, §92.

2. We can also think of the temperature dependence of the equilibrium constant (Your Turn 8D on page 302) as an instance of Le Châtelier's Principle. Dumping thermal energy into a closed system increases the temperature (thermal energy becomes more available). This change shifts the equilibrium toward the higher-energy side of the reaction, so the system absorbs thermal energy, making the actual temperature increase smaller than it would have been if no reaction had occurred. In other words, the reaction partially undoes our original disturbance.

 8.3.4′ Track 2

The discussion of electrophoresis in Section 8.3.4 is rather naïve; the full theory is quite involved. For an introductory discussion, see Benedek & Villars, 2000c, §3.1.D; for many details, see Viovy, 2000.

T_2' **8.6.1′ Track 2**

1. Section 8.6.1 argued that a bilayer membrane prefers to be flat. Strictly speaking, this argument only applies to artificial, pure lipid bilayers. Real plasma membranes have significant compositional differences between their two layers, with a corresponding spontaneous tendency to bend in one direction.

2. The logic given for the elastic energy of a membrane may be more familiar in the context of an ordinary spring. Here we find the elastic (potential) energy for a small deformation to be of the form $U = \frac{1}{2}k(\Delta x)^2$, where Δx is the change in length from its relaxed value. Differentiating to find the force gives $f = -k(\Delta x)$, which is the Hooke relation (compare with Equation 5.14 on page 172).

3. More realistically, bending a bilayer involves a combination of stretching the outer layer and *squeezing* the inner layer. In addition, the bilayer's elasticity also contains contributions from deformation of the tails of the amphiphilic molecules, not just the heads. These elaborations do not change the general form of the bending elasticity energy. (For many more details about bilayer elasticity, see for example Seifert, 1997.)

PROBLEMS

8.1 *Coagulation*

a. Section 8.6.3 described how the addition of acid can trigger the coagulation (clumping) of proteins in milk or egg. The suggested mechanism was a reduction of the effective charge on the proteins and a corresponding reduction in their mutual repulsion. The addition of *salt* also promotes coagulation, whereas sugar does not. Suggest an explanation for these facts.

b. Cheese-making dates from at least 2300 BCE. More recently (since ancient Roman times), cheese-makers have used a milk-curdling method that does not involve acid or salt. Instead, a proteolytic (protein-splitting) enzyme (chymosin, or rennin) is used to cut off a highly charged segment of the κ-casein molecule (residues 106–169). Suggest how this change could induce curdling and relate it to the discussion in Section 8.6.2.

8.2 *Isomerization*

Our example of buffalo as a two-state system (Figure 6.8 on page 220) may seem a bit fanciful. A more realistic example from biochemistry is the isomerization of a phosphorylated glucose molecule from its 1-P to its 6-P form (see Figure 8.12), with $\Delta G^0 = -1.74\,\text{kcal/mole}$. Find the equilibrium concentration ratio of glucose-P in the two isomeric states shown.

8.3 *pH versus temperature*

The pH of pure water is not a universal constant; rather, it depends on the temperature: At $0\,°C$, it's 7.5, whereas at $40\,°C$, it's 6.8. Explain this phenomenon and comment on why your explanation is numerically reasonable.

8.4 *Difference between F and G*

a. Consider a chemical reaction in which a molecule moves from gas to a water solution. At atmospheric pressure, each gas molecule occupies a volume of about $24\,\text{L/mole}$, whereas in solution, the volume is closer to the volume occupied by a water molecule, or $1/(55\,\text{mole/L})$. Estimate $(\Delta V)p$, expressing your answer in units of $k_{\mathrm{B}}T_{\mathrm{r}}$.

b. Consider a reaction in which two molecules in aqueous solution combine to form one. Compare an estimate of $(\Delta V)p$ with what you found in (a) and comment on why we usually don't need to distinguish between F and G for such reactions.

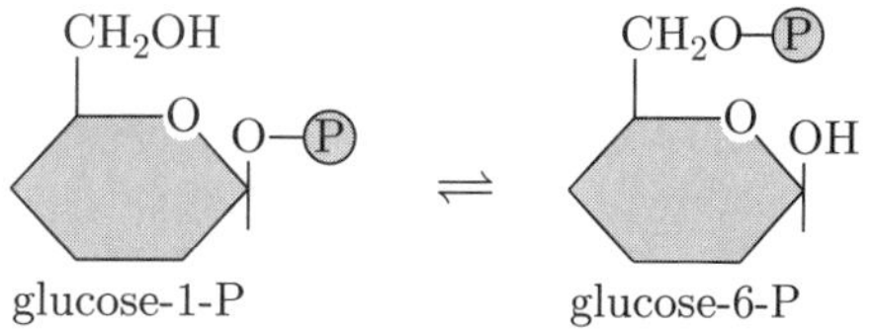

Figure 8.12: (Molecular structure diagrams.) Isomerization of glucose-P. [Adapted from Alberts et al., 1997.]

8.5 *Simple dissociation*

Section 8.3.2 gave the dissociation pK for acetic acid as 4.76. Suppose that we dissolve a mole of this weak acid in 10 L of water. Find the pH of the resulting solution. What fraction of acetic acid molecules is dissociated?

8.6 *Ionization state of inorganic phosphate*

Chapter 2 oversimplified somewhat in stating that phosphoric acid (H_3PO_4) ionizes in water to form the ion HPO_4^{2-}. In reality, all four possible protonation states, from three H's to none, exist in equilibrium. The three successive proton-removing reactions have the following approximate pK values:

$$H_3PO_4 \overset{pK_1=2}{\rightleftharpoons} H_2PO_4^- \overset{pK_2=7}{\rightleftharpoons} HPO_4^{2-} \overset{pK_3=12}{\rightleftharpoons} PO_4^{3-}.$$

Find the relative populations of all four protonation states at the pH of human blood, around 7.4.

8.7 *Electrophoresis*

In this problem, you will make a crude estimate of a typical value for the electrophoretic mobility of a protein.

a. Model the protein as a sphere of radius 3 nm, carrying a net electric charge $q = 10e$, in pure water. If we apply an electric field of $\mathcal{E} = 2\,\text{volt cm}^{-1}$, the protein will feel a force $q\mathcal{E}$. Write a formula for the resulting drift velocity and evaluate it numerically.[7]

b. In the experiment discussed in Section 8.3.4 on page 312, Pauling and coauthors used an electric field of $4.7\,\text{volt cm}^{-1}$, applied for up to 20 hours. For a mixture of normal and defective hemoglobin to separate into two distinguishable bands, they must travel different distances under these conditions. Estimate the separation of these bands for two species whose charges differ by just one unit and comment on the feasibility of the experiment.

8.8 $\boxed{T_2}$ *Grand partition function*

Review Section 8.1.2 on page 298.

a. Show that the distribution you found in Your Turn 8B is the one that minimizes the **grand potential** of system a at T, μ, defined by analogy with the usual free energy (Equation 6.32 on page 224) as

$$\Psi_a = \langle E_a - \mu N_a \rangle - TS_a. \tag{8.38}$$

b. Show that the minimal value of Ψ thus obtained equals $k_\mathrm{B} T \ln \mathcal{Z}$.

c. *Optional:* For the real gluttons, generalize your result in (a) and (b) to systems exchanging particles and energy, *and* changing volume as well (see Section 6.5.1).

---

[7] $\boxed{T_2}$ Actually, one uses a salt solution (buffer) instead of pure water. A more careful treatment would account for the screening of the particle's charge (Section 7.4.3′ on page 284); the result contains an extra factor of $(3/2)(\lambda_\mathrm{D}/a)$ relative to your answer.

PART III

Molecules, Machines, Mechanisms

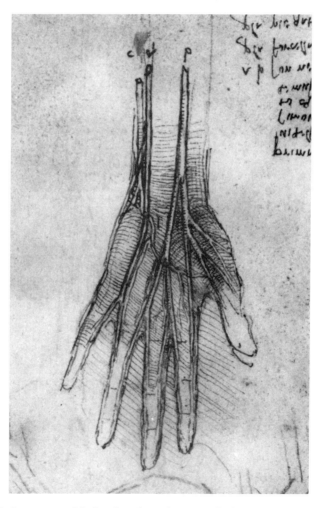

The median and ulnar nerves of the hand, as drawn by Leonardo da Vinci around 1504–1509. [From the Royal Library in Windsor Castle; Clark, *Catalog of the drawings of Leonardo da Vinci at Windsor Castle* (Cambridge University Press, Cambridge, UK, 1935).]

CHAPTER 9

Cooperative Transitions in Macromolecules

Hooke gave in 1678 the famous law of proportionality of stress and strain which bears his name, in the words "Ut tensio sic vis." This law he discovered in 1660, but did not publish until 1676, and then only under the form of an anagram, "ceiiinosssttuv."

— A. Love, *A treatise on the mathematical theory of elasticity*, 1906

The preceding chapters may have shed some light on particular molecular forces and processes, but they also leave us with a deeper sense of dissonance. On one hand, we have seen that the activity of individual small molecules is chaotic, leading to phenomena like Brownian motion. We have come to expect predictable, effectively deterministic behavior only when dealing with vast numbers of molecules, for example, the diffusion of a drop of ink or the pressure of air in a bicycle tire. On the other hand, Chapter 2 showed a gallery of exquisitely structured *individual* macromolecules, each engineered to do specific jobs reliably. So which image is right—should we think of macromolecules as being like gas molecules, or like tables and chairs?

More precisely, we'd like to know how individual molecules, held together by weak interactions, nevertheless retain their structural integrity in the face of thermal motion and, indeed, perform specific functions. The key to this puzzle is the phenomenon of cooperativity.

Chapter 8 introduced cooperativity, showing that it makes the micelle transition sharper than we would otherwise expect it to be. This chapter will extend the analysis and also deepen our understanding of macromolecules as brokers at the interface between the worlds of mechanical and chemical forces. Section 9.1 begins by studying how an external force affects the conformation of a macromolecule, first in a very simplified model and then in a second model adding the cooperative tendency of each monomer to do what its nearest neighbors are doing. The ideas of Chapter 6 and the partition function method for calculating entropic forces (from Section 7.1) will be very useful here. Next, Section 9.5 will extend the discussion to transitions induced by changes in the *chemical* environment. The final sections argue briefly that the lessons learned from simple model systems can help us understand qualitatively the sharp state transitions observed in biologically important systems, the allosteric proteins.

The Focus Questions for this chapter are

Biological question: Why aren't proteins constantly disrupted by thermal fluctuations? The cartoons in cell biology books show proteins snapping crisply between definite conformations as they carry out their jobs. Can a floppy chain of residues really behave in this way?

Physical idea: Cooperativity sharpens the transitions of macromolecules and their assemblies.

9.1 ELASTICITY MODELS OF POLYMERS

Roadmap The following sections introduce several physical models for the elasticity of DNA. Section 9.1.2 begins by constructing and justifying a physical picture of DNA as an elastic rod. Although physically simple, the elastic rod model is complex to analyze mathematically. Thus we work up to it with a set of reduced models, starting with the "freely jointed chain" (Section 9.1.3). Section 9.2 introduces experimental data on the mechanical deformation (stretching) of single molecules and interprets it, using the freely jointed chain model. Section 9.4 argues that the main feature neglected by the freely jointed chain is cooperativity between neighboring segments of the polymer. To redress this shortcoming, Section 9.4.1 introduces a simple model, the "one-dimensional cooperative chain." Later sections apply the mathematics of cooperativity to structural transitions within polymers, for example, the helix–coil transition.

Figure 2.15 on page 51 shows a segment of DNA. It's an understatement to say that this molecule has an elaborate architecture! Atoms combine to form bases. Bases bind into basepairs by hydrogen bonding; they also bond covalently to two outer backbones of phosphate and sugar groups. Worse, the beautiful picture in the figure is in some ways a *lie:* It doesn't convey the fact that a macromolecule is dynamic, with each chemical bond constantly flexing and involved in promiscuous, fleeting interactions with other molecules not shown (the surrounding water molecules, with their network of H-bonds, and so on). It may seem hopeless to seek a simple account of the mechanical properties of this baroque structure.

Before giving up on a simple description of DNA mechanics, though, we should pause to examine the length scales of interest. DNA is roughly a cylindrical molecule of diameter 2 nm. It consists of a stack of roughly flat plates (the basepairs), each about 0.34 nm thick. But the total *length* of a molecule of DNA (for example, in one of your chromosomes), can be 2 cm, or ten million times the diameter! Even a tiny virus such as the **lambda phage** has a genome 16.5 μm long, still far bigger than the diameter. We may hope that the behavior of DNA on such long length scales may not depend very much on the details of its structure.

9.1.1 Why physics works (when it does work)

There is plenty of precedent for such a hope. After all, engineers do not need to account for the detailed atomic structure of steel (nor, indeed, for the fact that steel is made of atoms at all) when designing bridges. Instead, they model steel as a continuum with a certain resistance to deformation, characterized by just *two numbers* (called the bulk modulus and shear modulus; see Section 5.2.3 on page 169). Similarly, the discussion of fluid mechanics in Chapter 5 made no mention of the detailed structure of the water molecule, its network of hydrogen bonds, and so on. Instead, we again summarized the properties of water relevant for physics on scales bigger than a couple of nanometers by just *two numbers,* mass density ρ_m and viscosity η. Any other Newtonian fluid, even with a radically different molecular structure, will flow like water if it matches the values of these two **phenomenological parameters**. What these two examples share is a deep theme running through all of physics:

> When we study a system with a **large number** of **locally interact-ing, identical** constituents on a far **bigger scale** than the size of the constituents, then we reap a huge simplification: Just a **few effective degrees of freedom** describe the system's behavior, with just a **few phenomenological parameters**. (9.1)

Thus the fact that bridges and pipes are *much bigger* than iron atoms and water molecules underlies the success of continuum elasticity theory and fluid mechanics.

Much of physics amounts to the systematic exploitation of Idea 9.1. A few more examples will help explain the statement of this principle. Then we'll try using it to address the questions of interest to this chapter.

Another way to express Idea 9.1 is to say that Nature is hierarchically arranged by length scale into levels of structure and that each successive level of structure forgets nearly everything about the deeper levels. It is no exaggeration to say that this principle explains why physics is possible at all. Historically, our ideas of the structure of matter have gone from molecules, to atoms, to protons, neutrons, electrons, and beyond this to the quarks composing the protons and neutrons, and perhaps to even deeper levels of substructure. Had it been necessary to understand *every* deeper layer of structure before making any progress, then the whole enterprise could never have started! Conversely, even now that we do know that matter consists of atoms, we would never make any progress understanding bridges (or galaxies) if we were obliged to consider them as collections of atoms. The simple rules emerging as we pass to each new length scale are examples of the emergent properties mentioned in Sections 1.2.3 and 6.3.2.

Continuum elasticity In elasticity theory, we pretend that a steel beam is a continuous object, ignoring the fact that it's made of atoms. To describe a deformation of the beam, we imagine dividing it into cells of, say, $1\,\mathrm{cm}^3$ (much smaller than the beam but much bigger than an atom). We label each cell by its position in the beam in its unstressed (straight) state. When we put a load on the beam, we can describe the resulting deformation by reporting the change in the position of each element relative to its neighbors, which is much less information than a full catalog of the positions of each atom. If the deformation is not too large, we can assume that its elastic energy cost per unit volume is proportional to the square of its magnitude (a Hooke-type relation; see Section 5.2.3 on page 169). The constants of proportionality in this relationship are examples of the phenomenological parameters mentioned in Idea 9.1. In this case, there are two of them, because a deformation can either stretch or shear the solid. We could try to predict their numerical values from the fundamental forces between atoms. But we can just as consistently take them to be experimentally measured quantities. As long as only one or a few phenomenological parameters characterize a material, we can get many falsifiable predictions after making only a few measurements to nail down the values of those parameters.

Fluid mechanics The flow behavior of a fluid can also be characterized by just a few numerical quantities. An isotropic Newtonian fluid, such as water, has no memory of its original (undeformed) state. Nevertheless, we saw in Chapter 5 that a fluid resists certain motions. Again dividing the fluid into imagined macroscopic cells, the effective degrees of freedom are each cell's velocity. Neighboring cells pull on one another via the viscous force rule (Equation 5.4 on page 164). The constant η appearing in that rule—the viscosity—relates the force to the deformation rate; it's one of the phenomenological parameters describing a Newtonian fluid.

Membranes Bilayer membranes have properties resembling both solids and fluids (see Section 8.6.1 on page 322). Unlike a steel beam or a thin sheet of aluminum foil, the membrane is a fluid: It maintains no memory of the arrangement of molecules within its plane, so it offers no resistance to a constant shear. But unlike sugar molecules dissolved in a drop of water, the membrane does remember that it prefers to lie in space as a continuous, flat sheet—its resistance to bending is an intrinsic phenomenological parameter (see Idea 8.37 on page 326). Once again, one constant, the bend stiffness κ, summarizes the complex intermolecular forces adequately, as long as the membrane adopts a shape whose radius of curvature is everywhere much bigger than the molecular scale.

Summary The preceding examples suggest that Idea 9.1 is a broadly applicable principle. But there are limits to its usefulness. For example, the individual monomers in a protein chain are *not* identical. As a result, the problem of finding the lowest-energy state of a protein is far more complex than the corresponding problem for, say, a jar filled with identical marbles. We need to use physical insights when they are helpful, while being careful not to apply them when inappropriate. Later sections of this chapter will find systems where simple models do apply and seem to shed at least qualitative light on complex problems.

$\boxed{T_2}$ *Section 9.1.1′ on page 384 discusses further the idea of phenomenological parameters and Idea 9.1.*

9.1.2 Four phenomenological parameters characterize the elasticity of a long, thin rod

Let's return to DNA and begin to think about what phenomenological parameters are needed to describe its behavior on length scales much longer than its diameter. Imagine holding a piece of garden hose by its ends. Suppose that the hose is naturally straight and of length L_{tot}. You can make it deviate from this geometry by applying forces and torques with your hands. Consider a little segment of the rod that is initially located a distance s from the end and of length ds. We can describe deformations of the segment by giving three quantities (Figure 9.1):

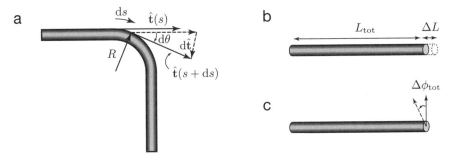

Figure 9.1: (Schematic.) Deformations of a thin elastic rod. (a) Definition of the bend vector, $\boldsymbol{\beta} = d\hat{\mathbf{t}}/ds$, illustrated for a circular segment of a thin rod. The parameter s is the contour length (also called arc length) along the rod. The tangent vector $\hat{\mathbf{t}}(s)$ at one point of the rod has been moved to a nearby point a distance ds away (*dashed arrow*), then compared with the tangent vector there, or $\hat{\mathbf{t}}(s+ds)$. The difference of these vectors, $d\hat{\mathbf{t}}$, points radially inward and has magnitude equal to $d\theta$, or ds/R. (b) Definition of stretch. For a uniformly stretched rod, $u = \Delta L / L_{\text{tot}}$. (c) Definition of twist density. For a uniformly twisted rod, $\omega = \Delta\phi_{\text{tot}}/L_{\text{tot}}$.

- The **stretch** $u(s)$ (or extensional deformation) measures the fractional change in length of the segment: $u = \Delta(ds)/ds$. The stretch is a dimensionless **scalar** (that is, a quantity with no spatial direction).
- The **bend** $\boldsymbol{\beta}(s)$ (or bend deformation) measures how the hose's unit tangent vector $\hat{\mathbf{t}}$ changes as we walk down its length: $\boldsymbol{\beta} = d\hat{\mathbf{t}}/ds$. Thus the bend is a vector with dimensions $\mathbb{L}^{-1}$.
- The **twist density** $\omega(s)$ (or torsional deformation) measures how each succeeding element has been rotated about the hose's axis relative to its neighbor. For example, if you keep the segment straight but twist its ends by a relative angle $d\phi$, then $\omega = d\phi/ds$. Thus the twist density is a scalar with dimensions $\mathbb{L}^{-1}$.

> **Your Turn 9A**
>
> Show that all three of these quantities are independent of the length ds of the small element chosen.

The stretch, bend, and twist density are local (they describe deformations near a particular location, s), but they are related to the overall deformation of the hose. For example, the total **contour length** of the hose (the distance a bug would have to walk to get from one end to the other) equals $\int_0^{L_{\text{tot}}} ds\,(1 + u(s))$. Note that the parameter s gives the contour length of the *unstretched* hose from one end to a given point, so it always runs from 0 to the total unstretched length, L_{tot}, of the rod.

In the context of DNA, we can think of the stretch as measuring how the contour length of a short tract of N basepairs differs from its natural (or "relaxed") value of $(0.34\,\text{nm}) \times N$ (see Figure 2.15 on page 51). We can think of the bend as measuring how each basepair lies in a plane tilted slightly from the plane of its predecessor. To visualize twist density, we first note that the relaxed double helix of DNA in solution makes one complete helical turn about every 10.5 basepairs. Thus we can think of the twist density as measuring the rotation $\Delta\psi$ of one basepair relative to its predecessor, minus the relaxed value of this angle. More precisely,

$$\omega = \frac{\Delta\psi}{0.34\,\mathrm{nm}} - \omega_0, \quad \text{where } \omega_0 = \frac{2\pi}{10.5\,\mathrm{bp}}\frac{1\,\mathrm{bp}}{0.34\,\mathrm{nm}} \approx 1.8\,\mathrm{nm}^{-1}.$$

Following Idea 9.1, we now write down the elastic energy cost E of deforming our cylindrical hose (or any long, thin elastic rod). Again divide the rod arbitrarily into short segments of length ds. Then E should be the sum of terms $dE(s)$ coming from the deformation of the segment at each position s. By analogy to the Hooke relation, we now argue that $dE(s)$ should be a quadratic function of the deformations, if these are small. The most general expression we can write is

$$dE = \frac{1}{2}k_{\mathrm{B}}T\left[A\boldsymbol{\beta}^2 + Bu^2 + C\omega^2 + 2Du\omega\right]ds. \tag{9.2}$$

The phenomenological parameters A, B, and C have dimensions $\mathbb{L}$, $\mathbb{L}^{-1}$, $\mathbb{L}$, respectively; D is dimensionless. The quantities $Ak_{\mathrm{B}}T$ and $Ck_{\mathrm{B}}T$ are called the rod's bend stiffness and twist stiffness at temperature T, respectively. It's convenient to express these quantities in units of $k_{\mathrm{B}}T$, which is why we introduced the **bend persistence length** A and the **twist persistence length** C. The remaining constants $Bk_{\mathrm{B}}T$ and $Dk_{\mathrm{B}}T$ are called the stretch stiffness and twist–stretch coupling, respectively.

It may seem as though we have forgotten some possible quadratic terms in Equation 9.2, for example, a twist–bend cross-term. But the energy must be a scalar, whereas $\boldsymbol{\beta}\omega$ is a vector; terms of this sort have the wrong geometrical status to appear in the energy.

In some cases, we can simplify Equation 9.2 still further. First, many polymers consist of monomers joined by single chemical bonds. The monomers can then rotate about these bonds, destroying any memory of the twist variable and eliminating twist elasticity: $C = D = 0$. In other cases (for example, the one to be studied in Section 9.2), the polymer is free to swivel at one of its attachment points, again leaving the twist variable uncontrolled; then ω again drops out of the analysis. A second simplification comes from the observation that the stretch stiffness $k_{\mathrm{B}}TB$ has the same dimensions as a *force*. If we pull on the polymer with an applied force much less than this value, the corresponding stretch u will be negligible, and we can forget about it, treating the molecule as an **inextensible rod**, that is, a rod having fixed total length. Making both these simplifications leads us to a *one-parameter* phenomenological model of a polymer, with elastic energy

$$E = \frac{1}{2}k_{\mathrm{B}}T\int_0^{L_{\mathrm{tot}}} ds\,A\boldsymbol{\beta}^2. \qquad \text{simplified \textbf{elastic rod model}} \tag{9.3}$$

Equation 9.3 describes a thin, inextensible rod made of a continuous, elastic material. Other authors call it the Kratky–Porod or wormlike chain model (despite the fact that real worms are highly extensible). It is certainly a simple, ultrareductive approach to the complex molecule shown in Figure 2.15! Nevertheless, Section 9.2 will show that it leads to a quantitatively accurate model of the mechanical stretching of DNA.

$\boxed{T_2}$ *Section 9.1.2′ on page 385 mentions some finer points about elasticity models of DNA.*

9.1.3 Polymers resist stretching with an entropic force

The freely jointed chain Section 4.3.1 on page 122 suggested that a polymer could be viewed as a chain of N freely jointed links and that it assumes a random-walk conformation in certain solution conditions. We begin to see how to justify this image when we examine Equation 9.3. Suppose that we bend a segment of our rod into a quarter-circle of radius R (see Figure 9.1 and its caption). Each segment of length ds then bends through an angle d$\theta = $ ds/R, so the bend vector $\boldsymbol{\beta}$ points inward, with magnitude $|\boldsymbol{\beta}| = $ d$\theta/$d$s = R^{-1}$. According to Equation 9.3, the total elastic energy cost of this bend is then one half the bend stiffness, times the circumference of the quarter-circle, times $\boldsymbol{\beta}^2$, or

$$\text{elastic energy cost of a } 90° \text{ bend} = \left(\frac{1}{2}k_\mathrm{B}TA\right) \times \left(\frac{1}{4}2\pi R\right) \times R^{-2} = \frac{\pi A}{4R}k_\mathrm{B}T.$$

(9.4)

The key point about this expression is that it gets smaller with increasing R. That is, a 90° bend can cost as little as we like, provided its radius is big enough. In particular, when R is much bigger than A, then the elastic cost of a bend will be negligible relative to the thermal energy $k_\mathrm{B}T$! In other words,

> *Any elastic rod immersed in a fluid will be randomly bent by thermal motion if its contour length exceeds its bend persistence length A.*

(9.5)

Idea 9.5 tells us that two distant elements will point in random, uncorrelated directions as long as their separation is much greater than A. This observation justifies the name "bend persistence length" for A: Only over separations less than A will the molecule remember which way it was pointing.[1]

A few structural elements in cells are extremely stiff, and so can resist thermal bending (Figure 9.2). But most biopolymers have persistence lengths much shorter than their total length. Because a polymer is rigid on the scale of a monomer, yet flexible on length scales much longer than A, it's reasonable to try the idealization that its conformation is a chain of *perfectly straight* segments, joined by *perfectly free* joints. We take the **effective segment length**, L_seg, to be a phenomenological parameter of the model. (Many authors refer to L_seg as the **Kuhn length**.) We expect L_seg

---

[1]The situation is quite different for two-dimensional elastic objects, for example, membranes. We already found in Section 8.6.1 that the energy cost to bend a patch of membrane into, say, a hemisphere, is $4\pi\kappa$, a constant *independent* of the radius. Hence membranes do *not* rapidly lose their planar character on length scales larger than their thickness.

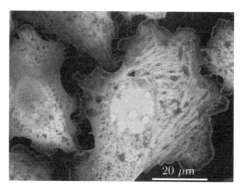

Figure 9.2: (Wet scanning electron micrograph.) Actin bundles in a stained CHO cell. Each bundle has a bend persistence length that is much larger than that of a single actin filament. The bundles are straight, not thermally bent, because their bend persistence length is longer than the cell's diameter. [Digital image kindly supplied by A. Nechushtan and E. Moses.]

to be roughly the same as A; because A is itself an unknown parameter, we lose no predictive power if we instead phrase the model in terms of L_{seg}.[2] We call the resulting model the **freely jointed chain** (or **FJC**). Section 9.2 will show that for DNA, $L_{seg} \approx 100\,\text{nm}$; conventional polymers like polyethylene have much shorter segment lengths, generally less than $1\,\text{nm}$. Because the value of L_{seg} reflects the bend stiffness of the molecule, DNA is often called a "stiff," or **semiflexible**, polymer.

The FJC model is a reduced form of the underlying elastic rod model (Equation 9.3). We will improve its realism later. But it at least incorporates the insight of Idea 9.5, and it will turn out to be mathematically simpler to solve than the full elastic rod model.

In short, we propose to study the conformation of a polymer as a random walk with step size L_{seg}. Before bringing any mathematics to bear on the model, let's first see if we find any qualitative support for it in our everyday experience.

The elasticity of rubber At first sight, the freely jointed chain may not seem like a promising model for polymer elasticity. Imagine pulling on the ends of the chain until it's nearly fully stretched, then releasing it. If you try this with a chain made of paperclips, the chain stays straight after you let go. But a rubber band, which consists of many polymer chains, will *recoil* when stretched and released. What have we missed?

The key difference between a macroscopic chain of paperclips and a polymer is scale: The thermal energy $k_B T$ is negligible for macroscopic paperclips but significant for the nanometer-scale monomers of a macromolecule. Suppose that we pull our paperclip chain out straight, then place it on a vibrating table, where it gets random kicks many times larger than $k_B T$: Its ends *will* spontaneously come closer together as its shape gradually becomes random. Indeed, we would have to place the ends of

---

[2] $\boxed{T_2}$ Section 9.1.3′ on page 386 shows that the precise relation is $L_{seg} = 2A$.

the chain under constant, gentle tension to *prevent* this shortening, just as we must apply a constant force to keep a rubber band stretched.

We can understand the retracting tendency of a stretched polymer by using ideas from Chapters 6 and 7. A long polymer chain of length L_{tot} can consist of hundreds (or millions) of monomers, with a huge number of possible conformations. If there's no external stretching, the vast majority of these conformations are spherelike blobs, with mean-square end-to-end length z much shorter than L_{tot} (see Section 4.3.1 on page 122). The polymer adopts these random-coil conformations because there's *only one way* to be straight but many ways to be coiled up. Thus, if we hold the ends a fixed distance z apart, the entropy decreases when z increases. According to Chapter 7, there must then be an entropic force opposing such stretching. That's why a stretched rubber band spontaneously retracts:

> *The retracting force supplied by a stretched rubber band is entropic in origin.* (9.6)

Thus the retraction of a stretched polymer, which increases disorder, is like the *expansion* of an ideal gas, which also increases disorder and can perform real work (see the heat engine Example, page 214). In either case, what must go down is not the elastic energy E of the polymer but the *free* energy, $F = E - TS$. Even if E *increases* slightly upon bending, still we'll see that the increase in entropy will more than offset the energy increase, driving the system toward the random-coil state. The free energy drop in this process can then be harnessed to do mechanical work, for example, flinging a wad of paper across the room.

Where does the energy to do this work come from? We already encountered some analogous situations while studying thermal machines in Sections 1.2.2, and Problem 6.3. As in those cases, the mechanical work done by a stretched rubber band must be extracted from the *thermal* energy of the surrounding environment. Doesn't the Second Law forbid such a conversion from disordered to ordered energy? No, because the disorder of the polymer molecules themselves increases upon retraction: Rubber bands are free energy transducers. (You'll perform an experiment to confirm this prediction and support the entropic force model of polymer elasticity in Problem 9.4.)

Could we actually build a heat engine based on rubber bands? Absolutely. To implement this idea, first notice a surprising consequence of the entropic origin of polymer elasticity. If the free energy increase upon stretching comes from a decrease in entropy, then the formula $F = E - TS$ implies that the free energy cost of a given extension will depend on the temperature. The tension in a stretched rubber band will thus increase with increasing T. Equivalently, if the imposed tension on the rubber is fixed, then the rubber will *shrink* as we heat it up—its coefficient of thermal expansion is negative, unlike, say, a block of steel.

To make a heat engine exploiting this observation, we need a cyclic process, analogous to the one symbolized by Figure 6.6 on page 216. Figure 9.3 shows one simple strategy.

The remainder of this chapter will develop heavier tools to understand polymers. But this section has a simple point: The ideas of statistical physics, which we have developed mainly in the context of ideal gases, are really of far greater applicability.

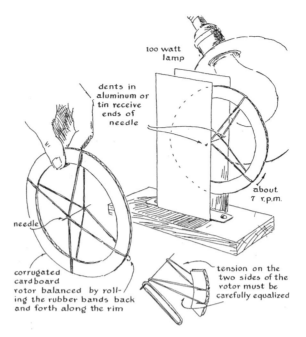

Figure 9.3: (Engineering sketch.) Rubber-band heat engine. The light bulb sequentially heats the rubber bands on one side of the disk, making them contract. The other side is shielded by the sheet metal screen; here the rubber bands cool. The resulting asymmetrical contraction unbalances the wheel, which turns. The turning wheel brings the warm rubber bands into the shaded region, where they cool; at the same time, cool rubber bands emerge into the warm region, making the wheel turn continuously. [From Stong, 1956.]

Even without writing any equations, these ideas have already yielded an immediate insight into a very different-seeming system, one with applications to living cells. Admittedly, your body is not powered by rubber-band heat engines, nor by any other sort of heat engine. Still, understanding the entropic origin of polymer elasticity is important for our goal of understanding cellular mechanics.

$\boxed{T_2}$ Section 9.1.3′ on page 386 gives a calculation showing that the bend stiffness sets the length scale beyond which a fluctuating rod's tangent vectors lose their correlation.

9.2 STRETCHING SINGLE MACROMOLECULES

9.2.1 The force–extension curve can be measured for single DNA molecules

We'll need some mathematics to calculate the free energy $F(z)$ as a function of the end-to-end length z of a polymer chain. Before doing this, let's look at some of the available experimental data.

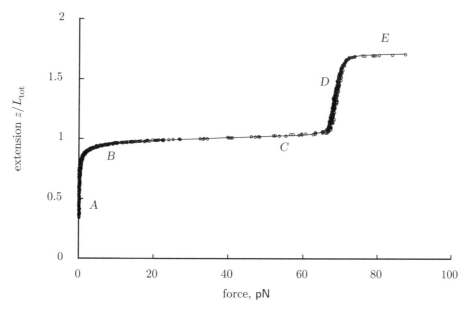

Figure 9.4: (Experimental data with fit.) Force f versus relative extension z/L_{tot} for a DNA molecule made of 10 416 basepairs, in high-salt solution. The regimes labeled *A, B, C, D,* and *E* are described in the text. The extension z was measured by video imaging of the positions of beads attached to each end; the force was measured by using the change of light momentum exiting a dual-beam optical tweezers apparatus (see Section 6.7 on page 226). L_{tot} is the DNA's total contour length in its relaxed state. The quantity z/L_{tot} becomes larger than 1 when the molecule begins to stretch, at around 20 pN. The *solid curve* shows a theoretical model obtained by a combination of the approaches in Sections 9.4.1′ and 9.5.1. [Experimental data kindly supplied by S. B. Smith; theoretical model and fit kindly supplied by C. Storm.]

To get a clear picture, we'd like to pass from pulling on rubber bands, with zillions of entangled polymer chains, to pulling on *individual* polymer molecules with tiny, precisely known forces. S. Smith, L. Finzi, and C. Bustamante accomplished this feat in 1992; a series of later experiments improved both the quality of the data and the range of forces probed, leading to the picture shown in Figure 9.4. Such experiments typically start with a DNA molecule of known length (for example, lambda phage DNA). One end is anchored to a glass slide, the other to a micrometer-sized bead, and the bead is then pulled by optical or magnetic tweezers (see Section 6.7 on page 226).

Figure 9.4 shows five distinct regimes of qualitative behavior as the force on the molecule increases:

A. At very low stretching force, $f < 0.01$ pN, the molecule is still nearly a random coil. Its ends then have a mean-square separation given by Equation 4.4 on page 115 as $L_{seg}\sqrt{N}$. For a molecule with 10 416 basepairs, Figure 9.4 shows that this separation is less than $0.3L_{tot}$, or 1060 nm, so we conclude that $L_{seg}\sqrt{L_{tot}/L_{seg}} < 0.3L_{tot}$, or $L_{seg} < (0.3)^2 L_{tot} \approx 300$ nm. (In fact, L_{seg} will prove to be much smaller than this upper bound—it's closer to 100 nm.)

B. At higher forces, the relative extension begins to level off as it approaches unity. At this point, the molecule has been stretched nearly straight. Sections 9.2.2–9.4.1 will discuss regimes A and B.

C. At forces beyond about 10 pN, the extension actually exceeds the total contour length of the relaxed molecule. Section 9.4.2 will discuss this "intrinsic stretching" phenomenon.

D. At around $f = 65$ pN, we find a remarkable jump, as the molecule suddenly extends to about 1.6 times its relaxed length. Section 9.5.5 briefly discusses this "overstretching transition."

E. Still higher forces again give elastic behavior, until eventually the molecule breaks.

9.2.2 A two-state system qualitatively explains DNA stretching at low force

The freely jointed chain model can help us understand regime A of Figure 9.4. We wish to compute the entropic force f exerted by an elastic rod subjected to thermal motion. This may seem like a daunting prospect. The stretched rod is constantly buffeted by the Brownian motion of the surrounding water molecules, receiving kicks in the directions perpendicular to its axis. Somehow all these kicks pull the ends closer together, maintaining a constant tension if we hold the ends a fixed distance z apart. How could we calculate such a force?

Luckily, our experience with other entropic forces shows how to sidestep a detailed dynamical calculation of each random kick: When the system is in thermal equilibrium, Chapter 7 showed that it's much easier to use the partition function method to calculate entropic forces. To use the method developed in Section 7.1.2, we need to elaborate the deep parallel between the entropic force exerted by a freely jointed chain and that exerted by an ideal gas confined to a cylinder:

• The gas is in thermal contact with the external world, and so is the chain.

• The gas has an external force squeezing it; the chain has an external force pulling it.

• The internal potential energy U_{int} of the gas molecules is independent of the volume. The chain also has fixed internal potential energy—the links are assumed to be free to point in any direction, with no potential-energy cost. In both systems, the kinetic energy is fixed by the ambient temperature, so it too is independent of the constraint. But, in both systems, the potential energy U_{ext} of the mechanism supplying the external force *will* vary.

In the polymer stretching system, U_{ext} goes up as the chain shortens:

$$U_{ext} = \text{const} - fz, \qquad (9.7)$$

where f is the applied external stretching force. The total potential $U_{int} + U_{ext}$ is what we need when computing the system's partition function.

The observations just made simplify our task greatly. Following the strategy leading to Equation 7.5 on page 247, we now calculate the average end-to-end distance of the chain at a given stretching force f directed along the $+\hat{z}$ axis.

In this section, we will work in one dimension for simplicity. (Section 9.2.2′ on page 389 extends the analysis to three dimensions.) Thus each link has a two-state variable σ, which equals $+1$ if the link points forward (along the applied force) or -1 if it points backward (against the force). The total extension z is then the sum of these variables:

$$z = L_{\text{seg}}^{(1d)} \sum_{i=1}^{N} \sigma_i. \tag{9.8}$$

(The superscript "1d" reminds us that this is the effective segment length in the *one-dimensional* FJC model.) The probability of a given conformation $\{\sigma_1, \dots, \sigma_N\}$ is then given by a Boltzmann factor:

$$P(\sigma_1, \dots, \sigma_N) = Z^{-1} e^{-\left(-f L_{\text{seg}}^{(1d)} \sum_{i=1}^{N} \sigma_i\right)/k_B T}. \tag{9.9}$$

Here Z is the partition function (see Equation 6.33 on page 224). The desired average extension is thus the weighted average of Equation 9.8 over all conformations, or

$$\langle z \rangle = \sum_{\sigma_1=\pm 1} \cdots \sum_{\sigma_N=\pm 1} P(\sigma_1, \dots, \sigma_N) \times z$$

$$= Z^{-1} \sum_{\sigma_1=\pm 1} \cdots \sum_{\sigma_N=\pm 1} e^{-\left(-f L_{\text{seg}}^{(1d)} \sum_{i=1}^{N} \sigma_i\right)/k_B T} \times \left(L_{\text{seg}}^{(1d)} \sum_{i=1}^{N} \sigma_i \right)$$

$$= k_B T \frac{d}{df} \ln \left[\sum_{\sigma_1=\pm 1} \cdots \sum_{\sigma_N=\pm 1} e^{-\left(-f L_{\text{seg}}^{(1d)} \sum_{i=1}^{N} \sigma_i\right)/k_B T} \right].$$

This looks like a formidable formula, until we notice that the argument of the logarithm is just the product of N independent, identical factors:

$$\langle z \rangle = k_B T \frac{d}{df} \ln \left[\left(\sum_{\sigma_1=\pm 1} e^{f L_{\text{seg}}^{(1d)} \sigma_1/k_B T} \right) \times \cdots \times \left(\sum_{\sigma_N=\pm 1} e^{f L_{\text{seg}}^{(1d)} \sigma_N/k_B T} \right) \right]$$

$$= k_B T \frac{d}{df} \ln \left(e^{f L_{\text{seg}}^{(1d)}/k_B T} + e^{-f L_{\text{seg}}^{(1d)}/k_B T} \right)^N$$

$$= N L_{\text{seg}}^{(1d)} \frac{e^{f L_{\text{seg}}^{(1d)}/k_B T} - e^{-f L_{\text{seg}}^{(1d)}/k_B T}}{e^{f L_{\text{seg}}^{(1d)}/k_B T} + e^{-f L_{\text{seg}}^{(1d)}/k_B T}}.$$

Recalling that $N L_{\text{seg}}^{(1d)}$ is just the total length L_{tot}, we have shown that

$$\boxed{\langle z/L_{\text{tot}} \rangle = \tanh(f L_{\text{seg}}^{(1d)}/k_B T). \quad \text{force versus extension for the 1d FJC}} \tag{9.10}$$

> **Your**   If you haven't yet worked Problem 6.5, do it now. Explain why this is mathe-
> **Turn**   matically the same problem as the one we just solved.
> **9B**

Solving Equation 9.10 for f shows that *the force needed to maintain a given extension z is proportional to the absolute temperature.* This property is the hallmark of any purely entropic force, for example, ideal-gas pressure or osmotic pressure; we anticipated it in Section 9.1.3.

The function in Equation 9.10 interpolates between two important limiting behaviors:

- At high force, $\langle z \rangle \to L_{\text{tot}}$. This behavior is what we expect from a flexible but inextensible rod: Once it's fully straight, it can't lengthen any more.
- At low force, $\langle z \rangle \to f/k$, where $k = k_{\text{B}}T/(L_{\text{tot}}L_{\text{seg}}^{(1d)})$.

The second point means that

> At low extension, a polymer behaves as a spring, that is, it obeys a
> Hooke relation, $f = k\langle z \rangle$. In the FJC model, the effective spring con-   (9.11)
> stant k is proportional to the temperature.

Figure 9.5 shows experimental data obtained by stretching DNA, together with the function in Equation 9.10 (top curve). The figure shows that taking $L_{\text{seg}}^{(1d)} = 35\,\text{nm}$ makes the curve pass through the first data point. Although the one-dimensional freely jointed chain correctly captures the qualitative features of the data, clearly it's not in good quantitative agreement throughout the range of forces shown. That's hardly surprising in the light of our rather crude mathematical treatment of the underlying physics of the elastic rod model. The following sections will improve the analysis, eventually showing that the simplified elastic rod model (Equation 9.3) gives a very good account of the data (see the black curve in Figure 9.5).

$\boxed{T_2}$ *Section 9.2.2′ on page 389 works out the three-dimensional freely jointed chain.*

9.3 EIGENVALUES FOR THE IMPATIENT

Section 9.4 will make use of some mathematical ideas not always covered in first-year calculus. Luckily, for our purposes only a few facts will be sufficient. Many more details are available in Shankar, 1995.

9.3.1 Matrices and eigenvalues

As always, it's best to approach this abstract subject through a familiar example. Look back at our force diagram for a thin rod being dragged through a viscous fluid (Figure 5.8 on page 175). Suppose, as shown in the figure, that the axis of the rod points in the direction $\hat{\mathbf{t}} = (\hat{\mathbf{x}} - \hat{\mathbf{z}})/\sqrt{2}$; let $\hat{\mathbf{n}} = (\hat{\mathbf{x}} + \hat{\mathbf{z}})/\sqrt{2}$ be the perpendicular unit vector.

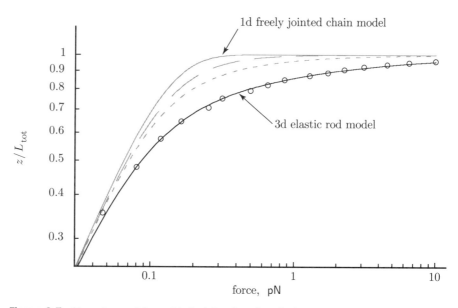

Figure 9.5: (Experimental data with fits.) Log-log plot of relative extension z/L_{tot} at low applied stretching force f for lambda phage DNA in 10 mM phosphate buffer. The points show experimental data corresponding to the regimes A–B in Figure 9.4. The curves show various theoretical models discussed in the text. For comparison, the value of L_{seg} has been adjusted in each model so that all the curves agree at low force. *Top curve:* One-dimensional freely jointed chain (Equation 9.10), with $L_{\mathrm{seg}}^{(1\mathrm{d})} = 35\,\mathrm{nm}$. *Long-dash curve:* One-dimensional cooperative chain (see Your Turn 9H(b)), with $L_{\mathrm{seg}}^{(1\mathrm{d})}$ held fixed at 35 nm and γ very large. *Short-dash curve:* Three-dimensional FJC (Your Turn 9O), with $L_{\mathrm{seg}} = 104\,\mathrm{nm}$. *Black curve through data points:* Three-dimensional elastic rod model (Section 9.4.1′ on page 390), with $A = 51\,\mathrm{nm}$. [Data kindly supplied by V. Croquette; see also Bouchiat et al., 1999.]

Section 5.3.1 stated that the drag force will be parallel to the velocity $\mathbf{v}$ if $\mathbf{v}$ is directed along either $\hat{\mathbf{t}}$ or $\hat{\mathbf{n}}$, but that the viscous friction coefficients in these two directions, $\zeta_\perp$ and $\zeta_\|$, are not equal: The parallel drag is typically $\frac{2}{3}$ as great as $\zeta_\perp$. For intermediate directions, we get a linear combination of a parallel force proportional to the parallel part of the velocity, plus a perpendicular force proportional to the perpendicular part of the velocity:

$$\mathbf{f} = \zeta_\| \hat{\mathbf{t}}(\hat{\mathbf{t}} \cdot \mathbf{v}) + \zeta_\perp \hat{\mathbf{n}}(\hat{\mathbf{n}} \cdot \mathbf{v}) = \zeta_\perp\!\left(\tfrac{2}{3}\hat{\mathbf{t}}(\hat{\mathbf{t}} \cdot \mathbf{v}) + \hat{\mathbf{n}}(\hat{\mathbf{n}} \cdot \mathbf{v})\right). \qquad (9.12)$$

This formula is indeed a linear function of v_x and v_z, the components of $\mathbf{v}$:

Your Turn 9C Use the preceding expressions for $\hat{\mathbf{t}}$ and $\hat{\mathbf{n}}$ to show that

$$\begin{bmatrix} f_x \\ f_z \end{bmatrix} = \zeta_\perp \begin{bmatrix} (\tfrac{1}{3} + \tfrac{1}{2})v_x + (-\tfrac{1}{3} + \tfrac{1}{2})v_z \\ (-\tfrac{1}{3} + \tfrac{1}{2})v_x + (\tfrac{1}{3} + \tfrac{1}{2})v_z \end{bmatrix}.$$

Expressions of this form arise so frequently that we introduce an abbreviation:

$$\left[\begin{array}{c} f_x \\ f_z \end{array} \right] = \zeta_\perp \left[\begin{array}{cc} (\frac{1}{3} + \frac{1}{2}) & (-\frac{1}{3} + \frac{1}{2}) \\ (-\frac{1}{3} + \frac{1}{2}) & (\frac{1}{3} + \frac{1}{2}) \end{array} \right] \left[\begin{array}{c} v_x \\ v_z \end{array} \right]. \tag{9.13}$$

Even though Equation 9.13 is nothing but an abbreviation for the formula above it, let's pause to put it in a broader context. *Any* linear relation between two vectors can be written as $\mathbf{f} = \mathsf{M}\mathbf{v}$, where the symbol M denotes a **matrix**, or rectangular array of numbers. In our example we are interested in only two directions $\hat{\mathbf{x}}$ and $\hat{\mathbf{z}}$, so our matrix is two-by-two:

$$\mathsf{M} \equiv \left[\begin{array}{cc} M_{11} & M_{12} \\ M_{21} & M_{22} \end{array} \right].$$

Thus the symbol M_{ij} denotes the entry in row i and column j of the matrix. Placing a matrix to the left of a vector, as in Equation 9.13, denotes an operation where we successively read across the rows of M, multiplying each entry we find by the corresponding entry of the vector $\mathbf{v}$ and adding to obtain the successive components of $\mathbf{f}$:

$$\mathsf{M}\mathbf{v} \equiv \left[\begin{array}{c} M_{11}v_1 + M_{12}v_2 \\ M_{21}v_1 + M_{22}v_2 \end{array} \right]. \tag{9.14}$$

The key question is now: Given a matrix M, *what are the special directions of* $\mathbf{v}$ (if any) that get transformed to vectors parallel to themselves under the operation symbolized by M? We already know the answer for the example in Equation 9.13: We constructed this matrix to have the special axes $\hat{\mathbf{t}}$ and $\hat{\mathbf{n}}$, with corresponding viscous friction coefficients $\frac{2}{3}\zeta_\perp$ and $\zeta_\perp$, respectively. More generally, though, we may not be given the special directions in advance, and there may not even be any. The special directions of a matrix M, if any, are called its **eigenvectors**; the corresponding multipliers are called the **eigenvalues**.[3] Let's see how to work out the special directions, and their eigenvalues, for a general 2×2 matrix.

Consider the matrix $\mathsf{M} = \left[\begin{smallmatrix} a & b \\ c & d \end{smallmatrix} \right]$. We want to know whether there is any vector $\mathbf{v}_*$ that turns into a multiple of itself after transformation by M:

$$\boxed{\mathsf{M}\mathbf{v}_* = \lambda\mathbf{v}_*. \qquad \text{eigenvalue equation}} \tag{9.15}$$

The notation on the right-hand side means that we multiply each entry of the vector $\mathbf{v}_*$ by the same constant λ. Equation 9.15 is actually *two* equations, because each side is a vector with two components (see Equation 9.14).

How can we solve Equation 9.15 without knowing in advance the value of λ? To answer this question, first note that there's always *one* solution, no matter what value

---

[3]Like "liverwurst," this word is a combination of the German *eigen* ("proper") and an English word. The term expresses the fact that the eigenvalues are intrinsic to the linear transformation represented by M. In contrast, the entries M_{ij} themselves *change* when we express the transformation in some other coordinate system.

we take for λ, namely, $\mathbf{v}_* = \begin{bmatrix} 0 \\ 0 \end{bmatrix}$. This is a boring solution. Regarding Equation 9.15 as two equations in the two unknowns v_1 and v_2, in general, we expect just one solution; in other words, *the eigenvalue equation, Equation 9.15, will in general have only the boring (zero) solution.* But for certain special values of λ, we may find a second, interesting solution after all. This requirement is what determines λ.

We are looking for solutions to the eigenvalue equation (Equation 9.15 with M = $\begin{bmatrix} a & b \\ c & d \end{bmatrix}$) in which v_1 and v_2 are not both zero. Suppose that $v_1 \neq 0$. Then we can divide both sides of the eigenvalue equation by v_1 and seek a solution of the form $\begin{bmatrix} 1 \\ \omega \end{bmatrix}$. The first of the two equations represented by Equation 9.15 then says that $a + \omega b = \lambda$, or $b\omega = \lambda - a$. The second equation says that $c + d\omega = \lambda\omega$. Multiplying by b and substituting the first equation lets us eliminate ω altogether, finding

$$bc = (\lambda - a)(\lambda - d). \qquad \text{(condition for } \lambda \text{ to be an eigenvalue)} \qquad (9.16)$$

Thus only for certain special values of λ—the eigenvalues—will we find any nonzero solution to Equation 9.15. The solutions are the desired eigenvectors.

Your Turn 9D

a. Apply Equation 9.16 to the matrix appearing in the frictional drag problem (Equation 9.13). Find the eigenvalues, and the corresponding eigenvectors, and confirm that they're what you expect for this case.

b. For some problems, it's possible that v_1 may be zero; in this case, we can't divide through by it. Repeat the preceding argument, this time assuming that $v_2 \neq 0$, and recover the same condition as Equation 9.16.

c. It's possible that Equation 9.16 will have no real solutions. Show that it will always have two real solutions if $bc \geq 0$.

d. Show that, furthermore, the two eigenvalues will be different (not equal to each other) if $bc > 0$.

Your Turn 9E

Continuing the previous problem, consider a *symmetric* 2×2 matrix, that is, one with $M_{12} = M_{21}$. Show that

a. It always has two real eigenvalues.

b. The corresponding eigenvectors are perpendicular to each other, if the two eigenvalues are not equal.

9.3.2 Matrix multiplication

Here is another concrete example. Consider the operation that takes a vector $\mathbf{v}$, rotates it through an angle α, and stretches or shrinks its length by a factor g. You can show that this operation is linear, that its matrix representation is $R(\alpha, g) = \begin{bmatrix} g\cos\alpha & g\sin\alpha \\ -g\sin\alpha & g\cos\alpha \end{bmatrix}$, and that it has *no* real eigenvectors (why not?).

Suppose we apply the operation R to a vector *twice*.

**Your
Turn
9F**
> **a.** Evaluate M(N**v**) for two arbitrary 2 × 2 matrices M and N. (That is, apply
> Equation 9.14 twice.) Show that your answer can be rewritten as Q**v**, where
> Q is a new matrix called the **product** N and M, or simply MN. Find Q.
> **b.** Evaluate the matrix product R(α, g)R(β, h), and show that it too can be
> written as a certain combination of rotation and scaling. That is, express it
> as R(γ, c) for some γ and c. Find γ and c and explain why your answers
> make sense.

$\boxed{T_2}$ *Section 9.3.2' on page 390 sketches the generalizations of some of the preceding
results to higher-dimensional spaces.*

9.4 COOPERATIVITY

9.4.1 The transfer matrix technique allows a more accurate treatment of bend cooperativity

Section 9.2.2 gave a provisional analysis of DNA stretching. To begin improving it,
let's recall some of the simplifications made so far:

- We treated a continuous elastic rod as a chain of perfectly stiff segments, joined by
perfectly free joints.
- We treated the freely jointed chain as being one-dimensional (Section 9.2.2' on
page 389 discusses the three-dimensional case).
- We ignored the fact that a real rod cannot pass through itself.

This section will consider the first of these oversimplifications.[4] Besides yielding a
slight improvement in our fit to the experimental data, the ideas of this section have
broader ramifications and go to the heart of this chapter's Focus Question.

Clearly, it would be better to model the chain, not as N segments with free joints,
but as, say, $2N$ shorter segments with some "peer pressure," a preference for neighboring units to point in the same direction. We'll refer to such an effect as a cooperative
coupling (or simply as **cooperativity**). In the context of DNA stretching, cooperativity is a surrogate for the physics of bending elasticity, but later we'll extend the concept to include other phenomena as well. To keep the mathematics simple, let's begin
by constructing and solving a one-dimensional version of this idea, which we'll call

---

[4] $\boxed{T_2}$ Section 9.4.1' on page 390 will tackle the first two together. Problem 7.9 discussed the effects of self-
avoidance; it's a minor effect for a stiff polymer (like DNA) under tension. The discussion in this section
will introduce yet another simplification, taking the rod to be infinitely long. Section 9.5.2 will illustrate
how to introduce finite-length effects.

the **1d cooperative chain model**. Section 9.5.1 will show that the mathematics of the one-dimensional cooperative chain is also applicable to another class of problems, the helix–coil transitions in polypeptides and DNA.

Just as in Section 9.2.2, we introduce N two-state variables σ_i, describing links of length ℓ. Unlike the FJC, however, the chain itself has an internal elastic potential energy U_{int}: When two neighboring links point in opposite directions ($\sigma_i = -\sigma_{i+1}$), we suppose that they contribute an extra $2\gamma k_B T$ to this energy, relative to when they point in parallel. We can implement this idea by introducing the term $-\gamma k_B T \sigma_i \sigma_{i+1}$ into the energy function; this term equals $\pm\gamma k_B T$, depending on whether the neighboring links agree or disagree. Adding contributions from all the pairs of neighboring links gives

$$U_{int}/k_B T = -\gamma \sum_{i=1}^{N-1} \sigma_i \sigma_{i+1}, \qquad (9.17)$$

where γ is a new, dimensionless phenomenological parameter (the **cooperativity parameter**). We are assuming that only next-door neighbor links interact with each other. The effective link length ℓ need not equal the FJC effective link length $L_{seg}^{(1d)}$; again we will find the appropriate ℓ by fitting the model to data.

We can again evaluate the extension $\langle z \rangle$ as the derivative of the free energy, computed by the partition function method (Equation 7.6 on page 248). Let $\alpha \equiv f\ell/k_B T$, a dimensionless measure of the energy term biasing each segment to point forward. With this abbreviation, the partition function is

$$Z(\alpha) = \sum_{\sigma_1=\pm1} \cdots \sum_{\sigma_N=\pm1} \left[e^{\alpha \sum_{i=1}^{N} \sigma_i + \gamma \sum_{i=1}^{N-1} \sigma_i \sigma_{i+1}} \right]. \qquad (9.18)$$

The first term in the exponential corresponds to the contribution U_{ext} to the total energy from the external stretching. We need to compute

$$\langle z \rangle = k_B T \frac{d}{df} \ln Z(f) = \ell \frac{d}{d\alpha} \ln Z(\alpha).$$

To make further progress, we must evaluate the summations in Equation 9.18. Sadly, the trick we used for the FJC doesn't help us this time: The coupling between neighboring links spoils the factorization of Z into N identical, simple factors. Nor can we have recourse to a mean-field approximation like the one that saved us in Section 7.4.3 on page 264. Happily, though, the physicists H. Kramers and G. Wannier found a beautiful end run around this problem in 1941. Kramers and Wannier were studying *magnetism*, not polymers. They imagined a chain of atoms, each a small permanent magnet that could point its north pole either parallel or perpendicular to an applied magnetic field. Each atom feels not only the applied field (analogous to the α term of Equation 9.18) but also the field of its nearest neighbors (the γ term). In a magnetic material like steel, the coupling tends to align neighboring atoms ($\gamma > 0$), just as in a stiff polymer the bending elasticity has the same effect. The fact that the solution to the magnet problem also solves interesting problems involving polymers is a beautiful example of the broad applicability of simple physical ideas.[5]

Suppose that there were just two links. Then the partition function Z_2 consists of just four terms; it's a sum over the two possible values for each of σ_1 and σ_2.

Your Turn 9G

a. Show that this sum can be written compactly as the matrix product $Z_2 = \mathbf{V} \cdot (\mathbf{TW})$, where $\mathbf{V}$ is the vector $\begin{bmatrix} e^\alpha \\ e^{-\alpha} \end{bmatrix}$, $\mathbf{W}$ is the vector $\begin{bmatrix} 1 \\ 1 \end{bmatrix}$, and $\mathbf{T}$ is the 2×2 matrix with entries

$$\mathsf{T} = \begin{bmatrix} e^{\alpha+\gamma} & e^{-\alpha-\gamma} \\ e^{\alpha-\gamma} & e^{-\alpha+\gamma} \end{bmatrix}. \tag{9.19}$$

b. Show that for N links, the partition function equals $Z_N = \mathbf{V} \cdot (\mathsf{T}^{N-1}\mathbf{W})$.

c. $\boxed{T_2}$ Show that for N links the average value of the middle link variable is $\langle \sigma_{N/2} \rangle = \left(\mathbf{V} \cdot \mathsf{T}^{(N-2)/2} \begin{pmatrix} 1 & 0 \\ 0 & -1 \end{pmatrix} \mathsf{T}^{N/2}\mathbf{W} \right) / Z_N$.

Just as in Equation 9.14, the notation in Your Turn 9G(a) is a shorthand way to write

$$Z_2 = \sum_{i=1}^{2} \sum_{j=1}^{2} V_i T_{ij} W_j,$$

where T_{ij} is the element in row i and column j of Equation 9.19. The matrix T is called the **transfer matrix** of our statistical problem.

Your Turn 9G(b) gives us an almost magical resolution to our difficult mathematical problem. To see this, we first notice that T has two eigenvectors, because its off-diagonal elements are both positive (see Your Turn 9D(c)). Let's call these eigenvectors $\mathbf{e}_\pm$ and their corresponding eigenvalues $\lambda_\pm$. Thus $\mathsf{T}\mathbf{e}_\pm = \lambda_\pm \mathbf{e}_\pm$.

Any other vector can be expanded as a combination of $\mathbf{e}_+$ and $\mathbf{e}_-$; for example, $\mathbf{W} = w_+ \mathbf{e}_+ + w_- \mathbf{e}_-$. We then find that

$$Z_N = p(\lambda_+)^{N-1} + q(\lambda_-)^{N-1}, \tag{9.20}$$

where $p = w_+ \mathbf{V} \cdot \mathbf{e}_+$ and $q = w_- \mathbf{V} \cdot \mathbf{e}_-$. This is a big simplification. It gets better when we realize that for very large N, we can forget about the second term of Equation 9.20, because one eigenvalue will be bigger than the other (Your Turn 9D(d)), and when raised to a large power, the bigger one will be *much* bigger. Moreover, we don't even need the numerical value of p: You are about to show that we need $N^{-1} \ln Z_N$, which equals $\ln \lambda_+ + N^{-1} \ln(p/\lambda_+)$. The second term is small when N is large.

---

[5] Actually, an ordinary magnet is a *three*-dimensional array of coupled spins, not a one-dimensional chain. The exact mathematical solution of the corresponding statistical physics problem remains unknown to this day.

Now finish the derivation:

Your Turn 9H

> a. Show that the eigenvalues are $\lambda_{\pm} = e^{\gamma} \left[\cosh \alpha \pm \sqrt{\sinh^2 \alpha + e^{-4\gamma}} \right]$.
>
> b. Adapt the steps leading to Equation 9.10 on page 353 to find $\langle z/L_{\text{tot}} \rangle$ as a function of f in the limit of large N.
>
> c. Check your answer by setting $\gamma \to 0$, $\ell \to L_{\text{seg}}^{(1d)}$, and showing that you recover the result of the FJC, Equation 9.10.

As always, it's interesting to check the behavior of your solution at very low force ($\alpha \to 0$). We again find that $\langle z \rangle \to f/k$, where now the spring constant is

$$ k = k_B T / (e^{2\gamma} \ell L_{\text{tot}}). \tag{9.21} $$

So at least we have not spoiled the partial success we had with the FJC: The low-force limit of the extension, where the FJC was successful, has the same form in the cooperative chain model, as long as we choose ℓ and γ to satisfy $\ell e^{2\gamma} = L_{\text{seg}}^{(1d)}$. We now ask whether the cooperative chain model can do a better job than the FJC of fitting the data at the *high*-force end.

The dashed curve in Figure 9.5 shows the function you found in Your Turn 9H. The cooperativity γ has been taken very large, while holding fixed $L_{\text{seg}}^{(1d)}$. The graph shows that the cooperative one-dimensional chain indeed does a somewhat better job of representing the data than the FJC.

Our 1d cooperative chain model is still not very realistic. The lowest curve on the graph shows that the three-dimensional cooperative chain (that is, the elastic rod model, Equation 9.3 on page 346) gives a very good fit to the data. This result is a remarkable vindication of the highly reductionist model of DNA as a uniform elastic rod. Adjusting just one phenomenological parameter (the bend persistence length A) gives a quantitative account of the relative extension of DNA, a very complex object (see Figure 2.15 on page 51). This success makes sense in the light of the discussion in Section 9.1.1: It is a consequence of the large difference in length scales between the typical thermal bending radius ($\approx 100\,\text{nm}$) and the diameter of DNA ($2\,\text{nm}$).

$\boxed{T_2}$ *Section 9.4.1' on page 390 works out the force–extension relation for the full, three-dimensional elastic rod model.*

9.4.2 DNA also exhibits linear stretching elasticity at moderate applied force

We have arrived at a reasonable understanding of the data in the low- to moderate-force regimes A and B shown in Figure 9.4. Turning to regime C, we see that at high force, the curve doesn't really flatten out as predicted by the inextensible rod model. Rather, the DNA molecule is actually *elongating*, not just straightening. In other words, an external force can induce structural rearrangements of the atoms in a macromolecule. We might have expected such a result—we arrived at the simple

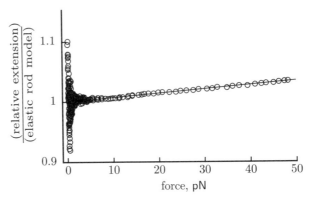

Figure 9.6: (Experimental data with fit.) Linear plot of the stretching of DNA in regime C of Figure 9.4. A DNA molecule with 38 800 basepairs was stretched with optical tweezers, in a buffer solution with pH 8.0. For each value of the force, the ratio of the observed relative extension and the prediction of the inextensible elastic rod model is plotted. The fact that this ratio is a linear function of applied force implies that the molecule has a simple elastic stretching response to the applied force. The *solid line* is a straight line through the point (0 pN, 1), with fitted slope $1/(Bk_{\mathrm{B}}T_{\mathrm{r}}) = 1/(1400\,\mathrm{pN})$. [Data kindly supplied by M. D. Wang; see Wang et al., 1997.]

model in Equation 9.3 on page 346 in part by neglecting the possibility of stretching, that is, by discarding the second term of Equation 9.2 on page 346. Now it's time to reinstate this term and in so doing, formulate an **extensible rod** model, due to T. Odijk.

To approximate the effects of this intrinsic stretching, we note that the applied force now has two effects: Each element of the chain aligns as before, but now each element also lengthens slightly. The relative extension is a factor of $1 + u$, where u is the stretch defined in Section 9.1.2 on page 344. Consider a *straight* segment of rod, initially of contour length Δs. Under an applied stretching force f, the segment will lengthen by $u \times \Delta s$, where u takes the value that minimizes the energy function $k_{\mathrm{B}}T[\frac{1}{2}Bu^2\,\Delta s - fu\,\Delta s]$ (see Equation 9.2). Performing the minimization gives $u = f/(k_{\mathrm{B}}TB)$. We will make the approximation that this formula holds also for the full fluctuating rod. In this approximation each segment of the rod again lengthens by a relative factor of $1 + u$, so $\langle z/L_{\mathrm{tot}}\rangle$ equals the inextensible elastic rod chain result, multiplied by $1 + \big(f/(k_{\mathrm{B}}TB)\big)$.

Figure 9.6 shows some experimental data on the stretching of DNA at moderate forces. Intrinsic stretching is negligible at low force, so the low-force data were first fit to the inextensible rod model, as shown in Figure 9.5. Next, all the extension data were divided by the corresponding points obtained by extrapolating the inextensible rod model to higher forces (corresponding to regime C of Figure 9.4). According to the previous paragraph, this residual extension should be a linear function of f — and the graph confirms this prediction. The slope lets us read off the value of the stretch stiffness as $Bk_{\mathrm{B}}T_{\mathrm{r}} \approx 1400\,\mathrm{pN}$ for DNA under the conditions of this particular experiment.

9.4.3 Cooperativity in higher-dimensional systems gives rise to infinitely sharp phase transitions

Equation 9.21 shows that the force-induced straightening transition becomes very sharp (the effective spring constant k becomes small) when γ is big. That is, cooperativity, a local interaction between neighbors in a chain, increases the sharpness of a global transition.

Actually, we are already familiar with cooperative transitions in our everyday, three-dimensional, life. Suppose that we take a beaker of water, carefully maintain it at a fixed, uniform temperature, and allow it to come to equilibrium. Then the water will be either all liquid or all solid ice, depending on whether the temperature is greater or less than 0°C. This sharp transition can again be regarded as a consequence of cooperativity. The interface between liquid water and ice has a surface tension, a free energy cost for introducing a *boundary* between these two phases, just as the parameter 2γ in the polymer stretching transition is the cost to create a boundary between a forward-directed domain and one pointing against the applied force. This cost disfavors a mixed water/ice state, making the water–ice transition discontinuous (infinitely sharp).

In contrast to the water/ice system, you found in Your Turn 9H(b) that the straightening of the one-dimensional FJC by applied tension is never discontinuous, no matter how large γ may be. We say that the freezing of water is a true **phase transition** but that such transitions are impossible in one-dimensional systems with local interactions.

We can understand qualitatively why the physics of a cooperative, one-dimensional chain is so different from analogous systems in three dimensions. Suppose that the temperature in your glass of water is just slightly below 0°C. There will certainly be occasional thermal fluctuations converting *small* pockets of the ice to water. But the probability of such a fluctuation is suppressed, both by the free energy difference between bulk ice and water and by the surface tension energy, which grows with the area of the small pockets of water. In one dimension, in contrast, the boundary of a domain of the energetically disfavored state is always just *two points*, no matter how large that domain may be. It turns out that this minor-seeming difference is enough to assure that in one dimension, a nonzero fraction of the sample will always be in the energetically disfavored state—the transition is never quite complete, just as in our polymer, $\langle z/L_{\text{tot}}\rangle$ is never quite equal to 1.

9.5 THERMAL, CHEMICAL, AND MECHANICAL SWITCHING

Section 9.4 introduced a conceptual framework—cooperativity—for understanding sharp transitions in macromolecules induced by externally applied forces. We saw how cooperativity sharpens the transition from random-coil to straight DNA. We found a simple interpretation of the effect in terms of a big increase in the effective segment length as we turn on cooperativity, from ℓ to $\ell e^{2\gamma}$ (see Equation 9.21).

Some important conformational transitions in macromolecules really are induced by mechanical forces. For example, the hair cells in your inner ear respond to pressure waves by a mechanically actuated ion channel. Understanding how such transitions can be sharp, despite the thermally fluctuating environment, was a major goal of this chapter. But other macromolecules function by undergoing conformational transitions in response to *chemical or thermal* changes. This section will show how these transitions, too, can become sharp by virtue of their cooperativity.

9.5.1 The helix–coil transition can be observed by using polarized light

A protein is a polymer; its monomers are the amino acids. Unlike DNA, whose large charge density gives it a uniform self-repulsion, the amino acid monomers of a protein have a rich variety of attractive and repulsive interactions. These interactions can stabilize definite protein structures.

For example, certain sequences of amino acids can form a right-handed helical structure, the alpha helix (Figure 2.17 on page 53). In this structure, the free energy gain of forming hydrogen bonds between monomers outweighs the entropic tendency of a chain to assume a random walk conformation. Specifically, H-bonds can form between the oxygen atom in the carbonyl group of monomer k and the hydrogen atom in the amide group on monomer $k + 4$, but only if the chain assumes the helical shape shown in Figure 2.17.[6]

Thus the question of whether a given polypeptide will assume a random coil or an alpha helix (ordered) conformation comes down to a competition between conformational entropy and H-bond formation. Which side wins this competition will depend on the polypeptide's composition, and on its thermal and chemical environment. The crossover between helix and coil as the environment changes can be surprisingly sharp, with nearly total conversion of a sample from one form to the other upon a temperature change of just a few degrees (see Figure 9.7). (To see why this is considered "sharp," recall that a change of a few degrees implies a *fractional* change of the thermal energy of a few degrees divided by 295 K.)

We can monitor conformational changes in polypeptides without having to look at them individually; instead, we look at changes in the bulk properties of their solutions. When studying the helix–coil transition, the most telling of these changes involves the solution's effect on polarized light.

Suppose that we pass a beam of polarized light rays through a suspension of perfectly spherical particles in water. The light will be scattered: It loses some of its perfect uniformity in direction and polarization, emerging with a slightly lower *degree* of polarization. This loss of purity can tell us something about the density of suspended particles. What we will not find, however, is any net rotation in the *direction* of polarization of the light. We can understand this important fact via a symmetry argument.

Suppose that our suspension rotated the angle of polarized light by an angle θ (Figure 9.8). Imagine a second solution, in which every atom of the first has been reflected through a mirror. Every particle of the second solution is just as stable as those in the first because the laws of atomic physics are invariant under reflection. And the second solution will rotate the polarization of incident light by $-\theta$, that is,

---

[6]Other ordered, H-bonded structures exist, for example, the beta sheet; this section will study only polypeptides whose main competing conformations are the alpha helix and random coil.

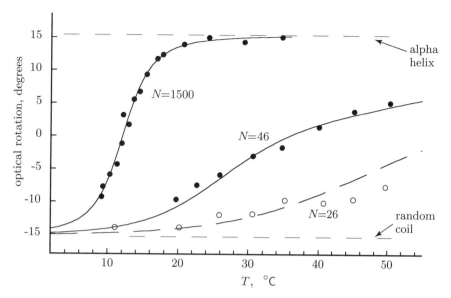

Figure 9.7: (Experimental data with fit.) Alpha helix formation as a function of temperature for solutions of poly-[γ-benzyl-L-glutamate] (an artificial polypeptide), dissolved in a mixture of dichloroacetic acid and ethylene dichloride. At low temperature, all samples displayed an optical rotation similar to that of isolated monomers; at high temperatures, the rotation changed, thus indicating alpha helix formation. *Top dots:* Polymer chains of weight-average length equal to 1500 monomers. *Middle dots:* 46-monomer chains. *Lower circles:* 26-monomer chains. The vertical axis gives the optical rotation; this value is linearly related to the fraction of all monomers in the helical conformation. *Top solid curve:* The large-N formula (Equations 9.25 and 9.24) obtained by fitting the values of the five parameters ΔE, T_m, γ, C_1, and C_2 to the experimental data. The lower two curves are then predictions of the model (see Sections 9.5.3 on page 369 and 9.5.3' on page 394), with no further fitting done. [Experimental data from Zimm et al., 1959.]

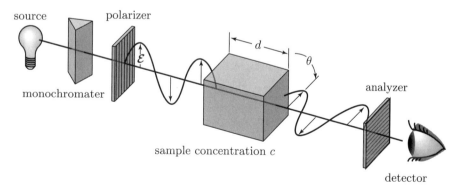

sample concentration c

Figure 9.8: (Schematic.) A polarimeter. The *arrows* represent the electric field vector $\mathcal{E}$ in a ray of light emerging from the source. They are shown rotating by an angle θ as the light passes through the sample; the rotation shown corresponds to the positive value $\theta = +\pi/2$. By convention, the plus sign means that an observer looking into the oncoming beam sees the electric field rotating in the clockwise direction as the beam advances through the medium. Try looking at this figure in a mirror to see that the optical rotation changes sign. [Adapted from Eisenberg & Crothers, 1979.]

opposite to the first solution's rotation. But, because a spherical object is *unchanged* upon reflection, then so is a random distribution (a suspension) of such objects. So we can also conclude that both solutions have the *same* value of θ. The only way that the second suspension could have an optical rotation that is equal to θ and to $-\theta$ is for θ to be zero, as claimed in the preceding paragraph.

Now consider a suspension of identical, but not necessarily spherical, molecules. If each molecule is equivalent to its mirror image (as is true of water, for example), then the argument just given again implies that $\theta = 0$—and that's what we observe in water. But most biological molecules are *not* equivalent to their mirror images—they are said to be **chiral** (see Figure 1.5 on page 25). To drive the point home, it's helpful to get a corkscrew and find its handedness, following the caption to Figure 2.17 on page 53. Next, look at the same corkscrew (or Figure 2.17) in a mirror and discover that its mirror image has the opposite handedness.[7] The two shapes are genuinely inequivalent: You cannot make the mirror image coincide with the original by turning the corkscrew end-over-end, nor by any other kind of rotation.

Solutions of chiral molecules really do rotate the polarization of incident light. Most interesting for our present purposes, a *single* chemical species may have different conformations with differing *degrees* of chirality (reflection asymmetry). Thus, whereas the individual amino acids of a protein may individually be chiral, the protein's ability to rotate polarized light at certain wavelengths changes dramatically when the individual monomers organize into the superstructure of the alpha helix. In fact,

> *The observed optical rotation of a solution of polypeptide is a linear function of the fraction of amino acid monomers in the alpha helix form.*   (9.22)

Observing θ thus lets us measure the degree of a polypeptide's conversion from random coil to alpha helix conformation. (This technique is used in the food industry, where θ is used to monitor the degree to which starches have been cooked.)

Figure 9.7 shows some experimental data obtained by P. Doty and K. Iso, together with the results of the analysis developed in Section 9.5.3. These experiments monitored the optical rotation of an artificial polypeptide in solution while raising its temperature. At a critical value of T, the rotation abruptly changed from the value typical for isolated monomers to some other value, signaling the self-assembly of alpha helices.

$\boxed{T_2}$ *Section 9.5.1' on page 393 defines the specific optical rotation, a more refined measurement of a solution's rotatory power.*

9.5.2 Three phenomenological parameters describe a given helix–coil transition

Let's try to model the data in Figure 9.7 by using the ideas set out in Section 9.4. Our approach is based on ideas pioneered by J. Schellman and extended by B. Zimm and J. Bragg.

---

[7]But don't look at your *hand* in the mirror while doing this! After all, the mirror image of your right hand looks like your left hand.

We can think of each monomer in an alpha helix-forming polypeptide as being in one of two states labeled by $\sigma = +1$ for alpha helix or $\sigma = -1$ for random coil. More precisely, we take $\sigma_i = +1$ if monomer number i is H-bonded to monomer $i + 4$, and -1 otherwise. The fraction of monomers in the helical state can be expressed as $\frac{1}{2}(\langle \sigma_{av} \rangle + 1)$. In this expression, σ_{av} denotes the average of σ_i over all the monomers, in one particular state of the chain; $\langle \sigma_{av} \rangle$ represents a further averaging over all allowed chain states. We suppose that each monomer makes a contribution to the overall optical rotation that depends only on its state. Then the total optical rotation (vertical axis of Figure 9.7) will be a linear function of $\langle \sigma_{av} \rangle$.

The three curves in Figure 9.7 show results obtained from three different samples of polymer, differing in their average length. The polymer was synthesized under three sets of conditions; the mean molar mass for each sample was then determined. We'll begin by studying the top curve, which was obtained with a sample of very long polymer chains. We need a formula for $\langle \sigma_{av} \rangle$ as a function of temperature. To get the required result, we adapt the analysis of Section 9.4.1, reinterpreting the parameters α and γ in Equation 9.18, as follows.

The helix-extension parameters   In the polymer stretching problem, we imagined an isolated thermodynamic system consisting of the chain, its surrounding solvent, and some kind of external spring supplying the stretching force. The bias parameter $2\alpha = 2\ell f / k_B T$ then described the reduction of the spring's potential energy when one link switched from the unfavorable (backward, $\sigma = -1$) to the favorable (forward, $\sigma = +1$) direction. The applied force f was known, but the effective link length ℓ was an unknown parameter to be fit to the data. In the present context, on the other hand, the link length is immaterial. When a monomer bonds to its neighbor, its link variable σ_i changes from -1 to $+1$ and an H-bond forms. We must remember, however, that the participating H and O atoms were already H-bonded to surrounding *solvent* molecules; to bond to each other, they must *break* these preexisting bonds, with a corresponding energy *cost*. The net energy change of this transaction, which we will call $\Delta E_{bond} \equiv E_{helix} - E_{coil}$, may therefore be either positive or negative, depending on solvent conditions.[8] The particular combination of polymer and solvent shown in Figure 9.7 has $\Delta E_{bond} > 0$. To see this, note that raising the temperature pushes the equilibrium toward the alpha helix conformation. Le Châtelier's Principle then says that forming the helix must cost energy (see Section 8.2.2 on page 301).

The formation and breaking of H-bonds also involves an entropy change, which we will call ΔS_{bond}.

There is a third important contribution to the free energy change when a tract of alpha helix extends by one more monomer. As mentioned in Section 9.5.1, the formation of intramolecular H-bonds requires the immobilization of all the intervening flexible links, so the participating H and O atoms stay within the very short range of the H-bond interaction. Each amino acid monomer contains two relevant flexible links. Even in the random-coil state, these links are not perfectly free, as a result of obstructions involving the atoms on either side of them; instead, each link flips between three preferred positions. But to get the alpha helix state, each link must oc-

---

[8] $\boxed{T_2}$ More precisely, we are discussing the enthalpy change, ΔH, but in this book we do not distinguish energy from enthalpy (see Section 6.5.1).

cupy just *one* particular position. Thus the change of conformational entropy upon extending a helix by one unit is roughly $\Delta S_{conf} \approx -k_B \ln(3 \times 3)$, with a corresponding contribution to the free energy change of about $+k_B T \ln 9$.

The statement that $\Delta E_{bond} > 0$ may seem paradoxical. If alpha helix formation is energetically unfavorable, and if it also reduces the conformational entropy of the chain, then why do helices ever form at *any* temperature? This paradox, like the related one involving depletion interactions (see the end of Section 7.2.2 on page 251), goes away when we consider all the actors on the stage. It is true that extending the helix brings a reduction in the polypeptide's conformational entropy, $\Delta S_{conf} < 0$. But the formation of an intramolecular H-bond also changes the entropy of the surrounding solvent molecules. If this entropy change ΔS_{bond} is positive and big enough that the *net* entropy change $\Delta S_{tot} = \Delta S_{bond} + \Delta S_{conf}$ is positive, then increasing the temperature can indeed drive helix formation because then $\Delta G_{bond} = \Delta E_{bond} - T\Delta S_{tot}$ will become negative at high enough temperature. We have already met a similar apparent paradox in the context of self-assembly: Tubulin monomers can be induced to polymerize into microtubules—lowering their entropy—by an *increase* in temperature (Section 7.5.2 on page 276). Again, the resolution of this paradox involved the entropy of the small, but numerous, water molecules.

Summarizing, we have identified two helix-extension parameters ΔE_{bond} and ΔS_{tot} describing a given helix–coil transition. We define the bias favoring the helical state as $\alpha \equiv (\Delta E_{bond} - T\Delta S_{tot})/(-2k_B T)$; extending an alpha helical stretch of the polypeptide by one unit changes the free energy by $-2\alpha k_B T$. (Some authors refer to the related quantity $e^{2\alpha}$ as the propagation parameter of the system.) Thus,

> *The free energy to extend the helix is a function of the polypeptide's temperature and chemical environment. A positive value of α means that extending a helical region is thermodynamically favorable.*   (9.23)

Clearly a first-principles prediction of α would be a very difficult problem, involving all the physics of the H-bond network of the solvent and so on. We will not attempt this level of prediction. But the ideas of Section 9.1.1 give us an alternative approach: We can view ΔE_{bond} and ΔS_{tot} as just two phenomenological parameters to be determined from experiment. If we get more than two nontrivial testable predictions out of the model, then we will have learned something. In fact, the complete shapes of all three curves in Figure 9.7 follow from these two numbers (plus one more, to be discussed momentarily).

It's convenient to rearrange the preceding expression for α slightly. Introducing the abbreviation $T_m \equiv \Delta E_{bond}/\Delta S_{tot}$ gives

$$\alpha = \frac{1}{2}\frac{\Delta E_{bond}}{k_B}\frac{T - T_m}{T T_m}. \tag{9.24}$$

The formula shows that T_m is the **midpoint temperature**, at which $\alpha = 0$. At this temperature, extending a helical section by one unit makes no change in the free energy.

The cooperativity parameter So far, each monomer has been treated as an independent, two-state system. If this were true, then we'd be done—you found $\langle \sigma \rangle$ in a two-state system in Problem 6.5. But so far, we have neglected an important feature of the physics of alpha helix formation: Extending a helical section requires the immobilization of two flexible bonds, but *creating* a helical section in the first place requires that we immobilize *all* the bonds between units i and $i + 4$. That is, the polymer must immobilize one full turn of its nascent helix before it gains any of the benefit of forming its first H-bond. The quantity $2\alpha k_B T$ introduced earlier thus exaggerates the decrease of free energy upon initiating a helical section. We define the cooperativity parameter γ by writing the true change of free energy upon making the first bond as $-(2\alpha - 4\gamma)k_B T$. (Some authors refer to the quantity $e^{-4\gamma}$ as the initiation parameter.)

> **Your Turn 9I**
>
> Use the previous discussion to find a rough numerical estimate of the expected value of γ.

The preceding discussion assumed that the extra free energy cost of initiating a tract of alpha helix is purely entropic in character. As you found in Your Turn 9I, this assumption implies that γ is a constant, independent of temperature. Although reasonable, this assumption is just an approximation. We will see, however, that it is quite successful in interpreting the experimental data.

9.5.3 Calculation of the helix–coil transition

Polypeptides, large N Having defined α and γ, we can now proceed to evaluate $\langle \sigma_{av} \rangle \equiv \langle N^{-1} \sum_{i=1}^{N} \sigma_i \rangle$, which we know is related to the observable optical rotation. We characterize conformations by listing $\{\sigma_1, \ldots, \sigma_N\}$ and give each such string a probability by the Boltzmann weight formula. The probability contains a factor of $e^{\alpha \sigma_i}$ for each monomer, which changes by $e^{2\alpha}$ when σ_i changes from -1 (unbonded) to $+1$ (H-bonded). In addition, we introduce a factor of $e^{\gamma \sigma_i \sigma_{i+i}}$ for each of the $N - 1$ links joining sites. Because introducing a single $+1$ into a string of -1's creates *two* mismatches, the total effect of initiating a stretch of alpha helix is a factor of $e^{-4\gamma}$, consistent with the definition of γ given earlier.

When N is very large, the required partition function is once again given by Equation 9.18, and $\langle \sigma_{av} \rangle = N^{-1} \frac{d}{d\alpha} \ln Z$. Adapting your result from Your Turn 9H and recalling that θ is a linear function of $\langle \sigma_{av} \rangle$ gives the predicted optical rotation as

$$\theta = C_1 + \frac{C_2 \sinh \alpha}{\sqrt{\sinh^2 \alpha + e^{-4\gamma}}}. \qquad (9.25)$$

In this expression, $\alpha(T)$ is the function given by Equation 9.24 and C_1, C_2 are two constants.

Your Turn 9J	Derive Equation 9.25, then calculate the maximum slope of this curve. That is, find $d\theta/dT$ and evaluate at the midpoint temperature, $T_m = \Delta E_{bond}/\Delta S_{tot}$ (see Equation 9.24). Comment on the role of γ.

The top curve of Figure 9.7 shows a fit of Equation 9.25 to Doty and Iso's experimental data. Standard curve-fitting software selected the values $\Delta E_{bond} = 0.78k_B T_r$, $T_m = 285\,K$, $\gamma = 2.2$, $C_1 = 0.08$, and $C_2 = 15$. The fit value of γ has the same general magnitude as your rough estimate in Your Turn 9I.

The ability of our highly reduced model to fit the large-N data is encouraging, but we allowed ourselves to adjust five phenomenological parameters to make Equation 9.25 fit the data! Moreover, only four combinations of these parameters correspond to the main visual features of the S-shaped (or **sigmoid**) curve in Figure 9.7, as follows:

- The overall vertical position and scale of the sigmoid fix the parameters C_1 and C_2.
- The horizontal position of the sigmoid fixes the midpoint temperature T_m.
- Once C_2 and T_m are fixed, the slope of the sigmoid at the midpoint fixes the combination $e^{2\gamma}\Delta E_{bond}$, according to your result in Your Turn 9J.

In fact, it is surprisingly difficult to determine the parameters γ and ΔE_{bond} separately from the data. We can see this in Figure 9.5 on page 355: There the top two curves, representing zero and infinite cooperativity, were quite similar once we adjusted ℓ to give them the same slope at the origin. Similarly, if we hold γ fixed to a particular value, then adjust ΔE_{bond} to get the best fit to the data, we find that we can get a visually good fit using *any* value of γ. To see this, compare the top curve of Figure 9.7 with the two curves in Figure 9.9a. It is true that unrealistic values of ΔE_{bond} are needed to fit the data with the alternative values of γ shown. But the point is that the large-N data alone do not really test the model—there are many ways to get a sigmoid.

Nevertheless, while our eyes would have a hard time distinguishing the curves in Figure 9.9a from the one in Figure 9.7, still there *is* a slight difference in shape, and numerical curve fitting says that the latter is the best fit. To test our model, we must now try to make some falsifiable *prediction* from the values we have obtained for the model's parameters. We need some situation in which the alternative values of ΔE_{bond} and γ shown in Figure 9.9a give wildly different results, so that we can see whether the best-fit values are really the most successful.

Polypeptides, finite N To get the new experimental situation we need, note that one more parameter of the system is available for experimental control: Different synthesis protocols lead to different *lengths* N of the polymer. In general, polymer synthesis leads to a mixture of chains with many different lengths, a polydisperse solution. But with care, it is possible to arrive at a rather narrow distribution of lengths. Figure 9.7 shows data on the helix–coil transition in samples with two different, finite values of N.

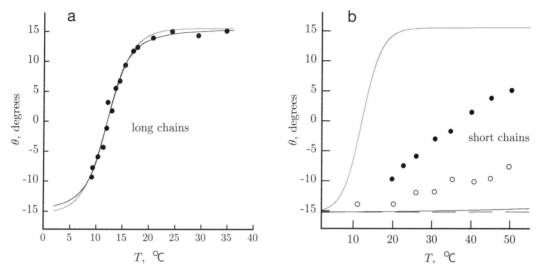

Figure 9.9: (Experimental data with fits.) Effect of changing the degree of cooperativity. (a) *Dots:* The same long-chain experimental data as Figure 9.7. *Dark curve:* Best fit to the data holding the cooperativity parameter γ fixed to the value 2.9 (too much cooperativity). The curve was obtained by setting $\Delta E_{bond} = 0.20 k_B T_r$, with the other three parameters the same as in the fit shown in Figure 9.7. *Light curve:* Best fit to the data fixing $\gamma = 0$ (no cooperativity). Here ΔE_{bond} was taken to be $57 k_B T_r$. (b) *Solid and open dots:* The same medium- and short-chain data as in Figure 9.7. The curves show the unsuccessful predictions of the same two alternative models shown in panel (a). *Top curve:* The model with no cooperativity gives no length dependence at all. *Lower curves:* In the model with too much cooperativity, short chains are influenced too much by their ends, so they stay overwhelmingly in the random-coil state. Solid line, $N = 46$; dashed line, $N = 26$.

Gilbert says: The data certainly show a big qualitative effect: The midpoint temperature is much higher for short chains. For example, $N = 46$ gives a midpoint at around 35°C (middle set of data points in Figure 9.7). Because we have no more free parameters in the model, it had better be able to predict that shift correctly.

Sullivan: Unfortunately, Equation 9.18 shows equally clearly that the midpoint is always at $\alpha = 0$, which we found corresponds to one fixed temperature, $T_m \equiv \Delta E_{bond}/\Delta S_{tot}$, independent of N. It looks like our model is no good.

Gilbert: How did you see that so quickly?

Sullivan: It's a symmetry argument. Suppose I define $\tilde{\sigma}_i = -\sigma_i$. Instead of summing over $\sigma_i = \pm 1$, I can equally well sum over $\tilde{\sigma}_i = \pm 1$. In this way, I show that $Z(-\alpha) = Z(\alpha)$; it's an "even function." Then its derivative must be an "odd function," so it must equal minus itself at $\alpha = 0$.

Gilbert: That may be good math, but it's bad physics! Why should there be any symmetry relating the alpha helix and a random coil conformations?

Putting Gilbert's point slightly differently, suppose that we have a tract of +1's starting all the way out at the end of the polymer and extending to some point in the middle. There is one junction at the end of this tract, giving a penalty of $e^{-2\gamma}$. In contrast, for a tract of $+1$'s starting *and* ending in the middle, there are *two* junctions, one at each end. But really, initiating a tract of helix requires that we immobilize several bonds, regardless of whether it extends to the end or not. So our partition function (Equation 9.18) underpenalizes those conformations with helical tracts extending all the way out to one end (or both). Such "end effects" will be more prominent for short chains.

We can readily cure this problem and incorporate Gilbert's insight. We introduce fictitious monomers in positions 0 and $N + 1$, but instead of summing over their values, we fix $\sigma_0 = \sigma_{N+1} = -1$. Now a helical tract extending to the end (position 1 or N) will still have two "junctions" and will get the same penalty as a tract in the middle of the polymer. Choosing -1 instead of $+1$ at the ends breaks the spurious symmetry between ± 1. That is, Sullivan's discouraging result no longer holds after this small change in the model. Let's do the math properly.

Your Turn 9K

a. Repeat Your Turn 9G(a) on page 360 with the modification just mentioned, showing that the partition function for $N = 2$ is $Z_2' = r \begin{bmatrix} 0 \\ 1 \end{bmatrix} \cdot \mathsf{T}^3 \begin{bmatrix} 0 \\ 1 \end{bmatrix}$, where T is the same as before and r is a quantity that you are to find.

b. Adapt your answer to Your Turn 9G(b) (general N) to the present situation.

c. Adapt Equation 9.20 on page 360 to the present situation.

Because N is not infinite, we can no longer drop the second term of Equation 9.20, nor can we ignore the constant p appearing in it. Thus we must find explicit eigenvectors of T. First we make the abbreviations

$$g_{\pm} = e^{\alpha - \gamma} \lambda_{\pm} = e^{\alpha} \left[\cosh \alpha \pm \sqrt{\sinh^2 \alpha + e^{-4\gamma}} \right].$$

Your Turn 9L

a. Show that we may write the eigenvectors as

$$\mathbf{e}_{\pm} = \begin{bmatrix} g_{\pm} - 1 \\ e^{2(\alpha - \gamma)} \end{bmatrix}.$$

b. Using (a), show that $\begin{bmatrix} 0 \\ 1 \end{bmatrix} = w_+ \mathbf{e}_+ + w_- \mathbf{e}_-$, where $w_{\pm} = \pm e^{2(\gamma - \alpha)}(1 - g_{\mp})/(g_+ - g_-)$.

c. $\boxed{T_2}$ Find an expression for the full partition function, Z_N', in terms of $g_{\pm}$.

The rest of the derivation is familiar, if involved: We compute $\langle \sigma_{av} \rangle_N = N^{-1} \frac{d}{d\alpha} \ln Z_N'$, using your result from Your Turn 9L(c). This calculation yields the lower two curves[9] in Figure 9.7.

---

[9] $\boxed{T_2}$ A small correction is discussed in Section 9.5.3′ on page 394 below.

Summary Earlier, we saw how fitting our model to the large-N data yields a satisfactory account of the helix–coil transition for long polypeptide chains. The result was slightly unsatisfying, though, because we adjusted several free parameters to achieve the agreement. Moreover, the data seem to underdetermine the parameters, including the most interesting one, the cooperativity parameter γ (see Figure 9.9a).

Nevertheless, we agreed to take seriously the value of γ obtained from the large-N data. We then successfully predicted, with no further fitting, the finite-N data. In fact, the finite-N behavior of our model does depend sensitively on the separate values of ΔE_{bond} and γ, as we see in Figure 9.9b: Both the noncooperative and the too-cooperative models, each of which seemed to do a reasonable job with the large-N data, fail miserably to predict the finite-N curves! It's remarkable that the large-N data, which seemed so indifferent to the separate values of ΔE_{bond} and γ, actually determine them well enough to predict successfully the finite-N data.

We can interpret our results physically as follows:

> ***a.*** *A two-state transition can be sharp either because its ΔE is large or because of cooperativity between many similar units.*
>
> ***b.*** *A modest amount of cooperativity can give as much sharpness as a very large ΔE, because it's e^γ that appears in the maximum slope (see Your Turn 9J). Thus cooperativity holds the key to giving sharp transitions between macromolecular states using only weak interactions (like H-bonds).*   (9.26)
>
> ***c.*** *A hallmark of cooperativity is a dependence on the system's size and dimensionality.*

Let's make point (c) quantitative. In the noncooperative case, each element behaves independently (light gray curves in Figure 9.9a,b), and so the sharpness of the transition is independent of N. With cooperativity, the sharpness goes down for small N (lower two curves of Figure 9.9b).

$\boxed{T_2}$ *Section 9.5.3′ on page 394 refines the analysis of the helix–coil transition by accounting for the sample's polydispersity.*

9.5.4 DNA also displays a cooperative "melting" transition

DNA famously consists of two strands wound around each other (Figure 2.15 on page 51); it's often called the DNA duplex. Each strand has a strong, covalently bonded backbone, but the two strands are only attached to each other by weak interactions, the hydrogen bonds between complementary bases in a pair. This hierarchy of interactions is crucial for DNA's function: Each strand must strictly preserve the linear sequence of the bases, but the cell frequently needs to unzip the two strands temporarily, to read or to copy its genome. Thus the marginal stability of the DNA duplex is essential for its function.

The weakness of the interstrand interaction leaves us wondering, however, why DNA's structure is so well-defined when it is *not* being read. We get a clue when we notice that simply heating DNA in solution to around 90°C does make it fall apart into two strands, or "melt." Other environmental changes, such as replacing the surrounding water by a nonpolar solvent, also destabilize the duplex. The degree of melting again follows a sigmoid (S-shaped) curve, similar to Figure 9.7 but with the disordered state at high, not low, temperature. That is, DNA undergoes a sharp transition as its temperature is raised past a definite melting point. Because of the general similarity to the alpha helix transition in polypeptides, many authors refer to DNA melting as another "helix–coil" transition.

To understand DNA melting qualitatively, we visualize each backbone of the duplex as a chain of sugar and phosphate groups, with the individual bases hanging off this chain like the charms on a bracelet. When the duplex melts, there are several contributions to the free energy change:

1. The hydrogen bonds between paired bases break.
2. The flat bases on each strand stop being neatly stacked like coins; that is, they **unstack**. Unstacking breaks some other energetically favorable interactions between neighboring bases, like dipole–dipole and van der Waals attractions.
3. The individual DNA strands are more flexible than the duplex, so the backbone's conformational entropy increases upon melting. The unstacked bases can also flop about on the backbone, giving another favorable entropic contribution to the free energy change.
4. Finally, unstacking exposes the hydrophobic surfaces of the bases to the surrounding water.

Under typical conditions,

- DNA melting is energetically unfavorable ($\Delta E > 0$). This fact mainly reflects the free energy contributions described in points (1) and (2) above. But,
- Unstacking is entropically favored ($\Delta S > 0$). This fact reflects the dominance of the contribution in point (3) over the entropic part of (4).

Thus, raising the temperature indeed promotes melting: $\Delta E - T\Delta S$ becomes negative at high temperature.

Now consider the reverse process, the annealing of single-stranded DNA. There will be a large entropic penalty when two flexible single strands of DNA come together and initiate a duplex tract. Thus we expect to find cooperativity, by analogy to the situation in Section 9.5.2. In addition, the unstacking energy is an interaction between neighboring basepairs, so it encourages the extension of an existing duplex tract more than the creation of a new one. The cooperativity turns out to be significant, leading to the observed sharp transition.

9.5.5 Applied mechanical force can induce cooperative structural transitions in macromolecules

Sections 9.2–9.4 showed how applying mechanical force can change the conformation of a macromolecule in the simplest way—by straightening it. Sections 9.5.1–

9.5.4 discussed another case, with a more interesting structural rearrangement. These two themes can be combined to study force-induced structural transitions:

> *Whenever a macromolecule has two conformations that differ in the distance between two points, then a mechanical force applied between those points will alter the equilibrium between the two conformations.* (9.27)

Idea 9.27 underlies the phenomenon of **mechanochemical coupling**. We saw this coupling in a simple context in our analysis of molecular stretching, via the external part of the energy function, $U_{ext} = -fz$ (Equation 9.7). This term altered the equilibrium between forward- and backward-pointing monomers from equally probable (at low force) to mainly forward (at high force). Section 6.7 on page 226 gave another example, where mechanical force altered the balance between the folded and unfolded states of a single RNA molecule. Here are three more examples of Idea 9.27.

Overstretching DNA DNA in solution normally adopts a conformation called the B-form (Figure 2.15), in which the H-bonded basepairs from the two chains stack on each other like the steps of a spiral staircase. The sugar-phosphate backbones of the two chains then wind around the periphery of the staircase. That is, the two backbones are far from being straight. The distance traveled along the molecule's axis when we take one step up the staircase is thus considerably shorter than it would be if the backbones were straight (vertical). Idea 9.27 then suggests that pulling on the two ends of a piece of DNA could alter the equilibrium between the B-form and some other, "stretched," form, in which the backbones are straightened. Figure 9.4 shows this **overstretching transition** as regime *D*. At a critical value of the applied force, DNA abandons the linear-elasticity behavior studied in Section 9.4.2 and begins to spend most of its time in a new state, about 60% longer than before. A typical value for f_{crit} in lambda phage DNA is 65 pN. The sharpness of this transition implies that it is highly cooperative.

Unzipping DNA It is even possible to tear the two strands of DNA apart without breaking them. F. Heslot and coauthors accomplished this in 1997 by attaching the two strands at one end of a DNA duplex to a mechanical stretching apparatus. They and later workers found the force needed to "unzip" the strands to be in the range 10–15 pN.

Unfolding titin Proteins, too, undergo massive structural changes in response to mechanical force. For example, titin is a structural protein found in muscle cells. In its native state, titin consists of a chain of globular domains. Under increasing tension, the domains pop open one at a time, somewhat like the RNA hairpin in Section 6.7 on page 226, thereby leading to a sawtooth-shaped force–extension relation. Upon release of the applied force, titin resumes its original structure, ready to be stretched again.

9.6 ALLOSTERY

So far, this chapter has focused on showing how nearest-neighbor cooperativity can create sharp transitions between conformations of rather simple polymers. It is a very big step to go from these model systems to proteins, with complicated, nonlocal interactions between residues that are distant along the chain backbone. Indeed, we will not attempt any more detailed calculations. Let's instead look at some biological consequences of the principle that cooperative effects of many weak interactions can yield definite conformations with sharp transitions.

9.6.1 Hemoglobin binds four oxygen molecules cooperatively

Returning to this chapter's Focus Question, first consider a protein critical to your own life, **hemoglobin**. Hemoglobin's job is to bind oxygen molecules on contact with air, then release them at the appropriate moment, in some distant body tissue. As a first hypothesis, one might imagine that

- Hemoglobin has a site where an O_2 molecule can bind.
- In an oxygen-rich environment, the binding site is more likely to be occupied, by Le Châtelier's Principle (Section 8.2.2 on page 301).
- In an oxygen-poor environment, the binding site is less likely to be occupied. Thus, when hemoglobin in a red blood cell moves from lungs to tissue, it first binds, then releases, oxygen as desired.

The problem with this tidy little scenario shows up when we try to model oxygen binding to hemoglobin quantitatively, using the Mass Action rule (Equation 8.17 on page 304) for the reaction $Hb + O_2 \rightleftharpoons HbO_2$. (The symbol "Hb" represents the whole hemoglobin molecule.) Let $Y \equiv [HbO_2]/([Hb] + [HbO_2])$ represent the fractional degree of oxygenation.

> **Your Turn 9M**
>
> Show that according to the preceding model, $Y = [O_2]/([O_2] + K_{eq}^{-1})$, where K_{eq} is the equilibrium constant of the binding reaction (see Equation 8.17 on page 304).

In a set of careful measurements, C. Bohr (father of the physicist Niels Bohr) showed in 1904 that the curve of oxygen binding versus oxygen concentration in solution (or pressure in the surrounding air) has a sigmoid form (open circles in Figure 9.10). The key feature of the sigmoid is its inflection point, the place where the graph switches from concave-up to concave-down. The data show such a point around $c_{O_2} = 8 \cdot 10^{-6}$ M. The formula you found in Your Turn 9M never gives such behavior, no matter what value we take for K_{eq}. Interestingly, though, the corresponding binding curve for **myoglobin**, a related oxygen-binding molecule, does have the form expected from simple Mass Action (solid dots in Figure 9.10).

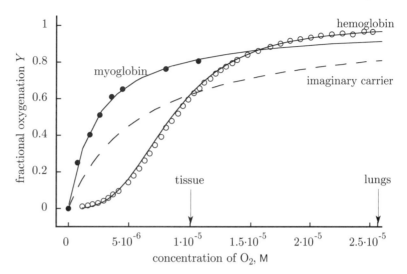

Figure 9.10: (*Experimental data with fits.*) Fractional degree of oxygen binding as a function of oxygen concentration. *Solid circles:* Data for myoglobin, an oxygen-binding molecule with a single binding site. The curve through the points shows the formula in Your Turn 9M with a suitable choice of K_{eq}. *Open circles:* Data for human hemoglobin, which has four oxygen-binding sites. The curve shows the formula in Your Turn 9N(a) with $n = 3.1$. *Dashed curve:* Oxygen binding for an imaginary, noncooperative carrier ($n = 1$), with the value of K_{eq} adjusted to agree with hemoglobin at the low oxygen concentration of body tissue (*left arrow*). The saturation at high oxygen levels (*right arrow*) is then much worse for the imaginary carrier than in real hemoglobin. [Data from Mills et al., 1976 and Rossi-Fanelli & Antonini, 1958.]

Archibald Hill found a more successful model for hemoglobin's oxygen binding in 1913: If we assume that hemoglobin can bind *several* oxygen molecules and does so in an all-or-nothing fashion, then the binding reaction becomes $\mathrm{Hb} + n\mathrm{O}_2 \rightleftharpoons \mathrm{Hb(O_2)}_n$. Hill's proposal is very similar to the cooperative model of micelle formation (see Section 8.4.2 on page 317).

Your *Turn* *9N*	**a.** Find the fractional binding Y in Hill's model. **b.** For what values of n and K_{eq} will this model give an inflection point in the curve of Y versus $[\mathrm{O}_2]$?

Fitting the data to both K_{eq} and n, Hill found the best fit for myoglobin gave $n = 1$, as expected, but $n \approx 3$ for hemoglobin.

These observations began to make structural sense after G. Adair established that hemoglobin is a **tetramer**: It consists of four subunits, each resembling a single myoglobin molecule and, in particular, each with its own oxygen binding site. Hill's result implies that the binding of oxygen to these four sites is highly cooperative. The cooperativity is not really all-or-none because the effective value of n is less than

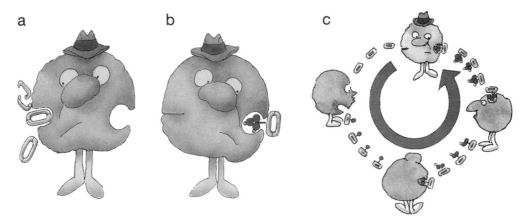

Figure 9.11: (Metaphor.) Allosteric feedback control. (a) An allosteric enzyme has an active site (*left*), at which it catalyzes the assembly of some intermediate product from substrate. (b) When a control molecule binds to the regulatory site (*right*), however, the active site becomes inactive. (c) In a simplified version of a synthetic pathway, an allosteric enzyme (*top*, in fedora) catalyzes the first step of the synthesis. Its product is the substrate for another enzyme, and so on. Most of the final product goes off on its errands in the cell, but some of it also serves as the control molecule for the initial enzyme. When the final product is present in sufficient concentration, it binds to the regulatory site, turning off the first enzyme's catalytic activity. Thus the final product acts as a messenger sent to the first worker on an assembly line, saying "stop production." [Cartoons by Bert Dodson, from Hoagland & Dodson, 1995.]

the number of binding sites (four). Nevertheless, *the binding of one oxygen molecule leaves hemoglobin predisposed to bind more.* After all, if each binding site operated independently, we would have found $n = 1$, because in that case, the sites might as well have been on completely separate molecules.

Cooperativity is certainly a good thing for hemoglobin's function as an oxygen carrier: It lets hemoglobin switch readily between accepting and releasing oxygen. Figure 9.10 shows that a noncooperative carrier would either have too high a saturation in tissue (like myoglobin) and hence fail to release enough oxygen, or have too low a saturation in the lungs (like the imaginary carrier shown as the dashed line) and hence fail to accept enough oxygen. Moreover, hemoglobin's affinity for oxygen can be modulated by other chemical signals besides the level of oxygen itself. For example, Bohr also discovered that the presence of dissolved carbon dioxide or other acids (produced in the blood by actively contracting muscle) promotes the release of oxygen from hemoglobin, delivering more oxygen when it is most needed. This **Bohr effect** fits with what we have already seen: Once again, binding of a molecule (CO_2) at one site on hemoglobin affects the binding of oxygen at another site, a phenomenon called **allostery**. More broadly, allosteric control is crucial to the feedback mechanisms regulating many biochemical pathways (Figure 9.11).

The puzzling aspect of all these interactions is simply that the binding sites for the four oxygen molecules (and for other regulatory molecules such as CO_2) are *not close* to one another. Indeed, M. Perutz's epochal analysis of the shape of the hemoglobin molecule in 1959 showed that the iron atoms in hemoglobin that bind the oxygens are 2.5 nm apart. Interactions between spatially distant binding sites on a macromolecule are called **allosteric**. At first it was difficult to imagine how such interactions could be possible at all. After all, we have seen that the main interactions responsible for molecular recognition and binding are of very short range. How, then, can one binding site communicate its occupancy to another one?

9.6.2 Allostery often involves relative motion of molecular subunits

A big clue to the allostery puzzle came in 1938, when F. Haurowitz found that crystals of hemoglobin had different morphologies when prepared with or without oxygen: The deoxygenated proteins took the form of scarlet needles, whereas crystals formed with oxygen present were purple plates. Moreover, crystals prepared without oxygen shattered upon exposure to air. (Crystals of myoglobin showed no such alarming behavior.) The crystals' loss of stability upon oxygenation suggested to Haurowitz that hemoglobin undergoes a *shape change* upon binding oxygen. Perutz's detailed structural maps of hemoglobin, obtained many years later, confirmed this interpretation: The quaternary (highest-order) structure of hemoglobin changes in the oxygenated form.

Today, many allosteric proteins are known, and their structures are being probed by an ever-widening array of techniques. For example, Figure 9.12 shows three-dimensional reconstructed electron micrographs of the motor protein kinesin. Each kinesin molecule is a **dimer**; that is, it consists of two identical subunits. Each subunit has a binding site that can recognize and bind a microtubule and another site

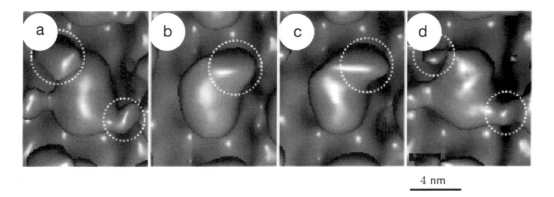

4 nm

Figure 9.12: (Image reconstructed from electron microscopy data.) Direct visualization of an allosteric change. The four panels show three-dimensional maps of a molecular motor (kinesin) attached to a microtubule. In each frame, the microtubule is in the background, running vertically and directed upward. A gold particle was attached to the neck linker region of the motor, enabling the microscope to show changes in the linker's position when the motor binds a small molecule. *Dotted circles* draw attention to the significant differences between the frames. (a) The motor has not bound any nucleotide in its catalytic domain; its neck linker flops between two positions (*circles*). (b,c) The motor has bound an ATP-like molecule (respectively, AMP–PNP and ADP–AlF$_4^-$, in the two frames). The position of the neck linker has changed. (d) When the motor has bound ADP, its conformation is much the same as in the unbound state (a). Each of the images shown was reconstructed from data taken on 10 000–20 000 individual molecules. [From Rice et al., 1999.]

that can bind the cellular energy-supply molecule ATP. The figure shows that one particular domain of the molecule, the **neck linker**, has two preferred positions when no ATP is bound. When ATP (or a similarly shaped molecule) binds to its binding site, however, the neck linker freezes into a definite third position. Thus kinesin displays a mechanochemical coupling. The function of this allosteric interaction is quite different from the one in hemoglobin. Instead of regulating the storage of a small molecule, Chapter 10 will show how the mechanical motion induced by the chemical event of ATP binding can be harnessed to create a single-molecule motor.

The observation of gross conformational changes upon binding suggests a simple interpretation of allosteric interactions:

- The binding of a molecule to one site on a protein can deform the neighborhood of that site. For example, the site's original shape may not precisely fit the target molecule, but the free energy gain of making a good fit may be sufficient to pull the binding site into tight contact.
- A small deformation can be amplified by a leverlike arrangement of the protein's subunits, then transmitted to other parts of the protein by mechanical linkage, and, in general, manipulated by the protein in ways familiar to us from macroscopic machinery.
- Distortions transmitted to a distant binding site in this way can alter the binding site's shape and hence its affinity for its own target molecule.

Although this purely mechanical picture of allosteric interactions is highly idealized, it has proved quite useful in understanding the mechanisms of motor proteins. More generally, we should view the mechanical elements in the picture just sketched (forces, linkages) as metaphors also representing more chemical mechanisms (such as charge rearrangements).

9.6.3 Vista: Protein substates

This chapter has emphasized the role of cooperative, weak interactions in giving macromolecules definite structures. Actually, however, it's an oversimplification to say that a protein has a unique native conformation. Although the native state is much more restricted than a random coil, nevertheless it consists of a very large number of closely related conformations.

Figure 9.13 summarizes one key experiment by R. Austin and coauthors on the structure of myoglobin. Myoglobin (abbreviated Mb) is a globular protein consisting of about 150 amino acids. Like hemoglobin, myoglobin contains an iron atom, which can bind either oxygen (O_2) or carbon monoxide (CO). The native conformation has a "pocket" region surrounding the binding site. To study the dynamics of CO binding, the experimenters took a sample of Mb·CO and suddenly dissociated all the carbon monoxide with an intense flash of light. At temperatures below about 200 K, the CO molecule remains in the protein's pocket, close to the binding site. Monitoring the optical absorption spectrum of the sample then let the experimenters measure the fraction $N(t)$ of myoglobin molecules that had rebound their CO, as a function of time.

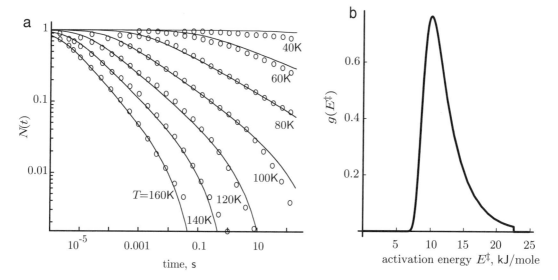

Figure 9.13: (Experimental data; theoretical model.) Rebinding of carbon monoxide to myoglobin after flash photodissociation. The myoglobin was suspended in a mixture of water and glycerol to prevent freezing. (a) Log-log plot of the fraction $N(t)$ of myoglobin molecules that have *not* rebound their CO by time t. *Circles:* Experimental data at various values of the temperature T. None of these curves is a simple exponential. (b) The distribution of activation barriers in the sample, inferred from just one of the data sets in (a) (namely, $T = 120$ K). The curves drawn in (a) were all computed by using this one fit function; thus, the curves at every temperature other than 120 K are all predictions of the model described in Section 9.6.3′ on page 394. [Data from Austin et al., 1974.]

We might first try to model CO binding as a simple two-state system, like those discussed in Section 6.6.2 on page 220. Then we'd expect the number of unbound myoglobin molecules to relax exponentially to its (very small) equilibrium value, following Equation 6.30 on page 222. This behavior was not observed, however. Instead, Austin and coauthors proposed that

- Each *individual* Mb molecule indeed has a simple exponential rebinding probability, reflecting an activation barrier E for the CO molecule to rebind, but
- The many Mb molecules in a bulk sample were each in slightly different **conformational substates**. Each substate is functional (it can bind CO), so it can be considered to be "native." But each differs subtly; for example, each has a different activation barrier to binding.

This hypothesis makes a testable prediction: It should be possible to deduce the probability of occupying the various substates from the rebinding data. More precisely, we should be able to find a distribution $g(\Delta E^{\ddagger})\mathrm{d}\Delta E^{\ddagger}$ of the activation barriers by studying a sample at one particular temperature and, from this function, predict the time course of rebinding at other temperatures. Indeed, Austin and coauthors found that the rather broad distribution shown in Figure 9.13b could account for all the data in Figure 9.13a. They concluded that a given primary structure (amino acid sequence) does *not* fold to a unique lowest-energy state; rather, it arrives at a group of closely related tertiary structures, each differing slightly in activation energy. These structures are the conformational substates. Pictorial reconstructions of protein structure from X-ray diffraction generally do not reveal this rich structure: They show only the one (or few) most heavily populated substates (corresponding to the peak in Figure 9.13b).

$\boxed{T_2}$ *Section 9.6.3′ on page 394 gives details of the functions drawn in Figure 9.13.*

THE BIG PICTURE

This chapter has worked through some case studies in which weak, nearest-neighbor couplings between otherwise independent actors created sharp transitions in the nanoworld of single molecules. Admittedly, we have hardly scratched the surface of protein structure and dynamics—our calculations involved only linear chains with nearest-neighbor cooperativity, whereas the allosteric couplings of greatest interest in cell biology involve *three*-dimensional protein structures. But as usual our goal was only to address the question "How could anything like that happen at all?" by using simplified but explicit models.

Cooperativity is a pervasive theme in both physics and cell biology, at all levels of organization. Thus, although this chapter mentioned its role in creating well-defined allosteric transitions in single macromolecules, Chapter 12 will turn to the cooperative behavior *between* thousands of proteins, the ion channels in a single neuron. Each channel has a sharp transition between "open" and "closed" states but makes that transition in a noisy, random way (see Figure 12.17). Each channel also communicates weakly with its neighbors, via its effect on the membrane's potential. We'll see how even such weak cooperativity leads to the reliable transmission of nerve impulses.

KEY FORMULAS

- *Elastic rod:* In the elastic rod model of a polymer, the elastic energy of a short segment of rod is $dE = \frac{1}{2}k_B T \left[A\boldsymbol{\beta}^2 + Bu^2 + C\omega^2 + 2Du\omega \right] ds$ (Equation 9.2). Here $Ak_B T$, $Ck_B T$, $Bk_B T$, and $Dk_B T$ are the bend stiffness, twist stiffness, stretch stiffness, and twist–stretch coupling, and ds is the length of the segment. (The quantities A and C are also called the bend and twist persistence lengths.) u, $\boldsymbol{\beta}$, and ω are the stretch, bend, and twist density. Assuming that the polymer is inextensible, and ignoring twist effects, led us to a simplified elastic rod model (Equation 9.3). This model retains only the first term of the elastic energy.

- *Stretched freely jointed chain:* The fractional extension $\langle z \rangle / L_{tot}$ of a one-dimensional, freely jointed chain is its mean end-to-end distance divided by its total unstretched length. If we stretch the chain with a force f, the fractional extension is equal to $\tanh(f L_{seg}^{(1d)} / k_B T)$ (Equation 9.10), where $L_{seg}^{(1d)}$ is the effective link length. The bending stiffness of the real molecule being represented by the FJC model determines the effective segment length $L_{seg}^{(1d)}$.

- *Alpha helix formation:* Let $\alpha(T)$ be the free energy change per monomer for the transition from alpha helix to random coil at temperature T. In terms of the energy difference ΔE_{bond} between the helical and coiled forms, and the midpoint temperature T_m, we found (Equation 9.24)

$$\alpha(T) = \frac{1}{2}\frac{\Delta E_{bond}}{k_B}\frac{T - T_m}{TT_m}.$$

The optical rotation of a solution of polypeptide is then predicted to be

$$\theta = C_1 + \frac{C_2 \sinh\alpha}{\sqrt{\sinh^2\alpha + e^{-4\gamma}}},$$

where C_1 and C_2 are constants and γ describes the cooperativity of the transition (Equation 9.25).

• *Simple binding:* The oxygen saturation curve of myoglobin is of the form

$$Y = [O_2]/([O_2] + K_{eq}^{-1})$$

(Your Turn 9M). Hemoglobin instead follows the formula you found in Your Turn 9N.

FURTHER READING

Semipopular:
Hemoglobin, science, and life: Perutz, 1998.
DNA: Frank-Kamenetskii, 1997; Austin et al., 1997; Calladine & Drew, 1997.
Allostery: Judson, 1995.

Intermediate:
Elasticity: Feynman et al., 1963b, §§38,39.
Physical aspects of biopolymers: Boal, 2001; Howard, 2001; Doi, 1996; Sackmann et al., 2002.
Allostery: Berg et al., 2002; Benedek & Villars, 2000b, §4.4.C.
Stacking free energy in DNA, and other interactions stabilizing macromolecules: van Holde et al., 1998.
In addition, many of the ideas in this chapter and Chapters 6–8 are applied to protein–DNA interactions in Bruinsma, 2002.

Technical:
Elasticity: Landau & Lifshitz, 1986.
Hemoglobin: Eaton et al., 1999.
DNA structure: Bloomfield et al., 2000.
Physical aspects of biopolymers: Grosberg & Khokhlov, 1994.
Single molecule manipulation and observation: Leuba & Zlatanova, 2001.
DNA stretching: Marko & Siggia, 1995.
Overstretching transition: Cluzel et al., 1996; Smith et al., 1996
Helix–coil transition: Poland & Scheraga, 1970 .

$\boxed{T_2}$ **9.1.1′ Track 2**

1. The idea of reducing the multitude of molecular details of a material down to just a few parameters may seem too ad hoc. How do we know that the viscous force rule (Equation 5.9 on page 168), which we essentially pulled from a hat, is complete? Why can't we add more terms, like

$$f/A = -\eta \frac{dv}{dx} + \eta_2 \frac{d^2 v}{dx^2} + \eta_3 \frac{d^3 v}{dx^3} + \cdots?$$

It turns out that the number of relevant parameters is kept small by *dimensional analysis* and by *symmetries* inherited from the microscopic world.

Consider, for example, the constant η_3 just mentioned. It is supposed to be an intrinsic property of the fluid, independent of the size R of its pipe. Clearly it has dimensions $\mathbb{L}^2$ times those of the ordinary viscosity. The only intrinsic length scale of a simple (Newtonian) fluid, however, is the average distance d between molecules. (Recall that the macroscopic parameters of a simple Newtonian fluid don't determine any length scale; see Section 5.2.1 on page 164.) Thus we can expect that η_3, if present at all, must turn out to be roughly d^2 as large as η. Because the gradient of the velocity is roughly R^{-1} as large as v itself (see Section 5.2.2 on page 166), we see that the η_3 term is less important than the usual η term by roughly a factor of $(d/R)^2$, a tiny number.

Turning to the η_2 term, it turns out that an even stronger result forbids it altogether: This term cannot be written in a way that is invariant under rotations. Thus it cannot arise in the description of an isotropic Newtonian fluid (Section 5.2.1 on page 164), which by assumption, is the same in every direction. In other words, symmetries of the molecular world restrict the number and types of effective parameters of a fluid. (For more discussion of these points, see Landau & Lifshitz, 1987; Landau & Lifshitz, 1986.)

The conclusions just given are not universal—hence the qualification that they apply to isotropic Newtonian fluids. They get replaced by more complicated rules in the case of non-Newtonian or **complex fluids** (for example, the viscoelastic ones mentioned in Section 5.2.3′ on page 188 or the liquid crystals in a wristwatch display).

2. Some authors refer to the systematic exploitation of Idea 9.1 as "generalized elasticity and hydrodynamics." Certainly there is some art involved in implementing the idea in a given situation, for example, in determining the appropriate list of effective degrees of freedom. Roughly speaking, a collective variable gets onto the list if it describes a disturbance to the system that costs very little energy or that relaxes very slowly, in the limit of long length scales. Such disturbances in turn correspond to broken symmetries of the lowest-energy state, or to conservation rules. For example,

- The centerline of a rod defines a line in space. Placing this line somewhere breaks two translation symmetries, singling out two corresponding collective modes, namely, the two directions in which we can bend the rod in its normal plane.

Section 9.1.2 shows that indeed bending costs very little elastic energy on long length scales.

- In diffusion, we assumed that the diffusing particles could neither be created nor destroyed—they're conserved. The corresponding collective variable of the system is the particle density, which indeed changes very slowly on long length scales according to the diffusion equation (Equation 4.20 on page 131).

(For examples of this approach in the context of soft condensed matter physics, see Chaikin & Lubensky, 1995.)

3. Much of the power of Idea 9.1 comes from the word *local*. In a system with local interactions, we can arrange the actors in such a way that each one interacts with only a few nearest neighbors, and the arrangement resembles a meshwork with only a few dimensions, typically two or three. For example, each point on a cubic lattice has just six nearest neighbors (see Problem 1.6).

Idea 9.1 is not strictly applicable to problems involving nonlocal interactions. In fact, one definition of a complex system is, "Many non-identical elements connected by diverse, nonlocal interactions." Many problems of biological and ecological organization do have this character; and indeed, general results have been harder to get in this domain than in the traditional fields of physics.

4. Section 9.1.1 stated that a fluid membrane has one elastic constant. This statement is a slight simplification: There are actually *two* elastic bending constants. The one discussed in Section 8.6.1 discourages "mean curvature," whereas the other involves "Gaussian curvature." (For more information and to see why the Gaussian stiffness doesn't enter many calculations, see for example Seifert, 1997.)

T_2 ### 9.1.2′ Track 2

1. Technically ω is called a *pseudo*scalar. The derogatory prefix *pseudo* reminds us that upon reflection through a mirror, ω changes sign (try viewing Figure 2.17 on page 53 in a mirror), whereas a true scalar like u does not. Similarly, the last term of Equation 9.2 on page 346 is also pseudoscalar, being the product of a true scalar times a pseudoscalar. We should expect to find such a term in the elastic energy of a molecule like DNA, whose structure is not mirror symmetric (see Section 9.5.1). The twist–stretch coupling in DNA has in fact been observed experimentally, in experiments that control the twist variable.

2. We implicitly used dimensional analysis reasoning (see Section 9.1.1′) to get the continuum rod-elasticity equation, Equation 9.2. Thus the only terms we retained were those with the fewest possible derivatives of the deformation fields (that is, none). In fact, *single*-stranded DNA is not very well described by the elastic rod model because its persistence length is *not* much bigger than the size of the individual monomers; so Idea 9.1 does not apply.

3. We also simplified our rod model by requiring that the terms have the symmetries appropriate to a uniform, cylindrical rod. Clearly DNA is not such an object. For example, at any position s along the molecule, it will be easier to bend in one direction than in the other: Bending in the easy direction squeezes the helical groove

in Figure 2.15 on page 51. Thus strictly speaking, Equation 9.2 is appropriate only for bends on length scales longer than the helical repeat of 10.5×0.34 nm because, on such scales, these anisotropies average out. (For more details, see Marko & Siggia, 1994.)

4. It is possible to consider terms in E of higher than quadratic order in the deformation. (Again see Marko & Siggia, 1994.) Under normal conditions, these terms have small effects because the large elastic stiffness of DNA keeps the local deformations small.

5. We need not, and should not, take our elastic rod imagery too literally as a representation of a macromolecule. When we bend the structure shown in Figure 2.15, some of the free energy cost indeed comes from deforming various chemical bonds between the atoms, roughly like bending a steel bar. But there are other contributions as well. For example, recall that DNA is highly charged—it's an acid, with two negative charges per basepair. This charge makes DNA a self-repelling object, adding a substantial contribution to the free energy cost of bending it. Moreover, this contribution depends on external conditions, such as the surrounding salt concentration (see Idea 7.28 on page 269). As long as we consider length scales longer than the Debye screening length of the salt solution, however, our phenomenological argument remains valid; we can simply incorporate the electrostatic effects into an effective value of the bend stiffness.

$\boxed{T_2}$ **9.1.3′ Track 2**

1. We can make the interpretation of A as a persistence length, and the passage from Equation 9.2 to a corresponding FJC model, more explicit. Recalling that $\hat{\mathbf{t}}(s)$ is the unit vector parallel to the rod's axis at contour distance s from one end, we first prove that for a polymer under no external forces,

$$\langle \hat{\mathbf{t}}(s_1) \cdot \hat{\mathbf{t}}(s_2) \rangle = e^{-|s_1 - s_2|/A}. \qquad \text{(to be shown)} \qquad (9.28)$$

Here s_1 and s_2 are two points along the chain; A is the constant appearing in the elastic rod model (Equation 9.3 on page 346). Once we prove it, Equation 9.28 will make precise the statement that the polymer "forgets" the direction $\hat{\mathbf{t}}$ of its backbone over distances greater than its bend persistence length A.

To prove Equation 9.28, consider three points A, B, C located at contour distances s, $s+s_{AB}$, and $s+s_{AB}+s_{BC}$ along the polymer. We will first relate the desired quantity $\hat{\mathbf{t}}(A) \cdot \hat{\mathbf{t}}(C)$ to $\hat{\mathbf{t}}(A) \cdot \hat{\mathbf{t}}(B)$ and $\hat{\mathbf{t}}(B) \cdot \hat{\mathbf{t}}(C)$. Set up a coordinate frame $\hat{\boldsymbol{\xi}}, \hat{\boldsymbol{\eta}}, \hat{\boldsymbol{\zeta}}$ whose $\hat{\boldsymbol{\zeta}}$-axis points along $\hat{\mathbf{t}}(B)$. (We reserve the symbols $\hat{\mathbf{x}}, \hat{\mathbf{y}}, \hat{\mathbf{z}}$ for a frame fixed in the laboratory.) Let (ϑ, ϕ) be the corresponding spherical polar coordinates, taking $\hat{\boldsymbol{\zeta}}$ as the polar axis. Writing the unit operator as $(\hat{\boldsymbol{\xi}}\hat{\boldsymbol{\xi}} + \hat{\boldsymbol{\eta}}\hat{\boldsymbol{\eta}} + \hat{\boldsymbol{\zeta}}\hat{\boldsymbol{\zeta}})$ gives

$$\hat{\mathbf{t}}(A) \cdot \hat{\mathbf{t}}(C) = \hat{\mathbf{t}}(A) \cdot (\hat{\boldsymbol{\xi}}\hat{\boldsymbol{\xi}} + \hat{\boldsymbol{\eta}}\hat{\boldsymbol{\eta}} + \hat{\boldsymbol{\zeta}}\hat{\boldsymbol{\zeta}}) \cdot \hat{\mathbf{t}}(C)$$

$$= \mathbf{t}_\perp(A) \cdot \mathbf{t}_\perp(C) + (\hat{\mathbf{t}}(A) \cdot \hat{\mathbf{t}}(B))(\hat{\mathbf{t}}(B) \cdot \hat{\mathbf{t}}(C)).$$

In the first term, the symbol $\mathbf{t}_\perp$ represents the projection of $\hat{\mathbf{t}}$ to the $\xi\eta$ plane. Choosing the $\hat{\boldsymbol{\xi}}$ axis to be along $\mathbf{t}_\perp(A)$ yields $\hat{\mathbf{t}}_\perp(A) = \sin\vartheta(A)\hat{\boldsymbol{\xi}}$, so

$$\hat{\mathbf{t}}(A) \cdot \hat{\mathbf{t}}(C) = \sin\vartheta(A)\sin\vartheta(C)\cos\phi(C) + \cos\vartheta(A)\cos\vartheta(C). \qquad (9.29)$$

So far we have done only geometry, not statistical physics. We now take the average of both sides of Equation 9.29 over all possible conformations of the polymer, weighted by the Boltzmann factor as usual. The key observations are that

- The first term of Equation 9.29 vanishes upon averaging. This result follows because the energy functional, Equation 9.2, doesn't care which direction the rod bends—it's isotropic. Thus, for every conformation with a particular value of $\phi(C)$, there is another with the same energy but a different value of $\phi(C)$, so our averaging over conformations includes integrating the right-hand side of Equation 9.29 over all values of $\phi(C)$. But the integral $\int_0^{2\pi} d\phi \cos\phi$ equals zero.[10]

- The second term is the product of two statistically independent factors. The shape of our rod between A and B makes a contribution ϵ_{AB} to its elastic energy and also determines the angle $\vartheta(A)$. The shape between B and C makes a contribution ϵ_{BC} to the energy and determines the angle $\vartheta(C)$. The Boltzmann weight for this conformation can be written as the product of $e^{-\epsilon_{AB}/k_\mathrm{B}T}$ (which does not involve $\vartheta(C)$), times $e^{-\epsilon_{BC}/k_\mathrm{B}T}$ (which does not involve $\vartheta(A)$), times other factors involving neither $\vartheta(A)$ nor $\vartheta(C)$. Thus the average of the product $\cos\vartheta(A)\cos\vartheta(C)$ equals the product of the averages, by the multiplication rule for probabilities.

Let's write $\mathcal{A}(x)$ for the **autocorrelation function** $\langle\hat{\mathbf{t}}(s)\cdot\hat{\mathbf{t}}(s+x)\rangle$; for a long chain this quantity does not depend on the starting point s chosen. Then the preceding logic implies that Equation 9.29 can be rewritten as

$$\mathcal{A}(s_{AB} + s_{BC}) = \mathcal{A}(s_{AB}) \times \mathcal{A}(s_{BC}). \qquad (9.30)$$

The only function with this property is the exponential, $\mathcal{A}(x) = e^{qx}$ for some constant q.

To finish the proof of Equation 9.28, then, we only need to show that the constant q equals $-1/A$. But for very small $\Delta s \ll A$, thermal fluctuations can hardly bend the rod at all (recall Equation 9.4 on page 347). Consider a circular arc in which $\hat{\mathbf{t}}$ bends by a small angle ψ in the $\xi\zeta$ plane as s increases by Δs. That is, suppose that $\hat{\mathbf{t}}$ changes from $\hat{\mathbf{t}}(s) = \hat{\boldsymbol{\zeta}}$ to $\hat{\mathbf{t}}(s+\Delta s) = (\hat{\boldsymbol{\zeta}} + \psi\hat{\boldsymbol{\xi}})/\sqrt{1+\psi^2}$, which is again a unit vector. Adapting Equation 9.4 for this situation yields the elastic energy cost as $(\frac{1}{2}k_\mathrm{B}TA) \times (\Delta s) \times (\Delta s/\psi)^{-2}$, or $(Ak_\mathrm{B}T/(2\Delta s))(\psi)^2$. This expression is a quadratic function of ψ. The equipartition of energy (Your Turn 6F on page 219) then tells us that the thermal average of this quantity will be $\frac{1}{2}k_\mathrm{B}T$, or that

$$\frac{A}{\Delta s}\langle\psi^2\rangle = 1.$$

---

[10]We used the same logic to discard the middle term of Equation 4.3 on page 115.

Repeating the argument for bends in the $\xi\eta$ plane, and remembering that ψ is small, we find

$$A(\Delta s) = \langle \hat{\mathbf{t}}(s) \cdot \hat{\mathbf{t}}(s + \Delta s) \rangle = \left\langle \hat{\boldsymbol{\zeta}} \cdot \frac{\hat{\boldsymbol{\zeta}} + \psi_{\xi\zeta}\hat{\boldsymbol{\xi}} + \psi_{\eta\zeta}\hat{\boldsymbol{\eta}}}{\sqrt{1 + (\psi_{\xi\zeta})^2 + (\psi_{\eta\zeta})^2}} \right\rangle$$

$$\approx 1 - \tfrac{1}{2}\langle(\psi_{\xi\zeta})^2\rangle - \tfrac{1}{2}\langle(\psi_{\eta\zeta})^2\rangle$$

$$= 1 - \Delta s/A.$$

Comparing this result with $A(x) = e^{qs} \approx 1 + qs + \cdots$ indeed shows that $q = -1/A$, finally establishing Equation 9.28, and with it, the interpretation of A as a persistence length.

2. To make the connection between an elastic rod model and its corresponding FJC model, we now consider the mean-square end-to-end length $\langle \mathbf{r}^2 \rangle$ of an elastic rod. Because the rod segment at s points in the direction $\hat{\mathbf{t}}(s)$, we have $\mathbf{r} = \int_0^{L_{\text{tot}}} ds\, \hat{\mathbf{t}}(s)$, so

$$\langle \mathbf{r}^2 \rangle = \left\langle \left(\int_0^{L_{\text{tot}}} ds_1\, \hat{\mathbf{t}}(s_1) \right) \cdot \left(\int_0^{L_{\text{tot}}} ds_2\, \hat{\mathbf{t}}(s_2) \right) \right\rangle$$

$$= \int_0^{L_{\text{tot}}} ds_1 \int_0^{L_{\text{tot}}} ds_2\, \langle \hat{\mathbf{t}}(s_1) \cdot \hat{\mathbf{t}}(s_2) \rangle = \int_0^{L_{\text{tot}}} ds_1 \int_0^{L_{\text{tot}}} ds_2\, e^{-|s_1 - s_2|/A}$$

$$= 2 \int_0^{L_{\text{tot}}} ds_1 \int_{s_1}^{L_{\text{tot}}} ds_2\, e^{-(s_2 - s_1)/A} = 2 \int_0^{L_{\text{tot}}} ds_1 \int_0^{L_{\text{tot}} - s_1} dx\, e^{-x/A},$$

where $x \equiv s_2 - s_1$. For a long rod, the first integral is dominated by values of s_1 far from the end, so we may replace the upper limit of the second integral by infinity:

$$\langle \mathbf{r}^2 \rangle = 2AL_{\text{tot}}. \qquad \text{(long, unstretched elastic rod)} \qquad (9.31)$$

This is a reassuring result: Exactly as in our simple discussion of random walks (Equation 4.4 on page 115), the mean-square end-to-end separation of a semi-flexible polymer is proportional to its contour length.

We now compute $\langle \mathbf{r}^2 \rangle$ for a freely jointed chain consisting of segments of length L_{seg} and compare it with Equation 9.31. In this case, we have a discrete sum over the $N = L_{\text{tot}}/L_{\text{seg}}$ segments:

$$\langle \mathbf{r}^2 \rangle = \sum_{i,j=1}^{N} \langle (L_{\text{seg}}\hat{\mathbf{t}}_i) \cdot (L_{\text{seg}}\hat{\mathbf{t}}_j) \rangle = (L_{\text{seg}})^2 \left[\sum_{i=1}^{N} \langle(\hat{\mathbf{t}}_i)^2\rangle + 2 \sum_{i<j}^{N} \langle \hat{\mathbf{t}}_i \cdot \hat{\mathbf{t}}_j \rangle \right].$$

$$(9.32)$$

By a now-familiar argument, we see that the second term equals zero: For every conformation in which $\hat{\mathbf{t}}_i$ makes a particular angle with $\hat{\mathbf{t}}_j$, there is an equally weighted conformation with the opposite value of $\hat{\mathbf{t}}_i \cdot \hat{\mathbf{t}}_j$, because the stretching force is zero and the joints are assumed to be free. The first term is also simple because, by definition, $(\hat{\mathbf{t}}_i)^2 = 1$ always, so we find

$$\langle \mathbf{r}^2 \rangle = N(L_{\text{seg}})^2 = L_{\text{seg}}L_{\text{tot}}. \qquad \text{(unstretched, three-dimensional FJC)} \qquad (9.33)$$

Comparing Equation 9.33 to Equation 9.31, we find that

> *The freely jointed chain model correctly reproduces the size of the underlying elastic rod's random-coil conformation, if we choose the* (9.34) *effective link length to be $L_{seg} = 2A$.*

3. This chapter regards a polymer as a stack of identical units. Such objects are called homopolymers. Natural DNA, in contrast, is a heteropolymer: It contains a message written in an alphabet of four different bases. But the effect of sequence on the large-scale elasticity of DNA is rather weak, essentially because the AT and GC pairs are geometrically similar to each other. Moreover, it is not hard to incorporate sequence effects into the results of the following sections. These homopolymer results also apply to heteropolymers when A is suitably interpreted as a combination of elastic stiffness and intrinsic disorder.

$\boxed{T_2}$ ### 9.2.2′ Track 2

1. One major weakness of the discussion in Section 9.2.2 is the fact that we used a *one-dimensional* random walk to describe the three-dimensional conformation of the polymer! This weakness is not hard to fix.

   The three-dimensional freely jointed chain has as its conformational variables a set of unit tangent vectors $\hat{t}_i$, which need not point only along the $\pm\hat{z}$ directions: They can point in any direction, as in Equation 9.32. We take $\mathbf{r}$ to be the end-to-end vector, as always; with an applied stretching force in the $\hat{z}$ direction, we know that $\mathbf{r}$ will point along $\hat{z}$. Thus the end-to-end extension z equals $\mathbf{r} \cdot \hat{z}$, or $(\sum_i L_{seg}\hat{t}_i) \cdot \hat{z}$. (The parameter L_{seg} appearing in this formula is not the same as the parameter $L_{seg}^{(1d)}$ in Section 9.2.2.) The Boltzmann weight factor analogous to Equation 9.9 on page 353 is then

$$P(\hat{t}_1, \ldots, \hat{t}_N) = Z^{-1}e^{-\left(-fL_{seg}\sum_i \hat{t}_i \cdot \hat{z}\right)/k_B T}. \qquad (9.35)$$

> **Your Turn 9O**
>
> If you haven't done Problem 6.9 yet, do it now. Then adapt Your Turn 9B on page 354 to arrive at an expression for the extension of a three-dimensional FJC. Again find the limiting form of your result at very low force.

The expression you just found is shown as the third curve from the top in Figure 9.5 on page 355. Your answer to Your Turn 9O shows why this time we took $L_{seg} = 104\,\text{nm}$. Indeed, the three-dimensional FJC gives a somewhat better fit to the experimental data than Equation 9.10 did.

2. The effective spring constant of a real polymer won't really be strictly proportional to the absolute temperature, as implied by Idea 9.11 on page 354, because the bend persistence length and hence L_{seg} themselves depend on temperature. Nevertheless, our qualitative prediction that the effective spring constant increases with temperature is observed experimentally (see Section 9.1.3).

9.3.2' Track 2

The ideas in Section 9.3.2 can be generalized to higher-dimensional spaces. A linear vector function of a k-dimensional vector can be expressed in terms of a $k \times k$ matrix M. The eigenvalues of such a matrix can be found by subtracting an unknown constant λ from each diagonal element of M, then requiring that the resulting matrix have determinant equal to zero. The resulting condition is that a certain polynomial in λ should vanish; the roots of this polynomial are the eigenvalues of M. An important special case is when M is real and symmetric; in this case, we are guaranteed that there will be k linearly independent real eigenvectors, so any other vector can be written as some linear combination of them. Indeed, in this case, all the eigenvectors can be chosen to be mutually perpendicular. Finally, if all the entries of a matrix are positive numbers, then one of the eigenvalues has greater absolute value than all the others; this eigenvalue is real, positive, and nondegenerate (the Frobenius–Perron theorem).

9.4.1' Track 2

Even though we found that the 1d cooperative chain fit the experimental data slightly better than the one-dimensional FJC, still it's clear that this is physically a very unrealistic model: We assumed a chain of straight links, each one joined to the next at an angle of either 0° or 180°! Really, each basepair in the DNA molecule is pointing in nearly the *same* direction as its neighbor. We did, however, discover one key fact: that the effective segment length L_{seg} is tens of nanometers long, much longer than the thickness of a single basepair (0.34 nm). This observation means that we can use our phenomenological elastic energy formula (Equation 9.3 on page 346) as a more accurate substitute for Equation 9.17 on page 359.

    Thus, "all" we need to do is to evaluate the partition function starting from Equation 9.3, then imitate the steps leading to Equation 9.10 on page 353 to get the force–extension relation of the three-dimensional elastic rod model. The required analysis was begun in the 1960s by N. Saito and coauthors, then completed in 1994 by J. Marko and E. Siggia, and by A. Vologodskii. (For many more details, see Marko & Siggia, 1995.)

    Unfortunately, the mathematics needed to carry out the program just sketched is somewhat more involved than in the rest of this book. But when faced with such beautifully clean experimental data as those in the figure, and with such an elegant model as Equation 9.3, we really have no choice but to go the distance and compare them carefully.

    We will treat the elastic rod as consisting of N discrete links, each of length ℓ. Our problem is more difficult than the one-dimensional chain because the configuration variable is no longer the discrete, two-valued $\sigma_i = \pm 1$; instead, it is the continuous variable $\hat{\mathbf{t}}_i$ describing the orientation of link number i. Thus the transfer matrix T has *continuous indices*.[11] To find T, we write the partition function at a fixed external

---

[11]This concept may be familiar from quantum mechanics. Such infinite-dimensional matrices are sometimes called kernels.

force f, analogous to Equation 9.18 on page 359:

$$Z(f) = \int d^2\hat{\mathbf{t}}_1 \cdots d^2\hat{\mathbf{t}}_N \, \exp\left[\sum_{i=1}^{N-1} \left(\frac{f\ell}{2k_B T}(\cos\theta_i + \cos\theta_{i+1}) - \frac{A}{2\ell}(\Theta_{i,i+1})^2 \right) \right.$$
$$\left. + \frac{f\ell}{2k_B T}(\cos\theta_1 + \cos\theta_N) \right]. \tag{9.36}$$

In this formula, the N integrals are over directions—each $\hat{\mathbf{t}}_i$ runs over the unit sphere. θ_i is the angle between link i's tangent and the direction $\hat{\mathbf{z}}$ of the applied force; in other words, $\cos\theta_i = \hat{\mathbf{t}}_i \cdot \hat{\mathbf{z}}$. Similarly, $\Theta_{i,i+1}$ is the angle between $\hat{\mathbf{t}}_i$ and $\hat{\mathbf{t}}_{i+i}$; Section 9.1.3' on page 386 showed that the elastic energy cost of a bend is $(Ak_B T/2\ell)\Theta^2$. Because each individual bending angle will be small, we can replace Θ^2 by the more convenient function $2(1 - \cos\Theta)$. Note that we have written every force term twice and divided by 2; the reason for this choice will become clear in a moment.

Exactly as in Your Turn 9G, we can reformulate Equation 9.36 as a matrix raised to the power $N-1$, sandwiched between two vectors. We again need the largest eigenvalue of the matrix T. Remembering that our objects now have continuous indices, a "vector" $\mathbf{V}$ in this context is specified by a *function*, $V(\hat{\mathbf{t}})$. The "matrix product" is the integral

$$(\mathsf{T}\mathbf{V})(\hat{\mathbf{t}}) = \int d^2\hat{\mathbf{n}} \, T(\hat{\mathbf{t}}, \hat{\mathbf{n}})V(\hat{\mathbf{n}}), \tag{9.37}$$

where

$$T(\hat{\mathbf{t}}, \hat{\mathbf{n}}) \equiv \exp\left[\frac{f\ell}{2k_B T}(\hat{\mathbf{t}} \cdot \hat{\mathbf{z}} + \hat{\mathbf{n}} \cdot \hat{\mathbf{z}}) + \frac{A}{\ell}(\hat{\mathbf{n}} \cdot \hat{\mathbf{t}} - 1) \right]. \tag{9.38}$$

The reason for our apparently perverse duplication of the force terms is that now T is a symmetric matrix, so the mathematical facts quoted in Section 9.3.2' on page 390 apply.

We will use a simple technique—the **Ritz variational approximation**—to estimate the maximal eigenvalue (see Marko & Siggia, 1995).[12] The matrix T is real and symmetric. As such, it must have a basis set of mutually perpendicular eigenvectors $\mathbf{e}_i$ satisfying $\mathsf{T}\mathbf{e}_i = \lambda_i \mathbf{e}_i$ with real, positive eigenvalues λ_i. *Any* vector $\mathbf{V}$ may be expanded in this basis: $\mathbf{V} = \sum_i c_i \mathbf{e}_i$. We next consider the "estimated maximal eigenvalue"

$$\lambda_{\text{max,est}} \equiv \frac{\mathbf{V} \cdot (\mathsf{T}\mathbf{V})}{\mathbf{V} \cdot \mathbf{V}} = \frac{\sum_i \lambda_i(c_i)^2 \mathbf{e}_i \cdot \mathbf{e}_i}{\sum_i (c_i)^2 \mathbf{e}_i \cdot \mathbf{e}_i}. \tag{9.39}$$

The last expression on the right cannot exceed the largest eigenvalue, $\lambda_{\max}$. It equals $\lambda_{\max}$ when we choose $\mathbf{V}$ to be equal to the corresponding eigenvector $\mathbf{e}_0$, that is, when $c_0 = 1$ and the other $c_i = 0$.

---

[12]A more general approach to maximal-eigenvalue problems of this sort is to find a basis of mutually orthogonal functions of $\hat{\mathbf{t}}$, expand $V(\hat{\mathbf{t}})$ in this basis, truncate the expansion after a large but finite number of terms, evaluate T on this truncated subspace, and use numerical software to diagonalize it.

Suppose that there is one maximal eigenvalue λ_0. Show that $\lambda_{\text{max,est}}$ is maximized precisely when $\mathbf{V}$ is a constant times $\mathbf{e}_0$. [*Hints:* Try the 2×2 case first; here you can see the result explicitly. For the general case, let $x_i = (c_i/c_0)^2$, $A_i = (\mathbf{e}_i \cdot \mathbf{e}_i)/(\mathbf{e}_0 \cdot \mathbf{e}_0)$, and $L_i = (\lambda_i/\lambda_0)$ for $i > 1$. Then show that the estimated eigenvalue has no maximum other than at the point where all the $x_i = 0$.]

To estimate $\lambda_{\max}$, then, we need to find the function $V_0(\hat{\mathbf{t}})$ that saturates the bound $\lambda_{\text{max,est}} \leq \lambda_{\max}$, or in other words, that maximizes Equation 9.39. This task may sound as hard as finding a needle in the infinite-dimensional haystack of functions $V(\hat{\mathbf{t}})$. The trick is to use physical reasoning to select a promising *family* of trial functions, $V_w(\hat{\mathbf{t}})$, depending on a parameter w. We substitute $V_w(\hat{\mathbf{t}})$ into Equation 9.39 and choose the value w_* that maximizes our estimated eigenvalue. The corresponding $V_{w_*}(\hat{\mathbf{t}})$ is then our best proposal for the true eigenvector, $V_0(\hat{\mathbf{t}})$. Our estimate for the true maximal eigenvalue $\lambda_{\max}$ is then $\lambda_* \equiv \lambda_{\text{max,est}}(w_*)$.

To make a good choice for the family of trial functions $V_w(\hat{\mathbf{t}})$, we need to think a bit about the physical meaning of $\mathbf{V}$. You found in Your Turn 9G(c) that the average value of a link variable can be obtained by sandwiching $\left[\begin{smallmatrix} 1 & 0 \\ 0 & -1 \end{smallmatrix}\right]$ between two copies of the dominant eigenvector $\mathbf{e}_0$. At zero force, each link of our chain should be equally likely to point in all directions, whereas at high force, it should be most likely to point in the $+\hat{\mathbf{z}}$ direction. In either case, the link should not prefer any particular azimuthal direction φ. With these considerations in mind, Marko and Siggia constructed a family of smooth trial functions, azimuthally symmetrical and peaked in the forward direction:

$$V_w(\hat{\mathbf{t}}) = \exp[w\hat{\mathbf{t}} \cdot \hat{\mathbf{z}}].$$

Thus, for each value of the applied force f, we must evaluate Equation 9.39, using Equations 9.37 and 9.38, then find the value w_* of the parameter w that maximizes $\lambda_{\text{max,est}}$, and substitute to get λ_*.

Let $v = \hat{\mathbf{t}} \cdot \hat{\mathbf{z}}$. We first need to evaluate

$$(\mathbf{T}V_w)(\hat{\mathbf{t}}) = e^{f\ell v/(2k_B T)} e^{-A/\ell} \int d^2\hat{\mathbf{n}} \, \exp\left[\frac{A}{\ell}\hat{\mathbf{n}} \cdot \hat{\mathbf{t}} + \frac{f\ell}{2k_B T}\hat{\mathbf{n}} \cdot \hat{\mathbf{z}} + w\hat{\mathbf{n}} \cdot \hat{\mathbf{z}}\right].$$

To do the integral, abbreviate $\zeta = w + \frac{f\ell}{2k_B T}$ and $\tilde{A} = A/\ell$. The integrand can then be written as $\exp[Q\hat{\mathbf{m}} \cdot \hat{\mathbf{n}}]$, where $\hat{\mathbf{m}}$ is the unit vector $\hat{\mathbf{m}} = (\tilde{A}\hat{\mathbf{t}} + \zeta\hat{\mathbf{z}})/Q$ and

$$Q = \|\tilde{A}\hat{\mathbf{t}} + \zeta\hat{\mathbf{z}}\| = \sqrt{\tilde{A}^2 + \zeta^2 + 2\tilde{A}\zeta v}.$$

We can write the integral using spherical polar coordinates ϑ and ϕ, choosing $\hat{\mathbf{m}}$ as the polar axis. Then $\int d^2\hat{\mathbf{n}} = \int_0^{2\pi} d\phi \int_{-1}^{1} d\mu$, where $\mu = \cos\vartheta$, and the integral becomes simply $\frac{2\pi}{Q}(e^Q - e^{-Q})$.

To evaluate the numerator of Equation 9.39, we need to do a second integral, over $\hat{\mathbf{t}}$. This time choose polar coordinates θ, φ with $\hat{\mathbf{z}}$ as the polar axis. Recalling that

$v \equiv \hat{\mathbf{t}} \cdot \hat{\mathbf{z}} = \cos\theta$, we find

$$\mathbf{V}_w \cdot (\mathsf{T}\mathbf{V}_w) = \int d^2\hat{\mathbf{t}}\, V_w(\hat{\mathbf{t}})(\mathsf{T}\mathbf{V}_w)(\hat{\mathbf{t}})$$

$$= e^{-\tilde{A}} 2\pi \int_{-1}^{+1} dv\, e^{\zeta v} \frac{2\pi}{Q}(e^Q - e^{-Q})$$

$$= e^{-\tilde{A}}(2\pi)^2 \int_{|\tilde{A}-\zeta|}^{\tilde{A}+\zeta} \frac{dQ}{\tilde{A}\zeta}\, e^{(Q^2 - \tilde{A}^2 - \zeta^2)/(2\tilde{A})}(e^Q - e^{-Q}). \qquad (9.40)$$

The last step changed variables from v to Q. The final integral in Equation 9.40 is not an elementary function, but you may recognize it as related to the error function (see Section 4.6.5′ on page 150).

Your Turn 9Q

Next evaluate the denominator of Equation 9.39 for our trial function V_w.

Having evaluated the estimated eigenvalue on the family of trial functions, it is now straightforward to maximize the result over the parameter w using mathematical software, to obtain λ_* as a function of A, ℓ, and f. For ordinary (double-stranded) DNA, the answer is nearly independent of the link length ℓ, as long as $\ell < 2\,\mathrm{nm}$. We can then finish the calculation by following the analysis leading to Equation 9.10 on page 353: For large $N = L_{\mathrm{tot}}/\ell$,

$$\langle z/L_{\mathrm{tot}} \rangle = \frac{k_{\mathrm{B}}T}{L_{\mathrm{tot}}} \frac{d}{df} \ln\big(\mathbf{W} \cdot (\mathsf{T}^{N-1}\mathbf{V})\big) \approx \frac{k_{\mathrm{B}}T}{\ell} \frac{d}{df} \ln \lambda_*(f). \qquad (9.41)$$

The force–extension curve given by the Ritz approximation (Equation 9.41) turns out to be practically indistinguishable from the exact solution (see Problem 9.7). The exact result cannot be written in closed form (see Marko & Siggia, 1995; Bouchiat et al., 1999). For reference, however, here is a simple expression that is very close to the exact result in the limit $\ell \to 0$:

$$\langle z/L_{\mathrm{tot}} \rangle = h(f) + 1.86h(f)^2 - 3.80h(f)^3 + 1.94h(f)^4,$$

$$\text{where } h(f) = 1 - \tfrac{1}{2}\left(\sqrt{\bar{f} + \tfrac{9}{4}} - 1\right)^{-1}. \qquad (9.42)$$

In this formula, $\bar{f} = fA/k_{\mathrm{B}}T$. Adjusting A to fit the experimental data gave the solid black curve shown in Figure 9.5.

$\boxed{T_2}$ **9.5.1′ Track 2**

The angle of optical rotation is not intrinsic to the molecular species under study: It depends on the concentration of the solution and so on. To cure this defect, biophysical chemists define the specific optical rotation as the rotation angle θ divided by

$\rho_m d/(100\,\text{kg m}^{-2})$, where ρ_m is the mass concentration of solute and d is the path length through solution traversed by the light. The data in Figure 9.7 show specific optical rotation (at the wavelength of the sodium D-line). With this normalization, the three different curves effectively all have the same total concentration of monomers and so may be directly compared.

9.5.3′ Track 2

1. Our discussion focused on hydrogen-bonding interactions between monomers in a polypeptide chain. Various other interactions are also known to contribute to the helix–coil transition, for example, dipole–dipole interactions. Their effects can be summarized in the values of the phenomenological parameters of the transition, which we fit to the data.

2. The analysis of Section 9.5.2 did not take into account the polydispersity of real polymer samples. We can make this correction in a rough way as follows.

    Suppose that a sample contains a number X_j of chains of length j. Then the fraction is $f_j = X_j/(\sum_k X_k)$, and the **number-averaged chain length** is defined as $N_n \equiv \sum_j (jf_j)$. Another kind of average can also be determined experimentally, namely, the **weight-averaged chain length** $N_w \equiv (1/N_n)\sum_j (j^2 f_j)$.

    Zimm, Doty, and Iso quoted the values ($N_n = 40, N_w = 46$) and ($N_n = 20, N_w = 26$) for their two short-chain samples. Let's model these samples as each consisting of *two* equal subpopulations, of lengths k and m. Then choosing $k = 55, m = 24$ reproduces the number- and weight-averaged lengths of the first sample; similarly, $k = 31, m = 9$ model the second sample. The lower two curves in Figure 9.7 actually show a weighted average of the result following from Your Turn 9L(c), assuming the two subpopulations just mentioned. Introducing the effect of polydispersity, even in this crude way, does improve the fit to the data somewhat.

9.6.3′ Track 2

Austin and coauthors obtained the fits shown in Figure 9.13 as follows: Suppose that each conformational substate has a rebinding rate related to its energy barrier $\Delta E^{\ddagger}$ by an Arrhenius relation, $k(\Delta E^{\ddagger}, T) = Ae^{-\Delta E^{\ddagger}/k_B T}$. The subpopulation in a given substate will relax exponentially, with this rate, to the bound state. We assume that the prefactor A does not depend strongly on temperature in the range studied. Let $g(\Delta E^{\ddagger})d\Delta E^{\ddagger}$ denote the fraction of the population in substates with barrier between $\Delta E^{\ddagger}$ and $\Delta E^{\ddagger} + d\Delta E^{\ddagger}$. We also neglect any temperature dependence in the distribution function $g(\Delta E^{\ddagger})$.

The total fraction in the unbound state at time t will then be

$$N(t, T) = N_0 \int d\Delta E^{\ddagger}\, g(\Delta E^{\ddagger})e^{-k(\Delta E^{\ddagger}, T)t}. \tag{9.43}$$

The normalization factor N_0 is chosen to get $N = 1$ at time zero. Austin and coauthors found that they could fit the rebinding data at $T = 120$ K by taking the population function $g(\Delta E^{\ddagger}) = \bar{g}(Ce^{-\Delta E^{\ddagger}/(k_B \times 120 \, \text{K})})$, where $C = 5.0 \cdot 10^8 \, \text{s}^{-1}$ and

$$\bar{g}(x) = \frac{(x/(67\,000\,\text{s}^{-1}))^{0.325} e^{-x/(67\,000\,\text{s}^{-1})}}{2.76 \,\text{kJ mole}^{-1}} \qquad \text{when } x < 23 \,\text{kJ mole}^{-1}. \qquad (9.44)$$

(The normalization constant has been absorbed into N_0.) Above the cutoff energy of 23 kJ mole^{-1}, $g(x)$ was taken to be zero. Equation 9.44 gives the curve shown in Figure 9.13b. Substituting $g(\Delta E^{\ddagger})$ into Equation 9.43 (the rebinding curve) at various temperatures gives the curves in Figure 9.13a.

Austin and coauthors also ruled out an alternative hypothesis, that all the protein molecules in the sample are identical but that each rebinds nonexponentially.

PROBLEMS

9.1   Big business

DNA is a highly *charged* polymer. That is, its neutral form is a salt, with many small positive counterions that dissociate and wander away in water solution. A charged polymer of this type is called a polyelectrolyte. A very big industrial application for polyelectrolytes is in the gels filling disposable diapers. What physical properties of polyelectrolytes do you think make them especially suitable for this critical technology?

9.2   Geometry of bending

Verify Equation 9.4 on page 347 explicitly as follows. Consider the circular arc in the xy plane defined by $\mathbf{r}(s) = (R\cos(s/R), R\sin(s/R))$ (see Figure 9.1). Show that s really is contour length, find the unit tangent vector $\hat{\mathbf{t}}(s)$ and its derivative, and thereby verify Equation 9.4.

9.3   Energy sleuthing

The freely jointed chain picture is a simplification of the real physics of a polymer: Actually, the joints are not quite free. Each polymer molecule consists of a chain of identical individual units, which stack best in a straight line (or in a helix with a straight axis). Thus Equation 9.2 on page 346 says that bending the chain into a tangle *costs* energy. And yet, a rubber band certainly can do work on the outside world as it retracts. Reconcile these observations qualitatively: Where does the energy needed to do mechanical work come from?

9.4   Thermodynamics of rubber

Take a wide rubber band. Hold it to your upper lip (moustache wearers may use some other sensitive, but public, part) and rapidly stretch it. Leave it stretched for a moment, then rapidly let it relax while still in contact with your lip. You will feel distinct thermal phenomena during each process.

a. Discuss what happened upon stretching, both in terms of energy and in terms of order.

b. Similarly discuss what happened upon release.

9.5   Simplified helix–coil transition

In this problem, you'll work through a simplified version of the cooperative helix–coil transition, assuming that the transition is *infinitely* cooperative. That is, each polypeptide molecule is assumed to be either all alpha helix or all random coil. The goal of the problem is to understand qualitatively a key feature of the experimental data shown in Figure 9.7 on page 365, namely, that longer chains have a sharper helix–coil transition. Let the chain have N amino acid units.

a. The observed optical rotation θ of the solution varies continuously from $\theta_{\min}$ to $\theta_{\max}$ as we change experimental conditions. How can an all-or-none model reproduce this observed behavior?

b. Section 9.5.1 argued that, for the conditions in Doty and Iso's experiments,

- The alpha helix form has greater energy per monomer than the random-coil form, or $\Delta E_{bond} > 0$.

- Forming the H-bond increases the entropy of the solvent, by an amount $\Delta S_{bond} > 0$.

- But forming a bond also *decreases* the molecule's conformational entropy, by $\Delta S_{conf} < 0$.

The total free energy change to extend a helical region is then $\Delta G = \Delta E_{bond} - T\Delta S_{bond} - T\Delta S_{conf}$. Suppose that the total free energy change for conversion of a chain were simply $N\Delta G$. What then would be the expected temperature dependence of θ? [*Hints:* Find the probability of being in the alpha helix form as a function of ΔE_{bond}, ΔS, N, and T and sketch a graph as a function of T. Don't forget to normalize your probability distribution properly. Make sure that the limiting behavior of your formula at very large and very small temperatures is physically reasonable.]

c. How does the sharpness of the transition depend on N? Explain that result physically.

d. The total free energy change for conversion of a chain is *not* simply $N\Delta G$, however, as a result of end effects. Instead suppose that the last two residues at each end are unable to benefit from the net free energy reduction of H-bonding. What is the physical origin of this effect? Again find the expected temperature dependence of θ. [*Hint:* Same hint as in (b).]

e. Continuing part (d), find the temperature T_m at which θ is halfway between θ_{min} and θ_{max}, including end effects. How does T_m depend on N? This is an experimentally testable qualitative prediction; compare it with Figure 9.7 on page 365.

9.6 $\boxed{T_2}$ *High-force limit*

The analysis of DNA stretching experiments in Sections 9.2.2–9.4.1 made a number of simplifications out of sheer expediency. Most egregious was working in one dimension: every link pointed either along $+\hat{\mathbf{z}}$ or $-\hat{\mathbf{z}}$, so every link angle was either 0 or π. In real life, every link points nearly (but not quite) *parallel* to the previous one. Section 9.4.1' on page 390 took this fact into account, but the analysis was very difficult. In this problem, you'll find a shortcut applicable to the high-force end of the stretching curve. You'll obtain a formula that, in this limit, agrees with the full elastic rod model.

In the high-force limit, the curve describing the rod's centerline is nearly straight. Thus at the point a distance s from the end, the tangent to the rod $\hat{\mathbf{t}}(s)$ is nearly pointing along the $\hat{\mathbf{z}}$ direction. Let $\mathbf{t}_\perp$ be the projection of $\hat{\mathbf{t}}$ to the xy plane; thus $\mathbf{t}_\perp$'s length is very small. Then

$$\hat{\mathbf{t}}(s) = M(s)\hat{\mathbf{z}} + \mathbf{t}_\perp(s), \tag{9.45}$$

where $M = \sqrt{1 - (\mathbf{t}_\perp)^2} = 1 - \frac{1}{2}(\mathbf{t}_\perp)^2 + \cdots$. The ellipsis denotes terms of higher order in powers of $\mathbf{t}_\perp$.

In terms of the small variables $\mathbf{t}_\perp(s) = (t_1(s), t_2(s))$, the bending term $\boldsymbol{\beta}^2$ equals $(\dot{t}_1)^2 + (\dot{t}_2)^2 + \cdots$. ($\dot{t}_i$ denotes dt_i/ds.) Thus the elastic bending energy of any configuration of the rod is (see Equation 9.3 on page 346)

$$E = \tfrac{1}{2} k_B T A \int_0^{L_{tot}} ds \, [(\dot{t}_1)^2 + (\dot{t}_2)^2] + \cdots . \qquad (9.46)$$

Just as in Section 9.2.2, we also add a term $-fz$ to Equation 9.46 to account for the external stretching force f. Work in the limit of very long $L_{tot} \to \infty$.

a. Rephrase E in terms of the Fourier modes of t_1 and t_2. [*Hint:* Write $-fz$ as $-f \int_0^{L_{tot}} ds \, \hat{t}(s) \cdot \hat{z}$ and use Equation 9.45. Express $M(s)$ in terms of t_1, t_2 as done after Equation 9.45.] Then E becomes the sum of a lot of decoupled quadratic terms, a little bit (not exactly!) like a vibrating string.

b. What is the mean-square magnitude of each Fourier component of t_1 and t_2? [*Hint:* Think back to Section 6.6.1 on page 218.]

c. We want the mean end-to-end distance $\langle z \rangle / L_{tot}$. Use the answer from (a) to write this in a convenient form. Evaluate it, using your answer to (b).

d. Find the force f needed to stretch the thermally dancing rod to a fraction $1 - \epsilon$ of its full length L_{tot}, where ϵ is small. How does f diverge as $\epsilon \to 0$? Compare your result with the 3d freely jointed chain (Your Turn 9O) and with the 1d cooperative chain (Your Turn 9H on page 361).

9.7   T₂ *Stretching curve of the elastic rod model*

We can get a useful simplification of the solution to the 3d cooperative chain (see Section 9.4.1′ on page 390) by taking the limit of small link length, $\ell \to 0$ (the elastic rod model).

a. Begin with Equation 9.40. Expand this expression in powers of ℓ, holding A, f, and w fixed and keeping the terms of order ℓ^1 and ℓ^2.

b. Evaluate the estimated eigenvalue $\lambda_{max,est}$ as a function of the quantity $\bar{f} \equiv Af/k_B T$, the variational parameter w, and other constants, again keeping only leading terms in ℓ. Show that

$$\ln \lambda_{max,est}(w) = \text{const} + \frac{\ell}{A}\left(-\frac{1}{2w} + \coth 2w\right)\left(\bar{f} - \frac{1}{2}w\right).$$

The first term is independent of f and w, so it won't contribute to Equation 9.41.

c. Even if you can't do (b), proceed using the result given there. Use some numerical software to maximize $\ln \lambda_{max,est}$ over w, and call the result $\ln \lambda_*(\bar{f})$. Evaluate Equation 9.41 and graph $\langle z/L_{tot} \rangle$ as a function of f. Also plot the high-precision result (Equation 9.42) and compare the plot with your answer, which used the Ritz variational approximation.

9.8   T₂ *Low-force limit of the elastic rod model*

a. If you didn't do Problem 9.7, take as given the result in (b). Consider only the case of very low applied force, $f \ll k_B T/A$. In this case, you can do the maximization analytically (on paper). Do it, find the relative extension by using Equation 9.41, and explain why you "had to" get a result of the form you did get.

b. In particular, confirm the identification $L_{seg} = 2A$ already found in Section 9.1.3′ on page 386 by comparing the low-force extension of the fluctuating elastic rod with that of the 3d freely jointed chain (see Your Turn 9O on page 389).

9.9 $\boxed{T_2}$ Twist and pop

A stretched macroscopic spring pulls back with a force f, which increases linearly with the extension z as $f = -kz$. Another familiar example is the *torsional spring:* It resists *twisting* by an angle θ, generating a *torque*

$$\tau = -k_t\theta. \tag{9.47}$$

Here k_t is called the torsional spring constant. To make sure you understand this formula, show that the dimensions of k_t are the same as those of energy.

It is possible to subject DNA to torsional stress, too. One way to accomplish this is by using an enzyme called **ligase**, which joins the two ends of a piece of DNA together. The two sugar-phosphate backbones of the DNA duplex then form two separate, closed loops. Each of these loops can bend (DNA is flexible), but they cannot break or pass through each other. Thus their degree of linking is fixed—it's a "topological invariant."

If we ligate a collection of identical, open DNA molecules at low concentration, the result is a mixture of various loop types (**topoisomers**), all chemically identical but topologically distinct.[13] Each topoisomer is characterized by a **linking number** M. If we measure M relative to the most relaxed possibility, then we can think of it as the number of extra turns that the DNA molecule had at the moment when it got ligated. M may be positive or negative; the corresponding total excess angle is $\theta = 2\pi M$. We can separate different topoisomers by electrophoresis, because a "supercoiled" shape (such as a figure-eight) is more compact, and hence will migrate more rapidly, than an open circular loop.

Normally, DNA is a right-handed helix, making one complete right-handed turn every 10.5 basepairs. This normal conformation is called B-DNA. Suppose that we overtwist our DNA; that is, suppose that we apply torsional stress tending to make the double helix tighter (one turn every J basepairs, where $J < 10.5$). Remarkably, it then turns out that the relation between torsional stress and excess linking number really does have the simple linear form shown in Equation 9.47, even though the DNA responds in a complicated way to the stress. The torsional spring constant k_t depends on the length of the loop: A typical value is $k_t = 56 k_B T_r / N$, where N is the number of basepairs in the loop.

When we *under*twist DNA, however, something more spectacular can happen. Instead of responding to the stress by supercoiling, a tract of the DNA loop can pop into a totally different conformation, a *left*-handed helix! This new conformation is called Z-DNA. No chemical bonds are broken in this switch. Z-DNA makes a left-handed turn every K basepairs, where K is a number you will find in a moment. Popping into the Z-form costs free energy, but it also partially relaxes the torsional stress on the rest of the molecule. That is, totally disrupting the duplex structure in a localized region allows a lot of the excess linking number to go there, instead

---

[13] At higher concentration, we may also get some double-length loops.

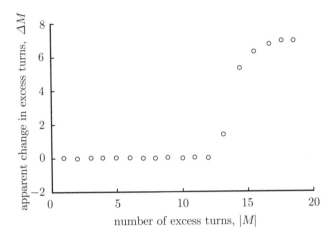

Figure 9.14: (Experimental data.) Evidence for the B–Z transition in a 40 basepair tract inserted into a closed circular loop of DNA (the plasmid pBR322). Each *circle* represents a particular topoisomer of DNA; the topoisomers were separated in a procedure called two-dimensional gel electrophoresis. In the horizontal direction, each circle is placed according to the topoisomer's number of excess turns (the linking number), relative to the most relaxed form. All circles shown correspond to negative excess linking number (tending to unwind the DNA duplex). Placement in the vertical direction reflects the apparent change of excess linking number after a change in the environment has allowed the B–Z transition to take place. [Data from Howell et al., 1996.]

of being distributed throughout the rest of the molecule as torsional strain (small deformations to the B-form helix).

Certain basepair sequences are especially susceptible to popping into the Z-form. Figure 9.14 shows some data taken for a loop of total length $N = 4300$ basepairs. The sequence was chosen so that a tract of length 40 basepairs was able to pop into the Z-form when the torsional stress exceeded some threshold.

Each point of the graph represents a distinct topoisomer of the 4300 basepair loop, with the absolute value of M on the horizontal axis. Only negative values of M (called negative supercoiling) are shown. Beyond a critical number of turns, suddenly the 40 basepair tract pops into the Z-form. As described earlier, this transition lets the rest of the molecule relax; it then behaves under electrophoresis as though $|M|$ had suddenly decreased to $|M| - \Delta M$. The quantity ΔM appears on the vertical axis of the graph.

a. From the data on the graph, find the critical torque τ_{crit} beyond which the transition occurs.

b. Find the number K mentioned earlier. That is, find the number of basepairs per left-handed turn of Z-DNA. Compare with the accepted value $K \approx 12$.

c. How much energy per basepair does it take to pop from B- to Z-form? Is this reasonable?

d. Why do you suppose the transition is so sharp? (Give a qualitative answer.)

CHAPTER 10

Enzymes and Molecular Machines

If ever to a theory I should say:
'You are so beautiful!' and 'Stay! Oh, stay!'
Then you may chain me up and say goodbye—
Then I'll be glad to crawl away and die.
— Delbrück and von Weizacker's update to *Faust*, 1932

A constantly recurring theme of this book has been the idea that living organisms transduce free energy. For example, Chapter 1 discussed how animals eat high-energy molecules and excrete lower-energy molecules, thereby generating not only thermal energy but also mechanical work. We have constructed a framework of ideas allegedly useful for understanding free energy transduction, and we have even presented some primitive examples of how it can work:

- Chapter 1 introduced the osmotic machine (Section 1.2.2); Chapter 7 worked through the details (Section 7.2).
- Section 6.5.3 introduced a motor driven by temperature differences.

Neither of the devices just mentioned is a very good analog of the motors we find in living organisms, however, because neither is driven by chemical forces. Chapter 8 set the stage for the analysis of more biologically relevant machines, developing the notion that chemical bond energy is just another form of free energy. For example, the change ΔG of chemical potential in a chemical reaction was interpreted as a force driving that reaction: The sign of ΔG determines in which direction a reaction will proceed. But we stopped short of explaining how a molecular machine can *harness* a chemical force to drive an otherwise unfavorable transaction, such as doing mechanical work on a load. Understanding how molecules can act as free energy brokers, sitting at the interface between the mechanical and chemical worlds, will be a major goal of this chapter.

Interest in molecular machines blossomed with the realization that much of cellular behavior and architecture depends on the active, directed transport of macromolecules, membranes, or chromosomes within the cell's cytoplasm. Just as disruption of traffic hurts the functioning of a city, so defective molecular transport can result in a variety of diseases.

The subject of molecular machines is vast. Rather than survey the field, this chapter will focus on showing how we can take some familiar mechanical ideas from the macroworld, add just one new ingredient (thermal motion), and obtain a rough picture of how molecular machines work. Thus many important biochemical details will be omitted; just as in Chapter 9, mechanical images will serve as metaphors for subtle chemical details.

This chapter has a character different from that of earlier ones because some of the stories are still unfolding. After outlining some general principles in Sections 10.2 and 10.3, Section 10.4 will look specifically at a remarkable family of real machines, the kinesins. A kinesin molecule's head region is just $4 \times 4 \times 8\,\mathrm{nm}$ in size (smaller than the smallest transistor in your computer) and is built from just 345 amino acid residues. Indeed, kinesin's head region is one of the smallest known natural molecular motors, and possibly the simplest. We will illustrate the interplay between models and experiments by examining two key experiments in some detail. Although the final picture of force generation in kinesin is still not known, still we will see how structural, biochemical, and physical measurements have interlocked to fill in many of the details.

The Focus Question for this chapter is

Biological question: How does a molecular motor convert chemical energy, a *scalar* quantity, into directed motion, a *vector*?

Physical idea: Mechanochemical coupling arises from a free energy landscape with a direction set by the geometry of the motor and its track. The motor executes a biased random walk on this landscape.

10.1 SURVEY OF MOLECULAR DEVICES FOUND IN CELLS

10.1.1 Terminology

This chapter will use the term **molecular device** to designate single molecules (or few-molecule assemblies) falling into two broad classes:

1. **Catalysts** enhance the rate of a chemical reaction. Catalysts created by cells are called **enzymes** (see Section 10.3.3).

2. **Machines** actively reverse the natural flow of some chemical or mechanical process by coupling it to another one. Machines can in turn be roughly divided:

   a. **One-shot** machines exhaust some internal source of free energy. The osmotic machine (Figure 1.3 on page 13) is a representative of this class.

   b. **Cyclic** machines process some external source of free energy such as food molecules, absorbed sunlight, a difference in the concentration of some molecule across a membrane, or an electrostatic potential difference. The heat engine in Section 6.5.3 on page 214 is a representative of this class; it runs on a temperature difference between two external reservoirs. Because cyclic machines are of greatest interest to us, let us subdivide them still further:

      i. **Motors** transduce some form of free energy into motion, either linear or rotary. This chapter will discuss motors abstractly, then focus on a case study, kinesin.

      ii. **Pumps** create concentration differences across membranes.

      iii. **Synthases** drive a chemical reaction, typically the synthesis of some product. An example is ATP synthase, to be discussed in Chapter 11.

A third broad class of molecular devices will be discussed in Chapters 11 and 12: Gated ion channels sense external conditions and respond by changing their permeability to specific ions.

Before embarking on the mathematics, Sections 10.1.2 through 10.1.4 describe a few representative classes of the molecular machines found in cells in order to have some concrete examples in mind as we begin to develop a picture of how such machines work. (Section 10.5 briefly describes still other kinds of motors.)

10.1.2 Enzymes display saturation kinetics

Chapter 3 noted that a chemical reaction, despite having a favorable free energy change, may proceed very slowly because of a large activation energy barrier (Idea 3.28 on page 87). Chapter 8 pointed out that this circumstance gives cells a convenient way to store energy, for example, in glucose or ATP, until it is needed. But what happens when it *is* needed? Quite generally, cells need to speed up the natural rates of many chemical reactions. The most efficient way to do this is with some reusable device—a catalyst.

Enzymes are biological catalysts. Most enzymes are made of protein, sometimes in the form of a complex with other small molecules (called coenzymes or prosthetic groups). Other examples include ribozymes, which consist of RNA. Complex catalytic organelles such as the ribosome (Figure 2.24) are complexes of protein with RNA.

To get a sense of the catalytic power of enzymes, consider the decomposition of hydrogen peroxide at room temperature, $H_2O_2 \rightarrow H_2O + \frac{1}{2}O_2$. This reaction is highly favorable energetically, with $\Delta G^0 = -41k_BT_r$, yet it proceeds very slowly in pure solution: With an initial concentration $1\,\mathrm{M}$ of hydrogen peroxide, the rate of spontaneous conversion at 25°C is just $10^{-8}\,\mathrm{M\,s^{-1}}$. This rate corresponds to a decomposition of just 1% of a sample after two weeks. Various substances can catalyze the decomposition, however. For example, the addition of $1\,\mathrm{mM}$ hydrogen bromide speeds up the reaction by a factor of 10. But the addition of the enzyme **catalase**, at a concentration of binding sites again equal to $1\,\mathrm{mM}$, increases the rate by a factor of 1 000 000 000 000!

Your Turn 10A

Reexpress this fact by giving the number of molecules of hydrogen peroxide that a *single* catalase molecule can split per second.

In your body's cells, catalase breaks down hydrogen peroxide generated by other enzymes (as a by-product of eliminating dangerous free radicals before they can damage the cell).

In the catalase reaction, hydrogen peroxide is called the **substrate** upon which the enzyme acts; the resulting oxygen and water are the **products**. The rate of change of the substrate concentration (here $10^4\,\mathrm{M\,s^{-1}}$) is called the **reaction velocity**. The reaction velocity clearly depends on how much enzyme is present. To get a quantity intrinsic to the enzyme itself, we divide the velocity by the concentration of enzyme[1] (taken to be 1 mM above). Even this number is not completely intrinsic to the enzyme but also reflects the availability (concentration) of the substrate. But most enzymes exhibit **saturation kinetics:** The reaction velocity increases up to a point as we increase substrate concentration, then levels off. Accordingly, we define the **turnover number** of an enzyme as the maximum velocity divided by the concentration of enzyme. The turnover number really is an intrinsic property: It reflects one enzyme molecule's competence at processing substrate when given as much substrate as it can handle. In the case of catalase, the numbers given in the previous paragraph reflect the saturated case, so the maximum turnover number is the quantity you found in Your Turn 10A.

Catalase is a speed champion among enzymes. A more typical example is fumarase, which hydrolyzes fumarate to L-malate,[2] with maximum turnover numbers somewhat above $1000\,\mathrm{s}^{-1}$. This is still an impressive figure, however: It means that a liter of 1 mM fumarase solution can process up to about a mole of fumarate per second, many orders of magnitude faster than a similar reaction catalyzed by an acid.

10.1.3 All eukaryotic cells contain cyclic motors

Section 6.5.3 made a key observation, that the efficiency of a free energy transduction process is greatest when the process involves small, controlled steps. Although we made this observation in the context of heat engines, it should seem reasonable in the chemically driven case as well, leading us to expect that Nature should choose to build even its most powerful motors out of many subunits, each made as small as possible. Indeed, early research on muscles discovered a hierarchy of structures on shorter and shorter length scales (Figure 10.1). As each level of structure was discovered, first by optical and then by electron microscopy, each proved to be not the ultimate force generator but instead a collection of smaller force-generating structures, right down to the molecular level. At the molecular scale, we find the origin of force residing in two proteins: **myosin** (golf club-shaped objects in Figure 10.1) and actin (spherical blobs in Figure 10.1). Actin self-assembles from its globular form (**G-actin**) into filaments (**F-actin,** the twisted chain of blobs in the figure), forming a track to which myosin molecules attach.

The direct proof that single actin and myosin molecules were capable of generating force came from a remarkable set of experiments, called single-molecule **motility assays.** Figure 10.2 summarizes one such experiment. A bead attached to a glass slide carries a small number of myosin molecules. A single actin filament attached at its ends to other beads is maneuvered into position over the stationary myosin by using optical tweezers. The density of myosin on the bead is low enough to ensure that at most one myosin engages the filament at a time. When the fuel molecule ATP is added to the system, the actin filament is observed to take discrete steps in one def-

---

[1] More precisely, we divide by the concentration of active sites, which is the concentration of enzyme times the number of such sites per enzyme molecule. Thus, for example, catalase has four active sites; the rates quoted here actually correspond to a concentration of catalase of 0.25 mM.

[2] Fumarase plays a part in the Krebs cycle (Chapter 11), splitting a water molecule and attaching the fragments to fumarate, thereby converting it to malate.

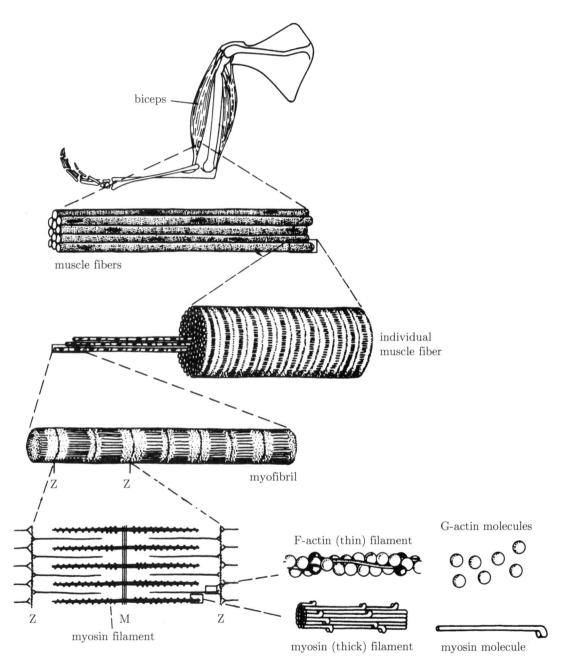

Figure 10.1: (Sketches.) Organization of skeletal muscle at successively higher magnifications. The ultimate generators of force in a myofibril (muscle cell) are bundles of myosin molecules, interleaved with actin filaments (also called F-actin). Upon activation, the myosins crawl along the actin fibers, pulling them toward the plane marked *M* and thus shortening the muscle fiber. [From McMahon, 1984.]

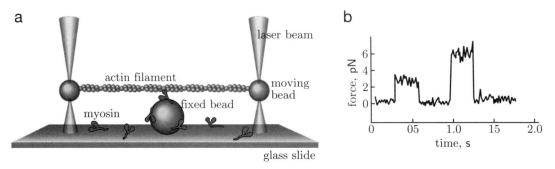

Figure 10.2: (Schematic; experimental data.) Force production by a single myosin molecule. (a) Beads are attached to the ends of an actin filament. Optical tweezers are used to manipulate the filament into position above another, fixed bead coated with myosin fragments. Forces generated by a myosin fragment pull the filament to the side, displacing the beads. The optical trap generates a known springlike force opposing this displacement, so the observed movement of the filament gives a measure of the force generated by the motor. (b) Force generation observed in the presence of 1 μM ATP. The trace shows how the motor takes a step, then detaches from the filament. [Adapted from Finer et al., 1994.]

inite direction away from the equilibrium position set by the optical traps; without ATP, no such stepping is seen. This directed, nonrandom motion occurs without any external macroscopic applied force (unlike, say, electrophoresis).

Muscles are obvious places to look for molecular motors because muscles generate macroscopic forces. Other motors are needed as well, however. In contrast with muscle myosin, many other motors work not in huge teams, but alone, generating tiny, piconewton-scale forces. For example, Section 5.3.1 described how locomotion in *E. coli* requires a rotary motor joining the flagellum to the body of the bacterium; Figure 5.9 on page 176 shows this motor as an assembly of macromolecules just a few tens of nanometers across. In a more indirect argument, Section 4.4.1 argued that passive diffusion alone could not transport proteins and other products synthesized at one place in a cell to the distant places where they are needed; instead some sort of "trucks and highways" are needed to transport these products actively. Frequently, the "trucks" consist of bilayer vesicles. The "highways" are visible in electron microscopy as long protein polymers called microtubules (Figure 2.18 on page 55). Somewhere between the truck and the highway, there must be an "engine."

One particularly important example of such an engine, **kinesin**, was discovered in 1985, in the course of single-molecule motility assays inspired by the earlier work on myosin. Unlike the actin/myosin system, kinesin molecules are designed to walk individually along microtubules (Figure 2.19 on page 56): Often just one kinesin molecule carries an entire transport vesicle toward its destination. Many other organized intracellular motions, for example, the separation of chromosomes during cell division, also imply the existence of motors to overcome viscous resistance to such directed motion. These motors too have been found to be in the kinesin family.[3]

---

[3] Actually, both "kinesin" and "myosin" are large families of related molecules; human cells express about 40 varieties of each. For brevity, we will use these terms to denote the best-studied members in each family: muscle myosin and "conventional" kinesin.

For a more elaborate example of a molecular machine, recall that each cell's genetic script is arranged linearly along a long polymer, the DNA. The cell must copy (or replicate) the script (for cell division) as well as transcribe it (for protein synthesis). An efficient way to perform these operations is to have a single readout machine through which the script is physically pulled. The pulling of a copy requires energy, just as a motor is needed to pull the tape across the read heads of a tape player. The corresponding machines are known as DNA or RNA polymerases for the cases of replication or transcription, respectively (see Section 2.3.4 on page 59). Section 5.3.5 has already noted that some of the chemical energy used by a DNA polymerase must be spent opposing rotational friction of the original DNA and the copy.

10.1.4 One-shot machines assist in cell locomotion and spatial organization

Myosin, kinesin, and polymerases are all examples of cyclic motors; they can take an unlimited number of steps without any change to their own structure, as long as "fuel" molecules are available in sufficient quantities. Other directed, nonrandom motions in cells do not need this property, and for them, simpler one-shot machines can suffice.

Translocation   Some products synthesized inside cells not only must be transported some distance inside the cell, but also must pass across a bilayer membrane to get to their destination. For example, mitochondria import certain proteins that are synthesized in the surrounding cell's cytoplasm. Other proteins need to be pushed outside the cell's outer plasma membrane. Cells accomplish this protein translocation by threading the chain of amino acids through a membrane pore.

Figure 10.3 shows some mechanisms that can help make translocation a one-way process. This motor's "fuel" is the free energy change of the chemical modification the protein undergoes upon emerging into the environment on the right. Once the protein has passed through the pore, there is no need for further activity: A one-shot machine suffices for translocation.

Polymerization   Many cells move, not by cranking flagella or waving cilia (Section 5.3.1), but by extruding their bodies in the direction of desired motion. Such extrusions are variously called pseudopodia, filopodia, or lamellipodia (see Figure 2.9 on page 44). To overcome the viscous friction opposing such motion, the cell's interior structure (including its actin cortex; see Section 2.2.4 on page 54) must push on the cell membrane. To this end, the cell stimulates the growth of actin filaments at the leading edge. At rest, the individual (or monomeric) actin subunits are bound to another small molecule, profilin, which prevents them from sticking to one another. Changes in intracellular pH trigger dissociation of the actin–profilin complex when the cell needs to move; the sudden increase in the concentration of actin monomers then causes them to assemble at the ends of existing actin filaments. To confirm that actin polymerization is capable of changing a cell's shape in this way, it's possible to recreate such behavior in vitro. A similar experiment, involving microtubules, is shown in Figure 10.4: Here the triggered assembly of just a handful of microtubules suffices to distend an artificial bilayer membrane.

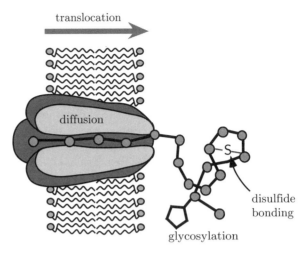

Figure 10.3: (Schematic.) Translocation of a protein through a pore in a membrane. Outside the cell (right side of figure), several mechanisms can rectify (make unidirectional) the diffusive motion of the protein through the pore, for example, disulfide bond formation and attachment of sugar groups (glycosylation). In addition, various chemical asymmetries between the cell's interior and exterior environment could enhance chain coiling outside the cell, thus preventing reentry. These asymmetries could include differences in pH and ion concentrations. [Adapted from Peskin et al., 1993.]

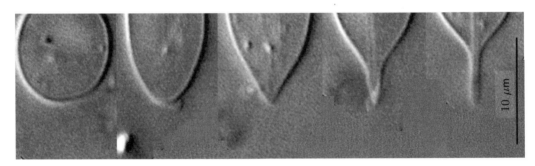

Figure 10.4: (Photomicrograph.) Microtubule polymerization distending an artificial bilayer membrane. Several microtubules gradually distort an initially spherical vesicle by growing inside it at about $2\,\mu$m per minute. [Digital image kindly supplied by D. K. Fygenson; see Fygenson et al., 1997.]

Actin polymerization can also get coopted by parasitical organisms. The most famous of these is the pathogenic bacterium *Listeria monocytogenes*, which propels itself through its host cell's cytoplasm by triggering the polymerization of the cell's own actin, thereby forming a bundle behind it. The bundle remains stationary, enmeshed in the rest of the host cell's cytoskeleton, so the force of the polymerization motor propels the bacterium forward. Figure 10.5 shows this scary process at work.

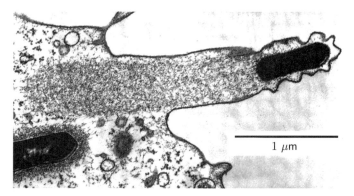

Figure 10.5: (Photomicrograph.) Polymerization from one end of an actin bundle provides the force that propels a *Listeria* bacterium (*black lozenge*) through the cell surface. The long tail behind the bacterium is the network of actin filaments whose assembly it stimulated. [From Tilney & Portnoy, 1989.]

Force generation by the polymerization of actin filaments or microtubules is another example of a machine, in the sense that the chemical binding energy of monomers turns into a mechanical force capable of doing useful work against the cell membrane (or invading bacterium). The machine is of the one-shot variety, because the growing filament is different (it's longer) after every step.[4]

10.2 PURELY MECHANICAL MACHINES

To understand the unfamiliar, we begin with the familiar. Accordingly, this section will examine some everyday macroscopic machines, show how to interpret them in the language of energy landscapes, and develop some terminology.

10.2.1 Macroscopic machines can be described by an energy landscape

Figure 10.6 shows three simple, macroscopic machines. In each panel, external forces acting on the machine are symbolized by weights pulled by gravity. Panel (a) shows a simple one-shot machine: Initially, cranking a shaft of radius R in the direction opposite that of the arrow stores potential energy in the spiral spring. When we release the shaft, the spring unwinds, thereby increasing the angular position θ. The machine can do useful work on an external load, for example, lifting a weight w_1, as long as Rw_1 is less than the torque τ exerted by the spring. If the entire apparatus is immersed in a viscous fluid, then the angular speed of rotation, $d\theta/dt$, will be proportional to $\tau - Rw_1$.

---

[4]Strictly speaking, living cells constantly recycle actin and tubulin monomers by depolymerizing filaments and microtubules and "recharging" them for future use, so perhaps we should not call this a one-shot process. Nevertheless, Figure 10.4 does show polymerization force generated in a one-shot mode.

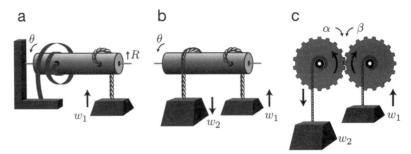

Figure 10.6: (Schematics.) Three simple macroscopic machines. In each case, the weights are not considered part of the machine proper. (a) A coiled spring exerting torque τ lifts weight w_1, driving an increase in the angular position θ. The spring is fastened to a fixed wall at one end and to a rotating shaft at the other; the rope holding the weight winds around the shaft. (b) A weight w_2 falls, lifting a weight w_1. (c) As (b), but the shafts to which w_1 and w_2 are connected are joined by gears. The angular variables α and β both decrease as w_2 lifts w_1.

Your Turn 10B   Explain that last assertion. [*Hint:* Think back to Section 5.3.5 on page 182.]

When the spring is fully unwound, the machine stops.

Figure 10.6b shows a cyclic analog of panel (a). Here the "machine" is simply the central shaft. An external source of energy (weight w_2) drives an external load w_1 against its natural direction of motion, as long as $w_2 > w_1$. This time the machine is a broker transducing a potential energy drop in its source to a potential energy gain in its load. Again, we can imagine introducing so much viscous friction that kinetic energy may be ignored.

Figure 10.6c introduces another level of complexity. Now we have two shafts, with angular positions α and β. The shafts are coupled by gears. For simplicity, suppose that the gears have a 1:1 ratio; so a full revolution of β brings a full revolution of α and vice versa. As in panel (b), we may regard (c) as a cyclic machine.

Our three little machines may seem so simple that they need no further explanation. But for future use, let us pause to extract from Figure 10.6 an abstract characterization of each one.

One-dimensional landscapes Figure 10.7a shows a potential energy graph, or **energy landscape**, for our first machine. The lower dotted line represents the potential energy of the spring. Adding the potential energy of the load (upper dashed line) gives a total (solid line) that decreases with increasing θ. The slope of the total energy is downward, so $\tau = -dU/d\theta$ is a positive net torque. In a viscous medium, the angular speed is proportional to this torque: We can think of the device as "sliding down" its energy landscape.

For the cyclic machine shown in Figure 10.6b, the graph is similar. Here U_{motor} is a constant, but there is a third contribution, $U_{\text{drive}} = -w_2 R\theta$, from the external driving weight, giving the same curve for $U_{\text{tot}}(\theta)$.

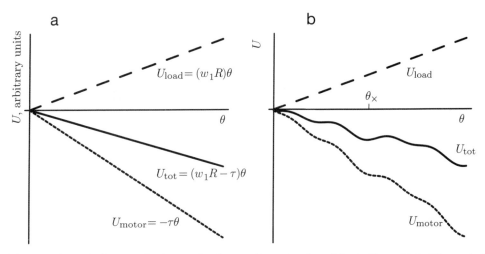

Figure 10.7: (Sketch graphs.) Energy landscapes for the one-dimensional machine in Figure 10.6a. The vertical scale is arbitrary. (a) *Lower dotted line:* The coiled spring contributes $U_{\mathrm{motor}} = -\tau\theta$ to the potential energy. *Upper dashed line:* The external load contributes $U_{\mathrm{load}} = w_1 R\theta$. *Solid line:* The total potential energy function $U_{\mathrm{tot}}(\theta)$ is the sum of these energies; it decreases in time, reflecting the frictional dissipation of mechanical energy into thermal form. (b) The same, but for an imperfect (slightly irregular) shaft. *Solid curve:* Under load, the machine will stop at the point $\theta_\times$. *Lower dotted curve:* Without load, the machine will slow down, but proceed, at $\theta_\times$.

Real machines are not perfect. Irregularities in the pivot may introduce bumps in the potential energy function, "sticky" spots where an extra push is needed to move forward. We can describe this effect by replacing the ideal potential energy $-\tau\theta$ by some other function $U_{\mathrm{motor}}(\theta)$ (lower dotted curve in Figure 10.7b). As long as the resulting total potential energy (solid curve) is everywhere sloping downward, the machine will still run. If a bump in the potential is too large, however, then a minimum forms in U_{tot} (point $\theta_\times$), and the machine will stop there. Note that the meaning of "too large" depends on the load: In the example shown, the unloaded machine *can* proceed beyond $\theta_\times$. Even in the unloaded case, however, the machine will slow down at $\theta_\times$: The net torque $-\mathrm{d}U_{\mathrm{tot}}/\mathrm{d}\theta$ is small at that point, as we see by examining the slope of the dotted curve in Figure 10.7b.

To summarize, the first two machines in Figure 10.6 operate by sliding down the potential energy landscapes shown in Figure 10.7. These landscapes give "height" (that is, potential energy) in terms of one coordinate θ, so we call them "one-dimensional."

Two-dimensional landscapes Our third machine involves gears. In the macroworld, the sort of gears we generally encounter link the angles α and β together rigidly: $\alpha = \beta$, or more generally $\alpha = \beta + 2\pi n/N$, where N is the number of teeth in each gear and n is any integer. But we could also imagine "rubber gears," which can deform and *slip* over each other under high load. Then the energy landscape for this machine will involve two *independent* coordinates, α and β. Figure 10.8 shows an imagined energy landscape for the internal energy U_{motor} of such gears with $N = 3$.

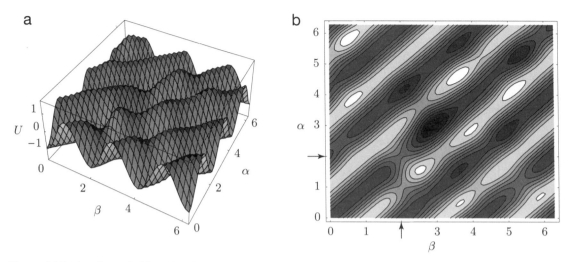

Figure 10.8: (Mathematical functions.) Imagined potential energy landscape for the gear machine in Figure 10.6c, with no load nor driving (but with some imperfections). For clarity, each gear is imagined as having only three teeth. (a) The two horizontal axes are the angles α, β in radians. The vertical axis is potential energy, with arbitrary scale. (b) The same, viewed as a contour map. The dark diagonal stripes are the valleys seen in panel (a). The valley corresponding to the main diagonal has a bump, seen as the light spot at $\beta = \alpha = 2$ (*arrows*).

The preferred motions are along any of the "valleys" of this landscape, that is, the lines $\alpha = \beta + 2\pi n/3$ for any integer n. Imperfections in the gears have again been modeled as bumps in the energy landscape; thus the gears don't turn freely even if we stay in one of the valleys. Slipping involves hopping from one valley to the next and is opposed by the energy ridges separating the valleys. Slipping is especially likely to occur at a bump in a valley, for example, the point ($\beta = 2, \alpha = 2$) (see the arrows in Figure 10.8b).

Now consider the effects of a load torque w_1R and a driving torque w_2R on the machine. Define the sign of α and β so that α increases when the gear on the left turns clockwise, whereas β increases when the other gear turns counterclockwise (see Figure 10.6). Thus the effect of the driving torque is to tilt the landscape downward in the direction of decreasing α. The effect of the load, however, is to tilt the landscape *upward* in the direction of decreasing β (see Figure 10.9). The machine *slides down the landscape*, following one of the valleys. The figure shows the case where $w_1 < w_2$; here α and β drive toward negative values.

Just as in the one-dimensional machine, our gears will get stuck if they attempt to cross the bump at ($\beta = 2, \alpha = 2$) under the load and driving conditions shown. Decreasing the load could get the gears unstuck. But if we instead increased the driving force, we'd find that our machine *slips* a notch at this point, sliding from the middle valley of Figure 10.9 to the next one closer to the viewer. That is, α can decrease without decreasing β.

Slipping is an important new phenomenon not seen in the one-dimensional idealization. Clearly it's bad for the machine's efficiency: A unit of driving energy gets

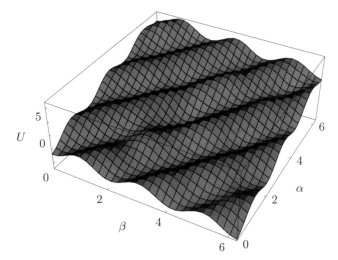

Figure 10.9: (Mathematical function.) Energy landscape for the driven, loaded, imperfect gear machine. The landscape is the same as the one in Figure 10.8, but tilted. The figure shows the case where the driving torque is larger than the load torque; in this case, the tilt favors motion to the front left of the graph. Again the scale of the vertical axis is arbitrary. The bump in the central valley (at $\beta = 2, \alpha = 2$) is now a spot where "slipping" is likely to occur. That is, the state of the machine can hop from one valley to the next lower one at such points.

spent (α decreases), but no corresponding unit of useful work is done (β does not decrease). Instead the energy all goes into viscous dissipation. In short,

> *The machine in Figure 10.6c stops doing useful work (that is, stops lifting the weight w_1) as soon as **either***
>
> ***a.*** *w_1 equals w_2, so the machine is in mechanical equilibrium (the valleys in Figure 10.9 become horizontal), or*   (10.1)
>
> ***b.*** *The slipping rate becomes large.*

10.2.2 Microscopic machines can step past energy barriers

The machines considered in Section 10.2.1 were deterministic: Noise, or random fluctuations, played no important role in their operation. But we wish to study molecular machines, which occupy a nanoworld dominated by such fluctuations.

Gilbert says: Some surprising things can happen in this world. For example, a machine need no longer stop when it encounters a bump in the energy landscape; after a while, a large enough thermal fluctuation will always arrive to push it over the bump. In fact, I have invented a simple way to translocate a protein, using thermal motion to my advantage. I've named my device the *G-ratchet* in honor of myself (Figure 10.10a). It's a shaft with a series of beveled bolts; they keep the shaft from taking steps to the left. Occasionally, a thermal fluctuation comes along and gives the shaft a kick with energy greater than ϵ, the energy needed to compress one of the little springs holding

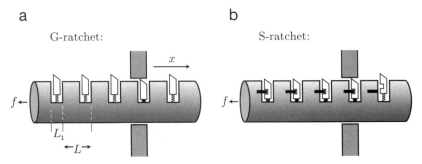

Figure 10.10: (Schematics.) Two thermally activated ratchets. (a) The G-ratchet. A rod (horizontal cylinder) makes a supposedly one-way trip to the right through a hole in a "membrane" (*shaded wall*), driven by random thermal fluctuations. It's prevented from moving to the left by sliding bolts, similar to those in a door latch. The bolts can move down to allow rightward motion; then they pop up as soon as they clear the wall. A possible external "load" is depicted as an applied force f directed to the left. The text explains why this device does *not* work. (b) The S-ratchet. Here the bolts are tied down on the left side, then released as they emerge on the right. This device is a mechanical model for protein translocation (Figure 10.3).

the bolts. Then the shaft takes a step to the right.

Sullivan: That certainly *is* surprising. I notice that you could even use your machine to pull against a load (the external force f shown in Figure 10.10).

Gilbert: That's right! It just slows down a bit, because now it has to wait for a thermal push with energy greater than $\epsilon + fL$ to take a step.

Sullivan: I have just one question: Where does the work fL done against the load come from?

Gilbert: I guess it must come from the thermal energy giving rise to the Brownian motion . . . .

Sullivan: Couldn't you wrap your shaft into a circle? Then your machine would go around forever, constantly doing work against a load.

Gilbert: Just what are you trying to tell me?

Yes, Sullivan is just about to point out that Gilbert's device would continuously extract mechanical work from the surrounding thermal motion, if it worked the way Gilbert supposes. Such a machine would spontaneously reduce the world's entropy and so violate the Second Law.[5] You can't convert thermal energy directly to mechanical energy without using up something else—think about the discussion of the osmotic machine in Section 1.2.2.

Sullivan continues: I think I see the flaw in your argument. It's not really so clear that your device takes only rightward steps. It cannot move at all unless the energy ϵ needed to retract a bolt is comparable to $k_{\mathrm{B}}T$. But if that's the case, then the bolts will *spontaneously* retract from time to time—they are thermally jiggling along with

---

[5]Unfortunately, it's already too late for Gilbert's financial backers, who didn't study thermodynamics.

everything else! If a leftward thermal kick comes along at just such a moment, then the rod will take a step to the left after all.

Gilbert: Isn't that an extremely unlikely coincidence?

Sullivan: Not really. The applied force will make the rod spend most of its time pinned at one of the locations $x = 0, L, 2L, \ldots$, at which a bolt is actually touching the wall. Suppose that now a thermal fluctuation momentarily retracts the obstructing bolt. If the rod then moves slightly to the right, the applied force will just pull it right back to where it was. But if the rod moves slightly to the left, the bolt will slip under the wall and f will pull the rod a full step to the left. That is, an applied force converts the random thermal motion of the rod to one-way, *left*ward, stepping. If $f = 0$, there will be no net motion at all, either to the right or left.

Sullivan continues: But I still like your idea. Let me propose a modification, the *S-ratchet* shown in Figure 10.10b. Here a latch keeps each bolt down as long as it's to the left of the wall; some mechaism releases the latch whenever a bolt moves to the right side.

Gilbert: I don't see how that helps at all. The bolts still never push the rod to the right.

Sullivan: Oh, but they do: They push on the wall whenever the rod tries to take a step to the left, and the wall pushes back. That is, they rectify its Brownian motion by bouncing off the wall.

Gilbert: But that's how my G-ratchet was supposed to work!

Sullivan: Yes, but now something is really getting used up: The S-ratchet is a one-shot machine, releasing potential energy stored in its compressed springs as it moves. In fact, it's a mechanical analog of the translocation machine (Figure 10.3). There's no longer any obvious violation of the Second Law.

Gilbert: Won't your criticism of my device (that it can make backward steps) apply to yours as well?

Sullivan: We can design the S-ratchet's springs to be so stiff that they rarely retract spontaneously, and hence leftward steps are rare. But thanks to the latches, rightward steps are still easy.

10.2.3 The Smoluchowski equation gives the rate of a microscopic machine

Qualitative expectations   Let's supply our protagonists with the mathematical tools they need to clear up their controversy. Panels (a) and (b) in Figure 10.11 show the energy landscapes of the G-ratchet, both without and with a load force, respectively. Rightward motion of the rod compresses a spring, increasing the potential energy. At $x = 0, L, 2L, \ldots$, the bolt clears the wall. It then snaps up, dissipating the spring's potential energy into thermal form. Panels (c) and (d) of the figure show the energy landscape of the S-ratchet, with small and large loads (f and f', respectively). Again each spring stores energy ϵ when compressed.

Note first that panel (d) is qualitatively similar to panel (b), and (a) is similar to the special case intermediate between (c) and (d), namely, the case in which $f = \epsilon/L$.

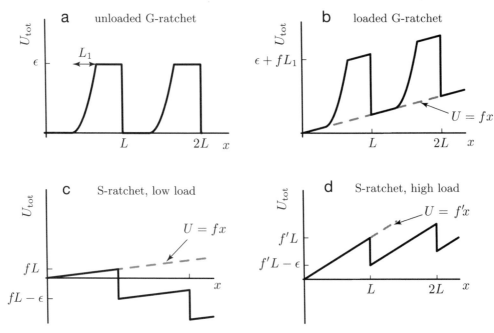

Figure 10.11: (Sketch graphs.) Energy landscapes. (a) The unloaded G-ratchet (see Figure 10.10a). Pushing the rod to the right compresses the spring on one of the bolts, raising the stored potential energy by an amount ϵ and giving rise to the curved part of the graph of U_{tot}. Once a bolt has been retracted, the potential energy is constant until it clears the wall; then the bolt pops up, releasing its spring, and the stored energy goes back down. (b) The loaded G-ratchet. Rightward motion now carries a net energy penalty, the work done against the load force f. Hence the graph of U_{tot} is tilted relative to (a). (c) The S-ratchet (see Figure 10.10b) at low load f. As the rod moves rightward, its potential energy progressively decreases, as more of its bolts get released. (d) The S-ratchet at high load, f'. The downward steps are still of fixed height ϵ, but the upward slope is greater, so rightward progress now carries a net energy penalty.

Thus we need only analyze the S-ratchet to find what's going on in both Gilbert's and Sullivan's devices. In brief, Sullivan has argued that

1. The unloaded G-ratchet will make no net progress in either direction; the situation is similar for the S-ratchet when $f = \epsilon/L$.
2. In fact, the loaded G-ratchet (or the S-ratchet with $f > \epsilon/L$) will move to the left.
3. The loaded S-ratchet, however, *will* make net progress to the right, if $f < \epsilon/L$.

Sullivan's remarks also imply that

4. The rate at which the loaded S-ratchet steps to the right will reflect the probability of getting a kick of energy at least fL, that is, enough to hop out of a local minimum of the potential shown in Figure 10.11c. The rate of stepping to the left will reflect the probability of getting a kick of energy at least ϵ.

Let's begin with Sullivan's third assertion. To keep things simple, assume, as he did, that ϵ is much bigger than $k_B T$. Thus, once a bolt pops up, it rarely retracts spontaneously; there is no backstepping. We'll refer to this special case of the S-ratchet as a **perfect ratchet**. First suppose that there's *no* external force: In our pictorial language, the energy landscape is a steep, descending staircase. Between steps, the rod wanders freely with some diffusion constant D. A rod initially at $x = 0$ will arrive at $x = L$ in a time given approximately by $t_{step} \approx L^2/(2D)$ (see Equation 4.5 on page 115). Once it arrives at $x = L$, another bolt pops up, thereby preventing return, and the process repeats. Thus the average net speed is

$$v = L/t_{step} \approx 2D/L, \qquad \text{speed of unloaded, perfect S-ratchet} \qquad (10.2)$$

which is indeed positive as Sullivan claimed.

We now imagine introducing a load f, still keeping the perfect ratchet assumption. The key insight is now Sullivan's observation that the fraction of time a rod spends at various values of x will depend on x, because the load force is always pushing x toward one of the local minima of the energy landscape. We need to find the probability distribution, $P(x)$, of being at position x.

Mathematical framework The motion of a single ratchet is complex, like any random walker. Nevertheless, Chapter 4 showed how a simple, deterministic equation describes the *average* motion of a large collection of such walkers: The averaging eliminates details of each individual walk, revealing the simple collective behavior. Let's adapt that logic to describe a large collection of M identical S-ratchets. To simplify the math further, we will also focus on just a few steps of the ratchet (say, four). We can imagine that the rod has literally been bent into a circle, so the point $x + 4L$ is the same as the point x. (To avoid Sullivan's criticism of the G-ratchet, we could also imagine that some external source of energy resets the bolts every time they go around.)

Initially, we release all M copies of our ratchet at the same point $x = x_0$, then let them walk for a long time. Eventually, the ratchets' locations form a probability distribution, like the one imagined in Figure 10.12. In this distribution, the individual ratchets cluster about the four potential minima (points just to the right of $x = -2L, \ldots L$; see Figure 10.11c), but all memory of the initial position x_0 has been lost. That is, $P(x)$ is a periodic function of x. In addition, eventually the probability distribution will stop changing in time.

The previous paragraphs should sound familiar: They amount to saying that our collection of ratchets will arrive at a steady, nonequilibrium state. We encountered such states in Section 4.6.1 on page 135 when studying diffusion through a thin tube joining two tanks with different concentrations of ink.[6] Shortly after set-

---

[6]The concept of a steady (or quasi-steady), nonequilibrium state also entered the discussion of bacterial metabolism in Section 4.6.2. Sections 10.4.1 and 11.2.2 will again make use of this powerful idea.

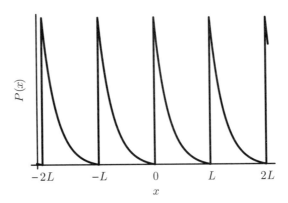

Figure 10.12: (Sketch graph.) The probability of being found at various positions x for a collection of S-ratchets, long after all were released at a common point. We imagine each ratchet to be circular, so the values $x = \pm 2L$ refer to the same point (see text). For illustration, the case of a perfect ratchet (large energy drop, $\epsilon \gg k_B T$) has been shown; see Your Turn 10C.

ting this system up, we found a steady flux of ink from one tank to the other. This state is not equilibrium—equilibrium requires that all fluxes equal *zero*. Similarly, in the ratchet case, the probability distribution $P(x, t)$ will come to a nearly time-independent form, as long as the external source of energy resetting the bolts remains available. The flux (net number of ratchets crossing $x = 0$ from left to right) need not be zero in this state.

To summarize, we have simplified our problem by arguing that we need only consider probability distributions $P(x, t)$ that are periodic in x and independent of t. Our next step is to find an equation obeyed by $P(x, t)$ and solve it with these two conditions. To do so, we follow the derivation of the Nernst–Planck formula (Equation 4.24 on page 140).

Note that in a time step Δt, each ratchet in our imagined collection gets a random thermal kick to the right or the left, in addition to the external applied force, just as in the derivation of Fick's law (Section 4.4.2 on page 128). Suppose first that there were no mechanical forces (no load and no bolts). Then we can just adapt the derivation leading to Equation 4.19 (recall Figure 4.10 on page 129):

- Subdivide each rod into imaginary segments of length Δx much smaller than L.
- The distribution contains $MP(a)\Delta x$ ratchets located between $x = a - \frac{1}{2}\Delta x$ and $x = a + \frac{1}{2}\Delta x$. About half of them step to the right in time Δt.
- Similarly, there are $MP(a + \Delta x)\Delta x$ ratchets located between $x = a + \frac{1}{2}\Delta x$ and $x = a + \frac{3}{2}\Delta x$, of which half step to the *left* in time Δt.
- Thus the net number of ratchets in the distribution crossing $x = a + \frac{1}{2}\Delta x$ from left to right is

$$\frac{1}{2}M[P(a) - P(a + \Delta x)]\Delta x \approx -\frac{1}{2}(\Delta x)^2 M \left.\frac{d}{dx}\right|_{x=a} P(x).$$

- We can compactly restate the last result as $-MD\frac{dP}{dx}\Delta t$, where D is the diffusion constant for the movement of the ratchet along its axis in the surrounding viscous medium. (Recall $D = (\Delta x)^2/(2\Delta t)$ from Equation 4.5b on page 115).

Now we add the effect of an external force:

- Each ratchet also drifts under the influence of the force $-dU_{tot}/dx$, where $U_{tot}(x)$ is the potential energy function sketched in Figure 10.11c.
- The average drift velocity of those ratchets located at $x = a$ is

$$v_{drift} = -\frac{D}{k_B T} \frac{d}{dx}\bigg|_{x=a} U_{tot}.$$

(To get this expression, write the force as $-dU_{tot}/dx$ and use the Einstein relation, Equation 4.16 on page 120, to express the viscous friction coefficient in terms of D.)

- The net number of ratchets crossing $x = a$ in time Δt from the left thus gets a second contribution, $M \times P(a)v_{drift}\Delta t$, or $-(MD/k_B T)(dU_{tot}/dx)P\,\Delta t$.

The arguments just given yielded two contributions to the number of systems crossing a given point in time Δt. Adding these contributions and dividing by Δt gives

$$j^{(1d)} \equiv \text{net number crossing per time} = -MD\Big(\frac{dP}{dx} + \frac{1}{k_B T}P\frac{dU_{tot}}{dx}\Big). \qquad (10.3)$$

(In this one-dimensional problem, the appropriate dimensions for a flux are $\mathbb{T}^{-1}$.) For the probability distribution $P(x)$ to be time independent, we now require that probability not pile up anywhere. This requirement means that the expression in Equation 10.3 must be independent of x. (A similar argument led us to the diffusion equation, Equation 4.20 on page 131.) In this context the resulting formula is called the **Smoluchowski equation**:

$$0 = \frac{d}{dx}\Big(\frac{dP}{dx} + \frac{1}{k_B T}P\frac{dU_{tot}}{dx}\Big). \qquad (10.4)$$

The equilibrium case We want to find some spatially periodic solutions to Equation 10.4 and interpret them. First suppose that the potential $U_{tot}(x)$ is itself periodic: $U_{tot}(x + L) = U_{tot}(x)$. This situation corresponds to the unloaded G-ratchet (Figure 10.11a) or to the S-ratchet (Figure 10.11c) with $f = \epsilon/L$.

Example: Show that in this case, the Boltzmann distribution is a solution of Equation 10.4, find the net probability per time to cross x, and explain why your result makes physical sense.

Solution: We expect that the system will just come to equilibrium, where it makes no net progress at all. Indeed, taking $P(x) = Ce^{-U_{tot}(x)/k_B T}$ gives a periodic, time-independent probability distribution. (C is a normalization constant.) Equation 10.3

then says that $j^{(1d)}(x) = 0$ everywhere. Hence this $P(x)$ is indeed a solution to the Smoluchowski equation with *no* net motion.

Because $j^{(1d)} = 0$, Sullivan's first claim was right (see page 416): The unloaded G-ratchet makes no net progress in either direction. We can also confirm Sullivan's physical reasoning for this claim: Indeed, the function $e^{-U_{tot}(x)/k_B T}$ peaks at the lowest-energy points, so each ratchet spends a lot of its time poised to hop *backward* whenever a chance thermal fluctuation permits this.

Beyond equilibrium The Boltzmann distribution only applies to systems at equilibrium. To tackle *non*equilibrium situations, begin with the perfect ratchet case (very large energy step ϵ). We already encountered the perfect ratchet when deriving the zero-force estimate Equation 10.2. Thus, as soon as a ratchet arrives at one of the steps in the energy landscape, it immediately falls down the step and cannot return; the probability $P(x)$ is thus nearly zero just to the left of each step, as shown in Figure 10.12.

Your Turn 10C

Verify that the function $P(x) = C(e^{-(x-L)f/k_B T} - 1)$ vanishes at $x = L$, solves the Smoluchowski equation with the potential energy $U_{tot}(x) = fx$, and resembles the curve sketched in Figure 10.12 between 0 and L. (Again C is a normalization constant.) Substitute into Equation 10.3 to find that $j^{(1d)}(x)$ is everywhere constant and positive.

You have just verified Sullivan's third claim (the loaded S-ratchet can indeed make net rightward progress), in the limiting case of a perfect ratchet. The constant C should be chosen to make $P(x)dx$ a properly normalized probability distribution, but we won't need its actual value. Outside the region between 0 and L, we make $P(x)$ periodic by just copying it (see Figure 10.12).

Let's find the average speed v of the perfect S-ratchet. First we need to think about what v means. Figure 10.12 shows the distribution of positions attained by a large collection of M ratchets. Even though the populations at each position are assumed to be constant in time, there can nevertheless be a net motion, just as we found when studying quasi-steady diffusion in a thin tube (Section 4.6.1 on page 135). To find this net motion, we count how many ratchets in the collection are initially located in a single period $(0, L)$, then find the average time Δt it takes for all of them to cross the point L from left to right, using the flux $j^{(1d)}$ found in Equation 10.3:

$$\Delta t = \text{(number)/(number/time)} = \left(\int_0^L dx\, MP(x) \right) \Big/ \left(j^{(1d)} \right). \qquad (10.5)$$

Then the average speed v is given by

$$v = L/\Delta t = \left(Lj^{(1d)} \right) \Big/ \left(\int_0^L dx\, MP(x) \right). \qquad (10.6)$$

The normalization constant C drops out of this result (and so does M).

Substituting the expressions in Your Turn 10C into Equation 10.5 gives

$$\Delta t = \frac{1}{Df/k_B T} \int_0^L dx \left(e^{-(x-L)f/k_B T} - 1\right),$$

or

$$v = \left(\frac{fL}{k_B T}\right)^2 \frac{D}{L} \left(e^{fL/k_B T} - 1 - fL/k_B T\right)^{-1}. \qquad \begin{array}{l}\text{speed of loaded,}\\ \text{perfect S-ratchet}\end{array} \qquad (10.7)$$

Although our answer is a bit complicated, it does have one simple qualitative feature: It's finite. That is, even though we took a very large energy step (the perfect ratchet case), the ratchet has a finite limiting speed.

Your Turn 10D

a. Show that in the case of *zero* external force, Equation 10.7 reduces to $2D/L$, agreeing with our rough analysis of the unloaded perfect ratchet (Equation 10.2).

b. Show that at *high* force (but still much smaller than ϵ/L), Equation 10.7 reduces to

$$v = \left(\frac{fL}{k_B T}\right)^2 \frac{D}{L} e^{-fL/k_B T}. \qquad (10.8)$$

The last result establishes Sullivan's fourth claim (forward stepping rate contains an exponential activation energy factor), in the perfect ratchet limit (backward stepping rate equals zero).

Although we only studied the perfect ratchet limit, we can now guess what will happen more generally. Consider the equilibrium case, where $f = \epsilon/L$. At this point, the activation barriers to forward and reverse motion are equal. Your result in Your Turn 10D(b) suggests that then the forward and reverse rates cancel, giving no net progress. This argument should sound familiar—it is just the kinetic interpretation of equilibrium (see Section 6.6.2 on page 220). At still greater force, $f > \epsilon/L$, the barrier to backward motion is actually smaller than the one for forward motion (see Figure 10.11d), and the machine makes net progress to the *left*. That was Sullivan's second claim.

Summary The S-ratchet makes progress to the right when $f < \epsilon/L$, then slows and reverses as we raise the load force beyond $f = \epsilon/L$.

The S-ratchet may seem rather artificial, but it illustrates some useful principles applicable to any molecular-scale machine:

1. Molecular-scale machines move by random-walking over their free energy landscape, not by deterministic sliding.

2. They can pass through potential energy barriers, with an average waiting time given by an exponential factor.

3. They can store potential energy (this is in part what creates the landscape) but not kinetic energy (because viscous dissipation is strong in the nanoworld, see Chapter 5).

Point (3) stands in contrast to familiar macroscopic machines like a pendulum clock, whose rate is controlled by the inertia of the pendulum. Inertia is immaterial in the highly damped nanoworld; instead the speed of a molecular motor is controlled by activation barriers.

Our study of ratchets has also yielded some more specific results:

> **a.** *A thermal machine can convert stored internal energy ϵ into directed motion if it is structurally asymmetrical.*
>
> **b.** *But structural asymmetry alone is not enough: A thermal machine won't go anywhere if it's in equilibrium (periodic potential, Figure 10.11a). To get useful work, we must push it out of equilibrium by arranging for a descending free energy landscape.*   (10.9)
>
> **c.** *A ratchet's speed does not increase without bound as we increase the drive energy ϵ. Instead, the speed of the unloaded ratchet saturates at some limiting value (Equation 10.7).*

You showed in Your Turn 10D that, with a load, the limiting speed gets reduced by an exponential factor relative to the unloaded $2D/L$. This result should remind you of the Arrhenius rate law (Section 3.2.4 on page 86). Chapter 3 gave a rather simple-minded approach to this law, imagining a single thermal kick carrying us all the way over a barrier. In the presence of viscous friction, such a one-kick passage may seem about as likely as a successful field goal in a football game played in molasses! But the Smoluchowski equation showed us the right way to derive the rate law for large molecules: Modeling the process as a random walk on an energy landscape gives qualitatively the same result as the naïve argument.

We could go on to implement these ideas for more complex microscopic machines, like the gears of Figure 10.6c. Rather than studying rolling on the potential energy surface (Figure 10.9), we would set up a *two-dimensional* Smoluchowski equation on the surface, again arriving at conclusions similar to Idea 10.1 on page 413. The following sections will not follow this program, however, instead seeking shortcuts to see the qualitative behavior without the difficult mathematics.

$\boxed{T_2}$ *Section 10.2.3′ on page 455 generalizes the preceding discussion to get the force–velocity relation for an imperfect ratchet.*

10.3 MOLECULAR IMPLEMENTATION OF MECHANICAL PRINCIPLES

The discussion of purely mechanical machines in Section 10.2 generated some nice formulas but still leaves us with many questions:

- Molecular-scale machines consist of one or a few molecules, unlike the macro-scopic machines sketched earlier. Can we apply our ideas to single molecules?
- We still have no candidate model for a cyclic machine that eats *chemical* energy. Won't we need some totally new ideas to create this?
- Most important of all, how can we make contact with experimental data?

To make progress on the first question, it's time to gather a number of ideas about single molecules developed in previous chapters.

10.3.1 Three ideas

First, the statistical physics of Chapter 6 was constructed to be applicable to single-molecule subsystems. For example, Section 6.6.3 on page 223 showed that such systems drive to minimize their free energy, just like macroscopic systems, although not necessarily in a one-way, deterministic fashion.

Second, Chapter 8 described how chemical forces are nothing but changes in free energy, in principle interconvertible with other changes involving energy (for example, the release of the little bolts in the S-ratchet). Chemical forces drive a reaction in a direction determined by its ΔG, a quantity involving the stoichiometry of the reaction but otherwise reflecting only the concentrations of molecules in the reservoir outside the reaction proper. (Idea 8.23 on page 307 expresses this conclusion succinctly.)

Third, Chapter 9 showed how even large, complex macromolecules, with thousands of atoms all in random thermal motion, can nevertheless act as though they had just a few discrete states. Indeed, macromolecules can snap crisply between those states, almost like a macroscopic light switch. We identified the source of this "multistable" behavior in the cooperative action of many weak physical interactions such as hydrogen bonds. Thus, for example, cooperativity made the helix–coil transition (Section 9.5.1) or the binding of oxygen by hemoglobin (Section 9.6.1) surprisingly sharp. Similarly, a macromolecule can be quite specific about what small molecules it binds, rejecting imposters by the cooperative effects of many charged or H-bonding groups in a precise geometrical arrangement (see Idea 7.17 on page 263).

10.3.2 The reaction coordinate gives a useful reduced description of a chemical event

The idea of multistability (the third point in Section 10.3.1) sometimes justifies us in writing extremely simple kinetic diagrams (or reaction graphs) for the reactions of huge, complex macromolecules, as if they were simple molecules jumping between just a few well-defined configurations. The reaction graphs we write will consist of discrete symbols (or **nodes**) joined by arrows, just like many we have already written in Chapter 8, for example, the isomerization reaction $A \rightleftharpoons B$ studied in Section 8.2.1 on page 299. A crucial point is that such reaction graphs are in general **sparsely connected**. That is, many of the arrows one could imagine drawing between nodes will in fact be missing, reflecting the fact that the corresponding rates are negligibly small (Figure 10.13). Thus, in many cases, reactions can proceed only in sequential steps,

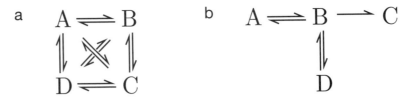

Figure 10.13: (Diagrams.) (a) A fully connected reaction diagram. (b) A sparsely connected reaction diagram.

rarely if ever taking shortcuts on the reaction graph. Usually we can't work out the details of the reaction graph from explicit calculations of molecular dynamics, but sometimes it's enough to frame guesses about a system from experience with similar systems, then look for quantitative predictions to test the guesses.

What exactly happens along those arrows in a reaction graph? To get from one configuration to the next, the atoms composing the molecule must rearrange their relative positions and angles. We could imagine listing the coordinates of every atom; then the starting and ending configurations are points on the many-dimensional space of these coordinates. In fact, they are special points, for which the free energy is much lower than elsewhere. This property gives those points a special, nearly stable status, entitling them to be singled out as nodes on the reaction graph. If we could reach in and push individual atoms around, we'd have to do work on the molecule to move it away from either of these points. But we can instead wait for thermal motion to do the pushing for us:

> *Chemical reactions reflect random walks on a free energy landscape in the space of molecular configurations.* (10.10)

Unfortunately, the size of the molecular configuration space is daunting, even for small molecules. To get a tractable example, consider an ultrasimple reaction: A hydrogen atom, called H_a, collides with a hydrogen molecule, picking up one of the molecule's two atoms, H_b. To describe the spatial relation of the three H atoms, we can specify the three distances between pairs of atoms. Consider, for example, the configurations in which all three atoms lie on a single line, so the two distances d_{ab} and d_{bc} fully specify the geometry. Then Figure 10.14 shows schematically the energy surface for the reaction. The energy is minimum at each end of the dashed line, where two H atoms are at the usual bond distance and the third is far away. The dashed line represents a path in configuration space that joins these two minima while climbing the free energy landscape as little as possible. The barrier that must be surmounted in such a walk corresponds to the bump in the middle of the dashed line, representing an intermediate configuration with $d_{ab} = d_{bc}$.

When a free energy landscape has a well-defined mountain pass, as in Figure 10.14, it makes sense to think of our problem approximately as just a one-dimensional walk *along this curve*, and to think in terms of the one-dimensional energy landscape seen along this walk. Chemists refer to the distance along the path as the **reaction coordinate**, and to the highest point along it as the **transition state**. We'll denote the height of this point on the graph by the symbol $\Delta G^{\ddagger}$.

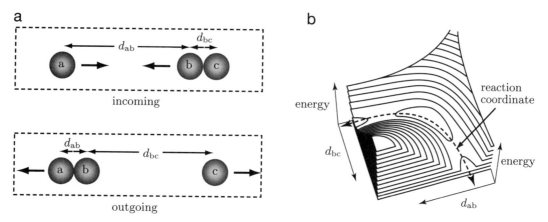

Figure 10.14: (Schematic; sketch graph.) (a) A simple chemical reaction: A hydrogen molecule transfers one of its atoms to a lone H atom, $H + H_2 \rightarrow H_2 + H$. (b) Imagined free energy landscape for this reaction, assuming that the atoms travel on one straight line. The dashed line is the lowest path joining the starting and ending configurations shown in (a); it's like a path through a mountain pass. The reaction coordinate can be thought of as distance along this path. The highest point on this path is called the transition state. [(b) Adapted from Eisenberg & Crothers, 1979.]

Remarkably, the utility of the reaction coordinate idea has proven not to be limited to small, simple molecules. Even macromolecules described by thousands of atomic coordinates often admit a useful reduced description with just one or two reaction coordinates. Section 10.2.3 showed how the rate of barrier passage for a random walk on a one-dimensional potential is controlled by an Arrhenius exponential factor, involving the activation barrier; in our present notation this factor takes the form $e^{-\Delta G^{\ddagger}/k_B T}$. To test the idea that a given reaction is effectively a random walk on a one-dimensional free energy landscape, we write[7] $\Delta G^{\ddagger}/k_B T = (\Delta E^{\ddagger}/k_B T) - (S^{\ddagger}/k_B)$. Then we predict that the reaction rate should depend on temperature as

$$\text{rate} \propto e^{-\Delta E^{\ddagger}/k_B T}. \tag{10.11}$$

Indeed, many reactions among macromolecules obey such relations (see Figure 10.15). Section 10.3.3 will show how these ideas can help explain the enormous catalytic power of enzymes.

$\boxed{T_2}$ *Section 10.3.2' on page 456 gives more details about the energy landscape concept.*

10.3.3   An enzyme catalyzes a reaction by binding to the transition state

Reaction rates are controlled by activation barriers, with a temperature dependence given roughly by an Arrhenius exponential factor (see Section 3.2.4 on page 86). Enzymes increase reaction rates but maintain that characteristic temperature depen-

---

[7] $\boxed{T_2}$ More precisely, we should use the enthalpy in place of $\Delta E^{\ddagger}$.

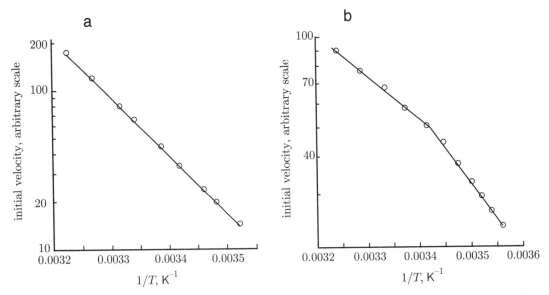

Figure 10.15: (Experimental data.) Rates of enzyme catalysis. (a) Semilog plot of initial reaction velocity versus inverse temperature for the conversion of L-malate to fumarate by the enzyme fumarase, at pH 6.35. (b) The same for the reverse reaction. The first reaction follows the Arrhenius rate law (Equation 10.11), as shown by the straight line. The line is the function $\log_{10} v_0 = \text{const} - (3650\,\text{K}/T)$, corresponding to an activation barrier of $29 k_B T_r$. The second reaction shows two different slopes; presumably an alternative reaction mechanism becomes available at temperatures above 294 K. [Data from Dixon & Webb, 1979.]

dence (Figure 10.15). Thus it's reasonable to guess that *enzymes work by reducing the activation barrier to a reaction.* What may not be so clear is *how* they could accomplish such a reduction.

Figure 10.16 summarizes a mechanism proposed by J. Haldane in 1930. Using the mechanical language of this chapter, let us imagine a substrate molecule S as an elastic body, with one particular chemical bond of interest shown in the figure as a spring. The substrate wanders at random until it encounters an enzyme molecule E. The enzyme molecule has been designed with a binding site whose shape is almost, but not quite, complementary to that of S. The site is assumed to be lined with groups that could make energetically favorable contacts with S (hydrogen bonds, electrostatic attractions, and so on), if only the shapes matched precisely.

Under these circumstances, states E and S may be able to lower their total free energy by deforming their shapes to make close contact and profit from the many weak physical attractions at the binding site.[8] In Haldane's words, E. Fischer's famous lock-and-key metaphor should be amended to say that "the key does not fit the lock quite perfectly, but rather exercises a certain strain on it." We will call the bound complex ES. But the resulting deformation on the particular bond of interest may push it closer to its breaking point or, in other words, reduce its activation barrier

---

[8]Other kinds of deformations besides shape changes are possible, for example, charge rearrangements. This chapter uses the mechanical idea of shape change as a metaphor for all sorts of deformations.

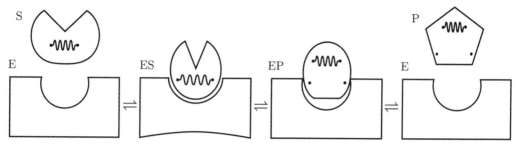

Figure 10.16: (Schematic.) Conceptual model of enzyme activity. (a) The enzyme E has a binding site with a shape and distribution of charges, hydrophobicity, and H-bonding sites approximately matching those presented by the substrate S. (b) To match perfectly, however, S (or both E and S) must deform. (Other, more dramatic conformational changes in the enzyme are possible, too.) One bond in the substrate (shown as a *spring* in S) stretches close to its breaking point. (c) From the ES state, then, a thermal fluctuation can readily break the stretched bond, giving rise to the EP complex. A new bond can now form (*upper spring*), stabilizing the product P. (d) The P state is not a perfect fit to the binding site either, so it readily unbinds, thereby returning E to its original state.

to breaking. Then ES will isomerize to a bound state of enzyme plus product, or EP, much more rapidly than S would spontaneously isomerize to P. If the product is also not a perfect fit to the enzyme's binding site, it can then readily detach, thereby leaving the enzyme in its original state. Each step in the process is reversible; the enzyme also catalyzes the reverse reaction P → S (see Figure 10.15).

Let us see how the little story just sketched actually implies a reduction in activation energy. Figure 10.17a sketches an imagined free energy landscape for a single molecule S to isomerize (convert) spontaneously to P (top curve). The geometrical change needed to make S fit the binding site of E is assumed to carry S along its reaction coordinate, with the tightest fit at the transition state $S^{\ddagger}$. The enzyme may also change its conformation to one different from its usual (lowest free energy) state (lower curve). These changes increase the self-energies of E and S, but they are partially offset by a sharp *decrease* in the interaction (or binding) energy of the complex ES (middle curve). Adding the three curves gives a total free energy landscape with a reduced activation barrier to the formation of the transition state $ES^{\ddagger}$ (Figure 10.17b).

The picture outlined in the preceding paragraph should not be taken too literally. For example, there's really no unambiguous way to divide the free energy into the three separate contributions shown in Figure 10.17a. Nevertheless, the conclusion is valid:

> *Enzymes work by reducing the activation energy for a desired reaction. To bring about this reduction, the enzyme is constructed to bind most tightly to the substrate's transition state.*　　　(10.12)

In effect, the enzyme–substrate complex *borrows* some of the free energy needed to form the transition state from the many weak interactions between the substrate and the enzyme's binding site. To return the enzyme to its original state, this borrowed energy must be paid back when the product unbinds. Thus,

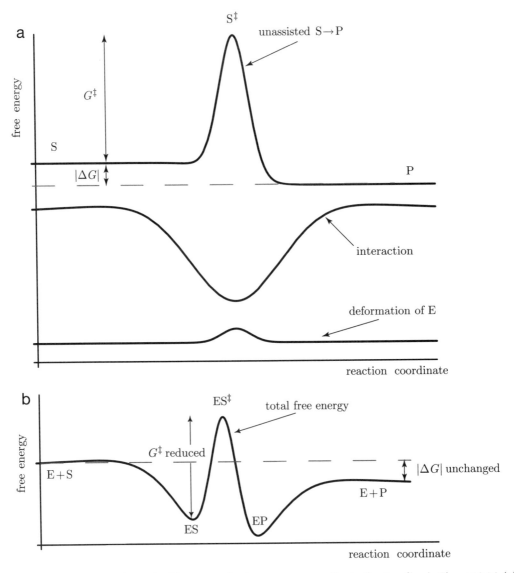

Figure 10.17: (Sketch graphs.) Imagined free energy landscapes corresponding to the story line in Figure 10.16. (a) *Top curve:* The substrate S can spontaneously convert to product P only by surmounting a large activation barrier $G^{\ddagger}$, which is the free energy of the transition state $S^{\ddagger}$ relative to S. *Middle curve:* The interaction free energy between substrate and product includes a large binding free energy (*dip*), as well as the entropic cost of aligning the substrate properly relative to the enzyme (slight *bumps* on either side of the dip). *Lower curve:* The binding free energy may be partly offset by a deformation of the enzyme upon binding, but still the net effect of the enzyme is to reduce the barrier $G^{\ddagger}$. All three curves have been shifted by arbitrary constants to show them on a single set of axes. (b) Imagined net free energy landscape obtained by summing the three curves in (a). The enzyme has reduced $G^{\ddagger}$, but it cannot change ΔG.

An enzyme cannot alter the net ΔG of the reaction.   (10.13)

An enzyme speeds up *both* the forward and backward reactions; the direction actually chosen is still determined by ΔG, a quantity external to the enzyme, as always (see Idea 8.15 on page 303).

Up to this point, we have been imagining a system containing just one molecule of substrate. With a simple modification, however, we can now switch to thinking of our enzyme as a *cyclic machine*, progressively processing a large batch of S molecules. When many molecules of S are available, then the net driving force for the reaction includes an entropic term of the form $k_B T \ln c_S$, where c_S is their concentration. (See Equation 8.3 on page 296 and Equation 8.14 on page 303.) The effect of a high concentration of S, then, is to pull the left end of the free energy landscape (Figure 10.17b) upward, reducing or eliminating any activation barrier to the formation of the complex ES and thus speeding up the binding of substrate. Similarly, an increase in product concentration c_P pulls up the right end of the free energy landscape, thereby slowing or halting the unbinding of product. Just as in any chemical reaction, a large enough concentration of P can even reverse the sign of ΔG, and hence reverse the direction of the net reaction (see Section 8.2.1 on page 299).

We can now make a simple but crucial observation: The state of our enzyme/substrate/product system depends on how many molecules of S have been processed into P. Although the enzyme returns to its original state after one cycle, still the whole system's free energy falls by ΔG every time it takes one net step. We can acknowledge this fact by generalizing the reaction coordinate to include the progress of the reaction, for example, the number N_S of remaining substrate molecules. Then *the complete free energy landscape consists of many copies of Figure 10.17b, each shifted downward by ΔG to make a continuous curve* (Figure 10.18). In fact, this curve looks qualitatively like one we have already studied, namely, Figure 10.11c! We identify the barrier $f L$ in that figure as $G^{\ddagger}$, and the net drop $f L - \epsilon$ as ΔG. In short,

> *Many enzymes can be regarded as simple cyclic machines; they work by random-walking down a one-dimensional free energy landscape. The net descent of this landscape in one forward step is the value of ΔG for the reaction S $\rightarrow$ P.*   (10.14)

Idea 10.14 gives an immediate qualitative payoff: We see at once why so many enzymes exhibit saturation kinetics (Section 10.1.2 on page 403). Recall what this means. The rate of an enzyme-catalyzed reaction S $\rightarrow$ P typically levels off at high concentration of S instead of being proportional to c_S as simple collision theory might have led us to expect. Viewing enzyme catalysis as a walk on a free energy landscape shows that saturation kinetics is a result we've already obtained, namely, our result for the speed of a perfect ratchet (Idea 10.9c on page 422). A large concentration of S pulls the left side of the free energy landscape upward. In other words, the step from E + S to ES in Figure 10.17b is steeply downhill. Such a steep downward step makes the process effectively irreversible, essentially forbidding backward steps; but after a certain point, it doesn't speed up the net progress, as seen in the analysis leading to Equation 10.7 on page 421. The reaction doesn't speed up because eliminating the first bump in Figure 10.17b doesn't affect the *middle* bump. Indeed, the activation

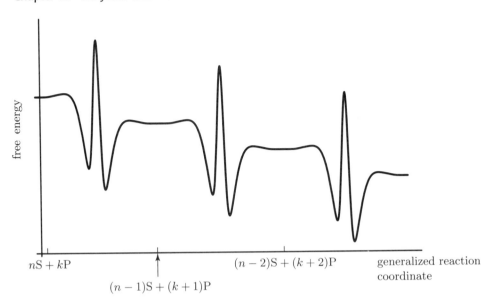

$nS + kP$

$(n-1)S + (k+1)P$

$(n-2)S + (k+2)P$

generalized reaction coordinate

Figure 10.18: (Sketch graph.) The free energy landscape of Figure 10.17b, duplicated and shifted to show three steps in a cyclic reaction. The reaction coordinate of Figure 10.17 has been generalized to include changes in the number of enzyme and substrate molecules; the curve shown connects the state with n substrate and k product molecules to the state with three fewer S (and three more P).

barrier controlling passage from ES to EP is insensitive to the availability of S, because the binding site is already occupied throughout this process.

We also see another way to make the catalytic cycle essentially irreversible: Instead of raising c_S, we can *lower* c_P, pulling the *right* side of the landscape steeply *down*. It makes sense—if there's no product, then the rate for E to bind P and convert it to S will be zero! Section 10.4 will turn all these qualitative observations into a simple, quantitative theory of enzyme catalysis rates, then apply the same reasoning to molecular machines.

Idea 10.14 also yields a second important qualitative prediction. Suppose that we find another molecule $\widetilde{S}$ similar to S, but whose relaxed state resembles the stretched (transition) state of S. Then we may expect that $\widetilde{S}$ will bind to E even more tightly than S itself, because it gains the full binding energy without having to pay any elastic-strain energy. Linus Pauling suggested in 1948 that introducing even a small amount of such a **transition state analog** $\widetilde{S}$ into a solution of E and S would *poison* the enzyme: E will bind $\widetilde{S}$ tightly and, instead of catalyzing a change in $\widetilde{S}$, will simply hold on to it. Indeed, today's protease inhibitors for the treatment of HIV infection were created by seeking transition state analogs directed at the active site of the HIV protease enzyme.

$\boxed{T_2}$ *Section 10.3.3′ on page 458 mentions other physical mechanisms that enzymes can use to facilitate reactions.*

10.3.4 Mechanochemical motors move by random-walking on a two-dimensional landscape

Idea 10.14 has brought chemical devices (enzymes) into the same conceptual frame-work as the microscopic mechanical devices studied in Section 10.2.2. This picture also lets us imagine how *mechano*chemical machines might work. Consider an enzyme that catalyzes the reaction of a substrate at high chemical potential, μ_S yielding a product with low μ_P. In addition, this enzyme has a second binding site, which can attach it to any point of a periodic "track." This situation is meant as a model of a motor like kinesin (see Section 10.1.3 on page 404), which converts ATP to ADP plus phosphate and can bind to periodically spaced sites on a microtubule.

The system just described has *two* markers of net progress, namely, the number of remaining substrate molecules and the spatial location of the machine along its track. Taking a step in either of these two directions will generally require surmounting some activation barrier; for example, stepping along the track involves first unbinding from it. To describe these barriers, we introduce a two-dimensional free energy landscape, conceptually similar to Figure 10.8 on page 412. Let β denote the spatial position of one particular atom on the motor. Imagine holding β fixed with a clamp, then finding the easiest path through the space of conformations at fixed β that accomplishes one catalytic step, finding a slice of the free energy landscape along a line of fixed β. Putting these slices together could, in principle, give the two-dimensional landscape.

If no external force acts on the enzyme and if the concentrations of substrate and product correspond to thermodynamic equilibrium ($\mu_S = \mu_P$), then we get a picture like Figure 10.8a, and no net motion. If, however, there are net chemical and mechanical forces, then we instead get a tilted landscape like Figure 10.9 on page 413, and the enzyme will move, exactly as in Section 10.2.2! The diagonal valleys in the landscape of Figure 10.9 implement the idea of a mechanochemical cycle:

> *A mechanochemical cycle amounts to a free energy landscape with directions corresponding to reaction coordinate and spatial displacement. If the landscape is not symmetrical under reflection in the mechanical (β) direction, and if the concentrations of substrate and product are out of equilibrium, then the cycle can yield directed net motion.* (10.15)

This result is just a restatement of Ideas 10.9a,b on page 422.

Figure 10.9 represents an extreme form of mechanochemical coupling, called **tight coupling**, in which a step in the mechanical (β) direction is nearly always linked to a step in the chemical (α) direction. There are well-defined valleys, well separated by large barriers, and so very little hopping takes place from one valley to the next. In such a situation it makes sense to eliminate altogether the direction perpendicular to the valleys, just as we already eliminated the many other configurational variables (Figure 10.14b). Thus, we can imagine reducing our description of the system to a *single* reaction coordinate describing the location along just one of the valleys. With this simplification, our motor becomes simple indeed: It's just another one-dimensional device, with a free energy landscape resembling the S-ratchet (Figure 10.11c on page 416).

We must keep in mind that tight coupling is just a hypothesis to be checked; indeed Section 10.4.4 will argue that tight coupling is not necessary for a motor to function usefully. Nevertheless, for now let us keep the image of Figure 10.9 in mind as our provisional, intuitive notion of how coupling works.

10.4 KINETICS OF REAL ENZYMES AND MACHINES

Certainly real enzymes are far more complicated than the sketches in the preceding sections might suggest. Figure 10.19 shows phosphoglycerate kinase, an enzyme playing a role in metabolism. (Chapter 11 will discuss the glycolysis pathway, to which this enzyme contributes.) The enzyme binds to phosphoglycerate (a modified fragment of

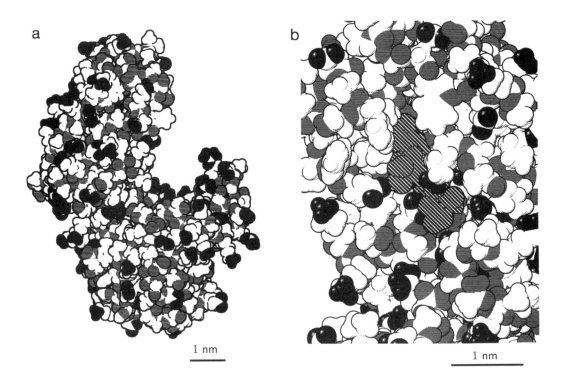

a

b

1 nm

1 nm

Figure 10.19: (Structure drawn from atomic coordinates.) (a) Structure of phosphoglycerate kinase, an enzyme composed of one protein chain of 415 amino acids. The chain folds into this distinctive shape, with two large lobes connected by a flexible hinge. The active site, where the chemical reaction occurs, is located between the two halves. The atoms are shown in a gray scale according to their hydrophobicity, with the most hydrophobic in white, the most hydrophilic in black. (b) Close-up of (a), showing the active site with a bound molecule of ATP (*hatched atoms*). This view is looking from the right in (a), centered on the upper lobe. Amino acids from the enzyme wrap around and hold the ATP molecule in a specific position. [From Goodsell, 1993.]

glucose) and transfers its phosphate group to an ADP molecule, forming ATP. If the enzyme were instead to bind phosphoglycerate and a *water* molecule, the phosphate could be transferred to the water, and no ATP would be made. The kinase enzyme is beautifully designed to solve this engineering problem. It is composed of two do-mains connected by a flexible hinge. Some of the amino acids needed for the reaction are in its upper half, some in the lower half. When the enzyme binds to phosphoglyc-erate and ADP, the energy of binding these substrate molecules causes the enzyme to close around them. Only then are all the proper amino acids brought into position; inside, sheltered from water by the enzyme, the reaction is consummated.

In short, phosphoglycerate kinase is complex because it must not only channel the flow of probability for molecular states into a useful direction but also *prevent* probability from flowing into *useless* processes. Despite this complexity, we can still see from its structure some of the general themes outlined in the preceding sections. The enzyme is much larger than its two substrate binding sites; it grips the substrates in a close embrace, making several weak physical bonds; optimizing these physical bonds constrains the substrates to a precise configuration, presumably corresponding to the transition state for the desired phosphate transfer reaction.

10.4.1 The Michaelis–Menten rule describes the kinetics of simple enzymes

The MM rule Section 10.2.3 gave us some experience calculating the net rate of a random walk down a free energy landscape. We saw that such calculations boil down to solving the Smoluchowski equation (Equation 10.4 on page 419) to find the ap-propriate quasi-steady state. However, we generally don't know the free energy land-scape. Even if we did, such a detailed analysis focuses on the specifics of one enzyme, whereas we would like to begin by finding some very broadly applicable lessons. Let's instead take an extremely reductionist approach.

First, focus on a situation where initially there is no product present, or hardly any. Then the chemical potential μ_P of the product is a large negative number. Thus, the third step of Figure 10.17b, EP $\rightarrow$ E + P, is steeply downhill, so we may treat this step as one-way forward—a perfect ratchet. We also make a related simplifying assumption, that the transition EP $\rightarrow$ E + P is so rapid that we may neglect EP alto-gether as a distinct step, lumping it together with E + P. Finally, we assume that the remaining quasi-stable states, E+S, ES, and E+P, are well separated by large barriers, so each transition may be treated independently. Thus the transition involving bind-ing of substrate from solution will also be supposed to proceed at a rate given by a first-order rate law, that is, the rate is proportional to the substrate concentration c_S (see Section 8.2.3 on page 306).

Now suppose that we throw a single enzyme molecule into a vat initially contain-ing substrate at concentration $c_{S,i}$ and a negligible amount of product.[9] This system is far from equilibrium, but it soon comes to a quasi-steady state: The concentra-tion of substrate remains nearly constant and that of product nearly zero, because substrate molecules enormously outnumber the one enzyme. The enzyme spends a

---

[9]Even if there are many enzyme molecules, we can expect the same calculations to hold as long as their concentration is much smaller than that of substrate.

certain fraction P_E of its time unoccupied, and the rest, $P_{ES} = 1 - P_E$ bound to substrate, and these numbers too are nearly constant in time. Thus the enzyme converts substrate at a constant rate, which we'd like to find.

Let us summarize the discussion so far in a reaction diagram:

$$E + S \underset{k_{-1}}{\overset{c_S k_1}{\rightleftharpoons}} ES \overset{k_2}{\rightharpoonup} E + P. \tag{10.16}$$

It's a cyclic process: The starting and ending states in this formula are different; but in each, the enzyme itself is in the same state. The notation associates rate constants to each process (see Section 8.2.3). We are considering only one molecule of E; thus the rate of conversion for $E + S \rightharpoonup ES$ is $k_1 c_S$, not $k_1 c_S c_E$.

In a short time interval dt, the probability P_E to be in the state E can change in one of three ways:

1. If the enzyme is initially in the unbound state E, it has probability per unit time $k_1 c_S$ of binding substrate and hence leaving the state E.

2. If the enzyme is initially in the enzyme–substrate complex state ES, it has probability per unit time k_2 of processing and releasing product, hence reentering the unbound state E.

3. The enzyme–substrate complex also has probability per unit time k_{-1} of losing its bound substrate, reentering the state E.

Expressing the preceding argument in a formula (see Idea 6.29 on page 222),

$$\frac{d}{dt} P_E = -k_1 c_S \times (1 - P_{ES}) + (k_{-1} + k_2) \times P_{ES}. \tag{10.17}$$

Make sure you understand the units on both sides of this formula.

The quasi-steady state is the one for which Equation 10.17 equals zero. Solving gives the probability to be in the state ES as

$$P_{ES} = \frac{k_1 c_S}{k_{-1} + k_2 + k_1 c_S}. \tag{10.18}$$

According to Equation 10.16, the rate at which a single enzyme molecule creates product is Equation 10.18 times k_2. Multiplying by the concentration c_E of enzyme then gives the reaction velocity v, defined in Section 10.1.2.

The preceding paragraph outlined how to get the initial reaction velocity as a function of the initial concentrations of enzyme and substrate, for a reaction with an irreversible step (Reaction 10.16). We can tidy up the formula by defining the **Michaelis constant** K_M and maximum velocity v_{max} of the reaction to be

$$K_M \equiv (k_{-1} + k_2)/k_1 \quad \text{and} \quad v_{max} \equiv k_2 c_E. \tag{10.19}$$

Thus K_M has the units of a concentration; v_{max} is a rate of change of concentration. In terms of these quantities, Equation 10.18 becomes the **Michaelis–Menten (or MM) rule**:

$$v = v_{max}\frac{c_S}{K_M + c_S}. \qquad \text{Michaelis–Menten rule} \qquad (10.20)$$

The MM rule displays saturation kinetics. At low substrate concentration, the reaction velocity is proportional to c_S, as we might have expected from naïve one-step kinetics (Section 8.2.3 on page 306). At higher concentration, however, the extra delay in escaping from the enzyme–substrate complex starts to modify that result: v continues to increase with increasing c_S but never exceeds v_{max}.

Let's pause to interpret the two constants v_{max} and K_M describing a particular enzyme. The maximum turnover number v_{max}/c_E, defined in Section 10.1.2, reflects the intrinsic speed of the enzyme. According to Equation 10.19, this quantity just equals k_2, which is indeed a property of a single enzyme molecule. To interpret K_M, we first notice that when $c_S = K_M$, then the reaction velocity is just one-half of its maximum. Suppose that the enzyme binds substrate rapidly relative to the rate of catalysis and the rate of substrate dissociation (that is, suppose that k_1 is large). Then even a low value of c_S will suffice to keep the enzyme fully occupied, or in other words, K_M will be small. The explicit formula (Equation 10.19) confirms this intuition.

The Lineweaver–Burk plot Our very reductionist model of a catalyzed reaction has yielded a testable result: We claim to predict the full dependence of v upon c_S, a *function*, using only two phenomenological fitting parameters, v_{max} and K_M. An algebraic rearrangement of the result shows how to test whether a given experimental data set follows the Michaelis–Menten rule. Instead of graphing v as a function of c_S, consider graphing the reciprocal $1/v$ as a function of $1/c_S$ (such a graph is called a **Lineweaver–Burk plot**). Equation 10.20 then becomes

$$\frac{1}{v} = \frac{1}{v_{max}}\left(1 + \frac{K_M}{c_S}\right). \qquad (10.21)$$

That is, the MM rule predicts that $1/v$ should be a linear function of $1/c_S$, with slope K_M/v_{max} and intercept $1/v_{max}$.

Remarkably, many enzyme-mediated reactions really do obey the MM rule, even though few are so simple as to satisfy our assumptions literally.

Example: Pancreatic carboxypeptidase cleaves amino acid residues from one end of a polypeptide. The table gives the initial reaction velocity versus c_S for this reaction for a model system, a peptide of just two amino acids:

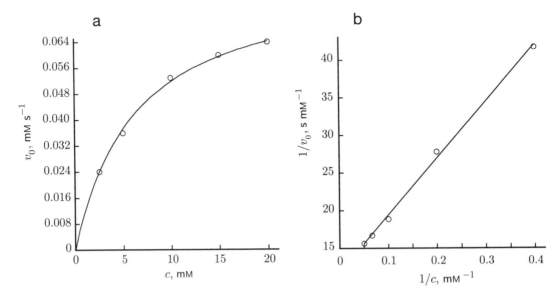

Figure 10.20: (Experimental data.) (a) Reaction velocity versus substrate concentration for the reaction catalyzed by pancreatic carboxypeptidase. (b) The same data, this time plotted in the Lineweaver–Burk form (see Equation 10.21). [Data from Lumry et al., 1951.]

substrate concentration, mM	initial velocity, mM s^{-1}
2.5	0.024
5.0	0.036
10.0	0.053
15.0	0.060
20.0	0.064

Find K_M and v_{max} by the Lineweaver–Burk method and verify that this reaction obeys the MM rule.

Solution: The graph in Figure 10.20b is indeed a straight line, as expected from the MM rule. Its slope equals 75 s, and the intercept is 12 mM^{-1} s. From the preceding formulas, then, $v_{max} = 0.085$ mM s^{-1} and $K_M = 6.4$ mM.

The key to the great generality of the MM rule is that some of the assumptions we made were not necessary. Problem 10.7 illustrates the general fact that *any* one-dimensional device (that is, one with a linear sequence of steps) effectively gives rise to a rate law of the form Equation 10.20, as long as the last step is irreversible.

10.4.2 Modulation of enzyme activity

Enzymes create and destroy molecular species. To keep everything working, the cell must regulate these activities. One strategy involves controlling the rate at which an enzyme is created, by regulating the gene coding for it (see Section 2.3.3 on page

58). For some applications, however, this strategy is not fast enough; instead, the cell adjusts the turnover numbers of the existing enzyme molecules. For example, an enzyme's activity may be slowed by the presence of another molecule that binds to, or otherwise directly interferes with, its substrate binding site (**competitive inhibition**; see Problem 10.5). Or a control molecule may bind to a second site on the enzyme, thereby altering activity at the substrate site by an allosteric interaction (**noncompetitive inhibition**; see Problem 10.6). One particularly elegant arrangement is a chain of enzymes, the first of which is inhibited by the presence of the last one's product to make a feedback loop (see Figure 9.11 on page 378).

10.4.3 Two-headed kinesin as a tightly coupled, perfect ratchet

Section 10.3.4 suggested that the kinetics of a tightly coupled molecular motor would be much the same as those of an enzyme. In the language of free energy landscapes (Figure 10.9 on page 413), we expect to find a one-dimensional random walk down a single valley, corresponding to the successive processing of substrate to product (shown as motion toward negative values of α), combined with successive spatial steps (shown as motion toward negative values of β). If the concentration of product is kept very low, then the random walk along α will have an irreversible step, and so will the overall motion along the valley. We therefore expect that the analysis of Section 10.4.1 should apply, with one modification: Because the average rate of stepping depends on the free energy landscape along the valley, in particular it will depend on the applied load force (the tilt in the β direction), just as in Sections 10.2.1–10.2.3. In short, then, we expect that

> *A tightly coupled molecular motor, with at least one irreversible step in its kinetics, should move at a speed governed by the Michaelis–Menten rule, with parameters $v_{\max}$ and K_M dependent upon the load force.* (10.22)

A real molecular motor will, however, have some important differences from the gear machine imagined in Section 10.2.1. One difference is that we expect an enzyme's free energy landscape to be even more rugged than the one shown in Figure 10.9. Activation barriers will give the most important limit on the rate of stepping, not the viscous friction imagined in Section 10.2.1. In addition, we have no reason to expect that the valleys in the energy landscape will be the simple diagonals imagined in Figure 10.9. More likely, they will zigzag from one corner to the other. Some substeps may follow a path nearly parallel to the α-axis (a "purely chemical step"). The landscape along such a substep is unaffected by tilting in the β direction, so its rate will be nearly independent of the applied load. Other substeps will follow a path at some nonzero angle to the α-axis (a "mechanochemical step"); their rate will be sensitive to load.

Physical measurements can reveal details about the individual kinetic steps in a motor's operation. This section will follow an analysis due to M. Schnitzer, K. Visscher, and S. Block. Building on others' ideas, these authors argued for a model of kinesin's cycle (see Figure 10.24). The rest of this section will outline the evidence leading to this model and describe the steps symbolized by the cartoons in the figure.

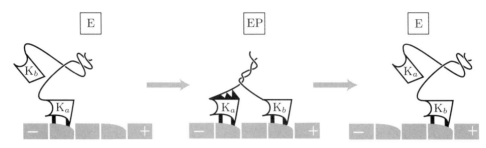

Figure 10.21: (Schematic.) One plausible model for directed motion of two-headed kinesin: the "hand-over-hand" scheme. After one cycle, the two heads of the kinesin dimer have exchanged roles, and the dimer has advanced along the microtubule (*gray*) by one step, or 8 nm. Figure 10.24 explains the symbols and gives more details about the intermediate biochemical steps between the illustrative states shown here.

Clues from kinetics Conventional (that is, two-headed) kinesin forms a homodimer, an association of two identical protein subunits. This structure lets kinesin walk along its microtubule track with a **duty ratio** of nearly 100%. The duty ratio is the fraction of the total cycle during which the motor is bound to its track and cannot slide freely along it; a high duty ratio lets the motor move forward efficiently even when an opposing load force is applied. One way for kinesin to achieve its high duty ratio could be by coordinating the detachment of its two identical heads in a "hand-over-hand" manner, so that at any moment one is always attached while the other is stepping (Figure 10.21).[10]

Kinesin is also highly **processive**. That is, it takes many steps (typically about 100) before detaching from the microtubule. Processivity is a very convenient property for the experimentalist seeking to study kinesin. Thanks to processivity, it's possible to follow the progress of a micrometer-size glass bead for many steps as it is hauled along a microtubule by a single kinesin molecule. Using optical tweezers and a feedback loop, experimenters can also apply a precisely known load force to the bead, then study the kinetics of kinesin stepping at various loads.

K. Svoboda and coauthors initiated a series of single-molecule motility assays of the type just described in 1993. Using an interferometry technique, they resolved individual steps of a kinesin dimer attached to a bead of radius $0.5\,\mu$m, finding that each step was 8 nm long, exactly the distance between successive kinesin binding sites on the microtubule track Some later data appear in Figure 10.22. As shown in the figure, kinesin rarely takes backward steps, even under a significant backward load force: In the terminology of Section 10.2.3, it is close to the perfect ratchet limit.

Further experiments showed that, in fact, two-headed kinesin is tightly coupled: It takes exactly one spatial step for each ATP molecule it consumes, even under moderate load. From the discussion at the start of this subsection, then, we may expect that two-headed kinesin would obey MM kinetics, with load-dependent parameters.

---

[10] $\boxed{T_2}$ Recent work has cast doubt on the hand-over-hand picture, in which the two kinesin heads execute identical chemical cycles (see Hua et al., 2002). Whatever the final model of kinesin stepping may be, however, it will have to be consistent with the experiments discussed in this section.

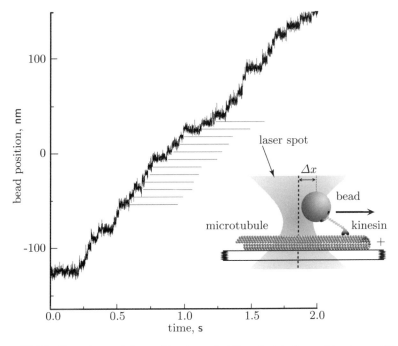

Figure 10.22: (Experimental data, with schematic.) Sample data from a kinesin motility assay. *Inset:* An optical tweezers apparatus pulls a 0.5 μm bead against the direction of kinesin stepping (not drawn to scale). A feedback circuit continuously moves the trap (*gray hourglass shape*) in response to the kinesin's stepping, maintaining a fixed displacement Δx from the center of the trap and hence a fixed backward load force (a procedure called force clamping). *Graph:* Stepping motion of the bead under a load force of 6.5 pN, with 2 mM ATP. The *gray lines* are separated by intervals of 7.95 nm; each corresponds to a plateau in the data. [Adapted from Visscher et al., 1999.]

Several experimental groups confirmed this prediction (Figure 10.23). Specifically, Table 10.1 shows that v_{max} decreases with increasing load, whereas K_M increases.

Your	The load forces tabulated in Table 10.1 reflect the force of the optical trap on the
Turn	bead. But the bead experiences another retarding force, namely, viscous drag
10E	friction. Shouldn't this force be included when we analyze the experiments?

Let's see what these results tell us about the details of the mechanism of force generation by kinesin.

One reasonable-sounding model for the stepping of kinesin might be the following: Suppose that binding of ATP is a purely chemical step, but its subsequent hydrolysis and release entail forward motion—a **power stroke.** Referring to Figure 10.17b on page 428, this proposal amounts to assuming that the load force pulls the second or third activation barrier up without affecting the first one; in the language of Equa-

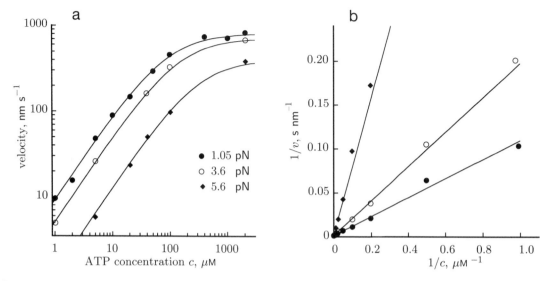

Figure 10.23: (Experimental data.) (a) Log-log plot of the speed v of kinesin stepping versus ATP concentration, at various loads (see legend). For each value of load, the data were fit to the Michaelis–Menten rule, yielding the solid curves with the parameter values listed in Table 10.1. (b) Lineweaver–Burk plot of the same data. [Data from Visscher et al., 1999.]

Table 10.1: Michaelis–Menten parameters for conventional kinesin stepping at fixed load force.

load force, pN	v_{max}, nm s^{-1}	K_M, μM
1.05	813 ± 28	88 ± 7
3.6	715 ± 19	140 ± 6
5.6	404 ± 32	312 ± 49

[Data from Schnitzer et al., 2000.]

tion 10.16, load reduces k_2 without affecting k_1 or k_{-1}. We already know how such a change will affect the kinetics: Equation 10.19 predicts that v_{max} will decrease with load (as observed), while K_M will also decrease (contrary to observation). Thus the data in Table 10.1 rule out this model.

Apparently there is another effect of load besides slowing down a combined hydrolysis/motion step. To explain their data, Schnitzer and coauthors proposed a model almost as simple as the unsuccessful one just described. Before discussing their proposed mechanism, however, we must digress to summarize some prior structural and biochemical studies.

Structural clues The microtubule itself also has a dimeric structure, with two alternating subunit types (see Figures 10.24, and 2.18 on page 55). One of the two subunits, called β, has a binding site for kinesin; these sites are regularly spaced at 8 nm intervals. The microtubule has a **polarity**; the subunits are all oriented in the same direction relative to one another, thus giving the whole structure a definite "front" and "back." We call the front the "+ end of the microtubule." Because protein binding is stereospecific (two matching binding sites must be oriented in a particular way), any bound kinesin molecule will point in a definite direction on the microtubule.

Each head of the kinesin dimer has a binding site for the microtubule and a second binding site for a nucleotide, such as ATP. Each kinesin head also has a short chain (15 amino acids) called the **neck linker**. The neck linkers in turn attach to longer chains. The two heads of the kinesin dimer are joined only by these chains, which intertwine, as shown schematically in Figure 10.24. The distance between the heads is normally too short for the dimer to act as a bridge between two binding sites on the microtubule, but under tension, the chains can stretch to allow such binding.

One further structural observation holds another clue to the mechanism of kinesin motility. Chapter 9 mentioned that the neck linker adopts strikingly different conformations, depending on the occupancy of the nucleotide-binding site (see Figure 9.12 on page 379). When the site is empty, or occupied by ADP, the neck linker flops between at least two different conformations. When the site contains ATP, however, the neck linker binds tightly to the core of the kinesin head in a single, well-defined orientation, pointing toward the "+" end of the microtubule. This allosteric change seems to be essential for motility: A modified kinesin, with its neck linker permanently attached to the head, was found to be unable to walk.

Biochemical clues We assign the abbreviations K, M, T, D for a single kinesin head, the microtubule, ATP, and ADP, respectively; DP represents the hydrolyzed combination ADP·P_i. In the absence of microtubules, kinesin binds ATP, hydrolyzes it, releases P_i, then stops—the rate of release for bound ADP is negligibly small. Thus kinesin alone has very little ATPase activity.

The situation changes if one removes the excess ATP and flows the solution of K·D (kinesin bound to ADP) onto microtubules. D. Hackney found in 1994 that in this case, single-headed (monomeric) kinesin rapidly releases all its bound ADP upon binding to the microtubules. Remarkably, Hackney also found that two-headed kinesin rapidly releases *half* of its bound ADP, retaining the rest. These and other results suggested that

- Kinesin binds ADP strongly, and
- Kinesin without bound nucleotide binds microtubules strongly, but
- The complex M·K·D is only weakly bound.

In other words, an allosteric interaction within one head of kinesin prevents it from binding strongly to both a microtubule and an ADP molecule at the same time. Thus the weakly bound complex M·K·D can readily dissociate. Hackney proposed an explanation for why only half of the ADP was released by kinesin dimers upon binding to microtubules: In the presence of ADP only, just one head at a time can reach a microtubule binding site (see state E of Figure 10.24).

It's hard to assess the ability of the complex K·T to bind microtubules because the ATP molecule is short lived (kinesin splits it). To overcome this difficulty, experimenters used an ATP analog molecule. This molecule, called AMP–PNP, has a shape and binding properties similar to those of ATP, but it does not split. Its complex with kinesin turned out to bind strongly to microtubules.

We can now state the key experimental observation. Suppose that we add two-headed (K·D)$_2$ to microtubules, thereby releasing half of the bound ADP as described earlier. *Adding ATP then causes the rapid release of the other half of the bound ADP!* Indeed, even the analog molecule AMP–PNP works: Binding, not hydrolysis, of nucleotide is sufficient. Somehow the unoccupied kinesin head, strongly bound to the microtubule, communicates the fact that it has bound an ATP to its partner head, stimulating the latter to release its ADP. This collaboration is remarkable, in the light of the rather loose connection between the two heads; it is not easy to imagine an allosteric interaction across such a floppy system.

In the rest of this section, we need to interpret these surprising phenomena and see how they can lead to a provisional model for the mechanochemical cycle of two-headed kinesin.

Provisional model: Assumptions Some of the following assumptions remain controversial. Still, we'll see that the model makes definite, and tested, predictions about the load dependence of kinesin's stepping kinetics.

We make the following assumptions, based on the clues listed earlier:

A1. We first assume that in the complexes M·K·T and M·K·DP, the kinesin binds (or "docks") its neck linker tightly, in a position that moves the attached chain forward, toward the "+" end of the microtubule. The other kinesin head in the dimer will then also move forward. The states M·K and M·K·D, in contrast, have the neck linker in a flexible state.

A2. When the neck linker is docked, the detached kinesin head will spend most of its time in front of the bound head. Nevertheless, the detached head will spend *some* of its time to the rear of its partner.

A3. We assume that kinesin with no nucleotide binds strongly to the microtubule, as does K·T. The weakly bound state M·K·D readily dissociates, either to M+K·D or to M·K+D.

Assumption A3 says that the free energy gain from ATP hydrolysis and phosphate release is partly spent on a conformational change that pulls the M·K·T complex out of a deep potential energy well to a shallower one. Similarly, A1 says that some of this free energy goes to relax the head's grip on its neck linker.

Provisional model: Mechanism The proposed mechanism is summarized graphically in Figure 10.24. This cycle is not meant to be taken literally; it just shows some of the distinct steps in the enzymatic pathway. Initially (top left panel of the figure), a kinesin dimer approaches the microtubule from solution and binds one head, releasing one of its ADPs. We name the subsequent states in the ATP hydrolysis cycle by abbreviations describing the state of the head that was initially bound.

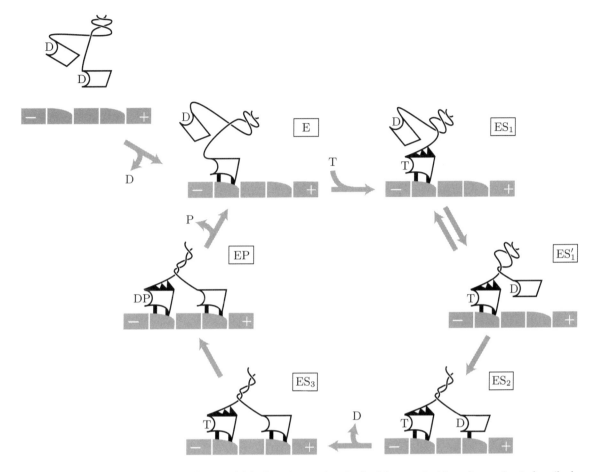

Figure 10.24: (Schematic.) Details of the model for kinesin stepping. Each of the steps in this cyclic reaction is described in the text, starting on page 442. Some elements of this mechanism are still under debate. The steps form a loop, to be read clockwise from upper left. The *gray symbols* represent a microtubule, with its "+" end at the right. Strong physical bonds are denoted by multiple lines, weak ones by single lines. The symbols T, D, and P denote ATP, ADP, and inorganic phosphate, respectively. The rapid isomerization step, $ES_1 \rightleftharpoons ES_1'$, is assumed to be nearly in equilibrium. The states denoted ES_2, ES_3, and EP are under internal strain, as described in the text. In the step from EP back to E, the roles of the two kinesin heads exchange. [Similar schemes, with some variations, appear in Gilbert et al., 1998; Rice et al., 1999; Schnitzer et al., 2000; Vale & Milligan, 2000; Schief & Howard, 2001; Mogilner et al., 2001; Uemura et al., 2002.]

E: This panel shows the dimer with one head strongly bound to the microtubule. The other, free head cannot reach any binding site because its tether is too short; the sites are separated by a fixed distance along the rigid microtubule.

ES_1,ES_1': The bound head binds an ATP molecule from solution. Its neck linker then docks onto its head, biasing the other head's random motion in the forward direction (assumption A2). Schnitzer and coauthors assumed that interactions with the microtubule effectively give the complex a weak energy landscape, making the unbound head hop between two distinct states ES_1 and ES_1'.

ES_2: The chains joining the two heads will have entropic elasticity, as discussed in Chapter 9. Being thrown forward by the bound head's neck linker greatly increases the probability that the tethers will momentarily stretch far enough for the free head to reach the next binding site. It may bind weakly, then detach, many times.

ES_3: Eventually, instead of detaching, the forward head releases its ADP and binds strongly to the microtubule. Its stretched tether now places the whole complex under sustained strain. Both heads are now tightly bound to the microtubule, however, so the strain does not pull either one off.

EP: Meanwhile, the rear head splits its ATP and releases the resulting phosphate. This reaction weakens its binding to the microtubule (assumption A3). The strain induced by the binding of the forward head then biases the rear head to unbind from the microtubule (rather than releasing its ADP).

E: The cycle is now ready to repeat, with the roles of the two heads reversed (see Figure 10.21). The kinesin dimer has made one 8 nm step and hydrolyzed one ATP.

The assumptions made earlier ensure that *free* kinesin (not bound to any microtubule) does not waste any of the available ATP, as observed experimentally. According to assumption A3, free kinesin will bind and hydrolyze ATP at each of its two heads, then stop, because the resulting ADPs are both tightly bound in the absence of a microtubule.

Our model is certainly more complicated than the S-ratchet imagined in Section 10.2! But we can see how our assumptions implement the necessary conditions for a molecular motor found there (Idea 10.9a,b on page 422):

- The forward flip induced by neck linker binding (assumption A1), together with the polarity of the microtubule, creates the needed spatial asymmetry.
- The tight linkage to the hydrolysis of ATP creates the needed out-of-equilibrium condition, since the cell maintains the reaction quotient $c_{ATP}/(c_{ADP}\, c_{P_i})$ at a level much higher than its equilibrium value.

Let's see how to make these ideas quantitative.

Kinetic predictions Let's simplify the problem by lumping all the states other than ES_1 and ES_1' into a single state called E, just as Equation 10.16 on page 434 lumped EP with E. The model sketched in Figure 10.24 then amounts to splitting the bound state ES of the Michaelis–Menten model into two substates, ES_1 and ES_1'. To extract predictions from this model, Schnitzer and coauthors proposed that the step $ES_1 \rightleftharpoons ES_1'$ is nearly in equilibrium. That is, they assumed that the activation barrier to this transition is small enough, and hence the step is rapid enough relative to the others, that the relative populations of the two states stay close to their equilibrium values.[11] We

---

[11] Some authors refer to this assumption as rapid isomerization.

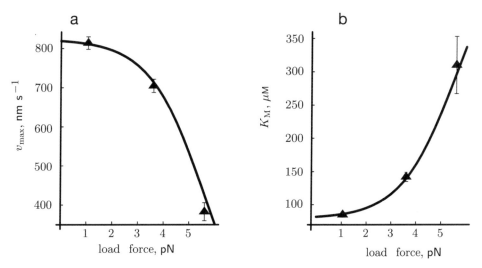

Figure 10.25: (Experimental data with fit.) Dependence of kinesin's MM parameters on applied load. Points denote the data derived from Figure 10.23 (see Table 10.1). The curves show (a) Equation 10.31 and (b) Equation 10.30, with the fit parameter values given in Section 10.4.3′ on page 459.

can consider these two states together, thinking of them jointly as a **composite state**. In the language of Equation 10.16, the fraction of time spent in ES_1 effectively lowers the rate k_2 of leaving the composite state in the forward direction. Similarly, the fraction of time spent in ES_1' effectively lowers the rate k_{-1} of leaving the composite state in the backward direction.

We wish to understand the effect of an applied load force, that is, an external force directed away from the "+" end of the microtubule. To do this, note that the step $ES_1 \rightarrow ES_1'$, besides throwing head K_b forward, also moves the common connecting chains to a new average position, shifted forward by some distance ℓ. All we know about ℓ is that it is greater than zero, but less than a full step of 8 nm. Because a spatial step does work against the external load, the applied load force will affect the composite state: It shifts the equilibrium away from ES_1' and toward ES_1. Schnitzer and coauthors neglected other possible load dependences, focusing only on this one effect.

We now apply the arguments of the previous two paragraphs to the definitions of the MM parameters (Equation 10.19 on page 434), finding that load reduces v_{max}, as observed, and moreover may increase K_M by effectively increasing k_{-1} by more than it reduces k_2. Thus we have the possibility of explaining the data in Table 10.1 with the proposed mechanism.

To test the mechanism, we must see whether it can model the actual data. That is, we must see whether we can choose the free energy change ΔG of the isomerization $ES_1 \rightleftharpoons ES_1'$, as well as the substep length ℓ, in a way that explains the numbers in Table 10.1. Some mathematical details are given in Section 10.4.3′ on page 459. A reasonably good fit can indeed be found (Figure 10.25). More important than the literal fit shown is the observation that the simplest power stroke model does not fit the data, but an almost equally simple model, based on structural and biochemical clues, reproduces the qualitative facts of Michaelis–Menten kinetics, with K_{M} rising and v_{max} falling as the load is increased.

The fit value of the equilibrium constant for the isomerization reaction is reasonable: It corresponds to a ΔG^0 of about $-5k_{\mathrm{B}}T_{\mathrm{r}}$. The fit value of ℓ is about 4 nm, which is also reasonable: It's half the full step length.

$\boxed{T_2}$ *Section 10.4.3′ on page 459 completes the analysis, obtaining the relation between speed, load, and ATP availability in this model.*

10.4.4 Molecular motors can move even without tight coupling or a power stroke

Section 10.4.3 argued that deep within the details of kinesin's mechanochemical cycle, there lies a simple mechanism: Two-headed kinesin slides down a valley in its free energy landscape. Even while admitting that the basic idea is simple, we can still marvel at the elaborate mechanism that evolution has had to create to implement it. For example, we saw that to have a high duty ratio, kinesin has been cunningly designed to coordinate the action of its two heads. How could such a complex motor have evolved from something simpler?

We could put the matter differently by asking, "Isn't there some simpler force-generating mechanism, perhaps not as efficient or as powerful as two-headed kinesin, which could have been its evolutionary precursor?" In fact, a single-headed (monomeric) form of kinesin, called KIF1A, has been found to have single-molecule motor activity. Y. Okada and N. Hirokawa studied a modified form of this kinesin, which they called C351. They labeled their motor molecules with fluorescent dye, then watched as successive motors encountered a microtubule, bound to it, and began to walk (see Color Figure 4).

Quantitative measurements of the resulting motion led Okada and Hirokawa to conclude that C351 operates as a **diffusing ratchet** (or D-ratchet). In this class of models, the operating cycle includes a step involving unconstrained diffusive motion, unlike the G- and S-ratchets. Also, in place of the unspecified agent resetting the bolts in the S-ratchet (see Section 10.2.3), the D-ratchet couples its spatial motion to a chemical reaction.

The free energy landscape of a single-headed motor cannot look like our sketch, Figure 10.9 on page 413. To make progress, the motor must periodically detach from its track; once detached, it's free to move along the track. In the gear metaphor (Figure 10.6c on page 410), the gears must disengage on every step, thereby allowing free slipping; in the landscape language, there are certain points in the chemical cycle (certain values of α) at which the landscape is *flat* in the β direction. Thus there are no well-defined diagonal valleys in the landscape. How can such a device make net progress?

The key observation is that, even though the grooved landscape of Figure 10.9 was convenient for us (it made the landscape effectively one-dimensional), still such a structure isn't really necessary for net motion. Idea 10.9a,b on page 422 gave the requirements for net motion as simply *a spatial asymmetry in the track* and *some out-of-equilibrium process coupled to spatial motion*. In principle, we should expect that solving the Smoluchowski equation on *any* two-dimensional free energy landscape will reveal net motion, as long as the landscape is tilted in the chemical (α) direction and asymmetrical in the spatial (β) direction.

As mentioned earlier, however, it's not easy to solve the Smoluchowski equation (Equation 10.4 on page 419) in two dimensions, nor do we even know a realistic free energy landscape for any real motor. To show the essence of the D-ratchet mechanism, then, we will as usual construct a simplified mathematical model. Our model motor will contain a catalytic site, which hydrolyzes ATP, and another site, which binds to the microtubule. We will assume that an allosteric interaction couples the ATPase cycle to the microtubule binding in a particular way:

1. The chemical cycle is autonomous—it's not significantly affected by the interaction with the microtubule. The motor snaps back and forth between two states, which we will call *s* (for "strong-binding") and *w* (for "weak-binding"). After entering the *s* state, it waits an average time t_s before snapping over to *w*; after entering the *w* state, it waits some other average time t_w before snapping back to *s*. (One of these states could be the one with the nucleotide-binding site empty, and the other one E·ATP, as drawn in Figure 10.26.) The assumption is that t_s and t_w are both independent of the motor's position *x* along the microtubule.

2. However, the binding energy of the motor to the microtubule does depend on the state of the chemical cycle. Specifically, we will assume that in the *s* state, the motor prefers to sit at specific binding sites on the microtubule, separated by a distance of 8 nm. In the *w* state, the motor will be assumed to have no positional preference at all—it diffuses freely along the microtubule.

3. In the strongly binding state, the motor feels an asymmetrical (that is, lopsided) potential energy $U(x)$ as a function of its position *x*. This potential is sketched as the sawtooth curve in Figure 10.26a; asymmetry means that this curve is not the same if we flip it end-for-end. Indeed, we do expect the microtubule, a polar structure, to give rise to such an asymmetrical potential.

In the D-ratchet model, the free energy of ATP hydrolysis can be thought of as entering the motion solely by an assumed allosteric conformational change, which alternately glues the motor onto the nearest binding site, then pries it off. To simplify the math, we will assume that the motor spends enough time in the *s* state to find a binding site, then binds and stays there until the next switch to the *w* state.

Let's see how the three assumptions listed above yield directed motion, following the left panels of Figure 10.26. As in Section 10.2.3, imagine a collection of *many* motor–microtubule systems, all starting at one position, $x = 0$ (panels (b1) and (b2)). At later times we then seek the probability distribution $P(x)$ to find the motor at various positions *x*. At time zero the motor snaps from *s* to *w*. The motor then diffuses freely along the track (panel (c1)), so its probability distribution spreads out into a Gaussian centered on x_0 (panel (c2)). After an average wait of t_w, the motor snaps back to its *s* state. Now it suddenly finds itself strongly attracted to the periodically spaced binding sites. Accordingly, it drifts rapidly down the gradient of $U(x)$ to the first such minimum, and we end up with the probability distribution symbolized by panel (d2) of Figure 10.26. The cycle then repeats.

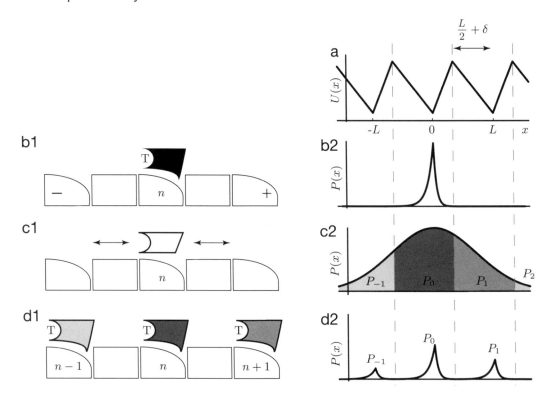

Figure 10.26: (Schematic; sketch graphs.) Diffusing ratchet (or D-ratchet) model for single-headed kinesin motility. *Left panels:* Bound ATP is denoted by T; ADP and P_i molecules are not shown. Other symbols are as in Figure 10.24. (b1) Initially, the kinesin monomer is strongly bound to site n on the microtubule. (c1) In the weakly bound state, the kinesin wanders freely along the microtubule. (d1) When the kinesin reenters the strongly bound state, it is most likely to rebind to its original site, somewhat likely to rebind to the next site, and least likely to bind to the previous site. Relative probabilities are represented by shading. *Right panels:* (a) A periodic but asymmetrical potential for the strong-binding (or s) state, as a function of position x along the microtubule track. The minimum of the potential is not midway between the maxima, but instead is shifted by a distance δ. The potential repeats every distance L ($L = 8$ nm for a microtubule). (b2) Quasi-equilibrium probability distribution for a motor in its s state, trapped in the neighborhood of the minimum at $x = 0$. The motor now suddenly switches to its w (or weak-binding) state. (c2) (Change of vertical scale.) The probability distribution just before the motor switches out of its w state. The *dark gray region* represents all the motors in an initial ensemble that are about to fall back into the microtubule binding site at $x = 0$; the area under this part of the curve is P_0. The *medium gray region* represents those motors about to fall into the site at $x = L$; the corresponding area is P_1. The *light gray regions* to the left and right have areas P_{-1} and P_2, respectively. (d2) (Change of vertical scale.) The probability distribution just before the motor switches back to the w state. The areas P_k from (c2) have each collapsed to sharp spikes. Because $P_1 > P_{-1}$, the mean position has shifted slightly to the right.

The key observation is that the average position of the motor after one cycle is now shifted relative to where it was originally. Some of this shift may arise from conformational changes, "power stroke" shifts analogous to those in myosin or two-headed kinesin. But the surprise is that there will be a net shift even without any power stroke! To see this, examine Figure 10.26 and its caption. The dark gray part of the curve in panel (c2) of the figure represents all the motors in the original collection that are about to rebind to the microtubule at their original position, $x = 0$. Thus the probability of taking *no* step is the area P_0 under this part of the curve. The two flanking parts of the curve, medium and light gray, represent respectively those motors about to rebind to the microtubule at positions shifted by $+L$ or $-L$, respectively. But the areas under these parts of the curve are not equal: $P_1 \neq P_{-1}$. The motor is more likely to diffuse over to the basin of attraction at $x = +L$ than to the one at $x = -L$, simply because the latter's boundary is farther away from the starting position.

Thus the diffusing ratchet model predicts that a one-headed molecular motor can make net progress. Indeed, we found that *it makes net progress even if no conformational change in the motor drives it in the x direction.* The model also makes some predictions about experiments. For one thing, we see that the diffusing ratchet can make backward steps;[12] P_{-1} is not zero, and can indeed be large if the motor diffuses a long way between chemical cycles. In fact, each cycle gives an independent displacement, with the same probability distribution $\{P_k\}$ for every cycle. Section 4.1.3 on page 117 analyzed the mathematics of such a random walk. The conclusion of that analysis, translated into the present situation, was that

> *The diffusing ratchet makes net progress uL per step, where $u = \langle k \rangle$.*
> *The variance (mean-square spread) of the total displacement in-*
> *creases linearly with the number of cycles, increasing by $L^2 \times$*   (10.23)
> *variance(k) per cycle.*

In our model, the steps come every $\Delta t = t_s + t_w$, so we predict a constant mean velocity $v = uL/\Delta t$ and a constant rate of increase in the variance of x given by

$$\langle (x(t) - vt)^2 \rangle = t \times \frac{L^2}{\Delta t} \times \text{variance}(k). \qquad (10.24)$$

Okada and Hirokawa tested these predictions with their single-headed kinesin construct, C351. Although the optical resolution of the measurements, $0.2\,\mu\text{m}$, was too large to resolve individual steps, still Figure 10.27 shows that C351 often made net backward progress (panel (a)), unlike conventional two-headed kinesin (panel (b)). The distribution of positions at a time t after the initial binding, $P(x, t)$, showed features characteristic of the diffusing ratchet model. As predicted by Equation 10.24, the mean position moved steadily to larger values of x, while the variance steadily increased. In contrast, two-headed kinesin showed uniform motion with very little increase in variance (panel (b)).

To make these qualitative observations sharp, Figure 10.27c plots the observed mean-square displacement, $\langle x(t)^2 \rangle$. According to Equation 10.24, we expect this quantity to be a quadratic function of time, namely, $(vt)^2 + t(L^2/\Delta t)\text{variance}(k)$. The figure shows that the data fit such a function well. Okada and Hirokawa concluded that, although monomeric kinesin cannot be tightly coupled, it makes net progress in the way predicted by the diffusing ratchet model.

---

[12]Compare with Problem 4.1 on page 153.

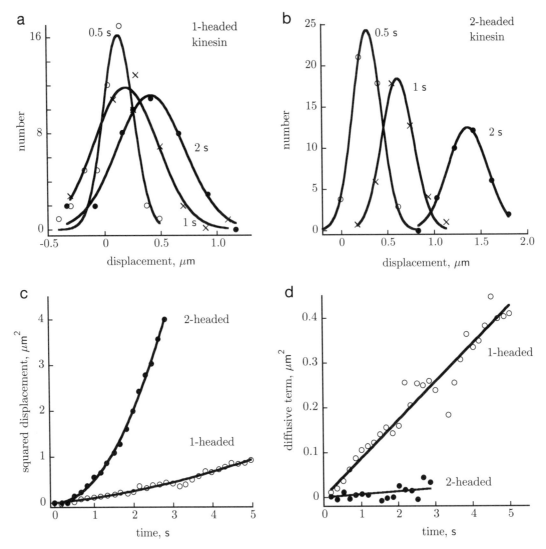

Figure 10.27: (*Experimental data.*) Analysis of the movement of single kinesin molecules. (a) Data for C351, a single-headed form of kinesin. The graphs give the observed distributions of displacement x from the original binding site, at three different times. The solid curves show the best Gaussian fit to each data set. Notice that even at 2 s, a significant fraction of all the kinesins has made net backward progress. (b) The same data as (a), but for conventional two-headed kinesin. None of the observed molecules made net backward progress. (c) Mean-square displacement, $\langle x(t)^2 \rangle$, as a function of time for single-headed (*open circles*) and two-headed (*solid circles*) kinesin. The curves show the best fits to the predicted random-walk law (see text). (d) The same data and fits as (c), after subtracting the $(vt)^2$ term (see text). [Data from Okada & Hirokawa, 1999.]

Subtracting away the $(vt)^2$ term to focus attention on the diffusive part reveals a big difference between one- and two-headed kinesin. Figure 10.27d shows that both forms obey Equation 10.24, but C351 had a far greater diffusion constant of proportionality, a difference reflecting the loosely coupled character of single-headed kinesin.

To end this section, let's return to the question that motivated it: How could molecular motors have evolved from something simpler? We have seen how the bare minimal requirements for a motor are simple, indeed:

- It must cyclically process some substrate like ATP, to generate out-of-equilibrium fluctuations.
- These fluctuations must in turn couple allosterically to the binding affinity for another protein.
- The latter protein must be an asymmetrical polymer track.

It's not so difficult to imagine how an ATPase enzyme could gain a specific protein-binding site by genetic reshuffling; the required allosteric coupling would arise naturally from the general fact that all parts of a protein are tied together. Indeed, a related class of enzymes is already known in eukaryotic cells, the GTP-binding proteins (or **G-proteins**); they play a number of intracellular signaling roles, including a key step in the detection of light in your retina. It seems reasonable to suppose that the first, primitive motors were G-proteins whose binding targets were polymerizing proteins, like tubulin. Interestingly, G-proteins have indeed turned out to have close structural links to both kinesin and myosin, perhaps reflecting a common evolutionary ancestry.

$\boxed{T_2}$ *Section 10.4.4′ on page 461 gives some quantitative analysis of the model and compares it with the experimental data.*

10.5 VISTA: OTHER MOLECULAR MOTORS

New molecular machines are constantly being discovered. Table 10.2 lists some of the known examples. Yet another class of machines transport ions across membranes against their electrochemical gradient; these "pumps" will play a key role in Chapter 11.

THE BIG PICTURE

Let's return to the Focus Question. This chapter has uncovered two simple requirements for a molecular device to transduce chemical energy into useful mechanical work: The motor and track must be asymmetrical in order to select a direction of motion, and they must couple to a source of excess free energy, for example, a chemical reaction that is far from equilibrium. The following chapter will introduce two other classes of molecular machines, ion pumps and the rotary ATP synthase.

Table 10.2: Examples of proteins that are believed to act as molecular motors.

motor	pushes on	energy source	motion	role
Cytoskeletal motors:				
kinesin	microtubule	ATP	linear	mitosis, organelle transport
myosin	actin	ATP	linear	muscle contraction, organelle transport
dynein	microtubule	ATP	linear	ciliary beating, organelle transport, mitosis
Polymerization motors:				
actin	none	ATP	extend/shrink	cell motility
microtubule	none	GTP	extend/shrink	mitosis
dynamin	membranes	GTP	pinching	endocytosis, vesicle budding
G-proteins:				
EfG	ribosome	GTP	lever	movement of peptidyl-tRNA and mRNA in ribosomes
Rotary motors:				
F0 motor	F1 ATPase	$\Delta[H^+]$	rotary	ATP synthesis
bacterial flagellar	peptidoglycan	$\Delta[H^+]$	rotary	propulsion
Nucleic acid motors:				
polymerases	DNA/RNA	ATP	linear	template replication
helicases	DNA/RNA	ATP	linear	opening of DNA duplex
phage portal motor	DNA	ATP	linear	packing virus capsid

[See Vale, 1999.]

A mechanochemical motor transduces chemical free energy to mechanical work. When the relative concentrations of fuel and waste differ from equilibrium, that's a form of order, analogous to the temperature differential that ran our heat engine in Section 6.5.3. It may seem surprising that the motors in this chapter can work inside a single, well-mixed chamber; in contrast, a heat engine must sit at the *junction* between a hot reservoir and a cold one. But if there is an activation barrier to the spontaneous conversion of fuel to waste, then even a well-mixed solution has an invisible wall separating the two, like a dam on a river. It's really not in equilibrium at all. The motor is like a hydroelectric plant on that dam: It offers a low-barrier pathway to the state of lower μ. Molecules will rush down that pathway, even if they are required to do some work along the way, just as water rushes to drive the turbine of the hydroelectric plant.

Cells contain a staggering variety of molecular motors. This chapter has made no attempt to capture Nature's full creative range, once again focusing on the humbler question, "How could anything like that happen at all?" Nor did we attempt even a survey of the many beautiful experimental results now available. Rather, the goal was simply to create some explicit mathematical models, anchored in simpler, known phenomena and displaying some of the behavior experimentally observed in real motors. Such conceptually simple models are the armatures upon which more detailed understanding must rest.

KEY FORMULAS

- *Perfect ratchet:* A perfect ratchet (that is, one with an irreversible step) at zero load makes progress at the rate $v = L/t_{\text{step}} = 2D/L$ (Equation 10.2).
- *Smoluchowski:* Consider a particle undergoing Brownian motion on a potential landscape $U(x)$. In a steady (not necessarily equilibrium) state, the probability $P(x)$ of finding the particle at location x is a solution to (Equation 10.4)

$$0 = \frac{d}{dx}\left(\frac{dP}{dx} + \frac{1}{k_B T}P\frac{dU}{dx}\right),$$

with appropriate boundary conditions.

- *Michaelis–Menten:* Consider the catalyzed reaction

$$E + S \underset{k_{-1}}{\overset{c_S k_1}{\rightleftharpoons}} ES \overset{k_2}{\rightarrow} E + P.$$

A steady, nonequilibrium state can arise when the supply of substrate S is much larger than the supply of enzyme E. The reaction velocity (rate of change of substrate concentration c_S) in this case is $v = v_{\text{max}}c_S/(K_M + c_S)$ (Equation 10.20), where the saturating reaction velocity is $v_{\text{max}} = k_2 c_E$ and the Michaelis constant is $K_M = (k_{-1} + k_2)/k_1$ (Equation 10.19).

FURTHER READING

Semipopular:
Enzymes: Dressler & Potter, 1991.

Intermediate:
Enzymes: Berg et al., 2002; Voet & Voet, 2003.
Chemical kinetics: Tinoco et al., 2001; Dill & Bromberg, 2002.
From actin/myosin up to muscle: McMahon, 1984.
Ratchets: Feynman et al., 1963a, §46.
Motors: Berg et al., 2002; Howard, 2001; Bray, 2001; Duke, 2002.

Technical:
Kramers theory: Kramers, 1940; Frauenfelder & Wolynes, 1985; Hänggi et al., 1990.
The abstract discussion of motors was largely drawn from the work of four groups around 1993. Some representative reviews by these groups include Jülicher et al., 1997; Astumian, 1997; Mogilner et al., 2002; Magnasco, 1996.
Single-molecule motility assays: Howard et al., 1989; Finer et al., 1994.
Myosin, kinesin, and G-proteins: general, Vale & Milligan, 2000; role of kinesin neck linker: Rice et al., 1999; Schnitzer et al., 2000; Mogilner et al., 2001.

RNA polymerase: Wang et al., 1998.

Polymerization ratchet, translocation ratchet: Mahadevan & Matsudaira, 2000; Borisy & Svitkina, 2000; Prost, 2002; for the shape assumed by a vesicle with a growing microtubule inside, see Powers et al., 2002.

T_2 ### 10.2.3′ Track 2

1. Strictly speaking, t_{step} in Equation 10.2 on page 417 should be computed as the mean time for a random walker to arrive at an absorber located at $x = L$, after being released at a reflecting wall at $x = 0$. Luckily, this time is given by the same formula, $t_{step} = L^2/(2D)$, as the naïve formula we used to get Equation 10.2! (See Berg, 1993, Equation 3.13.)

2. M. Smoluchowski foresaw many of the points made in this chapter around 1912. Some authors instead use the term *Fokker–Planck equation* for Equation 10.4 on page 419; others reserve that term for a related equation involving both position and momentum.

3. Equation 10.7 on page 421 was applicable only in the perfect-ratchet limit. To study the S-ratchet in the general nonequilibrium case, we first need the general solution to Equation 10.4 on the interval $(0, L)$ with $dU_{tot}/dx = f$, namely, $P(x) = C(b\,e^{-xf/k_BT} - 1)$ for any constants C and b. The corresponding probability flux is $j^{(1d)} = Mf DC/k_BT$.

   To fix the unknown constant b, we next show quite generally that the function $P(x)e^{U_{tot}(x)/k_BT}$ must have the same value just above and just below any discontinuity in the potential.[13] Multiply both sides of Equation 10.3 by $e^{U_{tot}(x)/k_BT}$, to find

$$\frac{d}{dx}\left(Pe^{U_{tot}/k_BT}\right) = -\frac{j^{(1d)}}{MD}e^{U_{tot}/k_BT}.$$

Integrating both sides of this equation from $L - \delta$ to $L + \delta$, where δ is a small distance, yields

$$P(L - \delta)e^{U_{tot}(L-\delta)/k_BT} = P(L + \delta)e^{U_{tot}(L+\delta)/k_BT}, \qquad (10.25)$$

plus a correction that vanishes as $\delta \to 0$. That is, $P(x)e^{U_{tot}(x)/k_BT}$ is continuous at L, as was to be shown.

   Imposing Equation 10.25 on our solution at $x = L$ and using the periodicity assumption, $P(L + \delta) = P(\delta)$, gives

$$P(L - \delta)e^{f L/k_BT} = P(0 + \delta)e^{(f L-\epsilon)/k_BT}$$

$$b(e^{-f L/k_BT} - e^{-\epsilon/k_BT}) = 1 - e^{-\epsilon/k_BT}$$

or

$$b = \frac{e^{\epsilon/k_BT} - 1}{e^{(-f L+\epsilon)/k_BT} - 1}.$$

---

[13] $P(x)$ itself will *not* be continuous; for example, in equilibrium, Example 10A on page 419 gives $P(x) \propto e^{-U(x)/k_BT}$, which is discontinuous whenever $U(x)$ is.

Proceeding as in the derivation of Equation 10.5, we find

$$\Delta t = \frac{MC}{j^{(1d)}}\left(-\frac{b}{f/k_BT}\left(e^{-fL/k_BT}-1\right)-L\right).$$

and hence

$$v = \frac{L}{\Delta t} = -\frac{D}{L}\left(\frac{fL}{k_BT}\right)^2\left[\frac{fL}{k_BT}-\frac{(1-e^{-\epsilon/k_BT})(1-e^{-fL/k_BT})}{e^{-fL/k_BT}-e^{-\epsilon/k_BT}}\right]^{-1}. \qquad (10.26)$$

You can verify from this formula that all four of Sullivan's claims listed on page 416 are correct:

Your Turn 10F

a. Check what happens when the load is close to the thermodynamic stall point, $f = \epsilon/L$.

b. What happens to Equation 10.26 at $f \to 0$? How can the ratchet move to the right, as implied by Sullivan's third point? Doesn't the formula $j^{(1d)} = MfDC/k_BT$, along with Equation 10.6, imply that $v \to 0$ when $f \to 0$?

c. Find the limit of very high drive, $\epsilon \gg k_BT$, and compare with the result in Equation 10.7.

d. Find the limit $\epsilon \gg fL \gg k_BT$, and comment on Sullivan's fourth assertion.

4. The physical discussion of Figure 10.12 on page 418 was subtle; an equivalent way to express the logic may be helpful. Rather than wrap the ratchet into a circle, we can take it to be straight and infinitely long. Then the probability distribution will *not* be periodic, nor will it be time independent. Instead, $P(x)$ will look like a broad bump (or envelope function), modulated by the spikes of Figure 10.12. The envelope function drifts with some speed v, whereas the individual spikes remain fixed at the multiples of L. To make contact with the discussion given in Section 10.2.3, we imagine sitting at the peak of the envelope function. After the system has evolved a long time, the envelope will be very broad and hence nearly flat at its peak. Therefore $P(x, t)$ will be approximately periodic and time independent. Our earlier analysis, leading to the Smoluchowski equation, is thus sufficient to find the average speed of advance.

T_2 **10.3.2′ Track 2**

1. The ultimate origin of the energy landscape lies in quantum mechanics. For the case of simple molecules in isolation (that is, in a gas), one can calculate this landscape explicitly. It suffices to treat only the electrons quantum-mechanically. Thus,

in the discussion of Section 10.3.2, the phrase "positions of atoms" is interpreted as "positions of nuclei." One imagines nailing the nuclei at specified locations, computing the ground-state energy of the electrons, and adding in the mutual electrostatic energy of the nuclei, to obtain the energy landscape. This procedure is known as the Born–Oppenheimer approximation. For example, it could be used to generate the energy landscape of Figure 10.14 on page 425.

For macromolecules in solution, more phenomenological approaches are widely used. Here one attempts to replace the complicated effects of the surrounding water (hydrophobic interaction and so on) by empirical interatomic potentials involving only the atoms of the macromolecule itself.

Many more sophisticated calculations than these have been developed. But quite generally the strategy of understanding chemical reactions as essentially classical random walks has proved successful for many biochemical processes. (An example of the exceptional, intrinsically quantum-mechanical processes is the detection of single photons by the retina.)

2. You may have noticed that in passing from Section 10.2.3 to Section 10.3.2, the word *energy* changed to *free energy*. To understand this shift, we first note that in a complex molecule, there may be *many* critical paths, each accomplishing the same reaction, not just one as shown in Figure 10.14. In this case, the reaction's rate gets multiplied by the number N of paths; equivalently, we can replace the barrier energy $\Delta E^{\ddagger}$ by an effective barrier $\Delta E^{\ddagger} - k_B T \ln N$. If we interpret the second term as the entropy of the transition state (and neglect the difference between energy and enthalpy), then we find that the reaction is really suppressed by $\Delta G^{\ddagger}$, not $\Delta E^{\ddagger}$. Indeed, we already knew that *equilibrium* between two complex states is controlled by free energy differences (Section 6.6.4 on page 225).

Further evidence that we should use the free energy landscape comes from a fact we already know about reaction rates. Suppose that the reaction involves binding a molecule that was previously moving independently, in a simple one-step process. In this case, we expect the rate of the reaction to increase with the concentration of that molecule in solution. The same conclusion emerges from our current picture, if we consider a walk on the free energy landscape. To see this, note that the bound molecule is being withdrawn from solution as it binds, so its initial entropy S_{in} makes a positive contribution to $\Delta G^{\ddagger} = \Delta E^{\ddagger} - TS^{\ddagger} - (E_{in} - TS_{in})$, decreases the Arrhenius exponential factor $e^{-\Delta G^{\ddagger}/k_B T}$, and therefore slows the predicted reaction rate. For example, if the bound molecule is present in solution at very low concentration, then its entropy loss upon binding will be large, and the reaction will proceed slowly, as we know it must. (Reactions are also slowed by the entropy loss implicit in *orienting* the reacting molecule properly for binding.)

More quantitatively, at small concentrations, the entropy per molecule is $S_{in} = -\mu/T = -k_B \ln c + \text{const}$ (see Equations 8.1 and 8.3), so its contribution to the exponential factor is a constant times c. This is just the familiar statement that simple binding leads to a first-order rate law (see Section 8.2.3 on page 306).

Finally, Section 10.2.3 argued that a molecular-scale device makes no net progress when its free energy landscape has zero average slope. But we saw in

Chapter 8 that a chemical reaction makes no net progress when its ΔG is zero, another way to see that the free energy, not ordinary energy, is the appropriate landscape to use. (For more on this important point, see Howard, 2001, Appendix 5.1.)

3. It's an oversimplification to say that we can simply ignore all directions in configuration space perpendicular to the critical path between two quasi-stable states. Certainly there will be excursions in these directions, with their own contribution to the entropy and so on. The actual elimination procedure involves finding the free energy by doing a partition sum over these directions, following essentially the methods of Section 7.1; the resulting free energy function is often called the potential of mean force. (See Grabert, 1982.)

  Besides modifying the free energy landscape, the mathematical step of eliminating all but one or two of the coordinates describing the macromolecule and its surrounding bath of water has a second important effect. The many eliminated degrees of freedom are all in thermal motion and are all interacting with the one reaction coordinate we have retained. Thus all contribute, not only to generating random motion along the reaction coordinate, but also to impeding directed motion. That is, the eliminated degrees of freedom give rise to *friction*, by an Einstein relation. (Again see Grabert, 1982.) H. Kramers pointed out in 1940 that this friction could be large and that, for complicated molecules in solution, the calculation of reaction rates via the Smoluchowski equation is more complete than the older Eyring theory. He reproduced Eyring's earlier results in a special (intermediate-friction) case, then generalized it to cover low and high friction. (For a modern look at some of the issues and experimental tests of Kramers' theory, see Frauenfelder & Wolynes, 1985.)

T_2 **10.3.3′ Track 2**

1. The discussion in Section 10.3.3 focused on the possibility that the lowest free energy state of the enzyme–substrate complex may be one in which the substrate is geometrically deformed to a conformation closer to its transition state. Figure 10.28 shows two other ways in which the grip of an enzyme can alter its substrate(s), accelerating a reaction. (For more biochemical details, see for example Dressler & Potter, 1991.)

2. The physical picture of walking down a free energy landscape (Idea 10.14 on page 429) is also helpful in understanding a new phenomenon occurring at extremely low concentrations of substrate. In this case, there will be large random variations in the arrival times of substrate molecules at E. We interpret these variations as the times to hop over the first bump in Figure 10.17 on page 428b. Because this bump is large when c_S is low, this contribution to the randomness of the process can be as important as the usual one (hopping over the middle bump of the figure). See Svoboda et al., 1994 for a discussion of this effect in the context of kinesin.

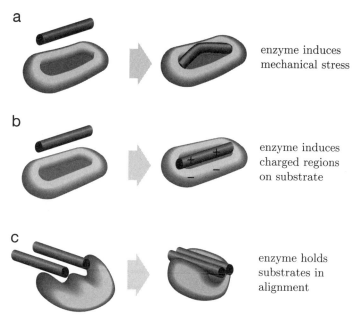

a — enzyme induces mechanical stress

b — enzyme induces charged regions on substrate

c — enzyme holds substrates in alignment

Figure 10.28: (Schematic.) Three mechanisms for an enzyme to assist a reaction. (a) The enzyme may exert mechanical forces on the substrate. (b) The enzyme may change the substrate's reactivity by altering its ionic environment. (c) The enzyme may hold two substrate molecules in the precise orientation needed for a joining bond to form, reducing the entropic part of the free energy barrier to the desired reaction. All these induced deviations from the substrate's normal distribution of states can be considered as forms of strain, pushing the substrate closer to its transition state. [Adapted from Karp, 2002.]

T_2

10.4.3′ Track 2

1. The assumptions outlined in Section 10.4.3 were chosen to discourage any short-cuts across the reaction diagram. For example, after state EP, the trailing head could, in principle, release its ADP, remain bound to the microtubule, then bind and split another ATP—a **futile hydrolysis**, as there would be no forward motion. The strain from binding the forward head makes this outcome less likely than the alternative shown (the head retains ADP but lets go of the microtubule), and so helps ensure tight coupling. Interestingly, a large, externally applied force in the backward direction could cancel the effect of strain, leading to a breakdown of tight coupling at a threshold load force. The motor would then "slip," as imagined in Figure 10.9 (see Idea 10.1 on page 413). Schnitzer and coauthors measured and analyzed the force at which the motor stalls and argued that stalling reflects slipping (or futile hydrolysis), not thermal equilibrium.

   It's also possible for the trailing kinesin head to hydrolyze ATP and release P_i prior to step ES_2, thereby allowing the entire kinesin dimer to detach from the microtubule and reducing its processivity. The transition to ES_2 (binding of the forward head) is normally rapid enough to make this process rare.

2. The discussion at the end of Section 10.4.3 simplified the reaction diagram of kinesin, replacing Figure 10.24 by[14]

$$E \underset{k_-}{\overset{c_{ATP}k_+}{\rightleftharpoons}} ES_1 \overset{equil}{\rightleftharpoons} ES_1' \cdots \overset{k_n}{\rightharpoonup} E.$$

Instead of writing ATP explicitly as a participant, we are thinking of E as spontaneously isomerizing to ES_1 with a rate proportional to c_{ATP}. (Some authors call the combination $c_{ATP}k_+$ a pseudo–first-order rate constant.) The dots represent possible other substeps, which we ignore; as usual, the last step (hydrolysis of ATP and release of P_i) is assumed to be effectively irreversible, as in Section 10.4.1.

Proceeding almost as in Section 10.4.1, we first note that each kinesin head must be in one of the three states E, ES_1, or ES_1'. We assumed near-equilibrium between the latter two states. The appropriate equilibrium constant will reflect an intrinsic free energy change, ΔG^0, plus a force-dependent term $f\ell$ (see Section 6.7 on page 226). Finally, although the state E is not in equilibrium with the others, we do assume that the whole reaction is in a quasi-steady state. All together, then, we are to solve three equations for the three unknown probabilities P_E, P_{ES_1}, and $P_{ES_1'}$:

$$1 = P_E + P_{ES_1} + P_{ES_1'} \quad \text{(normalization)} \tag{10.27}$$

$$P_{ES_1} = P_{ES_1'} e^{(\Delta G^0 + f\ell)/k_B T} \quad \text{(near-equilibrium)} \tag{10.28}$$

$$0 = \frac{d}{dt} P_E = -c_{ATP}k_+ P_E + k_- P_{ES_1} + k_n P_{ES_1'}. \quad \text{(quasi-steady state)} \tag{10.29}$$

Solving gives

$$v = k_n \times (8\,\text{nm}) \times \left[\frac{k_- e^{(\Delta G^0 + f\ell)/k_B T} + k_n}{k_+ c_{ATP}} + e^{(\Delta G^0 + f\ell)/k_B T} + 1 \right]^{-1}.$$

For any fixed value of load force f, this expression is of Michaelis–Menten form (Equation 10.20 on page 435), with load-dependent parameters analogous to Equation 10.19 given by

$$K_M = \frac{1}{k_+} \frac{k_- e^{(\Delta G^0 + f\ell)/k_B T} + k_n}{e^{(\Delta G^0 + f\ell)/k_B T} + 1} \tag{10.30}$$

and

$$v_{max} = k_n \times (8\,\text{nm}) \times [e^{(\Delta G^0 + f\ell)/k_B T} + 1]^{-1}. \tag{10.31}$$

Figure 10.25 on page 445 shows the kinetic data of Table 10.1, along with solid curves showing the preceding functions with the parameter choices $\ell = 3.7\,\text{nm}$, $\Delta G^0 = -5.1 k_B T_r$, $k_n = 103\,\text{s}^{-1}$, $k_+ = 1.3\,\mu\text{M}^{-1}\text{s}^{-1}$, and $k_- = 690\,\text{s}^{-1}$.

---

[14]See Problem 10.7 for another example of an enzymatic mechanism with a rapid-isomerization step.

Schnitzer and coauthors actually compared their model with a data set larger than the one shown here, including additional measurements of speed versus force at fixed c_{ATP}, and again found satisfactory agreement. Their model, however, is not the only one that fits the data. Other models also account for the observed statistical properties of kinesin stepping, stall forces, and the loss of processivity at high loads (see for example Fisher & Kolomeisky, 2001).

T_2 **10.4.4′ Track 2**

1. C351 is not a naturally occurring motor; it is a construct designed to have certain experimentally convenient properties. We nevertheless take it as emblematic of a class of natural molecular machines simpler than conventional kinesin.

2. Okada and Hirokawa also interpreted the numerical values of their fit parameters, showing that they were reasonable (see Okada & Hirokawa, 1999). Their data (Figure 10.27 on page 450) gave a mean speed v of $140\,\mathrm{nm\,s^{-1}}$ and a variance increase rate of $88\,000\,\mathrm{nm^2\,s^{-1}}$. To interpret these results, we must connect them with the unknown molecular quantities δ, t_s, t_w and the one-dimensional diffusion constant D for the motor as it wanders along the microtubule in its weak-binding state.

   Figure 10.29 shows again the probability distribution at the end of a weak-binding period. If we make the approximation that the probability distribution

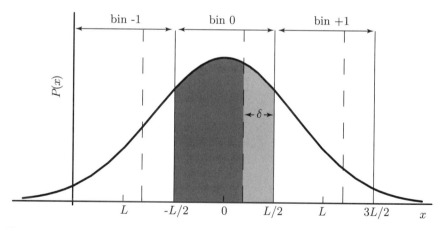

Figure 10.29: (Sketch graph.) Illustrating the calculation of the diffusing ratchet's average stepping rate. The *solid lines* delimit the bins discussed in the text. The *dashed lines* are the same as those on the right side of Figure 10.26 on page 448: They mark potential maxima, or "watersheds." Thus, a motor located in the region between two neighboring dashed lines will be attracted to whichever minimum $(0, L, 2L, \dots)$ lies between those lines. For example, the *dark gray region* is the part of bin 0 attracted to $x = 0$, whereas the *light gray region* is the part of bin 0 attracted to $x = L$. (The width of the bins has been exaggerated for clarity; actually the calculation assumes that the distribution $P(x)$ is roughly constant within each bin.)

was very sharply peaked at the start of this period, then the curve is just given by the fundamental solution to the diffusion equation (Equation 4.28 on page 143). To find the P_k, we must compute the areas under the various shaded regions in Figure 10.26, and from these, compute $\langle k \rangle$ and variance(k). This calculation is not difficult to do numerically, but there is a shortcut that makes it even easier.

We begin by dividing the line into bins of width L (solid lines on Figure 10.29). Suppose that Dt_w is much larger than L^2; so the motor diffuses many steps in each weak-binding time. Then $P(x)$ will be nearly a constant in the center bin of the figure, that is, the region between $\pm L/2$. As we move outward from the center, $P(x)$ will decrease. But we can still take it to be a constant in each bin, for example, the one from $L/2$ to $3L/2$. Focus first on the center bin. Those motors lying between $-L/2$ and $+L/2 - \delta$ (dark gray region of Figure 10.29) will fall back to the binding site at $x = 0$, whereas the ones from $L/2 - \delta$ to $L/2$ (light gray region) will land at $x = L$. For this bin, then, the mean position will shift by

$$\langle k \rangle_{\text{bin } 0} = \frac{P(0)\big((L - \delta) \times 0 + \delta \times 1\big)}{P(0) \times L} = \frac{\delta}{L}.$$

Your Turn 10G

Show that for the two flanking bins, centered at $\pm L$, the mean position also shifts by $\langle k \rangle_{\text{bin } \pm 1} = \delta/L$, and similarly for all the other pairs of bins, $\langle k \rangle_{\text{bin } \pm i}$.

We have divided the entire range of x into strips, in each of which the mean position shifts by the same amount δ/L. Hence the total mean shift per step, u, is also δ/L. According to Idea 10.23 on page 449, then, $v \equiv uL/\Delta t$ is given by $\delta/\Delta t$. You can also show using Idea 10.23 that the increase in the variance of x per cycle just equals the diffusive spread, $2Dt_w$.

The rate of ATP hydrolysis per motor under the conditions of the experiment was known to be $(\Delta t)^{-1} \approx 100\,\text{s}^{-1}$. Substituting the experimental numbers then yields

$$140\,\text{nm s}^{-1} \approx \delta \times (100\,\text{s}^{-1}) \quad \text{and} \quad 88\,000\,\text{nm}^2\,\text{s}^{-1} \approx (100\,\text{s}^{-1}) \times 2Dt_w,$$

or $\delta = 1.4\,\text{nm}$ and $Dt_w = 440\,\text{nm}^2$. The first of these gives a value for the asymmetry of the kinesin–microtubule binding that is somewhat smaller than the size of the binding sites. That's reasonable. The second result justifies a posteriori our assumption that $Dt_w \gg L^2 = 64\,\text{nm}^2$. That's good. Finally, biochemical studies imply that the mean duration t_w of the weak-binding state is several milliseconds; thus $D \approx 10^{-13}\,\text{m}^2\,\text{s}^{-1}$. This diffusion constant is consistent with measured values for other proteins that move passively along linear polymers. Everything fits.

3. It is not currently possible to apply a load force to single-headed kinesin molecules, as it is with two-headed kinesin. Nevertheless, the velocity calculation, corresponding to the result for the tightly coupled case (Equation 10.26 on page 456), is instructive. (See for example Peskin et al., 1994.) The motor will stall when its backward drift in the w state equals the net forward motion expected from the asymmetry of the potential.

But shouldn't the condition for the motor to stall depend on the chemical potential of the "food" molecule? This question involves the first assumption made when defining the diffusing ratchet model, that the hydrolysis cycle is unaffected by microtubule binding (page 447). This assumption is chemically unrealistic, but it is not a bad approximation when ΔG is very large compared with the mechanical work done on each step. If the chemical potential of ATP is too small, this assumption fails; the times t_s and t_w spent in the strong- and weak-binding states will start to depend on the location x along the track. Then the probabilities P_k to land at kL will not be given simply by the areas under the diffusion curve (see Figure 10.26 on page 448), and the stall force will be smaller for smaller ΔG. (For more details, see Jülicher et al., 1997; Astumian, 1997.)

More generally, suppose that a particle diffuses along an asymmetrical potential energy landscape, which is kicked by some external mechanism. The particle will make net progress only if the external kicks correspond to a nonequilibrium process. Such disturbances will have time correlations absent in pure Brownian motion. Some authors use the terms *correlation ratchet* or *flashing ratchet* instead of *diffusing ratchet* to emphasize this aspect of the physics. (Still other related terms in the literature include *Brownian ratchet*, *thermal ratchet*, and *entropic ratchet*.) For a general argument that asymmetry and an out-of-equilibrium step are sufficient to get net directed motion, see Magnasco, 1993 and Magnasco, 1994. This result is a particular case of the general result that whenever a reaction graph contains closed loops and is coupled to an out-of-equilibrium process, there will be circulation around one of the loops.

PROBLEMS*

10.1 Complex processes

Figure 10.30 shows the rate of firefly flashing as a function of the ambient temperature. (Insects do not maintain a fixed internal body temperature.) Propose a simple explanation for the behavior shown. Extract a quantitative conclusion and comment on why your answer is numerically reasonable.

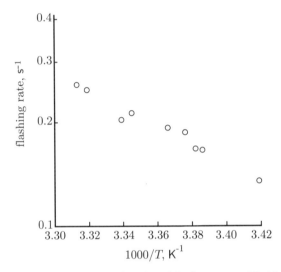

Figure 10.30: (Experimental data.) Semilog plot of the frequency of flashing of fireflies (arbitrary units), as a function of inverse temperature. [Data from Laidler, 1972.]

10.2 Scaling in muscle

Figure 10.1 on page 405 sketches the organization of vertebrate skeletal muscles. Assume that all creatures great and small have muscle tissues that are similar on the microscopic level; thicker muscles simply have more myofibrils in parallel, and longer muscles have longer myofibrils (or more copies laid end-to-end), than do smaller muscles.

Typically each end of a myosin filament (bottom left of the figure) has about 100 myosin molecules pulling in the same direction. Under physiological conditions, each myosin can exert a force of about 5.3 pN. We get an upper bound on the force the filament can exert by assuming that all of the myosins are simultaneously attached and exerting force. Each myosin filament occupies a cross-sectional area of about $1.8 \cdot 10^{-15}$ m^2 in the relaxed muscle.

a. Use these data to estimate how much force your biceps can exert. Is your estimate reasonable?

---

*Problems 10.7 and 10.8 (and Example 10B) are adapted with permission from Tinoco et al., 2001.

b. Let's also make the rough approximation that large and small creatures are geometrically similar; that is, that all dimensions in the large creature's body are obtained by a uniform rescaling of those of the small one. Which creature will be better able to lift its own body weight over its head? Does your answer agree with what you know about ants and elephants?

10.3 *Rescuing Gilbert*

Sullivan suggested one possible modification of the G-ratchet; here is another, which we will call the F-ratchet.

We imagine that the rod in Figure 10.10 extends far to the right, all the way into another chamber full of gas at temperature T'. The rod ends with a plate in the middle of the second chamber; gas molecules bounce against this plate, giving random kicks to the rod. We further suppose that T' is greater than the temperature T of the part of the mechanism containing the ratchet mechanism.

a. Suppose the external force $f = 0$. Will the F-ratchet make net progress? In which direction? [*Hint:* Think about the case where the temperature T equals absolute zero.]

b. Recall Sullivan's critique of the G-ratchet: "Couldn't you wrap your shaft into a circle? Then your machine would go around forever, violating the Second Law." Figure 10.31 shows such a device. Here the one-way mechanism on the left is a spring (the "pawl") that jams against the asymmetrical teeth on a wheel when it tries to rotate backward. Reply to Sullivan's remark in the context of this circular F-ratchet. [*Hint:* First review Section 6.5.3.]

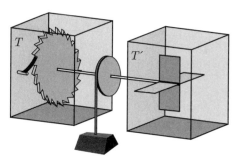

Figure 10.31: (Schematic.) The F-ratchet, an imagined motor. Two chambers are maintained at temperatures T and T', respectively; the right-hand chamber contains gas. Thermal motion in the right-hand chamber drives the shaft; its motion is rectified by the device in the left-hand chamber, perhaps lifting a weight attached to a pulley. [Adapted from Feynman et al., 1963a.]

10.4 *Ion pump energetics*

Textbooks quote the value $\Delta G'^0 = -7.3\,\text{kcal/mole}$ for the hydrolysis of ATP (Figure 2.12). Chapter 11 will introduce a molecular machine that uses one ATP per step and does useful work equal to $14k_B T_r$. Reconcile these statements, using the fact that typical intracellular concentrations are $[\text{ATP}] = 0.01$ (that is, $c_{\text{ATP}} = 10\,\text{mM}$), $[\text{ADP}] = 0.001$, and $[\text{P}_i] = 0.01$.

10.5　*Competitive inhibition*

Section 10.4.2 on page 436 described competitive inhibition as one strategy to control the activity of an enzyme; for example, the protease inhibitors used to treat HIV use this strategy. In this mechanism, an inhibitor molecule, which we will call I, binds to the active site of an enzyme E, blocking it from processing its substrate.

a. Write down the Mass Action rule for the reaction $I + E \rightleftharpoons EI$, with some equilibrium constant $K_{eq,I}$.

b. Now repeat the derivation of the Michaelis–Menten rule in Section 10.4.1, with the change that now E can be in any of three states: $P_E + P_{ES} + P_{EI} = 1$. Show that the reaction velocity can be written as

$$v = v_{max} \frac{c_S}{\alpha K_M + c_S}. \qquad \text{(competitive inhibition)} \qquad (10.32)$$

Here α is a quantity that you are to find; it involves the parameters of the uninhibited enzyme (K_M and v_{max}), $K_{eq,I}$, and the total concentration c_I of inhibitor.

c. Suppose that we measure the initial reaction velocity as a function of substrate concentration for two fixed values of c_I, then plot the two data sets in Lineweaver–Burk form. Describe the two curves we will get if I is a competitive inhibitor.

d. Ethanol and methanol are two similar, small molecules. Methanol is quite toxic: The liver enzyme alcohol dehydrogenase converts it to formaldehyde, which can cause blindness. The kidneys will eventually remove methanol from the blood, but not fast enough to avert this damage. Why do you suppose a therapy for methanol poisoning involves gradual intravenous injection of ethanol over several hours?

10.6　*Uncompetitive inhibition*

Modify the derivation of the Michaelis–Menten rule for enzyme kinetics (Section 10.4.1) to account for uncompetitive inhibition.[15] That is, augment the basic reaction diagram

$$E + S \underset{k_{-1}}{\overset{c_S k_1}{\rightleftharpoons}} ES \overset{k_2}{\rightarrow} E + P$$

by adding a second reaction (compare with Figure 10.13 on page 424),

$$ES + I \underset{k_{-3}}{\overset{c_I k_3}{\rightleftharpoons}} ESI.$$

Here E is the enzyme, S the substrate, P the product, and ES the enzyme–substrate complex. The inhibitor I is a second substance that, like E, is not used up. State I can bind to the enzyme–substrate complex ES to create a dead-end complex ESI, which cannot process substrate because of an allosteric interaction. Eventually, however, ESI spontaneously dissociates back to ES+I and the enzyme goes back to work. This

---

[15] $\boxed{T_2}$ Uncompetitive inhibition is a mathematical simplification of a more realistic situation called noncompetitive inhibition. For a full discussion, see Nelson & Cox, 2000.

dead-end branch slows down the reaction. As usual, assume a large reservoir of S, no product initially, and a small amount of enzyme.

a. Find the steady-state reaction rate v in terms of c_S, c_I, the total enzyme concentration $c_{E,tot}$, and the rate constants.

b. Consider the dependence of v on c_S, holding $c_{E,tot}$ and c_I fixed. Can you express your answer as a Michaelis–Menten function, where v_{max} and K_M are functions of c_I?

c. Regardless of your answer to (b), find the saturating value v_{max} as c_S increases at fixed $c_{E,tot}$ and c_I. Comment on why your answer is physically reasonable; if you did Problem 10.5, contrast to the case studied there.

10.7 $\boxed{T_2}$ Generality of MM kinetics

In this problem, you'll see that the MM formula is really more generally applicable than the discussion in the text may have made it seem.

The enzyme chymotrypsin catalyzes the hydrolysis of peptides (short protein fragments). We will denote the enzyme's original state as E–OH to emphasize one key hydroxyl group on one of its residues. We will also represent a peptide generically by the symbol R–CONH–R′, where the central atoms CONH indicate one particular peptide bond (see Figure 2.13 on page 48) and R, R′ denote everything to the left and right, respectively, of the bond in question.

The enzyme operates as follows: A noncovalent complex (E–OH·R–CONH–R′) forms rapidly between the enzyme E–OH and the peptide substrate R–CONH–R′, which we will call S. Next E–OH gives up a hydrogen and bonds covalently to one-half of the peptide, breaking the bond to the other half, which is released. Finally, the remaining enzyme–peptide complex splits a water molecule to restore E–OH to its original form and release the other half of the peptide:

$$\text{E–OH} + \text{S} + \text{H}_2\text{O} \overset{\text{equil}}{\rightleftharpoons} \text{E–OH} \cdot \text{S} + \text{H}_2\text{O} \overset{k_2}{\to} \text{E–OCO–R} + \text{NH}_2\text{–R}' + \text{H}_2\text{O}$$

$$\overset{k_3}{\to} \text{E–OH} + \text{R–CO}_2\text{H} + \text{NH}_2\text{–R}'.$$

Assume that the last step is irreversible, as indicated by the last arrow.

Suppose that the first reaction is so fast that it's practically in equilibrium, so that $c_{E-OH\cdot S}/c_{E-OH} = c_S K_{eq}$ for some constant K_{eq}. Apply the steady-state assumption to $c_{E-OCO-R}$ to show that the overall reaction velocity of the scheme just described is of Michaelis–Menten form. Find the effective Michaelis constant and maximum velocity in terms of K_{eq}, the rate constants k_2, k_3, and the total concentration $c_{E,tot}$ of enzyme.

10.8 $\boxed{T_2}$ Invertase

Earlier problems discussed two distinct forms of enzyme inhibition: competitive and uncompetitive. More generally, *noncompetitive* inhibition refers to any mechanism not obeying the rule you found in Problem 10.5. The enzyme invertase hydrolyzes sucrose (table sugar). The reaction is reversibly inhibited by the addition of urea, a small molecule. The initial rate of this reaction, for a certain concentration $c_{E,tot}$

of enzyme, is measured in terms of the initial sucrose concentration, both with and without a 2 M solution of urea:

$c_{sucrose}$, M	v (no urea), M s^{-1}	v (with urea), M s^{-1}
0.0292	0.182	0.083
0.0584	0.265	0.119
0.0876	0.311	0.154
0.117	0.330	0.167
0.175	0.372	0.192
0.234	0.371	0.188

Make the appropriate Lineweaver–Burk plots and determine whether the inhibition by urea is competitive in character. Explain.

CHAPTER **11**

Machines in Membranes

*In going on with these Experiments how many pretty Systems
do we build which we soon find ourselves oblig'd to destroy!
If there is no other Use discover'd of Electricity this however
is something considerable, that it may help to make a vain
man humble.*

— B. Franklin to P. Collinson, 1747

Chapter 12 will discuss the question of nerve impulses, the electric signals running along nerve fibers that make up the ghostly fabric of thought. Before we can discuss nerve impulses, however, this chapter must look at how living cells generate electricity in the first place. Chapter 4 skirted this question in the discussion of the Nernst formula; we are now ready to return to it as a matter of free energy transduction, armed with a general understanding of molecular machines. We will see how indirect, physical arguments led to the discovery of a remarkable class of molecular machines, the *active ion pumps,* long before the precise biochemical identity of these devices was known. The story may remind you of how Muller, Delbrück, and their colleagues characterized the nature of the genetic molecule, using physical experiments and ideas, many years before others identified it chemically as DNA (Section 3.3.3). The interplay of physical and biochemical approaches to life science problems will continue to bear fruit as long as both sets of researchers know about each others' work.

The Focus Question for this chapter is

Biological question: The cytosol's composition is very different from that of the outside world. Why doesn't osmotic flow through the plasma membrane burst (or shrink) the cell?

Physical idea: Active ion pumping by molecular machines can maintain a nonequilibrium, osmotically regulated state.

11.1 ELECTROOSMOTIC EFFECTS

11.1.1 Before the ancients

The separation of the sciences into disciplines is a modern aberration. Historically, there was a lively exchange between the study of bioelectric phenomena and the great project of understanding physically what electricity really was. For example, Benjamin Franklin's famous demonstration in 1752 that lightning was just a very big electric spark led to much speculation and experimentation on electricity in general. Lacking sophisticated measurement devices, it was natural for the scientists of the day to focus on the role of electricity in living organisms, in effect using them as their instruments. The physicians Albrecht von Haller and Luigi Galvani found that electricity, generated by physical means and stored in a capacitor, could stimulate strong contraction in animal muscles. Galvani published his observations in 1791 and speculated that muscles were also a *source* of electricity. After all, he reasoned, even without the capacitor he could evoke muscle twitches just by inserting electrodes between two points.

Alessandro Volta did not accept this last conclusion. He regarded muscles as electrically passive, receiving signals but not generating any electricity themselves. He explained Galvani's no-capacitor experiment by suggesting that an electrostatic potential could develop between two dissimilar metals in any electrolyte, alive or not. To prove his point, in 1800 he invented a purely nonliving source of electricity, merely placing two metal plates in an acid bath. Volta's device—the voltaic cell—led to decisive advances in our understanding of physics and chemistry. As technology, Volta's device also wins the longevity award: The batteries in your car, flashlight, and so on are voltaic cells.

But Volta was too quick to dismiss Galvani's idea that life processes could also generate electricity directly. Sections 11.1.2–11.2.3 will show how this can happen. Our discussion will rest upon many hard-won experimental facts. For example, after Galvani, decades would pass before E. DuBois Reymond, another physician, showed in the 1850s that living frog skin maintained a potential difference of up to 100 mV between its sides. And the concept of the cell membrane as an electrical insulator only a few nanometers thick remained a speculation until 1923, when H. Fricke measured quantitatively the capacitance of a cell membrane and thus estimated its thickness, essentially using Equation 7.26 on page 269.

To understand the origin of resting membrane potentials, we first return to the topic of ions permeating membranes, a story begun in Chapter 4.

11.1.2 Ion concentration differences create Nernst potentials

Figure 4.14 on page 140 shows a container of solution with two charged plates outside supplying a fixed external electric field. Section 4.6.3 calculated the concentration profile in equilibrium and, from this, the change in concentration of charged ions between the two ends of the container (Equation 4.26). We then noted that the potential drop needed to get a significant concentration jump across the container was roughly comparable to the difference in electrostatic potential across the membrane of most living cells. We're now in a position to see *why* the results of Section 4.6.3 should have anything to do with cells, starting with some ideas from Section 7.4.

Figure 11.1 shows the physical situation of interest. An uncharged membrane, shown as a long cylinder, separates the world into two compartments, *1* and *2*. Two electrodes, one inside and one outside, measure the electrostatic potential across the membrane. The figure is meant to evoke the long, thin tube, or axon, emerging from the body of a nerve cell. Indeed, one can literally insert a thin needlelike electrode into living nerve axons, essentially as sketched here, and connect them to an am-

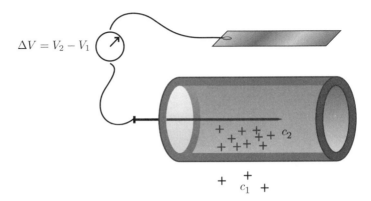

Figure 11.1: (Schematic.) Measurement of membrane potential. The bulk concentration c_2 of interior cations is greater than the exterior concentration, c_1, as shown; the corresponding bulk concentrations of negative charges follow the same pattern (*not shown*), as required by charge neutrality. The symbol on the left represents a voltmeter.

plifier. Historically, the systematic study of nerve impulses opened up only when a class of organisms was found with axons large enough for this delicate procedure: the cephalopods. For example, the **"giant" axon** of the squid *Loligo forbesi* has a diameter of up to a millimeter, much bigger than the typical axon diameter in your body, which is 5–20 μm.

Each compartment contains a salt solution, which for simplicity we'll take to be monovalent—say, potassium chloride. Imagine that the membrane is slightly permeable to K^+, but not at all to Cl^- (actually, squid axon membranes are about twice as permeable to K^+ as they are to Cl^-). For now, we will also ignore the osmotic flow of water (see Section 11.2.1). We imagine using different salt solutions on the inside and outside of the cell: Far from the membrane, the salt concentration in each compartment is uniform and equals c_2 on the inside and c_1 on the outside. Suppose that $c_2 > c_1$, as shown in Figure 11.1.

Let $c_+(r)$ denote the concentration of potassium ions at a distance r from the center of the inner compartment. After the system reaches equilibrium, $c_+(r)$ will not be uniform near the membrane, and neither will be the chloride concentration, $c_-(r)$ (see Figure 11.2a). To understand the origin of membrane potential, we must first explain these equilibrium concentration profiles.

The permeant K^+ ions face a dilemma: They could increase their entropy by crossing the membrane to erase the imposed concentration difference. Indeed, they will do this, up to a point. But their impermeant partners, the Cl^- ions, keep calling them back by electrostatic attraction. Thus, far from the membrane on both sides, the concentrations of K^+ and Cl^- will be equal, as required by overall charge neutrality. Only a few K^+ ions will actually cross the membrane, and even these won't travel far: They deplete a thin layer just inside the membrane and cling in a thin layer just outside (see the c_+ curve in Figure 11.2a).

The behavior shown in Figure 11.2 is just what we could have expected from our study of electrostatic interactions in Section 7.4.3 on page 264. To see the connection,

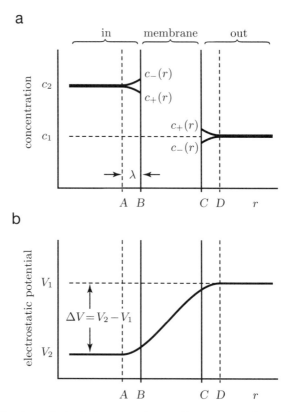

Figure 11.2: (Sketch graphs.) (a) Concentration profiles near a membrane, for the situation sketched in Figure 11.1. The radius r is the distance from the centerline of the cylindrical inner compartment. Far outside the membrane ($r \to \infty$), the concentrations $c_\pm$ of positive and negative ions must be equal, by charge neutrality; their common value c_1 is just the exterior salt concentration. Similarly, deep inside the cell, $c_+ = c_- = c_2$. The situation shown assumes that only the positive ions are permeant. Thus some positive ions leak out, enhancing c_+ in a layer of thickness λ just outside the membrane and depleting it just inside. c_- drops just outside the membrane because negative ions move away from the negatively charged cell. The concentrations in the membrane's hydrophobic interior (the region between B and C) are nearly zero. (b) The corresponding electrostatic potential V created by the charge distribution in (a). In equilibrium, ΔV equals the Nernst potential of the permeant species (in this case, the positive ions).

first consider the region to the right of point C in Figure 11.2. This region is a salt solution in contact with an "object" of net negative charge. The "object" consists of the membrane plus the interior of the cylinder in Figure 11.1; it's negatively charged because some of its positive ions have permeated the membrane and escaped. But a solution in contact with a negatively charged object develops a neutralizing positive layer, just as in Figure 7.8a on page 265. This layer is shown in Figure 11.2 as the region between points C and D. Its thickness λ is roughly analogous to x_0 in our discussion of the electric double layer (Equation 7.25 on page 268).[1] Unlike Figure 7.8a, however, we now have both positive and negative mobile charges in the solution. Hence, the layer of enhanced K^+ concentration is also *depleted* of Cl^-, because the negative region to the left of point C in the figure *repels* anions. The effect of both these disturbances is to create a layer of net positive charge just outside the membrane.

Just inside the membrane, the situation is reversed. Here we have a salt solution facing a *positive* object, namely, everything to the right of point B in the figure. Thus there is a region relatively depleted of K^+ and enriched in Cl^-, a layer of net negative charge just inside the membrane.

We can now turn to the question of finding the electrostatic potential jump across the membrane. One way to find it would be to solve the Gauss Law (Equation 7.20 on page 264) for the electric field $\mathcal{E}(x)$ given the charge density shown in Figure 11.2a, then integrate to find $V(x)$. Let's instead think physically (see Figure 11.2b). Suppose that we bring a positively charged test object in from outside (from the right of the figure). At first, everything to the left of our test object has net charge zero, so the net force on it is also zero and its potential energy is a constant. Once the test object enters the outer charge cloud, at point D, however, it starts to feel and be attracted to the net negative object to the left of point C. Its potential thus begins to decrease. The deeper it gets into the cloud, the more charge it sees: The slope of its potential curve increases.

The membrane itself was assumed to be uncharged. There will be very few permeant ions inside it, in transit. Thus, while traversing the membrane, the test charge feels a *constant* force attracting it toward the interior, from the charge of the region to the left of point B. Its potential thus falls linearly until it crosses point B, then levels off in the neutral interior of the cylinder.

The potential curve $V(r)$ sketched in Figure 11.2b summarizes the narrative in the preceding two paragraphs.

Your Turn 11A

Arrive at the same conclusion for the potential $V(r)$ by describing qualitatively the solution to the Gauss Law with the charge density $\rho_q(r) = e(c_+(r) - c_-(r))$, where $c_\pm(r)$ are as shown in Figure 11.2a.

Your Turn 11B

Repeat the discussion, again assuming that $c_2 > c_1$, but this time considering a fictitious membrane permeable to Cl^- but not to K^+. What changes?

To determine the potential drop $\Delta V = V_2 - V_1$ quantitatively, imagine replacing the voltmeter in Figure 11.1 by a battery of adjustable voltage and increasing the voltage until the current through the system just stops. The permeant ion species is then in equilibrium throughout the system. If we write its charge q as the proton charge

---

[1] $\boxed{T_2}$ Or more appropriately, to the Debye screening length λ_D (Equation 7.35 on page 285).

e times an integer z (the ion's **valence**), then its concentration must obey the Boltzmann distribution: $c(x) = \text{const} \times e^{-zeV(x)/k_B T}$. Taking the logarithm and evaluating on the inside and outside reproduces the Nernst relation:

$$\Delta V = \mathcal{V}^{\text{Nernst}} \text{ in equilibrium, where}$$

$$\Delta V \equiv V_2 - V_1 \text{ and } \mathcal{V}^{\text{Nernst}} \equiv -\frac{k_B T}{ze} \ln \frac{c_2}{c_1}. \tag{11.1}$$

In the language of Section 8.1.1 on page 295, the Nernst relation says that, in equilibrium, the electrochemical potential of any permeant ion species must be everywhere the same.

Notice that z in Equation 11.1 is the valence of the permeant species only (in our case, it's $+1$). In fact, the other (impermeant) species in the problem doesn't obey the Nernst relation at all, nor should it, because it's not at all in equilibrium. If we suddenly punched a hole through the membrane, the impermeant Cl^- would begin to rush out, whereas K^+ would not, because we adjusted the battery to exactly balance its electric force (to the left) against its entropic, diffusive force (to the right). Similarly, you just found in Your Turn 11B that switching the roles of the two species actually *reverses* the sign of the membrane's equilibrium potential drop.

$\boxed{T_2}$ *Section 11.1.2′ on page 501 gives some further comments involving ion permeation through membranes.*

11.1.3 Donnan equilibrium can create a resting membrane potential

Section 11.1.2 arrived at a simple conclusion:

> The Nernst relation gives the potential arising when a permeant species reaches equilibrium. Equivalently, it gives the potential that must be applied to **stop** the net flux of that species, given the concentration jump across a membrane. (11.2)

In this section, we begin to explore a slightly more complicated problem in which there are more than two ion species. The problem is relevant to living cells, where there are several important small permeant ions. We will simplify our discussion by considering only three species of small ions, with concentrations c_i, where the label i runs over Na^+, K^+, Cl^-.

Cells are also full of proteins and nucleic acids, huge macromolecules carrying net negative charge. The macromolecules are practically impermeant, so we expect a situation analogous to Figure 11.2, and a resulting membrane potential. Unlike the simpler case with just two species, however, the bulk concentrations are no longer automatically fixed by the initial concentrations and by the condition of charge neutrality: The cell can import some more Na^+ while still remaining neutral if, at the same time, it expels some K^+ or pulls in some Cl^-. Let's see what happens.

A typical value for the total charge density $\rho_{q,\text{macro}}$ of the trapped (impermeant) macromolecules is the equivalent of 125 mM of excess electrons. Just as in Section 11.1.2, small ions can and will cross the cell membrane, to reduce the total free energy of the cell. We will suppose that our cell sits in an infinite bath with exterior ion concentrations $c_{1,i}$. (It could be an algal cell in the sea or a bacterium in your blood.) These concentrations, like $\rho_{q,macro}$, are fixed and given; some illustrative values are $c_{1,Na^+} = 140\,\text{mM}$, $c_{1,K^+} = 10\,\text{mM}$, and $c_{1,Cl^-} = 150\,\text{mM}$. These values make sense, in that they imply that the exterior solution is neutral:

$$c_{1,Na^+} + c_{1,K^+} - c_{1,Cl^-} = 0.$$

The cell's interior is not infinite, so the concentrations there, $c_{2,i}$, are *not* fixed. Instead, they are all unknowns for which we must solve. Moreover, the membrane potential drop $\Delta V = V_2 - V_1$ is a fourth unknown. We therefore need to find four equations in order to solve for these four unknowns. First, charge neutrality in the bulk interior requires

$$c_{2,Na^+} + c_{2,K^+} - c_{2,Cl^-} + \rho_{q,macro}/e = 0. \tag{11.3}$$

(Section 12.1.2 will discuss neutrality in greater detail.) The other three equations reflect the fact that the same electrostatic potential function affects every ion species. Thus, in equilibrium, each permeant species must separately be in Nernst equilibrium at the same value of ΔV:

$$\Delta V = -\frac{k_B T}{e}\ln\frac{c_{2,Na^+}}{c_{1,Na^+}} = -\frac{k_B T}{e}\ln\frac{c_{2,K^+}}{c_{1,K^+}} = -\frac{k_B T}{-e}\ln\frac{c_{2,Cl^-}}{c_{1,Cl^-}}. \tag{11.4}$$

To solve Equations 11.3 and 11.4, we first notice that the latter can be rewritten as the **Gibbs–Donnan relations**:

$$\frac{c_{1,Na^+}}{c_{2,Na^+}} = \frac{c_{1,K^+}}{c_{2,K^+}} = \frac{c_{2,Cl^-}}{c_{1,Cl^-}} \quad \text{in equilibrium.} \tag{11.5}$$

Example:
a. Why is the chloride ratio in these relations inverted relative to the others?
b. Finish the calculation by using the illustrative values for $c_{1,i}$ and $\rho_{q,macro}$ given earlier in this section. That is, find $c_{2,i}$ and ΔV.

Solution:
a. The charge on a chloride ion is opposite to that on potassium or sodium, a situation leading to an extra minus sign in Equation 11.4. Upon exponentiating the formula, this minus sign turns into an inverse.
b. Let $x = [Na^+] = c_{2,Na^+}/1\,\text{M}$. Use Equation 11.5 and the given values of $c_{1,i}$ to express c_{2,K^+} and c_{2,Cl^-} in terms of x. Substitute into Equation 11.3 and multiply the equation by x to get

$$\left(1 + \frac{0.01}{0.14}\right)x^2 - 0.15 \times 0.14 - 0.125x = 0.$$

Solving with the quadratic formula gives $x = 0.21$, or $c_{2,\mathrm{Na}^+} = 210\,\mathrm{mM}$, $c_{2,\mathrm{K}^+} = 15\,\mathrm{mM}$, $c_{2,\mathrm{Cl}^-} = 100\,\mathrm{mM}$. Then Equation 11.4 gives $\Delta V = -10\,\mathrm{mV}$. (Appendix B gives $k_B T_r / e = \frac{1}{40}$ volt.)

The equilibrium state you just found is called the **Donnan equilibrium**; ΔV is called the **Donnan potential** for the system.

So we have found one realistic way in which a cell can maintain a permanent (resting) electrostatic potential across its membrane, simply as a consequence of the fact that some charged macromolecules are sequestered inside it. Indeed, the typical values of such potentials are in the tens of millivolts. No energy needs to be spent maintaining the Donnan potential—it's a feature of an equilibrium state, a state of minimum free energy. Notice that we could have arranged for charge neutrality by having only c_{2,Na^+} greater than the exterior value, with the other two concentrations the same inside and out. But that state is not the minimum of free energy; instead, all available permeant species share in the job of neutralizing $\rho_{\mathrm{q,macro}}$.

11.2 ION PUMPING

11.2.1 Observed eukaryotic membrane potentials imply that these cells are far from Donnan equilibrium

The sodium anomaly Donnan equilibrium appears superficially to be an attractive mechanism for explaining resting membrane potentials. But a little more thought reveals a problem. Let's return to the question of osmotic flow through our membrane, which we postponed at the start of Section 11.1.2. The macromolecules are not very numerous; their contribution to the osmotic pressure will be negligible. The small ions, however, greatly outnumber the macromolecules and pose a serious osmotic threat. To calculate the osmotic pressure in the Donnan equilibrium Example just given, we add the contributions from all ion species:

$$\Delta c_{\mathrm{tot}} = c_{2,\mathrm{tot}} - c_{1,\mathrm{tot}} \approx 25\,\mathrm{mM}. \tag{11.6}$$

The sign of our result indicates that small ions are more numerous inside the model cell than outside. To stop inward osmotic flow, the membrane thus would have to maintain an interior pressure of $25\,\mathrm{mM} \times k_B T_r \approx 6 \cdot 10^4\,\mathrm{Pa}$. But we know from Section 7.2.1 on page 248 that eukaryotic cells lyse (burst) at much smaller pressures than this!

Certainly our derivation is very rough. We have completely neglected the osmotic pressure of other, uncharged solutes (like sugar). But the point is still valid: The equations of Donnan equilibrium give a unique solution for electroosmotic equilibrium and neutrality. There is no reason why that solution should *also* coincidentally give small osmotic pressure! To maintain Donnan equilibrium, you've got to be strong. In fact, plant, algal, and fungal cells, as well as bacteria, surround their bilayer plasma membrane with a rigid wall; thus they can withstand significant osmotic pressures. Indeed, plant tissue actually uses the rigidity resulting from osmotic pressure for structural support and becomes limp when the plant dehydrates. (Think about

Table 11.1: Approximate ion concentrations inside and outside the squid giant axon. The second line illustrates the "sodium anomaly": The Nernst potential of sodium is nowhere near the actual membrane potential of $-60\,\mathrm{mV}$.

ion	valence z	interior $c_{2,i}$, mM	relation	exterior $c_{1,i}$, mM	Nernst potential $\mathcal{V}_i^{\mathrm{Nernst}}$, mV
K^+	$+1$	400	$>$	20	-75
Na^+	$+1$	**50**	$<$	**440**	$+54$
Cl^-	-1	52	$<$	560	-59

eating old celery.) But your own body's cells lack a strong wall. Why don't they burst from osmotic pressure?

Table 11.1 shows the actual (measured) concentration differences across one particular cell's membrane. Donnan equilibrium predicts that the presence of trapped, negative macroions will give $c_{2,\mathrm{Na}^+} > c_{1,\mathrm{Na}^+}$, $c_{2,\mathrm{K}^+} > c_{1,\mathrm{K}^+}$, $c_{2,\mathrm{Cl}^-} < c_{1,\mathrm{Cl}^-}$, and $\Delta V < 0$. These predictions make sense intuitively: The trapped negative macroions tend to push out negative permeant ions and pull in positive ones. But the table shows that of these four predictions, *the first one proves to be very wrong*. In thermodynamic equilibrium, all the entries in the last column would have to be the same, according to the Gibbs–Donnan relations. In fact, both the potassium and chloride ions roughly obey this prediction; and moreover, the measured membrane potential $\Delta V = -60\,\mathrm{mV}$ really is similar to each of their Nernst potentials. But the Gibbs–Donnan relation fails for sodium; and even for K^+, the quantitative agreement is not very successful.

To summarize:

> *The Nernst potential of sodium is much more positive than the actual membrane potential ΔV.* (11.7)

All animal cells (not just the squid axon) have a **sodium anomaly** of this type.[2]

One interpretation for these results might be that the sodium and other discrepant ions simply cannot permeate on the time scale of the experiment, so they need not obey the equilibrium relations. However, we are discussing the steady-state, or **resting**, potential; the "time scale" of this measurement is infinity. Any permeation at all would eventually bring the cell to Donnan equilibrium, contrary to the actual observed concentrations. More important, it's possible to measure directly the ability of sodium ions to pass through the axon membrane; the next section will show that this permeability, although small, is not negligible.

We are forced to conclude that the ions in a living cell are not in equilibrium. But why should they be? Equilibrium is not life; it's death. Cells at rest are constantly burning food, precisely to *combat* the drive toward equilibrium! If the metabolic cost of maintaining a nonequilibrium ion concentration is reasonable relative to the rest

---

[2] Many bacteria, plants, and fungi show a similar anomaly involving the concentration of protons; see Section 11.3.

of the cell's energy budget, then there's no reason not to do it. After all, the benefits can be great. We have already seen how maintaining electrostatic and osmotic equilibrium could place a cell under large internal pressure, bursting or at least immobilizing it.

We get a big clue that we're finally on the right track when we put our nerve cell in the refrigerator. Chilling a cell to just above freezing shuts down the cell's metabolism. Suddenly the cell loses its ability to maintain a nonequilibrium sodium concentration difference. Moreover, the shut-down cell also loses its ability to control its interior volume, or **osmoregulate**. When normal conditions are restored, the cell's metabolism starts up again and the interior sodium falls.

Certain genetic defects can also interfere with osmoregulation. For example, patients with hereditary spherocytosis have red blood cells whose plasma membrane is much more permeable to sodium than that of normal red cells. The affected cells must work harder than normal cells to pump sodium out. Hence they are prone to osmotic swelling, which in turn triggers their destruction by the spleen. Entropic forces can kill.

A look ahead   This section raised two puzzles: Eukaryotic cells maintain a far-from-equilibrium concentration drop of sodium, and they don't suffer from the immense osmotic pressure predicted by Donnan equilibrium. In principle, both these problems could be solved if, instead of being in equilibrium, cells could constantly *pump* sodium across their membranes by using metabolic energy. Such active pumping could create a nonequilibrium, but steady, state.

Here is a mechanical analogy: Suppose that you visit your friend and see a fountain in his garden (Figure 11.3). The fountain is supplied by a tank of water high above it. The water flows downhill, converting the gravitational potential energy it has in the tank to kinetic energy. You expect that eventually the water will run out of the tank and the fountain will stop, but this never happens. So you instead begin to suspect that your friend is recirculating the water with a *pump*, by using some external source of energy. In that case, the fountain is in a steady, but nonequilibrium, state.[3]

In the context of cells, we are exploring the hypothesis that that cells must somehow be using their metabolism to maintain resting ion concentrations far from equilibrium. To make this idea quantitative (that is, to see whether it's right), we now return to the topic of transport across membranes (introduced in different contexts in Sections 4.6.1 and 7.3.2).

11.2.2 The Ohmic conductance hypothesis

To begin exploring nonequilibrium steady states, first note that *the Nernst potential need not equal the actual potential jump* across a membrane, just as we found that the quantity $(\Delta c)k_B T$ need not equal the actual pressure jump Δp (Section 7.3.2 on page

---

[3]Similarly, Section 10.4.1 discussed the steady state of an enzyme presented with nonequilibrium concentrations of its substrate and product. We also encountered steady or quasi-steady nonequilibrium states in Sections 4.6.1, 4.6.2, 10.2.3, and 10.4.1.

Figure 11.3: (Metaphor.) If water is continuously pumped to the upper reservoir, the fountain will come to a nonequilibrium steady state. If not, it will come to a quasi-steady state, which lasts until the reservoir is empty.

259). If the actual pressure jump across a membrane differs from $(\Delta c)k_BT$, we found that there would be a *flux* of water across the membrane. Similarly, if the potential drop differs from the Nernst potential for some ion species, that species will be out of equilibrium and will permeate, thereby giving a net electric current. In this case, the potentials obtained from Equation 11.1 for different kinds of ions need not agree with one another.

   To emphasize the distinction, Equation 11.1 on page 474 introduced $\mathcal{V}_i^{\text{Nernst}}$ (read "the Nernst potential of ion species i") to mean precisely $-(k_BT/(ez_i))\ln(c_{2,i}/c_{1,i})$, reserving the symbol ΔV for the *actual* potential drop $V_2 - V_1$. Our sign convention assigns a positive Nernst potential to an entropic force driving positive ions into the cell.

   Prior experience (Sections 4.6.1 and 4.6.4) leads us to expect that the flux of ions through a membrane will be dissipative, and hence proportional to a net driving force, at least if the driving force is not too large. Furthermore, according to Idea 11.2 on page 474, the net driving force on ions of type i vanishes when $\Delta V = \mathcal{V}_i^{\text{Nernst}}$. Thus the net force is given by the sum of an energetic term, $z_ie\Delta V$ (from the electric fields), and an entropic term, $-z_ie\mathcal{V}_i^{\text{Nernst}}$ (from the tendency of ions to diffuse to erase any concentration difference).[4] This is just the behavior we have come to expect from our

---

[4]Equivalently, the net driving force acting on ions is the difference in electrochemical potential $\Delta\mu_i$ (see Section 8.1.1 on page 295).

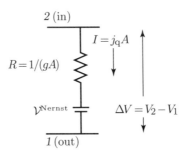

Figure 11.4: (Circuit diagram.) Equivalent circuit model for the electrical properties of a small patch of membrane of area A and conductance per area g, assuming the Ohmic hypothesis (Equation 11.8). The membrane patch is equivalent to a battery (*symbol* —|⊢) with potential drop $\mathcal{V}^{\mathrm{Nernst}}$, in series with a resistor (*symbol* —⋀⋀—) of resistance $R = 1/(gA)$. For a positive ion species ($z > 0$), a positive Nernst potential means that the ion concentration is greater outside the cell; in this case, an entropic force pushes ions upward in the diagram (into the cell). A positive applied potential ΔV has the opposite effect, pushing positive ions downward (out of the cell). Equilibrium is the state where these forces balance, or $\mathcal{V}^{\mathrm{Nernst}} = \Delta V$; then the net current I equals zero. The electric current is deemed positive when it is directed outward.

studies of osmotic flow (Section 7.3.2) and of chemical forces (see the gas chemical potential Example on page 296).

In short, we expect that

$$j_{\mathrm{q},i} = z_i e j_i = (\Delta V - \mathcal{V}_i^{\mathrm{Nernst}}) g_i. \qquad \text{Ohmic conductance hypothesis} \qquad (11.8)$$

Here as usual, the number flux j_i is the number of ions of type i per area per time crossing the membrane; the electric charge flux $j_{\mathrm{q},i}$ is this quantity times the charge $z_i e$ on one ion. We choose the sign convention that j is positive if the net flux is outward. The constant of proportionality g_i appearing in Equation 11.8 is called the **conductance per area** of the membrane to ion species i. It's always positive and has units[5] $\mathrm{m}^{-2}\Omega^{-1}$. A typical magnitude for the overall conductance per area of a resting squid axon membrane is about $5\,\mathrm{m}^{-2}\Omega^{-1}$.

Equation 11.8 is just another form of Ohm's law. To see this, note that the electric current I through a patch of membrane of area A equals $j_{\mathrm{q}}A$. If only one kind of ion can permeate, Equation 11.8 gives the potential drop across the membrane as $\Delta V = IR + \mathcal{V}^{\mathrm{Nernst}}$, where $R = 1/(gA)$. The first term is the usual form of Ohm's law. The second term corresponds to a battery of fixed voltage $\mathcal{V}^{\mathrm{Nernst}}$ connected in series with the resistor, as shown in Figure 11.4. The voltage across the terminals of this virtual battery is the Nernst potential of ion species i.

---

[5]Neuroscientists use the synonym **siemens** (symbol S) for inverse ohm; an older synonym is the mho (symbol ℧). We won't use either notation, instead writing Ω^{-1}. Note that conduc*tance* per area has units different from those of the conduc*tivity*, κ, of a bulk electrolyte (Section 4.6.4 on page 142): The latter has units $\mathrm{m}^{-1}\Omega^{-1}$.

We must bear in mind, though, that a membrane's regime of Ohmic behavior, where Equation 11.8 applies, may be very limited. First, Equation 11.8 is just the first term in a power series in $\Delta V - \mathcal{V}_i^{\text{Nernst}}$. Because we have seen that sodium is far from its equilibrium concentration difference (Table 11.1), we can't expect Equation 11.8 to give more than a qualitative guide to the resting electrical properties of cells. Moreover, the "constant" of proportionality g_i need not be constant at all; it may depend on environmental variables such as ion concentrations and ΔV itself. Thus, we can only use Equation 11.8 if both ΔV and the concentration of ion species i are close to their resting values. For other conditions, we'll have to allow for the possibility that the conductance per area changes, for example, writing $g_i(\Delta V)$. This section will consider only small deviations from the resting conditions; Section 12.2.4 will explore more general situations.

The conductance per area, g_i, is related to the ion's permeability $\mathcal{P}_s$ (see Equation 4.21 on page 135):

> **Your Turn 11C**
>
> Find a relation between the conductance per area and the permeability of a membrane to a particular ion species, assuming that the inside and outside concentrations are nearly equal. Discuss why your result is reasonable. [*Hint:* Remember that $c_{1,i} - c_{2,i}$ is small, and use the expansion $\ln(1 + \epsilon) \approx \epsilon$ for small ϵ.]

Notice that the conductances per area for various ion species, g_i, need not all be the same. Different ions have different diffusion constants in water; because they have different radii, they encounter different obstructions passing through different channels, and so on. Just as a membrane can be permeable to water but not to ions, so the conductances to different ions can differ. If a particular ion species is impermeant (like the Cl^- ions in the system imagined in Section 11.1.2), then its concentration needn't obey the Nernst relation. The impermeant species *are* important in determining the equilibrium membrane potential, however: They enter the system's overall charge neutrality condition.

$\boxed{T_2}$ *Section 11.2.2′ on page 501 mentions nonlinear corrections to the Ohmic behavior of membrane conductances.*

11.2.3 Active pumping maintains steady-state membrane potentials while avoiding large osmotic pressures

We can now return to the sodium anomaly in Table 11.1. To investigate nonequilibrium steady states using Equation 11.8, we need separate values of the conductances per area, g_i, of membranes to various ions. Several groups made such measurements around 1948 by using radioactively labeled sodium ions on one side of a membrane and ordinary sodium on the other side. They then measured the leakage of radioactivity across the membrane under various conditions of imposed potentials and concentrations. This technique yields the sodium current, separated from the contributions of other ions.[6] The result of such experiments was that nerve and muscle cells indeed behave ohmically (see Equation 11.8) under nearly resting conditions. The corresponding conductances are appreciable for potassium, chloride, *and sodium*; A. Hodgkin and B. Katz found that, for the squid axon,

$$g_{K^+} \approx 25 g_{Na^+} \approx 2 g_{Cl^-}. \qquad \text{(resting)} \qquad (11.9)$$

Thus the sodium conductance is small but not negligible, and certainly not zero.

Section 11.2.1 argued that a nonzero conductance for sodium implies that the cell's resting state is not in equilibrium. Indeed, in 1951 H. Ussing and K. Zehran found that living frog skin, with identical solutions on both sides, and membrane potential ΔV maintained at zero, nevertheless transported sodium ions, even though the net force in Equation 11.8 was zero. Apparently Equation 11.8 must be supplemented with an additional term describing the active ion pumping of sodium. The simplest modification we could entertain is

$$j_{Na^+} = \frac{g_{Na^+}}{e}(\Delta V - V_{Na^+}^{Nernst}) + j_{Na^+}^{pump}. \qquad (11.10)$$

The new, last term in this modified Ohm's law corresponds to a current source in parallel with the elements shown in Figure 11.4. This current source must do work if it's to push sodium ions "uphill" (against their electrochemical potential gradient). The new term distinguishes between the inner and outer sides of the membrane: It's positive, indicating that *the membrane pumps sodium outward*. The source of free energy needed to do that work is the cell's metabolism.

A more detailed study in 1955 by Hodgkin and R. Keynes showed that sodium is not the only actively pumped ion species: Part of the *inward* flux of potassium through a membrane also depends on the cell's metabolism. Intriguingly, Hodgkin and Keynes found that the outward sodium-pumping action stopped even in normal cells when they were deprived of any exterior potassium, a result suggesting that the pump couples its action on one ion to the other. Hodgkin and Keynes also found that metabolic inhibitors (such as dinitrophenol) reversibly stop the active pumping of both sodium and potassium in individual living nerve cells (Figure 11.5), leaving the passive, Ohmic part of the fluxes unchanged. Moreover, even with the cell's metabolism shut down, pumping resumes when one injects the cellular energy-storing molecule ATP into the cell.

To summarize, the results just described pointed to a hypothesis:

> *A specific molecular machine embedded in cell membranes hydrolyzes ATP, then uses some of the resulting free energy to pump sodium ions out of the cell. At the same time the pump imports potassium, partially offsetting the loss of electric charge from the exported sodium.*   (11.11)

---

[6]An alternative approach is to shut down the permeation of other ions by using specific **neurotoxins** (a class of poisons).

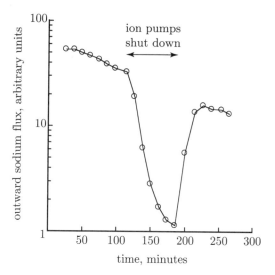

Figure 11.5: (Experimental data.) Flux of sodium ions out of a cuttlefish axon after electrical stimulation. At the beginning of the experiment, the axon was loaded with radioactive sodium, then placed in ordinary seawater; the loss of radioactivity was monitored. During the interval represented by the *arrow*, the axon was exposed to the toxin dinitrophenol, which temporarily shut down sodium pumping. Later the toxin was washed away with fresh seawater, and ion pumping spontaneously resumed. The horizontal axis gives the time after the end of electrical stimulation; the logarithmic vertical scale gives the rate at which radioactively labeled sodium left the axon. [Data from Hodgkin & Keynes, 1955.]

The pump operates only when sodium and ATP are available on its inner side and potassium is available on its outer side. If any of these are cut off, the cell slowly reverts to the ion concentrations appropriate for equilibrium.

Idea 11.11 amounts to a remarkably detailed portrait of the membrane pump, considering that in 1955 no specific membrane constituent was even known to be a candidate for this job. Clearly *something* was pumping those ions; but there are thousands of transmembrane proteins in a living cell membrane, and it was hard to find the right one. Then in 1957, the physiologist J. Skou isolated from crab leg neurons a single membrane protein with ATPase activity. By controlling the ion content of his solutions, Skou found that to hydrolyze ATP, his enzyme required both sodium and potassium, the same behavior Hodgkin and Katz had found for whole nerve axons (Figure 11.6). Skou concluded that his enzyme must have separate binding sites for both sodium and potassium. For this and other reasons, he correctly guessed that it was the anticipated sodium pump.

Additional experiments confirmed Skou's hypotheses: Remarkably, it is possible to prepare a pure lipid bilayer, introduce the purified pump protein, the necessary ions, and ATP, then watch as the protein self-assembles in the membrane and begins to function in this totally artificial system.

The fact that the pump's ATPase activity depends on the presence of the pumped ions has an important implication: The pump is a *tightly coupled* molecular machine,

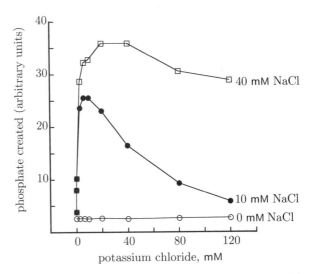

Figure 11.6: (Experimental data.) The rate of ATP hydrolysis catalyzed by the sodium–potassium pump, as a function of the available interior sodium and exterior potassium. The vertical axis gives the quantity of inorganic phosphate generated in a certain time interval. The data show that if either sodium or potassium is missing, ATP consumption, and hence P_i production, stop. [Data from Skou, 1957.]

wasting very little ATP on futile cycles. Later work showed that, in fact, the magnitude of the potassium current is always two-thirds as large as that of the sodium ions; the pump maintained this relation across a range of different ATP concentrations. In other words, the pump carries out **coupled transport** of sodium and potassium ions. We can think of the machine as a special kind of revolving door, which waits for three Na^+-binding sites to be occupied on its interior face. Then it pushes these ions out (or **translocates** them), releases them, and waits for two K^+-binding sites on the outer face to be occupied. Finally, it translocates the potassiums, releases them on the interior, and begins its cycle anew. Thus each cycle of this machine causes the net transport of one unit of charge out of the cell; we say that the pump is **electrogenic**.[7] Specific membrane pumps, or **active transporters**, of this sort are among the most important molecular machines in a cell.

Before concluding that the ATPase enzyme discovered by Skou really is (in part) responsible for resting membrane potentials, we should verify that the proposed pumping process is energetically reasonable.

Example: Compare the free energy gain from hydrolyzing one ATP molecule with the cost of running the pump through a cycle.

---

[7]Figure 2.21 on page 57 simplified the sodium–potassium pump, sketching only one of each kind of binding site. A *non*electrogenic pump would have had $j_{K^+}^{pump} + j_{Na^+}^{pump} = 0$. An example of this sort of behavior is the H^+/K^+ exchanger, found in the cells lining your stomach. In each cycle, it transports two protons out of the cell (helping to make your gastric fluid acidic) while importing two potassium ions.

Solution: To pump one sodium ion out of the cell costs both electrostatic potential energy $-e\Delta V$ and the free energy cost of enhancing the world's order (by incrementally increasing the difference in sodium concentration across the membrane). This entropy is what the Nernst potential measures. Consulting Table 11.1 on page 477, the total free energy cost to pump one sodium ion out is thus

$$-e(\Delta V - \mathcal{V}_{Na+}^{Nernst}) = e(60\,mV + 54\,mV) = e \times 114\,mV.$$

For inward pumping of potassium, the corresponding calculation gives

$$+e(\Delta V - \mathcal{V}_{K+}^{Nernst}) = e(-60\,mV - (-75\,mV)) = e \times 15\,mV,$$

which is also positive. The total cost of one cycle is then $3(e \times 114\,mV) + 2(e \times 15\,mV) = 0.0372\,eV = 15k_B T_r$. (The unit eV, or electron volt, is defined in Appendix A.) ATP hydrolysis, on the other hand, liberates about $19k_B T_r$ (see Problem 10.4). The pump is fairly efficient; only $4k_B T_r$ is lost as thermal energy.

Let's see how the discovery of ion pumping helps make sense of the data presented in Table 11.1 on page 477. Certainly the sodium–potassium pump's net effect of pushing one unit of positive charge out of the cell will drive the cell's interior potential down, away from the sodium Nernst potential and toward that of potassium. The net effect of removing one osmotically active ion from the cell per cycle also has the right sign to reduce the osmotic imbalance we found in Donnan equilibrium (Equation 11.6 on page 476).

To study pumping quantitatively, first note that a living cell is in a steady state because it maintains its potential and ion concentrations indefinitely (as long as it remains alive). Thus there must be no net flux of any ion; otherwise, some ion would pile up somewhere, eventually changing the concentrations. Every ion must be either impermeant (like the interior macromolecules), or in Nernst equilibrium, or actively pumped. Those ions that are actively pumped (Na^+ and K^+ in our simplified model) must separately have their Ohmic leakage exactly matched by their active pumping rates. Our model assumes that $j_{K+}^{pump} = -\frac{2}{3}j_{Na+}^{pump}$ and that $j_{Na+}^{pump} > 0$, because our convention is that j is the flux directed outward. In short, for steady state we must have $j_{Na+} = j_{K+} = 0$, or

$$j_{K+}^{pump} = -j_{K+}^{Ohmic} = -\frac{2}{3}j_{Na+}^{pump} = -\frac{2}{3}(-j_{Na+}^{Ohmic}).$$

In this model, chloride is permeant and not pumped, so its Nernst potential must agree with the resting membrane potential. Indeed, from Table 11.1, its Nernst potential really is in good agreement with the actual membrane potential $\Delta V = -60\,mV$. Turning to sodium and potassium, the previous paragraph implies that the Ohmic part of the corresponding ion fluxes must be in the ratio $-\frac{2}{3}$. The Ohmic hypothesis (Equation 11.8) says that

$$-\frac{2}{3}(\Delta V - \mathcal{V}_{Na+}^{Nernst})g_{Na+} = (\Delta V - \mathcal{V}_{K+}^{Nernst})g_{K+}.$$

Solving for ΔV gives

$$\Delta V = \frac{2g_{Na^+} \mathcal{V}^{Nernst}_{Na^+} + 3g_{K^+} \mathcal{V}^{Nernst}_{K^+}}{2g_{Na^+} + 3g_{K^+}}. \tag{11.12}$$

We now substitute the Nernst potentials appearing in Table 11.1 on page 477, and the measured relation between conductances (Equation 11.9), finding $\Delta V = -72\,\mathrm{mV}$. We can then compare our prediction with the actual resting potential, about $-60\,\mathrm{mV}$.

Our model is thus moderately successful at explaining the observed membrane potential. In part the inaccuracy stemmed from our use of the Ohmic (linear) hypothesis for membrane conduction, Equation 11.8, when at least one permeant species (sodium) was far from equilibrium. Nevertheless, we *have* qualitatively answered our paradox: The membrane potential predicted by Equation 11.12 lies *between* the Nernst potentials of sodium and potassium, and is much closer to the latter, as observed in experiments. Indeed, Equation 11.12 shows that

> *The ion species with the greatest conductance per area gets the biggest vote in determining the steady-state membrane potential. That is, the resting membrane potential ΔV is closer to the Nernst potential of the most permeant pumped species (here, $\mathcal{V}^{Nernst}_{K^+}$) than it is to that of the less permeant ones (here, $\mathcal{V}^{Nernst}_{Na^+}$).*   (11.13)

Our prediction for ΔV also displays experimentally verifiable trends as we change the ion concentrations on either side of the membrane.

Even more interesting, if our membrane could suddenly *switch* from conducting potassium better than sodium to the other way round, then Idea 11.13 predicts that its transmembrane potential would change drastically, switching suddenly from a negative value close to $\mathcal{V}^{Nernst}_{K^+}$ to a *positive* value closer to $\mathcal{V}^{Nernst}_{Na^+}$. And in fact, Chapter 12 will show that the measured membrane potential during a nerve impulse really does reverse sign and come close to $\mathcal{V}^{Nernst}_{Na^+}$. But this is idle speculation—isn't it? Surely the permeabilities of a membrane to various dissolved substances are fixed forever by its physical architecture and chemical composition—aren't they? Chapter 12 will come back to this point.

$\boxed{T_2}$ *Section 11.2.3′ on page 501 comments more about active ion pumping.*

11.3 MITOCHONDRIA AS FACTORIES

Like kinesin, studied in Chapter 10, the sodium–potassium pump runs on a fuel, the molecule ATP. Other molecular motors also run on ATP (or, in some cases, other NTPs). It takes a lot of ATP to run your body—some estimates are as high as $2 \cdot 10^{26}$ ATP molecules per day, all ultimately derived from the food you eat. That much ATP would weigh $160\,\mathrm{kg}$, but you don't need to carry such a weight around: Each ATP molecule gets recycled many times per minute. That is, ATP is a *carrier* for free energy.

ATP synthesis in eukaryotic cells also involves active ion pumping, although not of sodium or potassium. Instead, the last step in oxidizing your food (called **respiration**) pumps *protons* across a membrane. The next four sections will describe a remarkable molecular machine that accomplishes ATP synthesis starting from a proton gradient.

11.3.1 Busbars and driveshafts distribute energy in factories

Chapter 10 used the term *machine* to denote a relatively simple system, with few parts, doing just one job. Indeed, the earliest technology was of this sort: Turn a crank, and a rope lifts water out of the well.

As technology developed, it became practical to combine machines into a factory, a loose collection of several machines with specialized subtasks. The factory was flexible: It could be reconfigured as needed, individual machines could be replaced, all without disrupting the overall operation. Moreover, some of the machines could specialize in importing energy and converting it into a common currency to be fed into the other machines. The latter then made the final product, or perhaps yet another form of energy currency for export.

The drawing on page 1 shows such a factory, circa 1820. The waterwheel converts the weight of the incoming water to a torque on the driveshaft. The driveshaft runs through the mill, distributing mechanical energy to the various machines attached to it. Later, the invention of electric technology allowed a more flexible energy currency, the potential energy of electrons in a wire. With this system, the initial conversion of chemical energy (for example, in coal) to electricity could occur many kilometers away from the point of use in the factory. Within the factory, distribution could be accomplished by using a **busbar**, a large conducting bar running through the building, with various machines attached to it.

Figure 11.7 sketches a factory of a sort that could supply hydrogen-powered automobiles. Some high-energy substrate, like coal, comes in at the left. A series of transductions converts the incoming free energy to the potential energy of electrons for convenient transport (the electrons themselves are recirculated). In the factory, a busbar distributes the electricity to a series of electrolytic cells, which convert low-energy water molecules to high-energy hydrogen and oxygen. The hydrogen gets packaged and delivered to cars, which burn it (or convert it directly to electricity) and generate useful work. In winter, some of the electricity can instead be sent through a resistor, doing no mechanical work but warming up the factory for the comfort of those working inside it.

The next sections will discuss the close parallels between the industrial process just described and the activity of mitochondria.

11.3.2 The biochemical backdrop to respiration

The overall biochemical process we wish to study is one of **oxidation**. Originally this term referred to the chemical addition of oxygen to something else; and indeed, you breathe in oxygen, attach it to high-energy compounds containing carbon and hydrogen, and exhale low-energy H_2O and CO_2. Chemists have found it useful, however, to generalize the concept of oxidation in order to identify individual subreactions as oxidation or the opposite process, **reduction**. According to this generalization, the key fact about oxygen is the tremendous lowering of its internal energy when it acquires

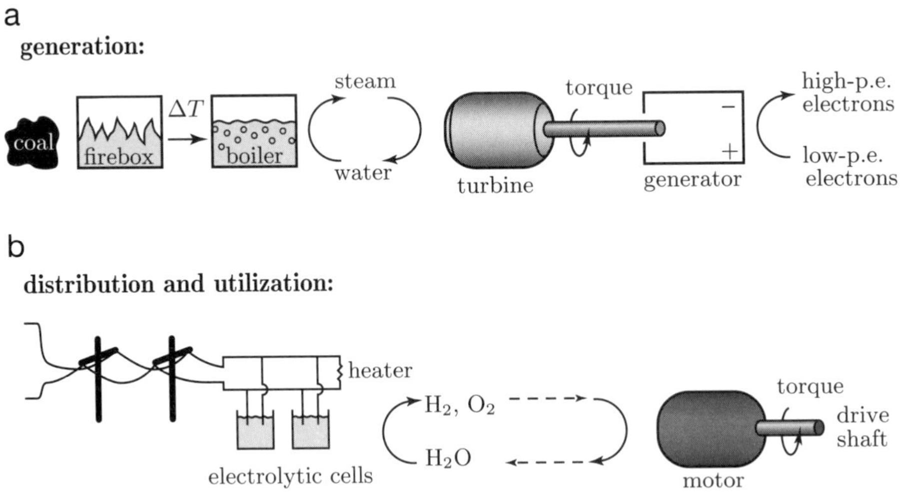

a

generation:

b

distribution and utilization:

Figure 11.7: (Schematic.) An imagined industrial process. (a) Chemical fuel is burned, ultimately creating a difference in the electrostatic potential of electrons across two wires. The difference is maintained by electrical insulation (in this case, air) between the wires on the far right. (b) Inside a factory, the electrons are used to drive an uphill chemical process, converting low-energy molecules to ones with high stored chemical energy. The latter can then be loaded into an automobile to generate torque and do useful work. If desired, some of the electrons' potential energy can be converted directly to thermal form by placing a resistor (the "heater") across the power lines.

an additional electron. Thus as mentioned in Chapter 7, in a water molecule, the hydrogen atoms are nearly stripped of their electrons, having given them almost entirely to the oxygen. Burning molecular hydrogen in the reaction $2H_2 + O_2 \rightarrow 2H_2O$ thus oxidizes it in the sense of removing electrons.

More generally, any reaction removing an electron from an atom or molecule is said to "oxidize" it. Because electrons are neither created nor destroyed in chemical reactions, any oxidation reaction must be accompanied by another reaction effectively *adding* an electron to something—a reduction reaction. For example, oxygen itself gets reduced when we burn hydrogen; indeed, adding a neutral hydrogen atom to anything is considered a reduction.

With this terminology in place, let's examine what happens to your food. The early stages of digestion break down complex fats and sugars to smaller molecules such as the simple sugar glucose, which then get transported to the body's individual cells. Once inside the cell, glucose undergoes **glycolysis** in the cytoplasm. We will not study glycolysis in detail, although it does generate a small amount of ATP (two molecules per glucose). Of greater interest to us is the fact that glycolysis splits glucose to two molecules of **pyruvate** (CH_3–CO–COO^-), another small, high-energy molecule.

In anærobic cells, glycolysis is essentially the end of the story. The pyruvate is a waste product, which typically gets converted to ethanol or lactate and excreted by the cell, thus leaving only the two ATP molecules per glucose as the useful product of metabolism. Prior to about 1.8 billion years ago, Earth's atmosphere lacked free oxygen, and living organisms had to manage with this anærobic metabolism. Even today, intense exercise can locally exhaust your muscle cells' oxygen supply, switching them to anærobic mode, with a resulting buildup of lactate.

With oxygen, however, a cell can synthesize *about 30 more* molecules of ATP per glucose. In 1948, E. Kennedy and A. Lehninger found that the site of this synthesis is the mitochondrion (Figure 2.6 on page 42). The mitochondrion carries out a process called **oxidative phosphorylation**: That is, it imports and oxidizes the pyruvate generated by glycolysis, coupling this energetically favorable reaction to the unfavorable one of attaching a phosphate group to ADP ("phosphorylating" it).

The mitochondrion is surrounded by an outer membrane, which is permeable to most small ions and molecules. Inside this membrane lies a convoluted inner membrane, whose interior is called the **matrix**. The matrix contains closed loops of DNA and its transcriptional apparatus, similar to those in a bacterium. The inner side of the inner membrane is densely studded with buttons visible in electron microscopy (sketched in Figure 2.6b). These are ATP synthase particles, to be discussed in Section 11.3.3.

Figure 11.8 shows in very rough form the steps involved in oxidative phosphorylation, discussed in this section and the next one. The figure has been drawn in a way

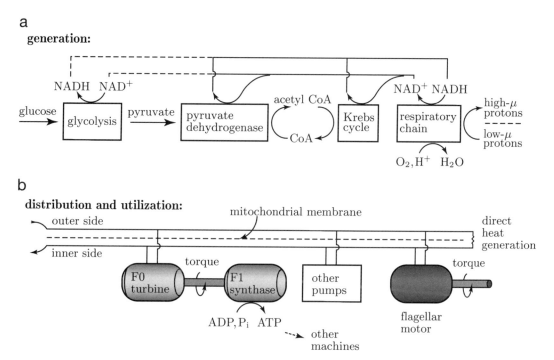

Figure 11.8: (Schematic.) Outline of the activity of a mitochondrion, emphasizing the parallels to Figure 11.7. (a) Metabolism of sugar generates a difference in the electrochemical potential of protons across the inner mitochondrial membrane. For simplicity, "NADH" represents both the carrier molecules NADH and $FADH_2$. The *dashed line* represents an indirect process of import into the mitochondrion. (b) The protons, in turn, drive a number of molecular machines. (Although mitochondria do not have flagella, bacteria such as *E. coli* have a similar arrangement, which does drive their flagellar motor.)

intended to stress the parallels between the mitochondrion and the simple factory in Figure 11.7.

Decarboxylation of pyruvate The first step in oxidative phosphorylation takes place in the mitochondrion's matrix. It involves the removal of the carboxyl (COO) group from pyruvate and its oxidation to CO_2, via a giant enzyme complex called pyruvate dehydrogenase (see Figure 2.4m on page 38). The remainder of the pyruvate is an acetyl group, CH_3–CO–; it gets attached to a carrier molecule called coenzyme A (abbreviated CoA) via a sulfur atom, thus forming acetyl-CoA. As mentioned earlier, a reduction must accompany the oxidation of the carbon. The pyruvate dehydrogenase complex couples the oxidation tightly to one *particular* reduction, that of the carrier molecule nicotinamide adenine dinucleotide (or **NAD$^+$**). The net reaction,

$$CH_3\text{–}CO\text{–}COO^- + HS\text{–}CoA + NAD^+ \rightarrow CH_3\text{–}CO\text{–}S\text{–}CoA + CO_2 + NADH,$$

$$(11.14)$$

adds two electrons (and a proton) to NAD$^+$, to yield NADH. Glycolysis also generates another molecule of NADH per pyruvate; this NADH enters the respiratory chain indirectly (dashed line in Figure 11.8).

Krebs cycle The second step also occurs in the mitochondrial matrix. A cycle of enzyme-catalyzed reactions picks up the acetyl-CoA generated in the previous step, oxidizing further the acetyl group and recovering coenzyme A. Corresponding to this oxidation, three more molecules of NAD$^+$ are reduced to NADH; in addition, a second carrier molecule, flavin adenine dinucleotide (abbreviated **FAD**), gets reduced to FADH$_2$. The net reaction,

$$CH_3\text{–}CO\text{–}S\text{–}CoA + 2H_2O + FAD + 3NAD^+ + GDP^{3-} + P_i^{2-}$$

$$\rightarrow 2CO_2 + FADH_2 + 3NADH + 2H^+ + GTP^{4-} + HS\text{–}CoA, \qquad (11.15)$$

thus adds eight electrons (and three protons) to the carriers FAD and NAD$^+$. It also generates one GTP, which is energetically equivalent to an ATP. This part of the reaction is called the **Krebs cycle**, or the **tricarboxylic acid cycle**.

Your
Turn
11D

Confirm that Reaction 11.15 is properly balanced.

Summary Reactions 11.14 and 11.15 oxidize pyruvate completely: Pyruvate's three carbon atoms each end up as molecules of carbon dioxide. Conversely, four molecules of the carrier NAD$^+$ and one of FAD get reduced to NADH and FADH$_2$. Because glycolysis also generates two molecules of pyruvate and two of NADH, the overall effect is to generate ten NADH and two FADH$_2$ per glucose. Two ATP per glucose have also been formed from glycolysis, and the equivalent of two more from the Krebs cycle.

11.3.3 The chemiosmotic mechanism identifies the mitochondrial inner membrane as a busbar

How does the chemical energy stored in the reduced carrier molecules get harnessed to synthesize ATP? Early attempts to solve this puzzle met with a frustrating inability to pin down the exact stoichiometry of the reaction: Unlike, say, Reaction 11.14, where each incoming pyruvate yields exactly one NADH, the number of ATP molecules generated by respiration did not seem to be any definite, integral number. This difficulty dispersed with the discovery of the **chemiosmotic mechanism**, proposed by Peter Mitchell in 1961.

According to the chemiosmotic mechanism, ATP synthesis is *indirectly* coupled to respiration via a power transmission system. Thus we can break the story down into the generation, transmission, and utilization of energy, just as in a factory (Figure 11.8).

Generation The final oxidation reaction in a mitochondrion (respiration) is

$$NADH + H^+ + \tfrac{1}{2}O_2 \rightarrow NAD^+ + H_2O. \qquad (11.16)$$

(FADH$_2$ undergoes a similar reaction.) This reaction has a standard free energy change of[8] $\Delta G'^0_{NAD} = -88k_B T_r$, but the enzyme complex that facilitates Reaction 11.16 couples it to the pumping of 10 protons across the inner mitochondrial membrane. The net free energy change of the oxidation reaction is thus partially offset by the difference in the electrochemical potential of a proton across the membrane (see Section 8.1.1 on page 295), times 10.

Your Turn 11E

a. Adapt the logic of the pump energetics Example (page 484) to find the difference in electrochemical potential for protons across the mitochondrial inner membrane. Use the following experimental input: The pH in the matrix minus that outside is $\Delta pH = 1.4$, whereas the corresponding electrostatic potential difference equals $\Delta V \approx -0.16$ volt.

b. The difference you just found is often expressed as a protonmotive force (or p.m.f.), defined as $(\Delta\mu_{H^+})/e$. Compute it, expressing your answer in volts.

c. Compute the total $\Delta G'^0_{NAD} + 10\Delta\mu_{H^+}$ for the coupled oxidation of 1 molecule of NADH and transport of 10 protons. Is it reasonable to expect this reaction to go forward? What information would you need to be sure?

Transmission Under normal conditions, the inner mitochondrial membrane is impermeable to protons. Thus by pumping protons out, the mitochondrion creates an electrochemical potential difference that spreads all over the surface of its inner mem-

---

[8]The actual ΔG is even greater in magnitude than $\Delta G'^0$ because the concentrations of the participating species are not equal to their standard values. We will nevertheless use the value given here as a rough guide.

brane. The impermeable membrane plays the role of the electrical insulation separating the two wires of an electric power cord: It maintains the potential difference
between the inside and outside of the mitochondrion. Any other machine embedded in the membrane can utilize the excess free energy represented by this $\Delta\mu$ to do
useful work, just as any machine can tap into the busbar along a factory.

Utilization The chemiosmotic mechanism requires a second molecular machine,
the **ATP synthase**, embedded in the inner membrane. These machines allow protons back inside the mitochondrion but couple their transport to the synthesis of
ATP. Under cellular conditions, the hydrolysis of ATP yields a ΔG_{ATP} of about $20k_B T_r$
(see Appendix B). This value is about 2.1 times the value you found for the proton's
$|\Delta\mu|$ in Your Turn 11E, so we conclude that at least 2.1 protons must cross back into
the mitochondrion per ATP synthesis. The actual value[9] is thought to be closer to 3.
Another proton is thought to be used by the active transporters that pull ADP and
P_i into, and ATP out of, the mitochondrion. As mentioned earlier, each NADH oxidation pumps 10 protons out of the mitochondrion. Thus we expect a maximum of
about 10/(3+1), or *roughly 2.5 ATP molecules synthesized per NADH.* This is indeed
the approximate stoichiometry measured in biochemical experiments. The related
molecule $FADH_2$ generates an average of another 1.5 ATP from its oxidation. Thus
the 10 NADH and 2 $FADH_2$ generated by the oxidation of 1 glucose molecule ultimately give rise to $10 \times 2.5 + 2 \times 1.5 = 28$ ATP molecules.

Adding to these 2 ATP generated directly from glycolysis and the 2 GTP from the
Krebs cycle yields a rough total of about *32 molecules of ATP or GTP from the oxidation
of a single glucose molecule.* This figure is only an upper bound because we assumed
high high efficiency (small dissipative losses) throughout the respiration/synthesis
system. Remarkably, the actual ATP production is close to this limit: The machinery
of oxidative phosphorylation is quite efficient. The schematic Figure 11.9 summarizes
the mechanism presented in this section.

$\boxed{T_2}$ *Section 11.3.3' on page 502 comments some more about ATP production.*

11.3.4 Evidence for the chemiosmotic mechanism

Several elegant experiments confirm the chemiosmotic mechanism.

Independence of generation and utilization Several of these experiments were designed to demonstrate that oxidation and phosphorylation proceed almost independently, linked only by the common value of the electrochemical potential difference,
$\Delta\mu$, across the inner mitochondrial membrane. For example, artificially changing
$\Delta\mu$ by preparing an acidic exterior solution was found to induce ATP synthesis in
mitochondria without any source of food. Similar results were obtained with chloroplasts in the absence of *light.* In fact, an external electrostatic potential can be directly applied across a cell membrane to operate other proton-driven motors—see
this chapter's Excursion.

---

[9]The precise stoichiometry of the ATP synthase is still under debate. Thus the numbers here are subject to
revision.

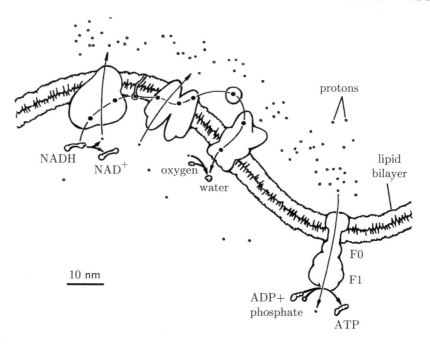

Figure 11.9: (Schematic.) Mechanism of oxidative phosphorylation. Electrons are taken from NADH molecules and transferred down a chain of carriers (*black dots*), ultimately ending up on an oxygen atom in water. Two of the membrane-bound enzymes shown couple this process to the pumping of protons across the inner mitochondrial membrane, seen in cross section. Protons then flow back through the F0F1 complex (*right*), which synthesizes ATP. See also the more realistic depiction of this crowded system in Figure 2.20 on page 57. [From Goodsell, 1993.]

    In a more elaborate experiment, E. Racker and W. Stoeckenius assembled a totally artificial system, combining artificial lipid bilayers with a light-driven proton pump (bacteriorhodopsin) obtained from a bacterium. The resulting vesicles generated a pH gradient when exposed to light. Racker then added an ATP synthase from beef heart to his preparation. Despite the diverse origins of the components, the combined system synthesized ATP when exposed to light, a result again emphasizing the independence of ATP synthase from any aspect of the respiratory cycle other than the electrochemical potential difference $\Delta\mu$.

Membrane as electrical insulation It is possible to rip apart the mitochondrial membrane into fragments (using ultrasound), without damaging the individual proteins embedded in it. Ordinarily these fragments would reassemble into closed vesicles, because of the high free energy cost of a bilayer membrane edge (see Section 8.6.1), but this reassembly can be prevented by adding a detergent. The detergent, a one-chain amphiphile, protects the membrane edges by forming a micellelike rim (Figure 8.8 on page 325). When such fragments were made from the mitochondrial inner membrane, they continued to oxidize NAD$^+$ *but lost the ability to synthesize ATP.* The loss of function makes sense in the light of the chemiosmotic mechanism: In a membrane fragment, the electrical transmission system is "short-circuited"; protons pumped to one side can simply diffuse to the other side.

Similarly, introducing any of a class of membrane channel proteins, or other lipid-soluble compounds known to transport protons short-circuits the mitochondrion, cutting ATP production. Analogous to the electric heater shown in Figure 11.7, such short-circuiting converts the chemical energy of respiration directly into heat. Some animals engage this mechanism in the mitochondria of "brown fat" cells when they need to turn food directly into heat (for example, during hibernation).

Operation of the ATP synthase We have seen that an elaborate enzymatic apparatus accomplishes the oxidation of NADH and the associated proton pumping. In contrast, the ATP synthase turned out to be remarkably simple. As sketched in Figure 11.10a, the synthase consists of two major units, called **F0** and **F1**. The F0 unit (shown as the elements a, b, and c in the figure) is normally embedded in the inner mitochondrial membrane, with the F1 unit (shown as the elements α, β, γ, δ, and ϵ in the figure) projecting into the matrix. Thus the F1 units are the round buttons (sometimes called lollipops) seen projecting from the inner side of the membrane in electron micrographs. They were discovered and isolated in the 1960s by H. Fernandez–Moran and by Racker, who found that, in isolation, they catalyzed the *breakdown* of ATP. This result seemed paradoxical: Why should the mitochondrion, whose job is to *synthesize* ATP, contain an ATPase?

To answer the paradox, we first must remember that an enzyme cannot alter the direction of a chemical reaction (see Ideas 8.15 on page 303 and 10.13 on page 429). ΔG sets the reaction's direction, regardless of the presence of enzyme. The only way an enzyme can implement an uphill chemical reaction ($\Delta G_{F1} > 0$ for ATP synthesis) is by coupling it to some downhill process ($\Delta G_{F0} < 0$), with the net process

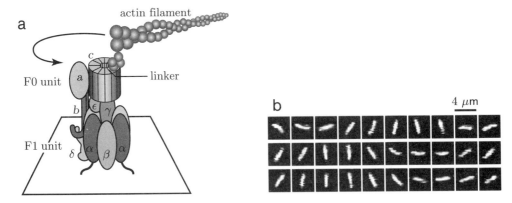

Figure 11.10: (Schematic; video micrograph frames.) Direct observation of the rotation of the c ring of the F0 proton turbine. (a) A complete ATP synthase from *E. coli* (both F0 and F1 units) is attached to a coverslip, and a long, fluorescently labeled filament of actin is attached. (b) Successive video frames showing the rotation of the actin filament in the presence of 5 mM ATP. The frames are to be read from left to right, starting with the first row; they show a counterclockwise rotation of the actin filament. [From Wada et al., 2000.]

being downhill ($\Delta G_{F1} + \Delta G_{F0} < 0$). So the F1 unit must somehow be coupled to the F0 unit; F0, being embedded in the membrane, is driven by the electrochemical potential difference of protons across the membrane. By isolating the F1 unit, the experimenters had inadvertently removed this coupling, thereby converting F1 from a synthase to an ATPase.

In 1979, P. Boyer proposed that both F0 and F1 are *rotary* molecular machines, mechanically coupled by a driveshaft. According to Boyer's hypothesis, we may think of F0 as a proton "turbine," driven by the chemical potential difference of protons and supplying *torque* to F1. Boyer also outlined a mechanochemical process by which F1 could convert rotary motion to chemical synthesis. Fifteen years later, J. Walker and coauthors gave concrete form to Boyer's model, finding the detailed atomic structure for F1 (sketched in Figure 11.10a). The elements labeled a, b, α, β, and δ in the figure remain fixed with respect to one another; c, γ, and ϵ rotate relative to them. Each time the driveshaft γ passes a β subunit, the F1 unit catalyzes the interconversion of ATP with ADP; the direction of rotation determines whether synthesis or hydrolysis takes place.

Although static atomic structures such as the one in Figure 11.10a can be highly suggestive, nevertheless they do not actually establish that one part moves relative to another. The look-and-see proof that F1 is a rotary machine came from an ingenious direct experiment by K. Kinosita, Jr., M. Yoshida, and coauthors. Figure 11.10 shows a second-generation version of this experiment.

With a diameter of less than 10 nm, F1 is far too small to observe directly by light microscopy. To overcome this problem, the experimenters attached a long, stiff actin filament to the c element, as sketched in Figure 11.10a. They labeled the filament with a fluorescent dye and anchored the α and β elements to a glass slide, so that relative rotary motion of the c element would crank the entire actin filament. The resulting motion pictures showed that the motor took random (Brownian) steps, with no net progress, until ATP was added. With ATP, it moved in one direction at speeds up to about six revolutions per second. The motion was not uniform; slowing the F1 motor by using low ATP levels showed discrete, 120° steps. Such steps are just what we would expect on structural grounds: The structure of F1 shows three β subunits, each spaced one-third of a revolution from the others. (Compare with the steps taken by kinesin, Figure 10.22 on page 439.) The subsequent experiment shown in Figure 11.10 used the entire F0F1 complex, not just F1, to confirm that the F0 really is rigidly connected to F1.

The experiments just described also allow an estimate of the torque generated by ATP hydrolysis (or the torque required for ATP synthesis), using ideas from low Reynolds-number flow. The experimenters found that an actin filament 1 μm long rotated at about 6 revolutions per second, or an angular velocity of $2\pi \times 6$ radians per second, when ATP was supplied. Section 5.3.1 on page 172 claimed that the viscous drag force on a thin rod, dragged sideways through a fluid, is proportional to its speed v and to the viscosity of water η. The force should also be proportional to the rod's length. Detailed calculation for a rod of length 1 μm, with the thickness of an actin filament, gave Kinosita and coauthors the constant of proportionality:

$$f \approx 3.0 \eta L v. \tag{11.17}$$

> **Your Turn 11F**
>
> Equation 11.17 gives the force needed to drag a rod at a given speed v. But we want the *torque* needed to crank a rod pivoted at one end at *angular* velocity ω.
>
> a. Work this out from Equation 11.17. Evaluate your answer for a rod of length $1\,\mu$m rotating at 6 revolutions per second.
> b. How much work must the F1 motor do for every one-third revolution of the actin filament?

More precisely, the rotation rate just quoted was achieved when ATP was supplied at a concentration $c_{ATP} = 2\,$mM, along with $c_{ADP} = 10\,\mu$M and $c_{P_i} = 10\,$mM.

> **Your Turn 11G**
>
> a. Find ΔG for ATP hydrolysis under these conditions (recall Section 8.2.2 on page 301 and Problem 10.4 on page 465).
> b. Each ATP hydrolysis cranks the γ element by one-third of a revolution. How efficiently does F1 transduce chemical free energy to mechanical work?

Thus F1 is a highly efficient transducer when operated in its ATPase mode. Under natural conditions, F1 operates in the opposite direction (converting mechanical energy supplied by F0 to ATP production) with a similarly high efficiency, contributing to the overall high efficiency of ærobic metabolism.

11.3.5 Vista: Cells use chemiosmotic coupling in many other contexts

Section 11.2 introduced ion pumping across membranes as a practical necessity, reconciling

- The need to segregate macromolecules inside a cellular compartment, so that they can do their jobs in a controlled chemical environment,
- The need to give macromolecules an overall net negative charge, to avert a clumping catastrophe (see Section 7.4.1 on page 261), and
- The need to maintain osmotic balance, or osmoregulate, to avoid excessive internal pressure (see Section 11.2.1).

This chain of logic may well explain why ion pumps evolved in the first place: to meet a challenge posed by the physical world.

But evolution is a tinkerer. Once a mechanism evolves to solve one problem, it's available to be pressed into service for some totally different need. Ion pumping implies that the resting, or steady, state of the cell is not in equilibrium and, hence, is not a state of minimal free energy. That is, the resting state is like a charged battery, with available free energy distributed all over the membrane. We should think of the ion pumps as a "trickle charger," constantly keeping the battery charged despite "current leaks" that tend to discharge it. Section 11.3.3 showed one useful cellular function that such a setup could perform: the transmission of useful energy among machines embedded in the mitochondrial membrane. In fact, the chemiosmotic mechanism is so versatile that it appears over and over in cell biology.

Proton pumping in chloroplasts and bacteria Chapter 2 mentioned a second class of ATP-generating organelles in the cell, the chloroplasts. These organelles capture sunlight and use its free energy to pump protons across their membrane. From this point on, the story is similar to that in Section 11.3.3: The proton gradient drives a "CF0CF1" complex similar to F0F1 in mitochondria.

Bacteria, too, maintain a proton gradient across their membranes. Some ingest and metabolize food, to drive proton pumps related to, though simpler than, those in mitochondria. Others, for example, the salt-loving *Halobacterium salinarium* contain a light-driven pump, bacteriorhodopsin. Again, whatever the source of the proton gradient, bacteria contain F0F1 synthases quite similar to those in mitochondria and chloroplasts. This high degree of homology, found at the molecular level, lends strong support to the theory that both mitochondria and chloroplasts originated as free-living bacteria. At some point in history, they apparently formed symbiotic relations with other cells. Gradually the mitochondria and chloroplasts lost their ability to live independently, for example, both losing some of their genome.

Other pumps Cells have an array of active pumps. Some are powered by ATP: For example, the calcium ATPase, which pumps Ca^{2+} ions out of a cell, plays a role in the transmission of nerve impulses (see Chapter 12). Others pull one molecule against its gradient by coupling its motion to the transport of a second species *along* its gradient. Thus, for example, the lactose permease allows a proton to enter a bacterial cell, but only at the price of bringing along a sugar molecule. Such pumps, where the two coupled motions are in the same direction, are generically called symports. A related class of pumps, coupling an inward to an outward transport, are called antiports. An example is the sodium–calcium exchanger, which uses sodium's electrochemical potential gradient to force calcium out of animal cells (see Problem 11.1).

The flagellar motor Figure 5.9 on page 176 shows the flagellar motor, another remarkable molecular device attached to the power busbar of *E. coli*. Like F0, the motor converts the electrochemical potential jump of protons into a mechanical torque; Section 5.3.1 on page 172 described how this torque turns into directed swimming motion. The flagellar motor spins at up to 100 revolutions per second; each revolution requires the passage of about 1000 protons. This chapter's Excursion describes a remarkable experiment showing directly the relation between protonmotive force and torque generation in this motor.

11.4 EXCURSION: "POWERING UP THE FLAGELLAR MOTOR" BY H. C. BERG AND D. FUNG

Flagellar rotary motors are driven by protons or sodium ions that flow from the outside to the inside of a bacterial cell. *Escherichia coli* uses protons. If the pH of the

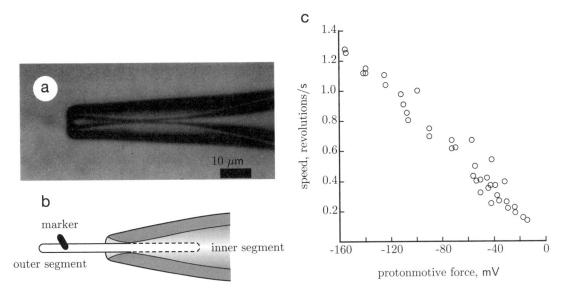

Figure 11.11: (Photomicrograph; schematic; experimental data.) Experiment to show that the flagellar motor runs on protonmotive force. (a) Micropipette tip used to study the bacterial flagellar motor. (b) Micropipette with a partially inserted bacterium. *Dashed lines* represent the part of the cell wall permeabilized by a chemical in the pipette. (c) Flagellar motor speed versus the protonmotive force across the part of the membrane containing the motor. [(a) Image kindly supplied by H. C. Berg; see Fung & Berg, 1995.]

external medium is lower than that of the internal medium, protons move inward by diffusion. If the electrostatic potential of the external medium is higher than that of the internal medium, they are driven in by a transmembrane electric field. We thought that it would be instructive to power up the flagellar motor with an external voltage source, for example, a laboratory power supply.[10] *Escherichia coli* is rather small, less than 1 μm in diameter by about 2 μm long. And its inner membrane, the one that needs to be energized, is enclosed by a cell wall and porous outer membrane. Thus, it is difficult to insert a micropipette into a cell. But one can put a cell into a micropipette.

First, we grew cells in the presence of a penicillin analog called cephalexin: This procedure suppresses septation (formation of new cell walls between the halves of a dividing cell). The cells then just grow longer without dividing—they become filamentous, like snakes. Then we attached inert markers (dead cells of normal size) to one or more of their flagella. We learned how to make glass micropipettes with narrow constrictions (Figure 11.11a). Then by suction, we pulled a "snake" about halfway into the pipette, as shown schematically in panel (b) of the figure. The pipette contained an ionophore, a chemical that made the inner segment of the cell permeable to ions, as indicated by the dashed lines. One electrode from the voltage clamp was placed in the external medium and the other was placed inside the pipette. At the beginning of the experiment, the largest resistance in the circuit between the elec-

---

[10] Actually, we used a voltage clamp; see Section 12.3.1 on page 532.

trodes was the membrane of the outer segment: The resistances of the fluid in the pipette and of the membrane of the inner segment were relatively small. Therefore, nearly all of the voltage drop was across the membrane of the outer segment, as desired. However, a substantial fraction of the current flowing between the electrodes leaked around the outside of the cell, so we could not measure the current flowing through the flagellar motors (or other membrane ion channels). The job of the voltage clamp circuitry was to supply whatever current was necessary to maintain a specified difference in potential.

When we turned up the control knob of the voltage clamp, the marker spun faster. When we turned it down, the marker spun more slowly. If we turned it up too far (beyond about 200 mV), the motor burned out. In between, the angular speed of the motor proved to be linearly proportional to the applied voltage, a satisfying result. When we reversed the sign of the voltage, the motor spun backward for a few revolutions and then stopped. When we changed the sign back again, the motor failed to start for several seconds, and then sped up in a stepwise manner, gaining speed in equally spaced increments. Evidently, the different force-generating elements of the motor—we think there are eight, as in a V-8 automobile engine—either were inactivated or came off of the motor when exposed to the reverse potential. They were reactivated or replaced, one after another, when the initial potential was restored! We did not expect to see this self-repair phenomenon.

The main difficulty with this experiment was that the ionophore used to permeabilize the inner segment soon found its way to the outer segment, destroying the preparation. Correction could be made for this, but only for a few minutes. We are still trying to find a better way to permeabilize the inner segment.

For more details See Blair & Berg, 1988 and Fung & Berg, 1995.
Howard Berg is Professor of Molecular and Cellular Biology and of Physics at Harvard University. Having studied chemistry, medicine, and physics, he began looking for a problem involving all these fields—and settled on the molecular biology of behavior. David Fung did his doctoral work on several aspects of the bacterial flagellar motor. He currently works on technology transfer at Memorial Sloan–Kettering Cancer Center in New York.

THE BIG PICTURE

Let's return to the Focus Question. This chapter gave a glimpse of how cells actively regulate their interior composition and, hence, their volume. We followed a trail of clues that led to the discovery of ion pumps in the cell membrane. In some ways, the story is reminiscent of the discovery of DNA (Chapter 3): A tour de force of indirect reasoning left little doubt that some kind of ion pump existed, years before the direct isolation of the pump enzyme.

We then turned to a second use for ion pumping, the transmission of free energy from the cell's respiration pathway to its ATP synthesis machinery. The following chapter will develop a third use: Ion pumps create a nonequilibrium state in which excess free energy is distributed over the cell's membrane. We will see how another class of molecular devices, the voltage-gated ion channels, can turn this "charged" membrane into an excitable medium, the resting state of a nerve axon.

KEY FORMULAS

- *Gibbs–Donnan:* If several ion species can all permeate a membrane, then to have equilibrium, their Nernst potentials must all agree with on another (and with the externally imposed potential drop, if any). For example, suppose that the ions are sodium, potassium, and chloride, and let $c_{1,i}$ and $c_{2,i}$ be the exterior and interior concentrations, respectively, of species i. Then (Equation 11.5)

$$\frac{c_{1,\mathrm{Na}^+}}{c_{2,\mathrm{Na}^+}} = \frac{c_{1,\mathrm{K}^+}}{c_{2,\mathrm{K}^+}} = \frac{c_{2,\mathrm{Cl}^-}}{c_{1,\mathrm{Cl}^-}}.$$

- *Pumps:* The effect of active ion pumping is to add an ATP-dependent current source to the membrane. Making the Ohmic hypothesis gives $j_{\mathrm{Na}^+} = \frac{g_{\mathrm{Na}^+}}{e}(\Delta V - V_{\mathrm{Na}^+}^{\mathrm{Nernst}}) + j_{\mathrm{Na}^+}^{\mathrm{pump}}$ (Equation 11.10). Here j_{Na^+} is the flux of sodium ions, g_{Na^+} is the membrane's conductance, $V_{\mathrm{Na}^+}^{\mathrm{Nernst}}$ is the Nernst potential, and ΔV is the actual potential difference across the membrane.

FURTHER READING

Semipopular:
History: Hodgkin, 1992.

Intermediate:
Section 11.2 follows in broad outline the approach of Benedek & Villars, 2000c. See also Katz's classic book: Katz, 1966.
Many biochemistry and cell biology texts describe the biochemical aspects of respiration, for example, Berg et al., 2002; Nelson & Cox, 2000; Voet & Voet, 2003; Karp, 2002.
Chemiosmotic mechanism: Atkins, 2001; Alberts et al., 1997.
Modeling of ion transport, cell volume control, and kidney function: Hoppensteadt & Peskin, 2002; Benedek & Villars, 2000c; Keener & Sneyd, 1998.

Technical:
General: Weiss, 1996.
Ion pumps: Läuger, 1991; Skou, 1989.
F0F1: Noji et al., 1997; Boyer, 1997; Oster & Wang, 2000.

11.1.2′ Track 2

1. To see why the charge density in the membrane is small, think of how permeation works:

   a. Some permeation occurs through channels; the volume of these channels is a small fraction of the total volume occupied by the membrane.

   b. Some permeation occurs by dissolving the ions in the membrane material. The corresponding partition coefficient (see Section 4.6.1 on page 135) is small because the ions have a large Born self-energy in the membrane interior, whose permittivity is low (see Section 7.4.1 on page 261).

2. We can get Equation 11.1 on page 474 more explicitly if we imagine membrane permeation literally as diffusion through a channel in the membrane. Applying the argument in Section 4.6.3 on page 139 to the channel gives

$$V_2' - V_1' = -\frac{k_B T}{ze} \ln \frac{c_2'}{c_1'}.$$

Here V' and c' refer to the potential and density at the mouth of the channel (at lines B or C in Figure 11.2). But we can write similar formulas for the potential drops across the charge layers themselves, for example, $V_2 - V_2' = -((k_B T/(ze)) \ln(c_2/c_2')$. Adding these three formulas again gives Equation 11.1.

   Actually, we needn't be so literal. The fact that the permeability of the membrane drops out of the Nernst relation means that any diffusive transport process will give the same result.

 11.2.2′ Track 2

Section 11.2.2 on page 478 mentioned that there will be nonlinear corrections to Ohmic behavior when $\Delta V - \mathcal{V}_i^{\text{Nernst}}$ is not small. Indeed, each of the many ion conductances has its own characteristic current-versus-potential relation, some of them highly nonlinear (or rectifying), others not. One simple model for a nonlinear current–voltage relation is the Goldman–Hodgkin–Katz formula. (See for example Appendix C of Berg, 1993.)

11.2.3′ Track 2

1. Adding up the columns of Table 11.1 on page 477 seems to show that even with ion pumping, there is a big osmotic imbalance across the cell membrane. We must remember, however, that even though the list of ions shown in the table is fairly complete for the extracellular fluid (essentially seawater), still the cytosol has many

other osmotically active solutes not listed in the table. The total of *all* interior solute species just balances the exterior salt, as long as active pumping keeps the interior sodium level small. If active pumping stops, the interior sodium level rises and an inward flow of water ensues.

2. The sodium–potassium pump can be artificially driven by external electric fields instead of by ATP. Even an oscillating field (which averages to zero) will induce a directed net flux of sodium in one direction and potassium in the other: The pump uses the nonequilibrium, externally imposed field to rectify the thermally activated barrier crossings of these ions, like the diffusing ratchet model of molecular motors (Section 10.4.4 on page 446). (See Astumian, 1997; Läuger, 1991.)

11.3.3′ Track 2

Section 11.3.3 mentioned that pyruvate and ADP enter the mitochondrial matrix, and ATP exits, via specialized transporters in the mitochondrial membrane. For details, see Berg et al., 2002.

PROBLEMS*

11.1   Heart failure

A muscle cell normally maintains a very low interior calcium concentration; Section 12.4.2 will discuss how a small increase in the interior $[Ca^{2+}]$ causes the cell to contract. To maintain this low concentration, muscle cells actively pump out Ca^{2+}. The pump used by cardiac (heart) muscle is an antiport (Section 11.3.5): It couples the extrusion of calcium ions to the entry into the cell of sodium.

The drug oubain suppresses the activity of the sodium–potassium pump. Why do you suppose this drug is widely used to treat heart failure?

11.2   Electrochemical equilibrium

Suppose we have a patch of cell membrane stuck on the end of a pipette (tube). The membrane is permeable to bicarbonate ions, HCO_3^-. On side A, we have a big reservoir with bicarbonate ions at a concentration of 1 M; on side B, there's a similar reservoir with a concentration of 0.1 M. Now we connect a power supply across the two sides of this membrane to create a fixed potential difference $\Delta V = V_A - V_B$.

a. What should ΔV be to maintain equilibrium (no net ion flow)?

b. Suppose $\Delta V = 100$ mV. Which way will bicarbonate ions flow?

11.3   Vacuole equilibrium

Here are data for the marine alga *Chætomorpha*. The extracellular fluid is seawater. The plasmalemma (outer cell membrane) separates the outside from the cytoplasm. A second membrane (the tonoplast) separates the cytoplasm from an interior organelle, the vacuole; see Section 2.1.1 on page 40. In this problem, pretend that there are no other small ions than those listed here:

ion	vacuole c_i, mM	cytoplasm c_i, mM	extracellular c_i, mM	$\mathcal{V}_i^{\text{Nernst}}$ (plasmalemma), mV	$\mathcal{V}_i^{\text{Nernst}}$ (tonoplast), mV
K^+	530	425	10	?	−5.5
Na^+	56	50	490	+57	?
Cl^-	620	30	573	−74	+76

a. The table gives some of the Nernst potentials across the two membranes. Fill in the missing ones.

b. The table does not list the charge density $\rho_{q,\text{macro}}$ arising from impermeant macroions in the cytoplasm. What is $-\rho_{q,\text{macro}}/e$ in mM?

c. The actual measured membrane potential difference across the tonoplast membrane is +76 mV. Which ion(s) must be actively pumped across the tonoplast membrane, and in which direction(s)?

d. Suppose that we selectively shut down the ion pumps in the tonoplast membrane but the cell metabolism continues to maintain the listed concentrations in the

---

*Problem 11.3 is adapted with permission from Benedek & Villars, 2000c.

cytoplasm. The system then relaxes to a Donnan equilibrium across the tonoplast membrane. What will be the approximate ion concentrations inside the vacuole, and what will be the final Donnan potential?

11.4 $\boxed{T_2}$ *Relaxation to Donnan equilibrium*

Explore what happens to the resting steady state (see Section 11.1.3 on page 474) after the ion pumps are suddenly turned off, as follows.

a. Table 11.1 on page 477 shows that sodium ions are far from equilibrium in the resting state. Find the conductance per area for these ions, using the value $5\,\Omega^{-1}\mathrm{m}^{-2}$ for the total membrane conductance per area and the ratios of individual conductances given in Equation 11.9 on page 482.

b. Using the Ohmic hypothesis, find the initial charge flux carried by sodium ions just after the pumps have been shut off. Reexpress your answer as charge per time per unit length along a giant axon, assuming its diameter to be 1 mm.

c. Find the total charge per unit length carried by all the sodium ions inside the axon. What would the corresponding quantity equal if the interior concentration of sodium matched the fixed exterior concentration?

d. Subtract the two values found in (c). Divide by the value you found in (b) to get an estimate for the time scale for the sodium to equilibrate after the pumps shut off.

e. Chapter 12 will describe a nerve impulse as an event that passes by one point on the axon in about a millisecond. Compare with the time scale you just found and comment.

CHAPTER 12

Nerve Impulses

In a series of five articles published in the *Journal of Physiology*, Alan Hodgkin, Andrew Huxley, and Bernard Katz described the results of experiments that determined how and when a cell membrane conducts ions. In the last of these papers, Hodgkin and Huxley presented experimental data on ion movement across electrically active cell membranes, a hypothesis for the mechanism of nerve impulse propagation, a fit of the model to their data, and a calculated prediction of the shape and speed of nerve impulses agreeing with experiment. Many biophysicists regard this work as one of the most beautiful and fruitful examples of what can happen when we apply the tools and ideas of physics to a biological problem.

Thinking about the problem in the light of this book's themes, living cells can do "useful work" not only in the sense of mechanical contraction but also in the sense of *computation.* Chapter 5 mentioned how single cells of *E. coli* make simple decisions that enable them to swim toward food. For more complex computations, multicellular organisms have had to evolve systems of specialist cells, the neurons. Like muscle cells, neurons are in the business of metabolizing food and, in turn, locally reducing disorder in an organism. Instead of generating organized mechanical motion, however, their job is *manipulating information* in ways useful to the organism. To give a glimpse of how they manage this task, this chapter will look at an elementary prerequisite for information processing, namely, information transmission.

One often hears a metaphorical description of the brain as a computer and individual nerve cells as the "wiring," but a little thought shows that this can't literally be true. Unlike, say, telephone wires, nerve cells are poorly insulated and bathed in a conductive medium. In an ordinary wire under such conditions, a signal would suffer serious degradation as a consequence of resistive losses—a form of dissipation. In contrast, even your longest nerve cells faithfully transmit signals without loss of amplitude or shape. We know the broad outlines of the resolution to this paradox from

Chapter 1: Living organisms constantly flush energy through themselves to combat dissipation. We'd like to see how nerve cells implement this program.

This chapter contains somewhat more historical detail than most of the others in this book. The aim is to show how careful biophysical measurements, aimed at answering the questions in the previous paragraph, disclosed the existence of yet another remarkable class of molecular devices, the voltage-gated channels, years before the specific proteins constituting those devices were identified.

The Focus Question for this chapter is

Biological question: How can a leaky cable carry a sharp signal over long distances?

Physical idea: Nonlinearity in a cell membrane's conductance turns the membrane into an excitable medium, which can transmit waves by continuously regenerating them.

12.1 THE PROBLEM OF NERVE IMPULSES

Roadmap Section 11.1 identified active ion pumps as the origin of the resting potential across the membranes of living cells. Section 12.1 attempts to use these ideas to understand nerve impulses, arriving at the linear cable equation (Equation 12.9). This equation does *not* have solutions resembling traveling impulses: Some important physical ingredient, not visible in the resting properties of cells, is missing. Section 12.2 argues that voltage gating is the missing ingredient, then shows how a modification to the linear cable equation (Equation 12.22) does capture some of the key phenomena we seek. Section 12.3 qualitatively sketches Hodgkin and Huxley's full analysis and the subsequent discovery of the molecular devices it predicted: voltage-gated ion channels. Finally, Section 12.4 sketches briefly how the ideas used so far to describe transmission of information have begun to yield an understanding of computation in the nervous system and of its interface to the outside world.

Some neurons surround their axon by a layer of electrical insulation called the myelin sheath. This chapter will study only neurons lacking this structure (those having unmyelinated axons). With appropriate changes, however, the analysis given here can be adapted to myelinated axons as well.

12.1.1 Phenomenology of the action potential

Section 2.1.2 on page 43 discussed anatomy: the shape and connectivity of neurons. The nerve cell's *function* can be summarized as three processes:

- Stimulation of the cell's inputs (typically the dendrite) from the preceding cells' outputs (typically axon terminals);
- Computation of the appropriate output signal; and
- Transmission of the output signal (nerve impulse) along the axon.

Sections 12.2 and 12.3 will discuss the last of these processes in some detail; Section 12.4 will discuss the other two briefly. (A fourth activity, the *adjustment* of synaptic properties, will also be mentioned in Section 12.4.3.)

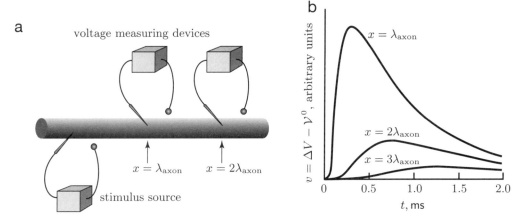

Figure 12.1: (Schematic; sketch graph.) (a) Schematic of an electrophysiology experiment. Stimuli at one point on an axon (shown as a cylinder) evoke a response, which is measured at distant points. (b) Responses to a short, weak, depolarizing pulse of current. The vertical axis represents the potential relative to its resting value. The response to a hyperpolarizing pulse looks similar, but the traces are inverted (*not shown*). The pulses observed at more distant points are weaker and more spread out than those observed up close, and they arrive later. The distance unit λ_{axon} is defined in Equation 12.8 on page 517.

Figure 12.1a shows a schematic of an experiment to examine the passage of nerve impulses. Measuring devices situated at various fixed positions along an axon all measure the time course of the membrane potential ΔV after the axon is stimulated. The axon could be attached to a living cell, or isolated. The external stimulus could be artificially applied, as shown, or could come from synapses to other neurons. Figure 12.1b sketches the results of an experiment in which a stimulating electrode suddenly injects positive charges into the interior of the axon (or removes negative charges). The effect of either change is to push the membrane potential at one point to a value less negative than the resting potential (that is, closer to zero); we say that the stimulus **depolarizes** the membrane. Then the external current source shuts off, allowing the membrane to return to its resting potential.

The sketch graphs in Figure 12.1b show that, for a weak depolarizing stimulus, a potential change at one point spreads to nearby regions; the response is weaker at more distant points. Moreover, the spread is not instantaneous. Another key point is that the peak height is proportional to the stimulus strength: We say that the response is **graded** (see Figure 12.2a). The response to a stimulus of the opposite sign—tending to drive V more negative or **hyperpolarize** the membrane—is qualitatively the same as that for weak depolarization. We just get Figure 12.1b turned upside down.

The behavior shown in Figure 12.1b is called **electrotonus**, or passive spread. Passive spread is not a "nerve impulse"; it dies out almost completely in a few millimeters. Something much more interesting happens, however, when we try larger depolarizing stimuli. Figure 12.2a shows the results of such an experiment. These graphs depict the response at a *single* location close to the stimulus, for stimuli of various strengths.

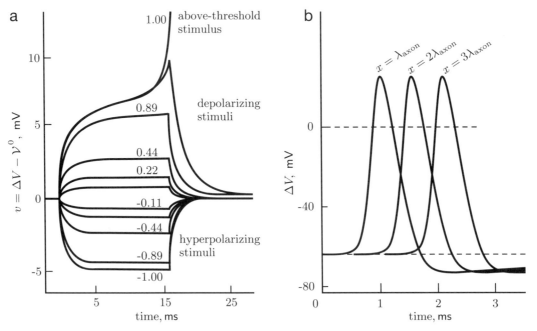

Figure 12.2: (Experimental data; sketch graph.) The action potential. (a) Response of a crab axon to long (15 ms) pulses of injected current. The vertical axis shows the membrane potential at a point close to the stimulus, measured relative to its resting value. The *lower traces* record the response to hyperpolarizing stimuli; the *upper traces* correspond to depolarizing stimuli. The threshold value of the stimulus has arbitrarily been designated as strength 1.0; the curves are labeled with their strength relative to this value. The *top trace* is just over threshold and shows the start of an action potential. (b) Sketch of the response to an above-threshold, depolarizing stimulus at three distances from the stimulation point (compare with Figure 12.1). The time courses (shapes) of the pulses assume a stereotyped form; each is shifted in time from its predecessor, reflecting a constant propagation speed. Note how the potential drops below its resting value (*lower dashed line*) after the pulse, then rises slowly. This is the phenomenon of afterhyperpolarization; see also Figure 12.6b. [(a) Data from Hodgkin & Rushton, 1946.]

The lower nine traces of Figure 12.2a correspond to small hyperpolarizing or depolarizing stimuli. They again show a graded response. The axon's response changes dramatically, however, when a depolarizing stimulus exceeds a threshold of about 10 mV. As shown in the top two traces in Figure 12.2a, such stimuli can trigger a massive response, called the **action potential**, in which the membrane potential shoots up. Figure 12.2b, drawn with a coarser vertical scale, shows schematically how the potential hits a peak (typically changing by 100 mV), then rapidly falls.

The action potential is the behavior we have been calling a nerve impulse. Experiments like the one sketched in Figure 12.1a show several remarkable features of the axon's electrical response (or **electrophysiology**):

- Instead of being graded, the action potential is an all-or-nothing response. That is, the action potential arises only when the membrane depolarization crosses a threshold; subthreshold stimuli give electrotonus, with no response far from the stimulating point. In contrast, above-threshold stimuli create a traveling wave of excitation, whose peak potential is independent of the strength of the initial stimulus.

- The action potential moves down the axon at a constant speed (see Figure 12.2b), which can be anywhere from 0.1 to $120\,\mathrm{m\,s^{-1}}$. This speed has nothing to do with the speed at which a signal moves down a copper wire (about a meter every three *nano*seconds, a billion times faster).

- When the progress of an action potential is measured at several distant points, as in Figure 12.2b, the peak potential is found to be independent of distance, in contrast to the decaying behavior for hyperpolarizing or subthreshold stimuli. A single stimulus suffices to send an action potential all the way to the end of even the longest axon.

- Indeed, the entire time course of the action potential is the same at all distant points (Figure 12.2b). That is, the action potential preserves its shape as it travels, and that shape is "stereotyped" (independent of the stimulus).[1]

- After the passage of an action potential, the membrane potential actually overshoots slightly, becoming a few millivolts more negative than the resting potential, and then slowly recovers. This behavior is called **afterhyperpolarization**.

- For a certain **refractory period** after transmitting an action potential, the neuron is harder to stimulate than it is at rest.

Our job in Sections 12.2 and 12.3 will be to explain all these remarkable qualitative features of the action potential from a simple physical model.

12.1.2   The cell membrane can be viewed as an electrical network

Iconography Section 11.2.2 on page 478 described the electrical properties of a small patch of membrane by using circuit diagram symbols from first-year physics (see Figure 12.3a). Before elaborating on this figure, we should pause to recall the meanings of the graphical elements of a schematic circuit diagram like this one and why they are applicable to our problem.

The figure shown consists of "wires," a resistor symbol, and a battery symbol. Schematic circuit diagrams like this one convey various implicit claims:

1. No significant net charge can pile up inside the individual circuit elements: The charge into one end of a symbol must always equal the charge flowing out the other end. Similarly,

2. A junction of three wires implies that the total current into the junction is zero.

3. The electrostatic potential is the same at either end of a wire and among any set of joined wires.

4. The potential changes by a fixed amount across a battery symbol.

---

[1] $\boxed{T_2}$ This statement requires a slight qualification. Close to the stimulating point, stronger stimuli indeed lead to a faster initial depolarization, because the membrane gets to threshold faster. These differences die out as the action potential travels down the axon, just as the *entire* response to a hyperpolarizing response dies out.

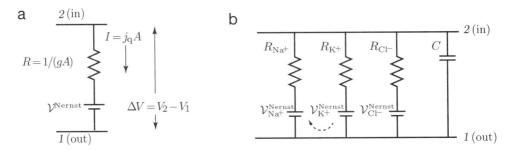

Figure 12.3: (Circuit diagrams.) Discrete-element models of a small patch of cell membrane of area A. (a) Duplicate of Figure 11.4, for reference. (b) A more realistic model. The orientations of the *battery symbols* (—⊦⊢—) reflect the sign convention in the text: A positive value of $\mathcal{V}^{\text{Nernst}}$ means that the upper wire entering the corresponding battery is at higher potential than the lower wire. Three representative ion species can flow between the interior and exterior of the cell, corresponding to $i = \text{Na}^+$, K^+, and Cl^-. Each species has its own resistance $R_i = 1/(g_i A)$ (*symbol* —⋀⋀—) and entropic driving force $\mathcal{V}_i^{\text{Nernst}}$. The capacitance $C = \mathcal{C}A$ (*symbol* —⊦⊢—) will be discussed later in this section. The *dashed arrow* depicts the circulating current flow expected from the data in Table 11.1. The effect of the sodium–potassium pumps described in Chapter 11 is not shown; see text.

5. The potential changes by the variable amount IR across a resistor symbol.

We'll refer to the first two of these statements as Kirchoff's first law. We prohibit charge buildup in ordinary circuits because of the prohibitive potential energy cost usually associated with it. In the cellular context, too, the separation of charge across micrometer-sized regions is energetically very costly (see the electrostatic self-energy Example on page 261 and Problem 12.2).

The rest of this section will adapt and extend Figure 12.3a to get a more realistic description of the resting cell membrane (Figure 12.3b).

Conductances as pipes The only "wire" in Figure 12.3a is the one joining the resistor to the battery. Thus items (3–5) in the preceding list amount to the statement that the total potential jump ΔV is the sum of two contributions, $\Delta V = IR + \mathcal{V}^{\text{Nernst}}$. This statement is just the Ohmic hypothesis (Equation 11.8 on page 480).

Thus, the electric circuit analogy appears to be useful for describing membranes. But Figure 12.3a describes the behavior of just *one* species of ion, just as in first-year physics you studied circuits with only one kind of charge carrier, the electron. Our situation is slightly different: We have at least *three* important kinds of charge carriers (Na^+, K^+, and Cl^-), each of whose numbers is separately fixed. Moreover, the conductance of a membrane will be different for different species (see for example Equation 11.9 on page 482). It might seem as though we would need to write circuit diagrams with three different kinds of wires, like the separate hot and cold plumbing in your house!

Fortunately, we don't need to go to this extreme. First note that there is only one kind of electrostatic potential V. Any charged particle feels the same force per charge, $-dV/dx$. Second, not only do all kinds of charged particles feel a common potential, they also all *contribute* to that potential in the same way. Thus the total electrostatic energy cost of a charge arrangement reflects only the spatial separation of net charge, without distinguishing between the types of charge. For example, pulling some sodium into a cell while at the same time pushing an equal number of potassium ions out (or an equal number of chloride ions in) creates no net separation of charge and carries no electrostatic energy cost.

Thus, when writing circuit diagrams, we can combine the various types of wires when dealing with elements, like the cell's cytoplasm, that do not discriminate between ion types.[2] We can think of these wires as representing one kind of pipe in which a *mixture* of different "fluids" (representing the various ion species) flows at a common "pressure" (the potential V). Kirchoff's first law then corresponds to the constraint that the total "volume of fluid" flowing in (total current) must equal that flowing out. Our wires must branch into different types when we describe the membrane, which has different resistances to different ion species in the mixture. In addition, each fluid will have a different entropic force driving it (the various Nernst potentials). We accommodate these facts by drawing the membrane as a compound object, with one resistor–battery pair in parallel for each ion species (Figure 12.3b). Notice that this figure does *not* imply that all three Nernst potentials are equal. Instead, the horizontal wires imply (by point (3)) that all three legs have the same value of

$$\Delta V = I_i R_i + \mathcal{V}_i^{\text{Nernst}}. \tag{12.1}$$

Here, as always, $\Delta V = V_2 - V_1$ is the interior potential relative to the outside; $R_i = 1/(g_i A)$ is the resistance; and $I_i = j_{\text{q},i} A$ are the currents through a patch of membrane of area A, considered positive if the ion current flows from inside to outside.

Quasi-steady approximation Figure 12.3b includes the effect of diffusive ion transport through the cell membrane, driven by entropic and electrostatic forces. However, the figure omits two important features of membrane physiology. One of these features (gated ion conductances) is not needed yet; it will be added in later sections. The other omitted feature is active ion pumping. This omission is a simplification that will be used throughout the rest of this chapter, so let's pause to justify it.

The situation sketched in Figure 12.3b cannot be a true steady state (see Sections 11.2.2 and 11.2.3). The dissipative flow of ions, shown by the dashed arrow in the figure, will eventually change the sodium and potassium concentrations until all three species come to Donnan equilibrium, obeying Equation 12.1 with all currents equal to zero.[3] To find a true steady state, we had to posit an additional element, the sodium–potassium pump. Setting the diffusive fluxes of both sodium and potassium equal to the pumped fluxes gave the steady state (Equation 11.12 on page 486).

But imagine that we begin in the steady state, then suddenly shut down the pumps. The ion concentrations will begin to drift toward their Donnan equilibrium values, but rather slowly (see Problem 11.4). In fact, the immediate effect on the membrane potential turns out to be rather small. We will denote the potential difference across the membrane shortly after shutting down the pumps (that is, the

---

[2] $\boxed{T_2}$ We are neglecting possible differences in bulk resistivity among the ion species.

[3] In this case, Equation 12.1 reduces to the Gibbs–Donnan relations, Equation 11.4.

quasi-steady value) by the symbol $\mathcal{V}^0$. To find it, note that, whereas the charge fluxes $j_{q,i}$ for each ion species need not separately be zero (as they must be in the true steady state), still they must add up to zero, to avoid net charge pileup inside the cell. The Ohmic hypothesis (Equation 12.1) then gives

$$\sum_i (\mathcal{V}^0 - \mathcal{V}_i^{\text{Nernst}}) g_i = 0. \tag{12.2}$$

Example: Find the value $\mathcal{V}^0$ of ΔV shortly after shutting off the pumps, assuming the initial ion concentrations in Table 11.1 on page 477 and the relative conductances per area given in Equation 11.9 on page 482. Compare with the estimated steady-state potential found in Section 11.2.3.

Solution: Collecting terms in Equation 12.2 and dividing by $g_{\text{tot}} \equiv \sum_i g_i$ gives the **chord conductance formula:**

$$\mathcal{V}^0 = \sum_i \frac{g_i}{g_{\text{tot}}} \mathcal{V}_i^{\text{Nernst}}. \tag{12.3}$$

Evaluating yields $\mathcal{V}^0 = -66\,\text{mV}$, only a few millivolts different from the true steady-state potential $-72\,\text{mV}$ found from Equation 11.12 on page 486.

In fact, the ion pumps *can* be selectively turned off, using drugs like oubain. The immediate effect of oubain treatment on the resting potential is indeed small (less than 5 mV), just as we found in the Example.

In summary, Equation 12.3 is an approximation to the resting potential difference (Equation 11.12).[4] Instead of describing a true steady state, Equation 12.3 describes the quasi-steady (slowly varying) state obtained immediately after shutting off the cell's ion pumps. We found that both approaches give roughly the same membrane potential. More generally, Equation 12.3 reproduces a key feature of the full steady-state formula: The ion species with the greatest conductance per area pulls ΔV close to its Nernst potential (compare with Idea 11.13 on page 486). Moreover, a nerve cell can transmit hundreds of action potentials after its ion pumps have been shut down. Both of these observations suggest that, for the purposes of studying the action potential, it's reasonable to simplify our membrane model by ignoring the pumps altogether and exploring fast disturbances to the slowly varying quasi-steady state.

Capacitors Figure 12.3b contains a circuit element not mentioned yet: a capacitor. This symbol acknowledges that some charge can flow toward a membrane without actually crossing it. To understand this effect physically, go back to Figure 11.2a on page 472. This time, imagine that the membrane is impermeable to both species but that external electrodes set up a potential difference ΔV. The figure shows how a net

---

[4] $\boxed{T_2}$ Actually, both these equations are rather rough approximations, because each relies on the Ohmic hypothesis, Equation 11.8.

charge density, $(c_+ - c_-)e$, piles up on one side of a membrane (and a correspond-ing deficit on the other side) whenever the potential difference ΔV is nonzero.[5] As ΔV increases, this pileup amounts to a net flow of charge into the cell's cytosol and another flow out of the exterior fluid, even though no charges actually cross the membrane. The constant of proportionality between the total charge q separated in this way and ΔV is called the **capacitance**, C:

$$q = C(\Delta V). \tag{12.4}$$

Gilbert says: Wait a minute. Doesn't charge neutrality (the first two points listed on page 509) say that charge can't pile up anywhere?

Sullivan replies: Yes, but look again at Figure 11.2a: The region just outside the membrane has acquired a net charge, but the region just *inside* has been *depleted* of charge, by an exactly equal amount. So the net charge between the dashed lines hasn't changed, as required by charge neutrality. As far as the inside and outside worlds are concerned, it looks as though current passed through the membrane!

Gilbert: I'm still not satisfied. The region between the dashed lines of the figure may be neutral overall, but the region from the dashed line on the left to the center of the membrane is not, nor is the region from the center to the dashed line on the right.

Sullivan: That's true. Indeed, Section 11.1.2 showed that it is this charge separation that creates any potential difference across a membrane.

Gilbert: So is charge neutrality wrong or right?

Gilbert needs to remember a key point in our discussion of Kirchoff's law. A charge imbalance over a micrometer-sized region will have an enormous electrostatic energy cost and is essentially forbidden. But the electrostatic self-energy Example on page 261 showed that over a nanometer-sized region, like the thickness of a cell membrane, such costs can be modest. We just need to acknowledge the energy cost of such an imbalance, which we do by using the notion of capacitance. Our assertion that the currents into the entire axon must balance, but that those in the immediate neighborhood of the membrane need not, really amounts to the quan-titative observation that the capacitance of the axon itself (an intermediate-scale object) is negligible relative to the much bigger capacitance of the cell membrane (a nanometer-scale object).

Unlike a resistor, whose potential drop is proportional to the *rate* of charge flow (current), Equation 12.4 says that ΔV across the membrane is proportional to the *total amount* of charge q that has flowed (the integral of current). Taking the time derivative of this equation gives a more useful form for our purposes:

$$\boxed{\frac{d(\Delta V)}{dt} = \frac{I}{C}. \qquad \textbf{capacitive current}} \tag{12.5}$$

---

[5] Actually, a larger contribution to a membrane's capacitance is the polarization of the interior insulator (the hydrocarbon tails of the constituent lipid molecules); see Problem 12.3.

So far, this section has considered steady- or quasi-steady situations, where the membrane potential is either constant, or nearly constant, in time. Equation 12.5 shows why we were allowed to neglect capacitive effects in such situations: The left-hand side equals zero. The following sections, however, will discuss transient phenomena such as the action potential; here, capacitive effects will play a crucial role.

Two identical capacitors in parallel will have the same ΔV as one when connected across a given battery because the electrostatic potential is the same among any set of joined lines (see point (3) in the list on page 509). Thus they will store twice as much charge as one capacitor (adding two copies of Equation 12.4). That is, they act as a single capacitor with *twice* the capacitance of either one. Applying this observation to a membrane, we see that a small patch of membrane will have capacitance proportional to its area. Thus $C = A\mathcal{C}$, where A is the area of the membrane patch and $\mathcal{C}$ is a constant characteristic of the membrane material. We will regard the capacitance per area $\mathcal{C}$ as a measured phenomenological parameter. A typical value for cell membranes is $\mathcal{C} \approx 10^{-2}\,\mathrm{F\,m^{-2}}$, more easily remembered as $1\,\mu\mathrm{F\,cm^{-2}}$.

In summary, we now have a simplified model for the electrical behavior of an individual small patch of membrane, pictorially represented by Figure 12.3b. Our model rests on the Ohmic hypothesis. The phrase *small patch* reminds us that we have been implicitly assuming that ΔV is uniform across our membrane, as implied by the horizontal wires in our idealized circuit diagram, Figure 12.3b. Our model involves several phenomenological parameters describing the membrane (g_i and $\mathcal{C}$) as well as the Nernst potentials ($\mathcal{V}_i^{\mathrm{Nernst}}$) describing the interior and exterior ion concentrations.

12.1.3 Membranes with Ohmic conductance lead to a linear cable equation with no traveling wave solutions

Although the membrane model developed in the previous section rests on some solid pillars (like the Nernst relation), nevertheless it contains other assumptions that are mere working hypotheses (like the Ohmic hypothesis, Equation 11.8). In addition, the analysis was restricted to either a small patch of membrane or a larger membrane maintained at a potential that was uniform along its length. This section will focus on lifting the last of these restrictions, to let us explore the behavior of an Ohmic membrane with a *non*uniform potential. We'll find that in such a membrane, external stimuli spread passively, giving behavior like that sketched in Figure 12.1b. Later sections will show that to understand nerve impulses (Figure 12.2b), we'll need to reexamine the Ohmic hypothesis.

When the potential is not uniform along the length of the axon, then current will flow axially (in the x direction, parallel to the axon). So far, we have neglected this possibility, considering only radial flow (in the r direction, through the membrane). In the language of Figure 12.3, axial flow corresponds to a current I_x flowing through the ends of the top and bottom horizontal wires. We will adopt the convention that I_x is called positive when positive ions flow in the $+\hat{\mathbf{x}}$ direction. If I_x is not zero, then the net radial current flow need not be zero, as assumed when deriving the chord conductance formula, Equation 12.3. Accordingly, we first need to generalize that result.

Your Turn 12A

Show that the three resistor–battery pairs in Figure 12.3b can equivalently be replaced by a *single* such pair, with effective conductance $g_{tot}A$ and battery potential $\mathcal{V}^0$ given by Equation 12.3.

We can now represent the axon as a *chain* of identical modules of the form you just found, each representing a cylindrical slice of the membrane (Figure 12.4). Current can flow axially through the interior fluid (representing the axon's cytoplasm, or **axoplasm**, represented by the upper horizontal line) or through the surround-

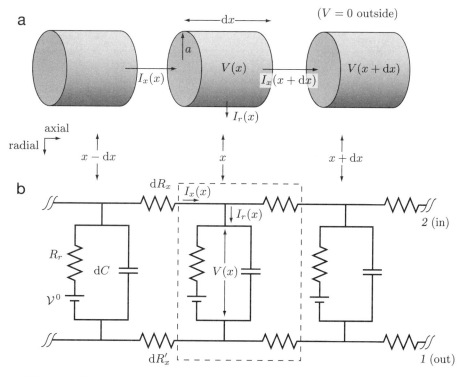

Figure 12.4: (Schematic; circuit diagram.) Distributed-element model of an axon. The axon is viewed as a chain of identical modules, labeled by their position x along the axon. (a) Modules viewed as cylindrical segments of length dx and radius a. Each one's surface area is thus $dA = 2\pi a\,dx$. (b) Modules viewed as electrical networks, each containing a battery of voltage $\mathcal{V}^0$ (recall that this quasi-steady state potential is negative). The "radial" resistor, with resistance $R_r = 1/(g_{tot}dA)$, represents passive ion permeation through the axon membrane; the associated capacitor has $dC = \mathcal{C}dA$. The "axial" resistors dR_x and dR'_x represent the fluid inside and outside the axon, respectively. We will make the approximation that $dR'_x = 0$, so the entire lower horizontal wire is at a common potential, which we define to be zero. The "radial" current, $I_r(x) \equiv j_{q,r}(x) \times dA$, reflects the net charge of all the ions leaving the axoplasm (that is, downward in (b)) at x; the axial current I_x represents the total current flowing to the right inside the axoplasm (that is, in the upper horizontal wire of (b)). $V(x)$ represents the potential inside the axon (and hence also the potential difference across the membrane, because we took the potential to be zero outside).

ing extracellular fluid (represented by the lower horizontal line). The limit $dx \to 0$ amounts to describing the membrane as a chain of infinitesimal elements, a **distributed network** of resistors, capacitors, and batteries.

To explore the behavior of such a network under the sort of stimuli sketched in Figure 12.1, we now take four steps:

a. Find numerical values for all the circuit elements in Figure 12.4, then

b. Translate the figure into an equation;

c. Solve the equation; and

d. Interpret the solution.

a. Values To find the interior axial resistance dR_x, recall that the resistance of a cylinder of fluid to axial current flow is proportional to the cylinder's length divided by its cross-sectional area, or $dR_x = dx/(\kappa\pi a^2)$, where κ is the fluid's electrical conductivity (see Section 4.6.4 on page 142). The conductivity of axoplasm can be measured in the lab. For squid axon, its numerical value is $\kappa \approx 3\,\Omega^{-1}\mathrm{m}^{-1}$, roughly what we would expect for the corresponding salt solution (see Problem 12.5).

To simplify the math, we will set the electrical resistance of the exterior fluid equal to zero: $dR_x' = 0$. This approximation is reasonable because the cross-sectional area available for carrying current outside the cylindrical axon is much larger than the area πa^2 of the interior. Thus we have the very convenient feature that the entire exterior of the axon is "short-circuited" and therefore is at a uniform potential, which we take to be zero: $V_1(x) \equiv 0$. The membrane potential difference is then $\Delta V(x) = V_2(x)$; to simplify the notation, we will abbreviate this quantity as $V(x)$.

The resistance R_r of the membrane surrounding the axon slice is just the reciprocal of its total conductance; according to Your Turn 12A, it equals $(g_{\mathrm{tot}} \times 2\pi a dx)^{-1}$, where g_{tot} is the sum of the g_i's. As mentioned in Section 11.2.2, a typical value for g_{tot} in squid axon is $\approx 5\,\mathrm{m}^{-2}\,\Omega^{-1}$.

Finally, Section 12.1.2 says that the membrane capacitance is $dC = (2\pi a dx) \times C$ and quoted a typical value of $C \approx 10^{-2}\,\mathrm{F}\,\mathrm{m}^{-2}$.

b. Equation To get the equation for the spread of an external stimulus, we write down the condition of charge neutrality for one cylindrical slice of the axon (Figure 12.4a). This condition says that the net current into the ends of the slice, $I_x(x) - I_x(x + dx)$, must balance the total rate at which charge flows radially out of the axoplasm. The radial current equals the sum of the charge permeating *through* the membrane, or $2\pi a dx \times j_{q,r}$, plus the rate at which charge *piles up* at the membrane, $(2\pi a dx) \times C\frac{dV}{dt}$ (see Equation 12.5). Thus

$$I_x(x) - I_x(x + dx) = -\frac{dI_x}{dx} \times dx = 2\pi a\left(j_{q,r}(x) + C\frac{dV}{dt}\right)dx. \qquad (12.6)$$

This equation is a good start, but we can't solve it yet: It's one differential equation in three unknown functions, namely, $V(x, t)$, $I_x(x, t)$, and $j_{q,r}(x, t)$. First let's eliminate I_x.

The axial current at a point x of our axon just equals the potential drop along a short distance, divided by the axial resistance dR_x:

$$I_x(x) = -\frac{V(x + \frac{1}{2}dx) - V(x - \frac{1}{2}dx)}{dx/(\pi a^2 \kappa)} = -\pi a^2 \kappa \frac{dV}{dx}.$$

To understand the minus sign, note that if V increases as we move to the right, then positive ions will be driven to the left. Substituting this result into Equation 12.6 yields our key formula:

$$\pi a^2 \kappa \frac{d^2 V}{dx^2} = 2\pi a \left(j_{q,r} + C\frac{dV}{dt} \right). \qquad \textbf{cable equation} \qquad (12.7)$$

(The cable equation also describes the propagation of signals along a wire, or "cable," partially short-circuited by a surrounding bath of salt water.)

Next, we write the membrane current in terms of the potential, using Your Turn 12A: $j_{q,r} = (V - V^0)g_{tot}$. We can also tidy up the cable equation some more by letting v be the difference between the interior potential and its quasi-steady value:

$$v(x, t) \equiv V(x, t) - V^0.$$

Also define the axon's **space constant** and **time constant** as

$$\lambda_{axon} \equiv \sqrt{a\kappa/2g_{tot}}; \qquad \tau \equiv C/g_{tot}. \qquad (12.8)$$

(Check that these expressions have the units of length and of time, respectively.) These abbreviations yield

$$(\lambda_{axon})^2 \frac{d^2 v}{dx^2} - \tau\frac{dv}{dt} = v. \qquad \textbf{linear cable equation} \qquad (12.9)$$

Equation 12.9 is a special form of the cable equation, embodying the extra assumption of the Ohmic hypothesis (see Your Turn 12A). As desired, it's one equation in one unknown, namely, $v(x)$. It has the pleasant feature of being a *linear* differential equation (every term is linear in v). And there's something very, very familiar about it: It's almost, but not quite, the diffusion equation (Equation 4.20 on page 131)!

c. Solution In fact, we can make the link to the diffusion equation complete by one last change of variables. Letting $w(x, t) \equiv e^{t/\tau}v(x, t)$, the linear cable equation becomes

$$\frac{(\lambda_{axon})^2}{\tau} \frac{d^2 w}{dx^2} = \frac{dw}{dt}.$$

We already know some solutions to this equation. Adapting the result of Section 4.6.5, we find that the response of our cable to a localized impulse is

$$v(x, t) = \text{const} \times e^{-t/\tau} t^{-1/2} e^{-x^2/(4t(\lambda_{\text{axon}})^2/\tau)}. \qquad \text{(passive-spread solution)} \quad (12.10)$$

In fact, *the linear cable equation has no traveling wave solutions* because the diffusion equation has no such solutions.

Some numerical values are revealing: Taking our illustrative values of $a = 0.5\,\text{mm}$, $g_{\text{tot}} \approx 5\,\text{m}^{-2}\,\Omega^{-1}$, $\mathcal{C} \approx 10^{-2}\,\text{F}\,\text{m}^{-2}$, and $\kappa \approx 3\,\Omega^{-1}\,\text{m}^{-1}$ (see step (a)) yields

$$\lambda_{\text{axon}} \approx 12\,\text{mm}\,, \qquad \tau \approx 2\,\text{ms}. \qquad (12.11)$$

d. Interpretation Our model axon is terrible at transmitting pulses! Besides the fact that it has no traveling wave solutions, we see that there is no threshold behavior, and stimuli die out after a distance of about 12 mm. Certainly a giraffe would have trouble moving its feet with neurons like this. Actually, though, these conclusions are not a complete disaster. Our model *has* yielded a reasonable account of electrotonus (passive spread, Section 12.1.1). Equation 12.10 does reproduce the behavior sketched in Figure 12.1; moreover, like the solution to any linear equation, ours gives a graded response to the stimulus. What our model lacks so far is any hint of the more spectacular action-potential response (Figure 12.2b).

12.2 SIMPLIFIED MECHANISM OF THE ACTION POTENTIAL

12.2.1 The puzzle

Following the Roadmap at the start of Section 12.1, this section will motivate and introduce the physics of voltage gating, in a simplified form, then show how it provides a way out of the impasse we just reached. The introduction to this chapter mentioned a key question whose answer will lead us to the mechanism we seek: The cellular world is highly dissipative, in the sense of electrical resistance (Equation 11.8) just as in the sense of mechanical friction (Chapter 5). How, then, can signals travel without diminution?

We found the beginning of an answer to this puzzle in Section 11.1. The ion concentrations inside a living cell are far from equilibrium (Section 11.2.1). When a system is not in equilibrium, its free energy is not at a minimum. When a system's free energy is not at a minimum, the system is in a position to do useful work. "Useful work" can refer to the activity of a molecular machine, but more generally, it can include the manipulation of information, as in nerve impulses. Either way, the resting cell membrane is poised to *do* something, like a beaker containing nonequilibrium concentrations of ATP and ADP.

In short, we'd like to see how a system with a continuous distribution of excess free energy can support traveling waves despite dissipation. The linear cable equation did not give this behavior, but in retrospect, it's not hard to see why: The value of $\mathcal{V}^0$ dropped out of the equation altogether, once we defined v as $V - \mathcal{V}^0$! This behavior is typical of any linear differential equation (it's called the superposition property of

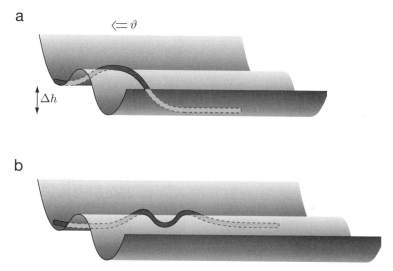

Figure 12.5: (Schematic.) Mechanical analog of the action potential. A heavy chain lies in a tilted channel, with two troughs at heights differing by Δh. (a) An isolated kink will move steadily to the left at a constant speed ϑ: successive chain elements are lifted from the upper trough, slide over the crest, and fall into the lower trough. (b) A disturbance can create a pair of kinks if it is above threshold. The two kinks then travel away from each other.

a linear equation).[6] Apparently, what we need to couple the resting potential to the traveling disturbance is some nonlinearity in the cable equation.

12.2.2 A mechanical analogy

We can imagine that a cell could somehow use the free energy stored along its membrane to *regenerate* the traveling action potential continuously as it passes, exactly compensating for dissipative losses so that the wave maintains its amplitude instead of dying out. These are easy words to say, but it may not be so easy to visualize how such a seemingly miraculous process could actually work, automatically and reliably. Before proceeding to the mathematics, we need an intuitive analogy to the mechanism we seek.

Figure 12.5 shows a molding such as you might find in a hardware store. The cross section of the molding is shaped like a rounded letter *w*. We hold the molding with its long axis parallel to the floor but with its cross section tilted, so that one of the two grooves is higher than the other. Call the height difference between the bottoms of the two troughs Δh.

Suppose that we lay a long, flexible chain in the higher groove and immerse everything in a viscous fluid. We pull on the ends of the chain, putting it under a

---

[6]Actually, a linear equation *can* have traveling wave solutions; the equations describing the propagation of light in vacuum are linear. What we cannot have is a traveling wave in a linear, *dissipative* medium. For example, light rays traveling through a smoke-filled room will get fainter and die out.

slight tension. In principle, the chain could lower its gravitational potential energy by hopping to the lower groove. The difference in height between the two grooves amounts to a certain stored potential energy density per length of chain. To release this energy, however, the chain would first have to move upward, which *costs* energy. What's more, the chain can't hop over the barrier all at once; it must first form a kink. The applied tension discourages the formation of a kink. Hence the chain remains stably in the upper groove. Even if we jiggle the apparatus gently, so that the chain wiggles a bit, it still stays up.

Next suppose that we begin laying the chain in the upper groove, starting from the far left end, but halfway along, we bring it over the hump and continue thereafter laying it in the lower groove (Figure 12.5a). We hold everything in place, then let go at time zero. We will then see the crossover region moving uniformly to the left at some velocity ϑ. Each second, a fixed length of chain $\vartheta \times (1\,\mathrm{s})$ rises over the hump, pulled upward by the weight of the falling segment to its right. That is, the system displays traveling wave behavior.

Each second the chain releases a fixed amount of its stored gravitational potential energy. The energy thus released gets spent overcoming frictional loss (dissipation).

> **Your Turn 12B**
>
> a. Suppose that the chain's linear mass density is $\rho_{\mathrm{m,chain}}^{(1d)}$. Find the rate at which gravitational potential energy gets released.
>
> b. The speed at which the chain moves is proportional to ϑ; hence, so is the retarding frictional force. Let the total retarding force be $\gamma\vartheta$, where γ is some constant. Find the rate at which mechanical work gets converted to thermal form.
>
> c. What sets the speed ϑ of the traveling wave?

Finally, let's begin again with the chain entirely in the upper channel. This time we grasp it in the middle, pull it over the hump, and let go (Figure 12.5b). If we pull too little over the hump, as shown in the figure, then both gravity and the applied tension act to pull it back to its initial state: No traveling wave appears, although the disturbance will spread before settling down. But if we drape a large enough segment over the hump initially, upon releasing the chain we'll see *two* traveling waves begin to spread from the center point, one moving in each direction.

Our thought experiment has displayed most of the qualitative features of the action potential, as described in Section 12.1.1! The chain's height roughly represents the deviation of concentrations from their equilibrium values; the friction represents electrical resistance. We saw how a dynamical system with continuously distributed stored potential energy, and dissipation, can behave as an **excitable medium**, ready to release its energy in a controlled way as a propagating wave of excitation:

- The wave requires a threshold stimulus.
- For subthreshold stimuli, the system gives a spreading, but rapidly decaying, response.

- Similarly, stimuli of any strength but the "wrong" sign give decaying responses (imagine lifting the rope up the far side of the higher trough in Figure 12.5b).
- Above-threshold stimuli create a traveling wave of excitation. The strength of the distant response does not depend on the stimulus strength. Although we did not prove this, it should be reasonable to you that its *form* will also be stereotyped (independent of the stimulus type).
- The traveling wave moves at constant speed. You found in Your Turn 12B that this speed is determined by a trade-off between the stored energy density and the dissipation (friction).

There will be numerous technical details before we have a mathematical model of the action potential rooted in verifiable facts about membrane physiology (Sections 12.2 and 12.3). In the end, though, the mechanism discovered by Hodgkin and Huxley boils down to the one depicted in Figure 12.5:

> *Each segment of axon membrane goes in succession from resisting change (like chain segments to the left of the kink in Figure 12.5a) to amplifying it (like segments immediately to the right of the kink) when pulled over a threshold by its neighboring segment.*   (12.12)

Although it's suggestive, our mechanical model has one very big difference from the action potential: It predicts one-shot behavior. We cannot pass a second wave along our chain. Action potentials, in contrast, are **self-limiting**: The passing nerve impulse stops itself before exhausting the available free energy, leaving behind it a state that is able to carry more impulses (after a short refractory period). Even after we kill a nerve cell, or temporarily suspend its metabolism, its axon can conduct thousands of action potentials before running out of stored free energy. This property is needed when a nerve cell is called upon to transmit rapid bursts of impulses in between quiescent periods of trickle charging by the ion pumps.

The following sections will explore a simplified, one-shot model for the action potential, starting with more details about membrane excitability. Section 12.3 will return to the question of how real action potentials can be self-limiting.

12.2.3 Just a little more history

After showing that living cells can maintain resting potentials, DuBois Reymond also undertook a systematic study of nerve impulses, showing around 1866 that they traveled along the axon at a constant speed. The physical origins of this behavior remained completely obscure.

It seemed natural to suppose that some process in the cell's interior was responsible for carrying nerve impulses. Thus, for example, when it became possible to see microtubules running in parallel rows down the length of the axon, most physiologists assumed that they were involved in the transmission. In 1902, however, Julius Bernstein set in motion a train of thought that ultimately overturned this expectation, locating the mechanism of the impulse in the cell's plasma membrane.

Bernstein correctly guessed that the resting membrane was selectively permeable to potassium. The discussion in Section 11.1 then implies that a cell's membrane potential should be around $V_{\mathrm{K+}}^{\mathrm{Nernst}} = -75\,\mathrm{mV}$, roughly as observed. Bernstein suggested that during a nerve impulse, the membrane temporarily becomes highly permeable to *all* ions, bringing it rapidly to a new equilibrium with no potential difference across the membrane. Bernstein's hypothesis explained the existence of a resting potential, its sign and approximate magnitude, and the observed fact that increasing the exterior potassium concentration changes the resting potential to a value closer to zero. It also explained roughly the depolarization observed during a nerve impulse.

Hodgkin was an early convert to the membrane-based picture of the action potential. He reasoned that if the passage of ions through the membrane was important to the mechanism (and not just a side effect), then changing the electrical properties of the exterior fluid should affect the propagation *speed* of the action potential. And indeed, Hodgkin found in 1938 that increasing the exterior resistivity gave slower-traveling impulses, whereas decreasing it (by laying a good conductor alongside the axon) almost doubled the speed.

Detailed tests of Bernstein's hypothesis had to await the technological advances made possible by electronics, which were needed to measure signals with the required speed and sensitivity. Finally, in 1938, K. Cole and H. Curtis succeeded in showing experimentally that the overall membrane conductance in a living cell indeed increased dramatically during a nerve impulse, as Bernstein had proposed. Hodgkin and Huxley, and independently Curtis and Cole, also managed to measure ΔV directly during an impulse by threading a tiny glass capillary electrode into an axon. Each group found to their surprise that, instead of driving to zero as Bernstein had proposed, *the membrane potential temporarily reversed sign*, as shown in Figure 12.6b. It seemed impossible to reconcile these observations with Bernstein's attractive idea.

Further examination of data like Figure 12.6b revealed a curious fact: The peak potential (about $+40\,\mathrm{mV}$ in the figure), although far from the potassium Nernst potential, is actually not far from the *sodium* Nernst potential (Table 11.1 on page 477). This observation offered an intriguing way to save Bernstein's selective-permeability idea:

> If the membrane could rapidly switch from being selectively permeable to potassium only to being permeable mainly to sodium, then the membrane potential would flip from the Nernst potential of potassium to that of sodium, explaining the observed polarization reversal (see Equation 12.3). (12.13)

Idea 12.13 is certainly a falsifiable hypothesis. It predicts that changing the exterior concentration of sodium, and hence the sodium Nernst potential, will alter the peak of the action potential.

At this exciting moment, most of British civilian science was interrupted for several years by the needs of the war effort. Picking up the thread in 1946, Katz prepared axons with the exterior seawater replaced by a solution containing no sodium.[7] Although this change did nothing to the interior of the axons, and indeed did not alter the resting potential very much, Katz found that eliminating exterior sodium

---

[7]In this and other modified-solution experiments, it's important to introduce some other solute to match the overall osmotic pressure across the cell membrane.

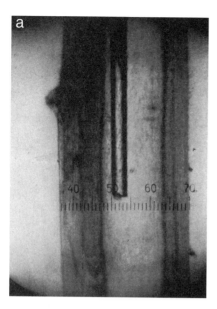

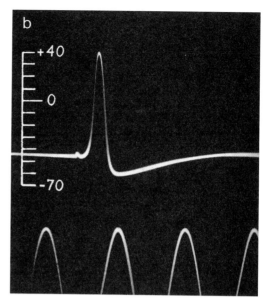

Figure 12.6: (Photomicrograph; oscilloscope trace.) Hodgkin and Huxley's historic 1939 result. (a) A recording electrode (a glass capillary tube) inside a giant axon, which shows as a clear space between divisions marked 47 and 63 on the scale. (The axon, in turn, is contained in a larger glass tube.) One division of the horizontal scale equals $33\,\mu$m. (b) Action potential and resting potential recorded between the inside and outside of the axon. Below the trace appears a time marker, showing reference pulses every 2 ms. The vertical scale indicates the potential of the internal electrode in millivolts, the seawater outside being taken as zero potential. Note that the membrane potential actually changes sign for a couple hundred microseconds; note also the overshoot, or afterhyperpolarization, before the potential settles back to its resting value. [Both panels from Hodgkin & Huxley, 1939.]

completely abolished the action potential, just as predicted by the hypothesis in Idea 12.13. Later Hodgkin and Katz showed in more detail that reducing the external sodium to a fraction of its usual concentration gave action potentials with reduced peak potentials (Figure 12.7), whereas increasing it increased the peak, all in quantitative accord with the Nernst equation. Rinsing out the abnormal solution and replacing it with normal seawater restored the normal action potential, as seen in Figure 12.7.

Hodgkin and Katz then managed to get a quantitative estimate of the changes of the individual conductances per area during an action potential. They found that they could explain the dependence of the action potential on the sodium concentration if g_{Na^+} increased about 500-fold from its resting value. That is, the resting values, $g_{K^+} \approx 25 g_{Na^+} \approx 2 g_{Cl^-}$ (Equation 11.9 on page 482), momentarily switch to

$$g_{K^+} \approx 0.05 g_{Na^+} \approx 2 g_{Cl^-}. \qquad \text{(at the action potential peak)} \qquad (12.14)$$

This is a dramatic result; but how exactly do the membrane permeabilities change, and how does the membrane know to change them in just the right sequence to create a traveling, stereotyped wave? Sections 12.2.4–12.3.2 will address these questions.

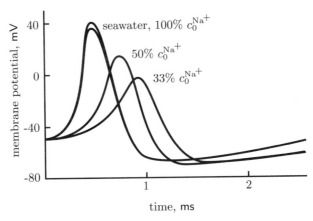

Figure 12.7: (Experimental data.) The role of sodium in the conduction of an action potential. One of the top traces was taken on a squid axon in normal seawater before exposure to low sodium. In the middle trace, external sodium was reduced to one-half that in seawater, and in the bottom trace, to one-third. (The other top trace was taken after normal seawater was restored to the exterior bath.) The data show that the peak of the action potential tracks the sodium Nernst potential across the membrane, an observation supporting the idea that the action potential is a sudden increase in the axon membrane's sodium conductance. [Data from Hodgkin & Katz, 1949.]

12.2.4 The time course of an action potential suggests the hypothesis of voltage gating

The previous sections have foreshadowed what is about to come. We must abandon the Ohmic hypothesis, which states that all membrane conductances are fixed, in favor of something more interesting: The temporary reversal of the sign of the membrane potential reflects a sudden increase in g_{Na^+} (Equation 12.14 instead of Equation 11.9), so g_{tot} temporarily becomes dominated by the sodium contribution instead of by potassium. The chord conductance formula (Equation 12.3 on page 512) then implies that this change drives the membrane potential away from the potassium Nernst potential and toward that of sodium, thus creating the temporary reversed polarization characteristic of the action potential.

In fact, the cable equation shows quite directly that the Ohmic hypothesis breaks down during a nerve impulse. We know that the action potential is a traveling wave of fixed shape, moving at some speed ϑ. For such a traveling wave, the entire history $V(x, t)$ is completely known once we specify its speed and its time course at *one* point:[8] We then have $V(x, t) = \tilde{V}(t - (x/\vartheta))$, where $\tilde{V}(t) \equiv V(0, t)$ is the curve shown in Figure 12.6b. Hence,

$$\frac{\mathrm{d}V}{\mathrm{d}x} = -\frac{1}{\vartheta} \frac{\mathrm{d}\tilde{V}}{\mathrm{d}t'}\bigg|_{t'=t-(x/\vartheta)},$$

---

[8]Recall the image of traveling waves as snakes under the rug (Figure 4.12b on page 134).

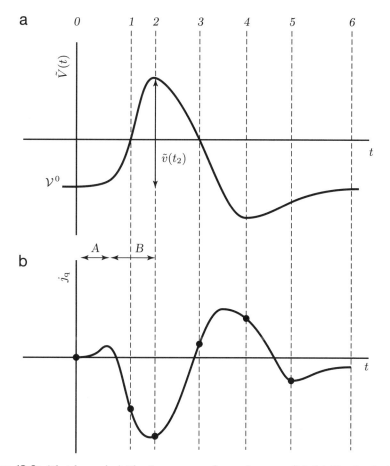

Figure 12.8: (Sketch graphs.) The time course of an action potential. (a) The sketch shows the membrane potential $\tilde{V}(t)$, measured at a fixed location $x = 0$. $\tilde{v}(t)$ refers to the difference between the membrane potential and its resting value $\mathcal{V}^0$. The *dashed lines* are six particular moments of time discussed in the text. (b) Reconstruction of the total membrane current from (a), using Equation 12.15. An Ohmic stage A gives way to another stage B. In B, the membrane potential continues to rise but the current falls and then reverses; this is non-Ohmic behavior. [Adapted from Benedek & Villars, 2000c.]

by the chain rule of calculus. Rearranging the cable equation (Equation 12.7) then gives us the total membrane current $j_{q,r}$ from the measured time course $\tilde{V}(t)$ of the membrane potential at a fixed position:

$$j_{q,r} = \frac{a\kappa}{2\vartheta^2} \frac{\mathrm{d}^2 \tilde{V}}{\mathrm{d}t^2} - \mathcal{C}\frac{\mathrm{d}\tilde{V}}{\mathrm{d}t}. \tag{12.15}$$

Applying Equation 12.15 to the measured time course of an action potential, sketched in Figure 12.8a, gives us the corresponding time course for the membrane current (Figure 12.8b). We can understand this result graphically, without any calculations. Note that the membrane current is particularly simple at the inflection points of panel (a) (the dashed lines labeled *1, 3,* and *5*): Here the first term of Equation 12.15 equals zero, and the sign of the current is opposite to that of the slope of $\tilde{V}(t)$. Similarly, at the extrema of panel (a) (the dashed lines labeled *2* and *4*), we find that the *second* term of Equation 12.15 vanishes: Here the sign of the current is that of the *curvature* of $\tilde{V}(t)$, as shown in panel (b). With these hints, we can work out the sign of $j_{q,r}$ at the points *0–6*; joining the dots gives the curve sketched in panel (b).

Comparing the two panels of Figure 12.8 shows what is happening during the action potential. Initially (stage *A*), the membrane conductance is indeed Ohmic: The cell's interior potential begins to rise above its resting value, thereby driving an outward current flux, as predicted from your calculation of the potential of three resistor–battery pairs (Your Turn 12A on page 514). But when the membrane has depolarized by about 10 mV, something strange begins to happen (stage *B*): The potential continues to rise, but the net current falls.

Idea 12.13 made the key point needed for understanding the current reversal, in terms of a switch in the membrane's permeabilities to various ions. Net current flows across a membrane whenever the actual potential difference *V* deviates from the "target" value given by the chord formula (Equation 12.3 on page 512). But the target value itself depends on the membrane conductances. If these suddenly change from their resting values, so will the target potential; if the target switches from being more negative than *V* to more positive, then the membrane current will change sign. Because the target value is dominated by the Nernst potential of the most permeant ion species, we can explain the current reversal by supposing that the membrane's permeability to sodium increases suddenly during the action potential.

So far, we have done little more than restate Idea 12.13. To go further, we must understand what *causes* the sodium conductance to increase. Because the increase does not begin until after the membrane has depolarized significantly (Figure 12.8, stage *B*), Hodgkin and Huxley proposed that

> *Membrane depolarization itself is the trigger that causes the sodium conductance to increase.*   (12.16)

That is, they suggested that some collection of unknown molecular devices in the membrane allow the passage of sodium ions, with a conductance depending on the membrane potential. Idea 12.16 introduces an element of **positive feedback** into our picture: Depolarization begins to open the sodium gates, a process that increases the degree of depolarization. The increased depolarization opens still more sodium gates; and so on.

The simplest way to implement Idea 12.16 is to retain the Ohmic hypothesis, but with the modification that each of the membrane's conductances may depend on *V*:

$$j_{q,r} = \sum_i (V - \mathcal{V}_i^{\text{Nernst}}) g_i(V). \qquad \text{simplified \textbf{voltage-gating hypothesis}}$$

(12.17)

In this formula, the conductances $g_i(V)$ are unknown (but positive) functions of the membrane potential. Equation 12.17 is our proposed replacement for the Ohmic hypothesis, Equation 11.8.[9]

The proposal Equation 12.17 certainly has a lot of content, even though we don't yet know the precise form of the conductance functions appearing in it. For example, it implies that the membrane's ion currents are still Ohmic (linear in $\ln(c_1/c_2)$) if we hold V fixed while changing the concentrations. However, the membrane current is now a *nonlinear* function of V, a crucial point for the following analysis.

Before proceeding to incorporate Equation 12.17 into the cable equation, let's place it in the context of this book's other concerns. We are accustomed to positive ions moving along the electric field, which then does work on them; they dissipate this work as heat as they drift against the viscous drag of the surrounding water. This migration has the net effect of reducing the electric field: Organized energy (stored in the field) has been degraded to disorganized (thermal) energy. But stage B of Figure 12.8b shows ions moving inward, that is, in a direction *opposite* to that of the potential drop. The energy needed to drive them can only have come from the thermal energy of their surroundings. Can thermal energy really turn back into organized (electrostatic) energy? Previous chapters have argued that such unintuitive energy transactions are possible, as long as they reduce the *free* energy of the system. And in fact, the axon started out with excess free energy, in the form of its nonequilibrium ion concentrations. Chapter 11 identified the source of this stored free energy as the cell's metabolism, via the membrane's ion pumps.

Note that Equation 12.17 implies that the conductances track changes in potential instantaneously. Section 12.2.5 will show how this simplified conductance hypothesis already accounts for much of the phenomenology of the action potential. Section 12.3.1 will then describe how Hodgkin and Huxley managed to measure the conductance functions and how they were forced to modify the simplified voltage-gating hypothesis somewhat.

12.2.5 Voltage gating leads to a nonlinear cable equation with traveling wave solutions

We can now return to the apparent impasse reached in our discussion of the linear cable equation (Section 12.2.1): There seemed to be no way for the action potential to gain access to the free energy stored along the axon membrane by the ion pumps. The previous section motivated a proposal for how to get the required coupling, namely, the simplified voltage-gating hypothesis. However, it left unanswered the question posed at the end of Section 12.2.3: Who orchestrates the orderly, sequential increases in sodium conductance as the action potential travels along the axon? The full answer to this question is mathematically rather complex. Before describing it qualitatively in Section 12.3, this section will implement a simplified version, in which we can actually solve an equation and see the outline of the full answer.

---

[9]The symbol ΔV appearing in Equation 11.8 is abbreviated as V in this chapter (see Section 12.1.3a on page 516).

Let's first return to our mechanical analogy, a chain that progressively shifts from a higher to a lower groove (Figure 12.5 on page 519a). Section 12.2.2 argued that this system can support a traveling wave of fixed speed and definite waveform. Now we must translate our ideas into the context of axons, and do the math.

Idea 12.12 said that the force needed to pull each successive segment of chain over its potential barrier came from the *previous* segment of chain. Translating into the language of our axon, this idea suggests that even though the resting state is a stable steady state of the membrane,

- *Once one segment depolarizes, its depolarization spreads passively to the neighboring segment;*
- *Once the neighboring segment depolarizes by more than 10 mV, the positive feedback phenomenon described in the previous section sets in, triggering a massive depolarization; and*   (12.18)
- *The process repeats, spreading the depolarized region.*

Let's begin by focusing only on the initial sodium influx. Thus we imagine only one voltage-gated ion species, say, Na^+. We also suppose that the membrane's conductance for this ion, $g_{Na^+}(v)$, depends only on the momentary value[10] of the potential disturbance $v \equiv V - V^0$.

A detailed model would use an experimentally measured form of the conductance per area $g_{Na^+}(v)$, as imagined in the dashed line of Figure 12.9a. We will instead use a mathematically simpler form (solid curve in the figure), namely, the function

$$g_{Na^+}(v) = g^0_{Na^+} + Bv^2. \tag{12.19}$$

Here $g^0_{Na^+}$ represents the resting conductance per area; as usual, we lump this in with the other conductances and call the sum g^0_{tot}. B is a positive constant. Equation 12.19 incorporates the key feature of increasing upon depolarization; moreover, it is always positive, as a conductance must be.

The total charge flux through the membrane is then the sum of the sodium contribution, plus Ohmic terms from the other ions:

$$j_{q,r} = \left(\sum_i (V - V_i^{\text{Nernst}})g_i^0\right) + (V - V_{Na^+}^{\text{Nernst}})Bv^2. \tag{12.20}$$

As in Your Turn 12A on page 514, the first term in Equation 12.20 can be rewritten as $g^0_{tot}v$. Letting H denote the constant $V_{Na^+}^{\text{Nernst}} - V^0$, we can also rewrite the last term as $(v - H)Bv^2$, obtaining

$$j_{q,r} = vg^0_{tot} + (v - H)Bv^2. \tag{12.21}$$

Figure 12.9b helps us understand graphically the behavior of our model. There are three important points on the curve of current versus depolarization, namely,

---

[10] As mentioned earlier, these assumptions are not fully realistic; thus our simple model will not capture all the features of real action potentials. Section 12.3.1 will discuss an improved model.

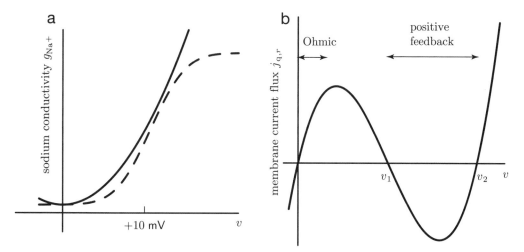

Figure 12.9: (Sketch graphs.) Voltage-gating hypothesis. (a) *Dashed curve:* The conductance g_{Na^+} of an axon membrane to sodium ions, showing an increase as the membrane potential increases from its resting value ($v = 0$). *Solid curve:* Simplified form for membrane sodium conductance (Equation 12.19). This form captures the relevant feature of the dashed curve, namely, that it increases as v increases and is positive. (Even the dashed line is not fully realistic: Real membrane conductances do not respond instantly to changes in membrane potential; rather they reflect the past history of v. See Section 12.3.1.) (b) Current-voltage relation resulting from the conductance model in (a) (Equation 12.21). The special values v_1 and v_2 are defined in the text.

the points where the membrane current $j_{q,r}$ is zero. Equation 12.21 shows that these points are the roots of a cubic equation. We write them as $v = 0$, v_1, and v_2, where v_1 and v_2 equal $\frac{1}{2}(H \mp \sqrt{H^2 - 4g_{tot}^0/B})$, respectively. At small depolarization v, the sodium permeability stays small, so the last term of Equation 12.21 is negligible. In this case, a small positive v gives small positive (outward) current, as expected: We are in the Ohmic regime (stage A of Figure 12.8). The outward flow of charge tends to reduce v back toward zero. A further increase of v, however, opens the voltage-gated sodium channels, eventually reducing $j_{q,r}$ to zero, and then below zero as we pass the point v_1. Now the net inward flow of charge tends to *increase* v, giving positive feedback—an avalanche. Instead of returning to zero, v drives toward the other root, v_2.[11] At still higher v, we once again get a positive (outward) current, as the large outward electric force on all the ions finally overcomes the entropic tendency for sodium to drift inward.

In short, our model displays threshold behavior: Small disturbances get driven back to $v = 0$, but above-threshold disturbances drive to the other stable fixed point v_2. Our program is now to repeat the steps in Section 12.1.3, starting from step (b) on page 516 (step (a) is unchanged).

b′. Equation We first substitute Equation 12.21 into the cable equation (Equation 12.7 on page 517). Some algebra shows that $v_1 v_2 = g_{tot}^0/B$, so the cable equation

---

[11]This bistability is reminiscent of the one studied in Problem 6.7c on page 241.

becomes

$$(\lambda_{\text{axon}})^2 \frac{d^2v}{dx^2} - \tau \frac{dv}{dt} = \frac{v(v - v_1)(v - v_2)}{(v_1 v_2)}. \qquad \textbf{nonlinear cable equation}$$

(12.22)

Unlike the linear cable equation, Equation 12.22 is not equivalent to the diffusion equation. In general, it's very difficult to solve nonlinear, many-variable differential equations like this one. But we can simplify things, because our main interest is in finding whether there are any traveling wave solutions to Equation 12.22. Following the discussion leading to Equation 12.15, we can represent a wave traveling at speed ϑ by a function $\tilde{v}(t)$ of *one* variable, via $v(x, t) = \tilde{v}(t - (x/\vartheta))$ (see Figure 4.12b on page 134). Substituting into Equation 12.22 leads to an *ordinary* (one-variable) differential equation:

$$\left(\frac{\lambda_{\text{axon}}}{\vartheta} \right)^2 \frac{d^2\tilde{v}}{dt^2} - \tau \frac{d\tilde{v}}{dt} = \frac{\tilde{v}(\tilde{v} - v_1)(\tilde{v} - v_2)}{v_1 v_2}. \qquad (12.23)$$

We can tidy up the equation by defining the dimensionless quantities $\bar{v} \equiv \tilde{v}/v_2$, $y \equiv -\vartheta t/\lambda_{\text{axon}}$, $s \equiv v_2/v_1$, and $Q \equiv \tau\vartheta/\lambda_{\text{axon}}$, finding

$$\frac{d^2\bar{v}}{dy^2} = -Q\frac{d\bar{v}}{dy} + s\bar{v}^3 - (1 + s)\bar{v}^2 + \bar{v}. \qquad (12.24)$$

c'. Solution You could enter Equation 12.24 into a computer-math package, substitute some reasonable values for the parameters Q and s, and look at its solutions. But it's tricky: The solutions are badly behaved (they blow up) unless you take Q to have one particular value (see Figure 12.10). This behavior is actually not surprising in the light of Section 12.2.2, which pointed out that our mechanical analog system selects one definite value for the pulse speed (and hence Q). You'll find in Problem 12.6 that choosing

$$\vartheta = \pm\frac{\lambda_{\text{axon}}}{\tau}\sqrt{\frac{2}{s}}\left(\frac{s}{2} - 1\right) \qquad (12.25)$$

yields a traveling wave solution (the solid curves in Figure 12.10).

d'. Interpretation The hypothesis of voltage gating, embodied in the nonlinear cable equation, has led to the appearance of traveling waves of definite speed and shape. In particular, the amplitude of the traveling wave is fixed: It smoothly connects two of the values of v for which the membrane current is zero, namely, 0 and v_2 (Figure 12.9). We cannot excite such a wave with a very small disturbance. Clearly, for small enough v, the nonlinear cable equation is essentially the same as the linear one

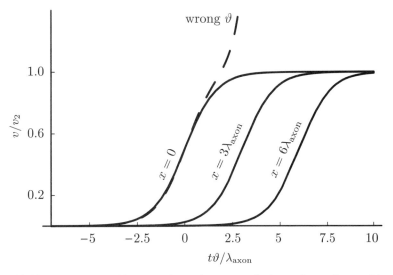

Figure 12.10: (*Mathematical functions.*) Traveling wave solution to the nonlinear cable equation (see Problem 12.6). The membrane potential relative to rest, $v(x, t)$, is shown as a function of time at three different fixed locations (*three solid curves*). Points at larger x see the wave go by at later times, so this wave is traveling in the $+\hat{x}$ direction. The parameter $s \equiv v_2/v_1$ has been taken equal to 3 for illustration. Comparison with Figure 12.2b on page 508 shows that this simplified model qualitatively reproduces the leading edge of the action potential. The *dashed line* shows a solution to Equation 12.23 with a value of the front velocity ϑ different from that in Equation 12.25; this solution is singular. Time is measured in units of $\lambda_{\mathrm{axon}}/\vartheta$. The potential relative to resting is measured in units of v_2 (see text).

(Equation 12.9 on page 517), whose solution we have already seen corresponds to passive, diffusive spreading (electrotonus), not an action potential. Thus

> **a.** *Voltage gating leads to a graded, diffusive response for stimuli below some threshold, but above-threshold, depolarizing stimuli yield a large, fixed-amplitude response.* (12.26)
>
> **b.** *The above-threshold response takes the form of a traveling wave of fixed shape and speed.*

Our model, a mathematical embodiment of Idea 12.18, has captured many of the key features of real nerve impulses, listed at the end of Section 12.1.1. We didn't prove that the wave rapidly forgets the precise nature of its initial stimulus, remembering only whether it was above threshold or not, but such behavior should seem reasonable in the light of the mechanical analogy (see Section 12.2.2). We also get a quantitative prediction. The velocity ϑ is proportional to $\lambda_{\mathrm{axon}}/\tau = \sqrt{a\kappa g_{\mathrm{tot}}/(2\mathcal{C}^2)}$ times a factor independent of the axon's radius a. Thus the model predicts that if we examine a family of unmyelinated axons of the same general type, with the same ion concentrations, we should find that the pulse speed varies with axon radius as $\vartheta \propto \sqrt{a}$. This prediction is roughly borne out in experimental data. Moreover, the overall magnitude of the pulse speed is approximately λ_{axon}/τ. For the squid giant axon, our estimates give this quantity as about $12\,\text{mm}/2\,\text{ms} = 6\,\text{m s}^{-1}$, a value within an order of magnitude of the measured action potential speed of about $20\,\text{m s}^{-1}$.

Our result also makes sense in the light of the mechanical analogy (Section 12.2.2). In Your Turn 12B(c), you found that the wave speed was proportional to the density of stored energy divided by a friction constant. Examining our expression for ϑ, we notice that both κ and g_{tot} are inverse resistances, so $\sqrt{\kappa g_{tot}}$ is indeed an inverse "friction" constant. In addition, the formula $E/A = \frac{1}{2}q^2/(\mathcal{C}A)$ for the electrostatic energy density stored in a charged membrane of area A shows that the stored energy is proportional to $1/\mathcal{C}$. Thus our formula for ϑ has essentially the structure expected from the mechanical analogy.

$\boxed{T_2}$ *Section 12.2.5′ on page 552 discusses how the nonlinear cable equation determines the speed of its traveling wave solution.*

12.3 THE FULL HODGKIN–HUXLEY MECHANISM AND ITS MOLECULAR UNDERPINNINGS

Section 12.2.5 showed how the hypothesis of voltage-gated conductances leads to a nonlinear cable equation, with self-sustaining, traveling waves of excitation reminiscent of actual action potentials. This is an encouraging preliminary result, but it makes us want to see whether axon membranes really do have the remarkable properties of voltage-dependent, ion-selective conductance we attributed to them. In addition, the simplified voltage-gating hypothesis has not given us any understanding of how the action potential *terminates;* Figure 12.10 shows the ion channels opening and staying open, presumably until the concentration differences giving rise to the resting potential have been exhausted. Finally, while voltage gating may be an attractive idea, we do not yet have any idea how the cell could implement it with molecular machinery. This section will address all these points.

12.3.1 Each ion conductance follows a characteristic time course when the membrane potential changes

Hodgkin, Huxley, Katz, and others confirmed the existence of voltage-dependent, ion-selective conductances in a series of elegant experiments, which hinged on three main technical points.

Space clamping The conductances g_i determine the current through a patch of membrane held at a fixed, uniform potential drop. But during the normal operation of an axon, deviations from the resting potential are highly *non*uniform along the axon—they are localized pulses. Cole and G. Marmont addressed this problem by developing the **space clamp** technique. The technique involved threading an ultra-fine wire down the inside of an axon (Figure 12.11). The metallic wire was a much better conductor than the axoplasm, so its presence forced the entire interior to be at

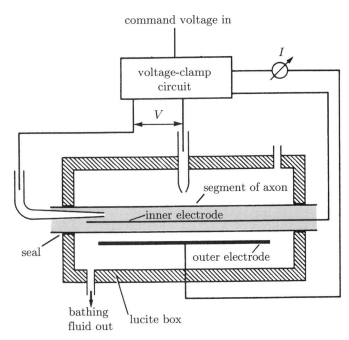

Figure 12.11: (Schematic.) An electrophysiology experiment. The long wire threaded through the axon maintains its interior at a spatially uniform electrostatic potential (space clamping). A feedback circuit monitors the transmembrane potential V and sends whatever current is needed to keep V fixed at a "command" value chosen by the experimenter (voltage clamping). The corresponding current I is then recorded. Typically the axon is 30–40 mm long. [From Läuger, 1991.]

a fixed, uniform potential. Introducing a similar long exterior electrode then forces $V(x)$ itself to be uniform in x.

Voltage clamping One could imagine forcing a given current across the membrane, measuring the resulting potential drop, and attempting to recover a relation like the one sketched in Figure 12.9b. There are a number of experimental difficulties with this approach, however. For one thing, the figure shows that a given $j_{q,r}$ can be compatible with *multiple* values of V. More important, we are exploring the hypothesis that the devices regulating conductance are themselves regulated by V, not by current flux, so V is the more natural variable to fix. For these and other reasons, Hodgkin and Huxley set up their apparatus in a **voltage clamp** mode. In this arrangement, the experimenter chooses a "command" value of the membrane potential; feedback circuitry supplies whatever current is needed to maintain V at that command value and reports the value of that current.

Separation of ion currents Even with space and voltage clamping, electrical measurements yield only the total current through a membrane, not the individual currents of each ion species. To overcome this problem, Hodgkin and Huxley extended Katz's technique of ion substitution (Section 12.2.3). Suppose we adjust the exterior concentration of ion species i so that $\mathcal{V}_i^{\text{Nernst}}$ equals the clamped value of V. Then this ion's contribution to the current equals zero, regardless of what its conductance $g_i(V)$ may be (see Equation 12.17 on page 526). Using an elaboration of this idea, Hodgkin and Huxley managed to dissect the full current across the membrane into its components at any V.

Results Hodgkin and Huxley systematized a number of observations made by Cole and Marmont. Figure 12.12 sketches some results from their voltage clamp apparatus (Figure 12.11). The command potential was suddenly stepped up from the membrane's resting potential to $V = -9\,\text{mV}$, then held there. One striking feature of these data is that the membrane conductance does *not* track the applied potential instantaneously. Instead, we have the following sequence of events:

1. Immediately after the imposed depolarization (Figure 12.12a), there is a very short spike of outward current (panel b), lasting a few microseconds. This is not really current through the membrane; rather, it is the discharge of the membrane's capacitance (a capacitive current), as discussed in Section 12.1.2.
2. A brief, inward sodium current develops in the first half-millisecond. Dividing by $V - \mathcal{V}_{\text{Na}^+}^{\text{Nernst}}$ gives the sodium conductance, whose peak value was found to depend on the selected command potential V.
3. After peaking, however, the sodium conductance drops to zero, even though V is held constant (Figure 12.12c).
4. Meanwhile, the potassium current rises slowly (in a few milliseconds, Figure 12.12d). Like g_{Na^+}, the potassium conductance rises to a value that depends on V. Unlike g_{Na^+}, however, g_{K^+} holds steady indefinitely at this value when V is held fixed.

Thus, the simplified voltage-gating hypothesis describes reasonably well the initial events following membrane depolarization (points (1) and (2)), which is why it gave a reasonably adequate description of the leading edge of the action potential. In the later stages, however, our simplified picture breaks down (points (3) and (4)) and indeed, here our solution deviated from reality (compare the mathematical solutions in Figure 12.10 with the experimental trace in Figure 12.6b on page 523). The results in Figure 12.12 show us what changes we should expect in our solutions when we introduce more realistic gating functions:

- After half a millisecond, the spontaneous drop in sodium conductance begins to drive V back down to its resting value.
- Indeed, the slow increase in potassium conductance after the main pulse implies that the membrane potential will temporarily *overshoot* its resting value, instead arriving at a value closer to $\mathcal{V}_{\text{K}^+}^{\text{Nernst}}$ (see Equation 12.3 on page 512 and Table 11.1 on page 477). This observation explains the phenomenon of afterhyperpolarization, mentioned in Section 12.1.1.

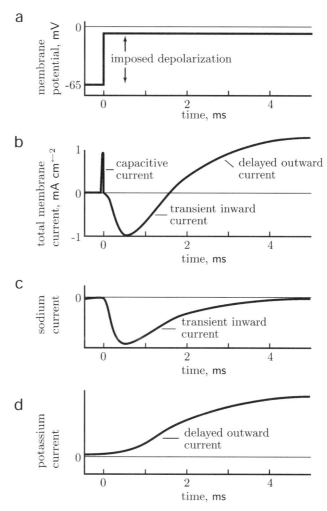

Figure 12.12: (Sketch graphs of experimental data.) Membrane currents produced by depolarizing stimuli. (a) Applied stimulus, a 56 mV depolarization of a squid axon membrane imposed by a voltage clamp apparatus. (b) Currents measured during the stimulus. The observed current consists of a brief positive pulse as the membrane's capacitance discharges, followed by a short phase of inward current, and then finally a delayed outward current. The inward and delayed outward currents are shown separately in (c) and (d). (c) The transient inward current is caused by sodium entry. (d) Potassium movement out of the axon gives the longer outward current. Dividing the traces in (c,d) by the imposed $V - \mathcal{V}_i^{\text{Nernst}}$ yields the corresponding conductances, $g_i(V, t)$, which depend on time. [Adapted from Hodgkin & Huxley, 1952a.]

- Once the membrane has repolarized, another slow process resets the potassium conductance to its original, lower value, and the membrane potential returns to its resting value.

Hodgkin and Huxley characterized the full time course of the potassium conductance by assuming that for every value of V, there is a corresponding saturation value of the potassium conductance, $g_{K+}^{\infty}(V)$. The rate at which g_{K+} relaxes to its saturation value was also taken to be a function of V. These two functions were taken as phenomenological membrane properties and were obtained by repeating experiments like Figure 12.12 with command voltage steps of various sizes. Thus the actual conductance of a patch of membrane at any time is not simply determined by the instantaneous value of the potential at that time, as implied by the simple voltage-gating hypothesis. Instead, the *recent history* of the potential (in this case, the time since V was stepped from V^0 to its command value) affects g_i. A similar, but slightly more elaborate, scheme successfully described the rise/fall structure of the sodium conductance.

Substituting the conductance functions just described into the cable equation led Hodgkin and Huxley to an equation more complicated than our Equation 12.24. Obtaining the solutions was a prodigious effort, originally taking weeks to compute on a hand-cranked, desktop calculator. But the solution correctly reproduced all the relevant aspects of the action potential, including its entire time course, speed of propagation, and dependence on changes of exterior ion concentrations.

There is an extraordinary postscript to this story. The model described in this chapter implies that as far as the action potential is concerned, the sole function of the cell's interior machinery is to supply the required nonequilibrium resting concentration differences of sodium and potassium across the membrane. P. Baker, Hodgkin, and T. Shaw confirmed this rather extreme conclusion by the extreme measure of emptying the axon of all its axoplasm, replacing it by a simple solution containing potassium but no sodium. Although it was almost entirely gutted, the axon continued to transmit action potentials indistinguishable from those in its natural state (Figure 12.13)!

12.3.2 The patch clamp technique allows the study of single ion channel behavior

Hodgkin and Huxley's theory of the action potential was phenomenological in character: They measured the behavior of the membrane conductances under space and voltage clamped conditions, then used these measurements to explain the action potential. Although they suspected that their membrane conductances arose by the passage of ions through discrete, molecular-scale **ion channels**, their data could not confirm this picture. Indeed, the discussion of this chapter so far leaves us with several questions:

a. What is the molecular mechanism by which ions pass through a membrane? The simple scheme of diffusion through the lipid bilayer cannot be the answer (see Section 4.6.1 on page 135) because the conductance of pure bilayer membranes is

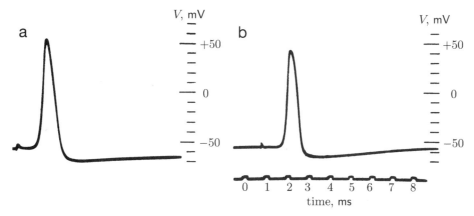

Figure 12.13: (Oscilloscope traces.) Perhaps the most remarkable experiment described in this book. (a) Action potential recorded with an internal electrode from an axon whose internal contents have been replaced by potassium sulfate solution. (b) Action potential of an intact axon, with same amplification and time scale. [From Baker et al., 1962.]

several orders of magnitude less than the value for natural membranes (see Section 11.2.2 on page 478).

b. What gives this mechanism its specificity for ion types? We have seen that the squid axon membrane's conductances to potassium and to sodium are quite different and are gated differently.

c. How do ion channels sense and react to the membrane potential?

d. How do the characteristic time courses of each conductance arise?

This section will briefly sketch the answers to these questions, starting with observations made in the 1970s.

Hodgkin and Huxley could not see the molecular mechanisms for ion transport across the axon membrane because they were observing the collective behavior of thousands of ion channels, not the behavior of any individual channel. The situation was somewhat like that of statistical physics at the turn of the twentieth century: The ideal gas law made it easy to measure the product $N_{mole}k_B$, but the individual values of N_{mole} and k_B remained in doubt until Einstein's analysis of Brownian motion (Chapter 4). Similarly, measurements of g_i in the 1940s gave only the product of the conductance G_i of an individual channel times the number of channels per unit area of membrane. Katz succeeded in the early 1970s in estimating the magnitude of G_i by analyzing the statistical properties of aggregate conductances. But others' (inaccurate) estimates disagreed with his, and confusion ensued.

The systematic study of membrane conductance at the single-channel level had to await the discovery of cell biology techniques capable of isolating individual ion channels and electronic instrumentation capable of detecting the tiny currents they carry. E. Neher developed the necessary electronic techniques in experiments with ion channel proteins embedded in artificial bilayers. The real breakthrough came in 1975, when Neher and B. Sakmann developed the **patch clamp** technique

a

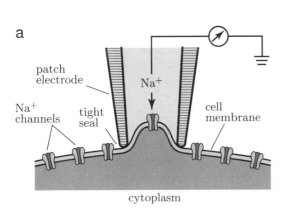

patch
electrode

Na$^+$

Na^+
channels

tight
seal

cell
membrane

cytoplasm

25 μm

Figure 12.14: (Schematic; optical micrograph.) The patch clamp technique. (a) A small patch of membrane containing only a single voltage-gated sodium channel (or a few) is electrically isolated from the rest of the cell by a patch electrode. The current entering the cell through these channels is recorded by a monitor connected to the patch electrode. (b) Patch clamp manipulation of a single, live photoreceptor cell from the retina of a salamander. The cell is secured by partially sucking it into a glass micropipette (*bottom*), and the patch clamp electrode (*upper left*) is sealed against a small patch of the cell's plasma membrane. [(a) Adapted from Kandel et al., 2000. (b) Digital image kindly supplied by T. D. Lamb; see Lamb et al., 1986.]

(Figure 12.14), thereby enabling the measurement of ion currents across single channels in intact, living cells. Neher and Sakmann's work helped launch an era of dynamical measurements on single-molecule devices.

One of the first results of patch clamp recording was an accurate value for the conductance of individual channels: A typical value is $G \approx 25 \cdot 10^{-12}\,\Omega^{-1}$ for the open sodium channel. Using the relations $V = IR$ and $R = 1/G$, we find that at a driving potential of $V - \mathcal{V}_{Na^+}^{\text{Nernst}} \approx 100\,\text{mV}$, the current through a single open channel is 2.5 pA.

> **Your Turn 12C**
>
> Express this result in terms of sodium ions passing through the channel per second. Is it reasonable to treat the membrane electric current as the flow of a continuous quantity, as we have been doing?

a. Mechanism of conduction The simplest imaginable model for ion channels has proved to be essentially correct: Each one is a barrel-shaped array of protein subunits inserted in the axon's bilayer membrane (Figure 2.21a on page 57), creating a hole through which ions can pass diffusively. (Problem 12.8 tests this idea for reasonableness with a simple estimate.)

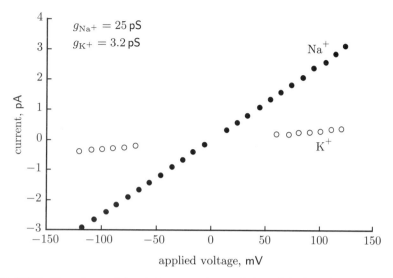

Figure 12.15: (Experimental data.) Current–voltage relation of single sodium channels re-constituted into an artificial bilayer in solutions of NaCl and KCl. The vertical axis gives the current observed when the channel was in its open state. The channels were kept open (that is, channel inactivation was suppressed) by adding batrachotoxin, the neurotoxin found in the skin of the poison dart frog. The slopes give the channel conductances shown in the legend; the unit pS equals $10^{-12}\,\Omega^{-1}$. The data show that this channel is highly selective for sodium. [Data from Hartshorne et al., 1985.]

b. Specificity The channel concept suggests that the independent conductances of the axon membrane arise through the presence of two (actually, several) subpopulations of channels, each carrying only one type of ion and each with its own voltage-gating behavior. Indeed, the patch clamp technique revealed the existence of distinct, specific channels. Figure 12.15 illustrates the great specificity of the sodium channel: The conductance of a single sodium channel to sodium is nearly ten times the conductance to other similar cations. The potassium channel is even more precise, admitting potassium 50 times as readily as sodium.

It's not hard to imagine how a channel can accept smaller ions, like sodium, while rejecting larger ones, like potassium and rubidium: We can just suppose that the channel is too small for the larger ions to pass. (More precisely, this geometrical constraint applies to the hydrated ions; see Section 8.2.2 on page 301.) It's also not hard to imagine how a channel can pass positive ions in preference to neutral or negative objects: A negative charge somewhere in the middle can reduce the activation barrier for a positive ion to pass, thereby increasing the rate of cation passage (Section 3.2.4 on page 86), while having the opposite effect on anions. Real sodium channels seem to employ both these mechanisms.

What *is* hard to imagine is how a channel could specifically pass a *large* cation, rejecting smaller ones, as the potassium channel must do! In the early 1970s, C. Armstrong and B. Hille proposed models exploring this idea. The idea is that the channel could contain a constriction so narrow that ions, normally hydrated, would have to "undress" (lose some of their bound water molecules) to pass through. The energy needed to break the corresponding hydration interactions will create a large activation barrier, thus disfavoring ion passage, unless some other interaction forms compensating (favorable) bonds at the same time. The first crystallographic reconstructions of a potassium channel, obtained by R. Mackinnon and coauthors in 1998, indeed showed such a constriction, exactly fitting the potassium ion (diameter 0.27 nm) and lined with negatively charged oxygen atoms from carbonyl groups in the protein making up the channel. Thus, just as the potassium ion is divested of its companion water molecules, it picks up similar attractive interactions to these oxygens and hence can pass without a large activation barrier. The smaller sodium ion (diameter 0.19 nm), however, does not make a tight fit, so it cannot interact as well with the fixed carbonyl oxygens. Nevertheless, it too must lose its hydration shell, thus incurring a net energy barrier. (In addition, because of its smaller size, sodium holds its hydration shell more tightly than does potassium.)

c. Voltage gating Already in 1952, Hodgkin and Huxley were imagining voltage-gated channels as devices similar to the fanciful valve sketched in Figure 12.16a: A net positive charge embedded in a movable part of the channel gets pulled by an external field. An allosteric coupling then converts this motion into a major conformational change, which opens a gate. Panel (b) of the figure shows a more realistic sketch of this idea, based in part on Mackinnon's crystallographic data.

The mechanism just outlined leaves open the question of whether the conformational change is continuous, as implied in Figure 12.16a, or discrete. The two possibilities respectively give rise to analog gating of the membrane current (as in the transistors of an audio amplifier) or digital, on/off, behavior (as in computer circuitry). Our experience with allostery in Chapter 9 shows that the latter option is a real possibility; and indeed, patch clamp recording showed that most ion channels have just two (or a few) discrete conductance states. For example, the traces in Figure 12.17b each show a single channel jumping between a closed state with zero current and an open state, which always gives roughly the same current.

The observation of digital (all-or-nothing) switching in single ion channels may seem puzzling in the light of our earlier discussion. Didn't our simple model for voltage gating require a *continuous* response of the membrane conductance to V (Figure 12.9a)? Didn't Hodgkin and Huxley find a continuous time course for their conductances, with a continuously varying saturation value of $g_{K+}^{\infty}(V)$? To resolve this paradox, we need to recall that there are many ion channels in each small patch of membrane (see Problem 12.7), each switching independently. Thus the values of g_i measured by the space clamp technique reflect not only the conductances of individual open channels (a discrete quantity) and their density σ_{chan} in the membrane (a constant) but also the average fraction of all channels that are open. The last factor mentioned *can* change in a nearly continuous manner if the patch of membrane being studied contains many channels.

We can test the idea just stated by noticing that the fraction of open channels should be a particular function of V. Suppose that the channel really is a simple two-state device. We studied the resulting equilibrium in Section 6.6.1 on page 218, arriving at a formula for the probability of one state ("channel open") in terms of the

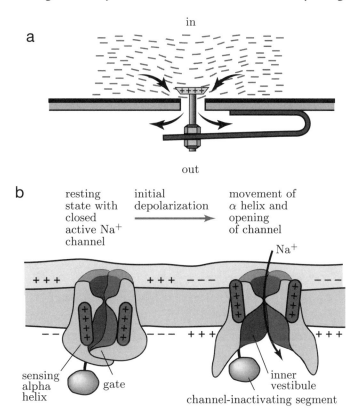

Figure 12.16: (Schematic; sketch based on structural data.) (a) Conceptual model of a voltage-gated ion channel. A spring normally holds a valve closed. An electric field pointing upward lifts the positively charged valve, letting water flow downward. (b) Sketch of the sodium channel. *Left:* In the resting state, positive charges in the channel protein's four "sensing" alpha helices are pulled downward, toward the negative cell interior. The sensing helices in turn pull the channel into its closed conformation. *Right:* Upon depolarization, the sensing helices are pulled upward. The channel now relaxes toward a new equilibrium, in which it spends most of its time in the open state. The *lower blob* depicts schematically the channel-inactivating segment. This attached object can move into the channel, thereby blocking ion passage even though the channel itself is in its open conformation. [(b) Adapted from Armstrong & Hille, 1998.]

free energy difference ΔF for the transition closed→open (Equation 6.34 on page 225):

$$P_{\text{open}} = \frac{1}{1 + e^{\Delta F/k_{\text{B}}T}}. \qquad (12.27)$$

We cannot predict the numerical value of ΔF without detailed molecular modeling of the channel. But we can predict the *change* in ΔF when we change V. Suppose that the channel's two states, and their internal energies, are almost unchanged by V. Then

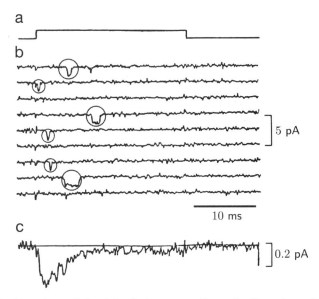

Figure 12.17: (Experimental data.) Patch clamp recordings of sodium channels in cultured muscle cells of rats, showing the origin of the inward sodium current from discrete channel-opening events. (a) Time course of a 10 mV depolarizing voltage step, applied across the patch of membrane. (b) Nine individual current responses elicited by the stimulus pulses, showing six individual sodium channel openings (*circles*). The potassium channels were blocked. The patch contained 2–3 active channels. (c) Average of 300 individual responses like those shown in (b). If a region of membrane contains many channels, all opening independently, we would expect its total conductance to resemble this curve; and indeed it does (see Figure 12.6c). [Data from Sigworth & Neher, 1980.]

the only change to ΔF comes from the fact that a few charges in the voltage-sensing region move in the external field, as shown in Figure 12.16b.

Suppose that upon switching, a total charge q moves a distance ℓ in the direction perpendicular to the membrane. The electric field in the membrane is $\mathcal{E} \approx V/d$, where d is the thickness of the membrane (see Section 7.4.3 on page 264). The external electric field then makes a contribution to ΔF equal to $-q\mathcal{E}\ell$, or $-qV\ell/d$; so our model predicts that

$$\Delta F(V) = \Delta F_0 - qV\ell/d \, , \text{ or } \; P_{\text{open}} = \frac{1}{1 + Ae^{-qV\ell/(k_{\mathrm{B}}Td)}}, \qquad (12.28)$$

where ΔF_0 is an unknown constant (the internal part of ΔF), and $A \equiv e^{\Delta F_0/k_{\mathrm{B}}T}$.

Equation 12.28 gives our falsifiable prediction: It makes a definite prediction about the sigmoidal shape of the opening probability, in terms of two phenomenological fit parameters (A and $q\ell/d$). Figure 12.18a shows experimental patch clamp data giving P_{open} as a function of V. To see whether it obeys our prediction, panel (b) shows the quantity $\ln((P_{\text{open}})^{-1} - 1) = \Delta F/k_{\mathrm{B}}T$. According to Equation 12.28, this quantity should be a constant minus $qV\ell/(k_{\mathrm{B}}Td)$. Figure 12.18b shows that it

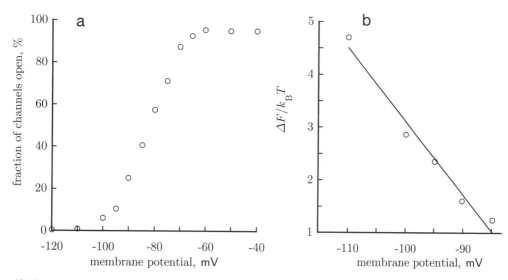

Figure 12.18: (Experimental data with fit.) Voltage dependence of sodium channel opening. (a) The current through a single sodium channel reconstituted in an artificial bilayer membrane was measured under voltage clamp conditions while increasing the voltage from hyperpolarized to depolarized (this particular channel opened at $\Delta V \approx -80\,\text{mV}$). Channel inactivation was suppressed; see Figure 12.15. (b) The free energy difference $\Delta F / k_B T$ between open and closed states, computed from (a) under the hypothesis of a two-state switch (Equation 12.28). The curve is nearly linear in the applied voltage, as we would expect if the channel snapped between two states with different, well-defined, spatial distributions of charge. The slope is $-0.15\,\text{mV}^{-1}$. [(a) Data from Hartshorne et al., 1985.]

is indeed a linear function of V. From the slope of this graph, Equation 12.28 gives $(q\ell)/(k_B T_r d) = 0.15\,\text{mV}^{-1}$.

> **Your Turn 12D**
>
> Interpret the last result. Using the fact that ℓ cannot exceed the membrane thickness, find a bound for q and comment.

d. Kinetics Section 6.6.2 on page 220 also drew attention to the implications of the two-state hypothesis for *non*equilibrium processes: If initially the probabilities of occupation are not equal to their equilibrium values, then they will approach those values exponentially, following the experimental relaxation formula (Equation 6.30). The situation with ion channels is somewhat complicated; most have more than two relevant states. Nevertheless, in many circumstances, one relaxation time dominates, and we do find nearly exponential relaxation behavior. Figure 12.19 shows the results of such an experiment. The figure also illustrates the similarities between the voltage-gated channels studied so far in this chapter and **ligand-gated ion channels**, which open in response to a *chemical* signal.

The channels studied in Figure 12.19 are sensitive to the presence of the molecule **acetylcholine**, a neurotransmitter. At the start of each trial, a sudden release of acetylcholine opens a number of channels simultaneously. The acetylcholine rapidly diffuses away, leaving the channels still open but ready to close. That is, the experiment

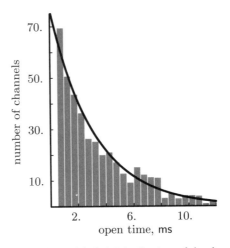

Figure 12.19: (Experimental data with fit.) Distribution of the durations of channel open times in a ligand-gated ion channel (frog synaptic channels exposed to acetylcholine). The histogram shows how many individual ion channels stayed open for various times following brief exposure to the activating neurotransmitter. The curve shows an exponential probability distribution with time constant $\tau = 3.2\,\mathrm{ms}$; the curve has been normalized appropriately for the total number of observations (480) and the bin width (0.5 ms). The first bin of data is not shown. Compare with the kinetics of RNA unfolding in Figure 6.10 on page 228. [Data from Colquhoun & Hawkes, 1983.]

prepares an initial nonequilibrium population of ion channel states. Each channel has a fixed probability per unit time of jumping to the closed state. The experimenters followed the time course of the membrane current, flagging individual channel-closing events. Repeating the experiment to build a large data set yielded the histogram of open times shown, which matches the exponential curve $e^{-t/3.2\,\mathrm{ms}}$. Although each channel is either fully open or fully shut, adding the conductances of many channels gives a total membrane current that roughly approximates a continuous exponential relaxation, analogous to that found in Hodgkin and Huxley's experiments for the potassium conductance upon sudden depolarization.

The complex, open-then-shut dynamics of the sodium channel is not a simple exponential, but it too arises from the all-or-nothing opening and closing of individual sodium channels. Figure 12.17 makes this point graphically. The nine traces in panel (b) show successive trials in which a single sodium channel, initially in its resting state, was suddenly depolarized. The individual traces show only digital behavior. To simulate the behavior of a large patch of membrane, containing many channels, the experimenters averaged 300 such single-channel time courses, obtaining the trace in panel (c). Remarkably, the result resembles closely the time course of the sodium current in space clamp experiments (Figure 12.12c on page 535).

Today we attribute the observed two-stage dynamics of the sodium conductance under sustained depolarization to two independent, successive obstructions that a sodium ion must cross. One of these obstructions opens rapidly upon depolarization, whereas the other closes slowly. The second process, called **inactivation**, involves a **channel-inactivating segment**. According to a model due to Armstrong and F. Bezanilla, the channel-inactivating segment is loosely attached to the sodium channel by a flexible tether (Figure 12.16b). Under sustained depolarization, this segment eventually enters the open channel, physically blocking it. Upon repolarization, the segment wanders away, and the channel is ready to open again.

Several ingenious experiments supported this model. For example, Armstrong found that he could cleave away the channel-inactivating segment with enzymes, thereby destroying the inactivation process but leaving the fast opening process unchanged. Later R. Aldrich and coauthors manufactured channels in which the flexible linker chain joining the inactivating segment to the channel was shorter than usual. The modified channels inactivated faster than their natural counterparts: Shortening the chain made it easier for the inactivation segment to find its docking site by diffusive motion.

12.4 NERVE, MUSCLE, SYNAPSE

Another book the size of this one would be needed to explore the ways in which neurons accept sensory information, perform computation, and stimulate muscle activity. This short section will at best convey a survey of such a survey, emphasizing links to our discussion of action potentials.

12.4.1 Nerve cells are separated by narrow synapses

Most body tissues consist of cells with simple, compact shapes. In contrast, nerve cells are large, have complex shapes, and are intertwined with one another to such an extent that by the late nineteenth century, many anatomists still thought of the brain as a continuous mass of fused cells and fibers, and not as a collection of distinct cells. The science of neuroanatomy could not begin until 1873, when Camillo Golgi developed a silver-impregnation technique that stained only a few nerve cells in a sample (typically 1%) but stained the selected cells completely. Thus the stained cells stood out from the intertwining mass of neighboring cells, and their full extent could be mapped.

Improving Golgi's technique, Santiago Ramón y Cajal drew meticulous and breathtaking pictures of entire neurons (Figure 12.20). Cajal argued in 1888 that neurons were, in fact, distinct cells. Golgi himself never regarded this "neuron doctrine" as proved.[12] Indeed, the definitive proof required the development of electron microscopy to see the narrow synapse separating adjoining neurons. Figure 12.21 shows a modern view of this region. One nerve cell's axon ends at another's dendrite (or on a dendritic spine attached to the dendrite). The cells interact when an impulse travels to the end of the axon and stimulates the next cell's dendrite across a narrow (10–30 nm wide) gap, the **synaptic cleft** (Figure 12.21). Thus, information flows from the **presynaptic** (axon) side of the cleft to the **postsynaptic** (dendrite) side. A

---

[12]Golgi was right to be cautious: His method does *not* always stain whole neurons; it often misses fine processes, especially axons.

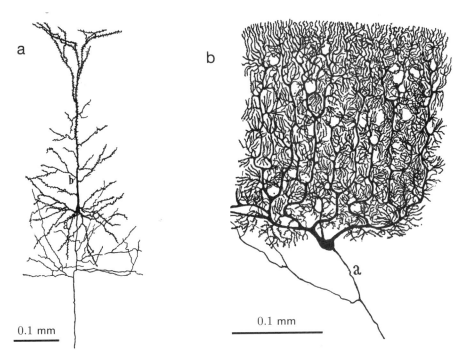

a

b

0.1 mm

0.1 mm

Figure 12.20: (Anatomical drawings.) Two classes of human neurons from the pioneering work of S. Ramón y Cajal. (a) A pyramidal cell from the rabbit cerebral cortex. The axon divides near the cell body (or soma, *dark blob* between *a* and *b*), sending branches to connect with nearby cells as well as a main axon (*bottom*) projecting to distant parts of the brain. The other branched lines extending from the soma are dendrites (input lines). (b) A Purkinje cell, with its extensive dendritic (input) system (*top*). The axon is labeled *a*. [From Ramón y Cajal, 1995.]

similar synapse joins a motor nerve axon to the muscle fiber whose contraction it controls.

12.4.2 The neuromuscular junction

The best-studied synapse is the junction between a motor neuron and its associated muscle fiber. As sketched in Figure 12.21, the axon terminals contain many synaptic vesicles filled with the neurotransmitter acetylcholine. In the quiescent state, the vesicles are mostly awaiting release, although a few release spontaneously each second.

As an action potential travels down an axon, it splits into multiple action potentials if the axon branches, finally arriving at one or more axon terminals. The terminal's membrane contains voltage-gated calcium channels; in response to depolarization, these channels open. The external concentration of Ca^{2+} is in the millimolar range, but active pumping maintains a much smaller (micromolar) concentration inside (see Section 11.3.5 on page 496). The resulting influx of calcium catalyzes the fusion of about 300 synaptic vesicles with the presynaptic membrane in about a millisecond (see Figure 2.7 on page 43). The vesicles' contents then diffuse across the synaptic cleft between the neuron and the muscle fiber.

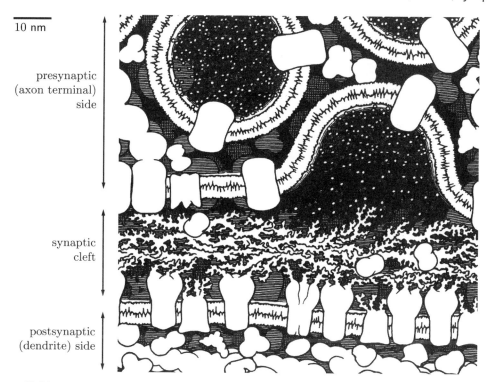

10 nm

presynaptic
(axon terminal)
side

synaptic
cleft

postsynaptic
(dendrite) side

Figure 12.21: (Drawing based on structural data.) Cross section of a chemical synapse (see also Figure 2.7 on page 43). The end of an axon is shown at the top, with two synaptic vesicles full of neurotransmitter molecules inside and one in the process of fusing with the axon's plasma membrane and dumping its contents into the synaptic cleft. The receiving (or postsynaptic) dendrite is shown at the bottom. Neurotransmitters diffusing across the cleft bind to receptor proteins embedded in the dendrite's membrane. Typically these receptors are ligand-gated ion channels. [From Goodsell, 1993.]

On the other side of the synapse, the muscle cell contains ligand-gated ion channels sensitive to acetylcholine (Figure 12.19). The release of a single synaptic vesicle generates a measurable, subthreshold depolarization in the muscle cell. S. Kuffler and coauthors showed that an identical response could be generated by manually injecting a tiny quantity of acetylcholine (fewer than 10 000 molecules) into the neuromuscular junction. The arrival of an action potential, however, releases many vesicles at once. The ensuing large depolarization triggers an action potential in the muscle cell, ultimately activating the myosin molecules that generate muscle contraction (Figure 10.1 on page 405).

Thus the neuromuscular connection involves two distinct steps: presynaptic release of acetylcholine, followed by its postsynaptic activity. One way to separate these steps experimentally involves the alkaloid curare, which paralyzes skeletal muscle. Stimulating a motor neuron in the presence of curare leads to normal action potentials and the release of normal amounts of acetylcholine, but no muscle contraction. It turns out that curare competes with acetylcholine for binding to the postsynaptic ligand-gated ion channels, inhibiting their normal action in a way analogous to competitive inhibition in enzymes (see Problem 10.5). Other neurotoxins, for example, the one in cobra venom, work similarly.

To stop muscle contraction after the neuron stops supplying action potentials, an enzyme called **acetylcholinesterase** is always present in the synaptic cleft, breaking down neurotransmitter molecules shortly after their release into their components, acetate and choline. Meanwhile, the neuron is constantly replenishing its supply of synaptic vesicles. It does this by actively transporting choline back inside to be used for acetylcholine synthesis, and by actively recovering the lipid bilayer that fused with the neuron's outer membrane and repackaging the acetylcholine into new vesicles.

12.4.3 Vista: Neural computation

The situation with synapses between neurons is similar to that just described for the neuromuscular junction. An axon terminal can release a variety of neurotransmitters, thereby altering the local membrane potential in another neuron's dendritic tree. The main difference between the neuromuscular and neuron–neuron junctions is that the former acts as a simple relay, transmitting an impulse without fail, whereas the latter are used to perform more subtle computations.

The effect of an arriving presynaptic action potential can either depolarize or hyperpolarize the postsynaptic dendrite, depending on the details of what neurotransmitter is released and the nature and state of the receiving point's ion channels.[13] In the depolarizing case, the synapse is **excitatory**; in the hyperpolarizing case, it is **inhibitory**.

If the total depolarization in the soma near the axon (the axon hillock) exceeds a threshold, the neuron will "fire," that is, generate an action potential. (Section 12.2.5 outlined how threshold behavior can arise in the context of the axon.) In many neurons, the arrival of a single action potential at a dendrite is not enough to make the cell fire. Instead, each incoming presynaptic impulse generates a localized, temporary disturbance to the membrane potential, similar to electrotonus (Section 12.1.3 on page 514). If enough of these disturbances arrive close enough in space and in time, however, they can add up to an above-threshold stimulus. With this **integrate-and-fire model** of neuron activity, we can begin to understand how a neuron can perform some simple computations:

- Adding up those disturbances that overlap in time lets a cell measure the rate of incoming action potentials at a particular synapse. Thus, although all action potentials along a given axon are stereotyped (identical), nevertheless your nervous system can encode quantitative signals as *rates* of action-potential firing, a "rate-coding scheme."
- Adding up those disturbances that overlap in *space*, that is, those arriving in the same neighborhood of the dendritic tree, lets a cell determine whether two different signals arrive together.

---

[13] $\boxed{T_2}$ It's also possible for a neurotransmitter to have an indirect effect on the postsynaptic membrane; for example, it can alter a voltage-dependent conductance that is not currently active, thereby modulating the response to other synaptic inputs.

One model for neural computation supposes that the cell sums its input signals with particular weights that correspond to the excitatory or inhibitory character of each component synapse. The cell fires if the sum exceeds a threshold.

A crucial aspect of the scenario just sketched is that a neuron can adjust its synaptic couplings—for example, altering the numbers of ligand-gated channels at a dendritic spine—and thereby alter the computation it performs. Neurons can also modulate their connections by adjusting the amount of neurotransmitter released in response to an action potential, and in other ways as well. Taken together, such reconfigurations allow a network of neurons to "learn" new behavior.

Connecting even a few dozen of such simple computational devices can yield a system with sufficiently complex behavior to operate a simple organism, like a mollusk. Connecting a hundred billion of them, as your body has done, can lead to very complex behavior indeed.

THE BIG PICTURE

Let's return to the Focus Question. This chapter has developed a picture of the unmyelinated nerve axon as an excitable medium, capable of transmitting nonlinear waves of excitation over long distances without loss of signal strength or definition. Hodgkin and Huxley's insight had an immense impact, even on applied mathematics, helping to launch the theory of such waves. In biology, too, the notion of nonlinear waves in an excitable medium has led to an understanding of systems as diverse as the cooperative behavior of individual slime mold cells (as they spontaneously coalesce into the multicellular fruiting body) and the cells in your heart (as they contract synchronously).

We located the source of the axon's excitability in a class of allosteric ion channels. Channels of the voltage-gated superfamily are the target of drugs widely used against pain, epilepsy, cardiac arrhythmias, cardiac failure, hypertension, and hyperglycemia. These advances are all rooted in Hodgkin and Huxley's biophysical measurements—which contained no direct evidence for individual ion channels!

KEY FORMULAS

- *Capacitors:* The potential across a capacitor is $V = q/C$. So the current flowing into a capacitor is $I = dq/dt = C(dV/dt)$ (Equation 12.5).

    For capacitors in parallel, the total capacitance is $C = C_1 + C_2$ because both share the same V; this observation explains why the capacitance of a membrane patch is proportional to its area.

- *Membrane conductance:* The symbol j_q always denotes net electric charge per time per area from inside the cell to outside (also called charge flux). The charge flux through the membrane due to ion species i is $j_{q,i}$; thus $j_q = \sum_i j_{q,i}$. In this chapter, V denotes the electrostatic potential inside the membrane (in our model, the potential is zero everywhere outside). V^0 denotes the quasi-steady value of V, and $v = V - V^0$. So $v = 0$ is the quasi-steady state. Also $V_i^{\text{Nernst}} = -(k_B T/(z_i e)) \ln(c_{i,2}/c_{i,1})$ is the Nernst potential for ion species i.

We studied three increasingly realistic models of membrane conductance:

- *Ohmic:* The fluxes are $j_{q,i} = (V - V_i^{\text{Nernst}})g_i$, where g_i are positive constants (Equation 11.8). Thus the current of ion type i flows in the direction tending to bring V back to that ion's equilibrium value, V_i^{Nernst}. Because equilibrium is the state of maximum disorder, this is a dissipative process converting free energy to heat, like a resistor.
- *Simplified voltage gating:* One or more of the conductances are not constant but instead depend on the instantaneous value of v. We explored a model with $j_{q,r} = vg_{\text{tot}}^0 + B(v - H)v^2$ (Equation 12.21). Here B and H are positive constants.
- *Hodgkin–Huxley:* Some conductances depend not only on the value of v but also on its recent past history via relaxation-type relations.

- *Chord:* If we neglect ion pumping, the Ohmic hypothesis yields the chord formula (Equation 12.3):

$$V^0 = \sum_i \frac{g_i}{g_{\text{tot}}} V_i^{\text{Nernst}}, \quad \text{where} \quad g_{\text{tot}} = \sum_i g_i.$$

V^0 describes a quasi-steady potential approximating the true resting potential. The formula shows that this value is a compromise between the various Nernst potentials and is dominated by the ion species with the greatest conductance.

If V is maintained at some value $V^0 + v$ other than its quasi-steady value, the Ohmic hypothesis says we get a net current $j_q = vg_{\text{tot}}$ (see Your Turn 12A). The voltage-gating hypothesis agrees with this prediction at small v; but at larger depolarization, it instead gives positive feedback (Figure 12.9).
- *Cable:* For a cylindrical axon of radius a filled with axoplasm of conductivity κ, with the approximation that the resistance of the exterior fluid is zero, the membrane current $j_{q,r}$ and potential V are related by (Equation 12.7)

$$\pi a^2 \kappa \frac{d^2 V}{dx^2} = 2\pi a \left(j_{q,r} + C\frac{dV}{dt}\right).$$

Here C is the capacitance per area of the membrane. Taking $j_{q,r}$ to be given by one of the three hypotheses listed earlier gives a closed equation (a cable equation), which in principle can be solved. In the Ohmic model, this equation is essentially a diffusion equation. Introducing voltage gating leads to a nonlinear traveling wave solution. The full Hodgkin–Huxley conductance model gives a cable equation with a realistic, self-limiting, traveling wave solution.

FURTHER READING

Semipopular:
Historical: Hodgkin, 1992; Neher & Sakmann, 1992.

Intermediate:

This chapter again follows the approach of Benedek & Villars, 2000c.

See also, for example, Dowling, 2001; Nicholls et al., 2001; Koch, 1999; and Katz's classic book: Katz, 1966.

Computer modeling of electrophysiology: Hoppensteadt & Peskin, 2002.

Technical:

General: Kandel et al., 2000.

Membrane electrophysiology: Aidley, 1998.

Action potentials: Hodgkin & Huxley, 1952b; Keener & Sneyd, 1998.

Nonlinear waves in excitable media: Murray, 2002.

Ion channels: Hille, 2001.

Synapses: Cowan et al., 2001.

$\boxed{T_2}$ **12.2.5′ Track 2**

The main qualitative feature of our formula for the speed of an action potential (Equation 12.25 on page 530) is that $\vartheta \propto \lambda_{\text{axon}}/\tau$; we could have guessed such a result from dimensional analysis. But how does the nonlinear cable equation select any velocity in the first place? To answer the question, notice that Equation 12.24 has a familiar form. Interpreting y as a fictitious "time" and $\bar{v}$ as "position," the equation resembles Newton's law of motion for a particle sliding with friction in a position-dependent force field. Such mathematical analogies allow us to apply intuition from a familiar system to a new one.

Write the right-hand side as

$$-Q\frac{d\bar{v}}{dy} - \frac{dU}{d\bar{v}}, \quad \text{where } U(\bar{v}) \equiv -\frac{s}{4}\bar{v}^4 + \frac{1+s}{3}\bar{v}^3 - \frac{1}{2}\bar{v}^2.$$

Then we can think of our fictitious particle as sliding on an energy landscape defined by U. The landscape has two peaks and a valley in between.

The waveform we are seeking must go smoothly from $\bar{v} = 0$ at $y \to \infty$ (resting potential, with channels closed at $t \to -\infty$) to $\bar{v} = 1$ at $y \to -\infty$ (channels open at $t \to +\infty$). In the language of our particle analogy, we want a solution in which the particle starts out just below $\bar{v} = 1$ at large negative y, rolls slowly off one of the two peaks of the potential U, gains speed, then slows down and approaches the other peak (at $\bar{v} = 0$) when $y \to \infty$.

Now $\bar{v}$ must pass through the value $\frac{1}{2}$ at some intermediate value y_*. Without loss of generality, we may take this point to be $y_* = 0$ (any solution can be shifted in y to make another solution). We now choose the "velocity" $d\bar{v}/dy$ at y_* to be just large enough that the particle comes to rest at the top of the $\bar{v} = 0$ peak. This value is unique: With a smaller push, the particle would stall, then slide back and end up at the bottom of the valley, at $\bar{v} = s^{-1}$, whereas with a bigger push, it would run off to $\bar{v} = -\infty$.

We have now used up all our freedom to choose constants of integration in our solution. Looking at our solution for large negative y, it is very unlikely that our solution will be perfectly at rest right at the top of the other peak, at $\bar{v} = 1$. The only way to arrange for this it to adjust some parameter in the equation of motion. The only available free parameter is the "friction" constant Q: We must tune Q so that the solution does not over- or undershoot. Thus Equation 12.24 has a well-behaved solution only for one particular value of Q (namely, the one quoted in Problem 12.6). The dashed line in Figure 12.10 on page 531 shows the result of attempting to find a solution by the procedure just outlined using a different value of Q: We can at best make one asymptotic region satisfy its boundary condition, but not both.

PROBLEMS*

12.1   *Conduction velocity*
The Chippendale Mupp is a mythical creature who bites his tail before going to sleep. As the poets have sung, his tail is so long that he doesn't feel the pain until it's time to wake up, eight hours after going to sleep. Suppose that a single unmyelinated axon connects the Mupp's tail to its spinal cord. Use axon parameters appropriate to squid. Given that the range of axon diameters in real animals is 0.2–1000 μm, estimate how long the Mupp's tail must be.

12.2   *Discharging the battery*
Imagine the resting axon membrane as a capacitor, an insulating layer that separates charge and hence creates the resting membrane potential difference.

a. How much charge per unit area must pass through the membrane to discharge the capacitor (that is, bring V from $V^0 = -60$ mV to zero)?

b. Reexpress your answer to (a) by giving the surface area per excess proton charge needed to maintain $V^0 = -60$ mV. Then express it a third time, as the charge per unit length of axon, taking the squid giant axon to be a cylinder of radius 0.5 mm.

c. We saw that depolarization is largely the result of the passage of sodium ions. Estimate the effect on the interior ion concentration of a charge transfer of the sort just described, as follows. Again imagine the giant axon as a cylinder filled with salt solution, with ion concentrations given by the data in Table 11.1 on page 477. Find the total number of interior sodium ions per length. Find the corresponding number if the interior sodium concentration matched the exterior value. Subtract these two numbers and compare with the total number of sodium ions passing through the membrane as estimated in (b).

d. Comment on your answer in the light of the observation that an axon can continue to transmit many action potentials after its ion pumps have been shut down.

12.3   *Contributions to capacitance*
a. Estimate the capacitance per area of a lipid bilayer. Consider only the electrically insulating part of the bilayer, the lipid tails, as a layer of oil about 2 nm thick. The dielectric constant of oil is $\varepsilon_{oil}/\varepsilon_0 \approx 2$.

b. $\boxed{T_2}$ As mentioned in Section 12.1.2, the charge-screening layers in the water on either side of the membrane also contribute to its capacitance (see Section 7.4.3' on page 284). In physiological salt concentrations, these layers are each roughly 0.7 nm thick. Use the formula for capacitors in series from first-year physics and your result in (a) to estimate the contribution to the total capacitance from these layers.

12.4   *Afterhyperpolarization*
a. The quasi-steady membrane potential Example on page 512 showed how the resting membrane conductances (Equation 11.9 on page 482) predict a membrane

---

*Problem 12.2 is adapted with permission from Benedek & Villars, 2000c.

potential in rough agreement with the actual resting potential. Repeat the calculation, using the conductances measured during an action potential (Equation 12.14 on page 523), and interpret in the light of Figure 12.6b on page 523.

b. Hodgkin and Katz also found that the membrane conductances immediately *after* the passage of an action potential did not return immediately to their resting values. Instead, they found that g_{Na^+} fell to essentially zero, whereas $g_{K^+} \approx 4g_{Cl^-}$. Repeat your calculation using these values, and again interpret in the light of Figure 12.6b.

12.5 *Conduction as diffusion*

Section 4.6.4 on page 142 argued that the conduction of electricity by a salt solution was just another diffusive process. Mobile charged objects (sodium and chloride ions) get pulled by an external force (from the imposed electrostatic potential gradient) and drift at a net speed v_{drift} much smaller than their thermal velocity. Let's rederive the result of that section and see how successful this claim is. We'll study fully dissociated salts with singly charged ions, like table salt, NaCl. In this problem, take all diffusion constants to have the approximate value for a generic small molecule in water, $D \approx 1\,\mu m^2/ms$.

The resistance of a column of conductor with cross-sectional area A and length L is not intrinsic to the material in the column, but a related quantity, the electrical conductivity, is. We defined conductivity[14] by $\kappa = L/(AR)$.

Not surprisingly, the conductivity of a salt solution goes up when we add more salt. For low concentrations, one finds experimentally at room temperature

$$\kappa = (12.8\,\Omega^{-1}\,m^{-1}) \times (c/1\,M), \tag{12.29}$$

where $c/1\,M$ is the salt concentration expressed in mole/L. We want to understand the magnitude of the numerical prefactor.

A potential difference V across the cell gives an electric field $\mathcal{E} = V/L$. Each mole of salt gives $2N_{mole}$ ions (one Na^+ and one Cl^- per NaCl molecule). Each ion drifts under the applied electric field. Suppose that the solution is dilute, so we can use ideal-solution formulas.

a. Write the force on each ion in terms of V, L, and known constants.

b. Write the resulting drift velocity v_{drift} in terms of V, L, and known constants.

c. Get a formula for the number of Na^+ ions crossing the centerline of the cell in time dt.

d. Write the resulting current I in the cell in terms of V, A, L, c, and known constants.

e. Write a formula for κ in terms of c and known constants. Discuss every factor in this formula and its physical meaning.

f. Put in the numbers and compare with experiment (Equation 12.29).

---

[14]You may be more familiar with the **resistivity**, which is $1/\kappa$. Because the resistance R has the units of ohms (denoted Ω), and an ohm is a $J\,s\,coul^{-2}$, κ has the SI units $coul^2\,J^{-1}m^{-1}s^{-1}$.

g. Now evaluate the conductivity for the ion concentrations characteristic of squid axoplasm (see Table 11.1 on page 477; pretend that you can use the dilute-solution formulas and ignore ions not listed in the table). Compare your answer with the measured value of $\kappa \approx 3\,\Omega^{-1}\mathrm{m}^{-1}$.

h. What would you expect for a solution of magnesium chloride? You can suppose that $(c/1\,\mathrm{M})$ moles of $MgCl_2$ dissociates completely into Mg^{2+} and Cl^- in 1 L of water.

12.6 Analytical solution for simplified action potential

Show that the function $\bar{v}(y) = (1 + e^{\alpha y})^{-1}$ solves Equation 12.24, if we take the parameter Q to be given by $\sqrt{2/s}\left(\frac{s}{2} - 1\right)$. Hence derive the speed of the action potential (Equation 12.25). α is another constant, which you are to find.

12.7 Discrete versus continuous

a. Use the overall membrane conductance per area g_{tot}^0 of the resting squid axon membrane, the 500-fold increase in total conductance during the action potential, and the conductance of a single open sodium channel to estimate the density of sodium channels in the squid axon membrane. Compare with the accepted value of roughly $300\,\mu\mathrm{m}^{-2}$ in squid.

b. For a cylindrical axon of diameter 1 mm, how many channels per unit length does your estimate yield? Comment on the continuous appearance of Hodgkin and Huxley's conductance curves in the light of your estimate.

12.8 Estimate for channel conductivity

a. Model a sodium channel as a cylindrical tube about 0.5 nm in diameter (the diameter of a hydrated ion) and 4 nm long (the thickness of a bilayer membrane). Use the discussion of Section 4.6.1 on page 135 to estimate the permeability of a membrane studded with such channels at an area density σ_{chan}.

b. Use your result from Your Turn 11C on page 481 to estimate the corresponding conductance per area. Take the concentration of ions to be $c = 250\,\mathrm{mM}$.

c. Convert your result to conductance per channel; σ_{chan} will drop out of your answer. Get a numerical answer and compare with the experimental value $G_{\mathrm{Na}+} = 25 \cdot 10^{-12}\,\Omega^{-1}$ quoted in Appendix B.

[Remark: Certainly the result you obtained is very rough: We cannot expect the results of macroscopic diffusion theory to apply to a channel so narrow that ions must pass through it single file! Nevertheless, you'll see that the idea of a water-filled channel can give the magnitude of real conductances observed in experiments.]

12.9 Mechanotransduction

Review Problem 6.7 on page 241. How could the arrangement shown in Figure 6.13b help your ear to transduce sound (mechanical stimulation) into electric signals (action potentials)?

12.10 $\boxed{T_2}$ Extracellular resistance

Repeat our derivation of the nonlinear cable equation, but this time don't set the external fluid's resistivity equal to zero. Instead, let Γ_1 denote the electrical resistance per unit length of the extracellular fluid (we found that the axoplasm's resistance per length is $\Gamma_2 = (\pi a^2 \kappa)^{-1}$). Get a new estimate of the propagation speed ϑ and see how it depends on Γ_1. Compare your answer qualitatively with Hodgkin's 1938 result (Section 12.2.3 on page 521).

Epilogue

Farewells should be brief. Instead of a lengthy repetition of what we have done, here is a short outline of what we *didn't* manage to do in this long book. (For what we *did* do, you may wish to reread the chapter openings and closings in one sitting.)

Put this book down, go outside, and look at an ant. After reading this book, you now have some detailed ideas about how the ant gets the energy needed to move around incessantly, how its nervous system controls the muscles, and so on. You can also write some simple estimates to understand how the ant can carry loads several times its body weight, whereas an elephant cannot. And yet reading this book has given you no insight into the fantastic choreography of muscles needed simply to walk, the interpersonal communication needed to tell other ants about sources of food, nor the complex sociology of the ant's nest. Even the equally fantastic choreography of the biochemical pathways in a *single cell,* to say nothing of cellular control and decision networks, have exceeded our grasp.

Nor could we touch on the ecological questions—why do some ants lovingly tend their host trees, but others intentionally stunt their host's reproduction, to make it a better home for ants? Clearly there is much, much more to biology than molecules and energy. I hope that by uncovering just one corner of this tapestry, I have heightened, not dulled, your sense of awe and wonder at the living world around us.

The master key for addressing all these questions is evolution by natural selection. Originally a modest proposal for understanding the origin of species, this principle has become an organizing paradigm for attacking problems as diverse as the development in cells of an array of self-folding protein sequences, the self-organization of metabolic networks, the self-wiring (and self-training) of the brain, the spontaneous development of human language and culture, and the very origins of life from its precursors.

Many scientists believe that the parallels between these problems go deeper than just words and that a common modularity underlies them all. Following up on this idea will require skills from many disciplines. Indeed, in its small way, this book has sought to weave together many threads, including biochemistry, physiology, physical chemistry, statistical physics, neuroscience, fluid mechanics, materials science, cell biology, nonlinear dynamics, the history of science, and yes, even French cooking. Our unifying theme has been to look at complex phenomena via simple model building. Now it's Your Turn to apply this approach to your own questions.

APPENDIX A

Global List of Symbols and Units

"What's the good of Mercator's North Poles and Equators
Tropics, Zones, and Meridian Lines?"
So the Bellman would cry: and the crew would reply
"They are merely conventional signs!"
— Lewis Carroll, *The Hunting of the Snark*

Notation is a perennial problem for scientists. We can give each quantity whatever symbolic name we choose, but chaos would ensue if every writer chose completely different names for familiar quantities. On the other hand, using standard names unavoidably leads to the problem of too many different quantities all having the same name. The following notation tries to walk a line between these extremes; when the same symbol has been pressed into service for two different quantities, the aim has been to ensure that they are *so* different that context will make it clear which is meant in any given formula.

Notation

Mathematics Vectors are denoted by boldface: $\mathbf{v} = (v_x, v_y, v_z)$. The symbols $\mathbf{v}^2$, or $\mathbf{v} \cdot \mathbf{v}$, or $|\mathbf{v}|^2$, refer to the total length-squared of $\mathbf{v}$, that is, $(v_x)^2 + (v_y)^2 + (v_z)^2$. Vectors of length equal to 1 are flagged with a circumflex, for example, the three unit vectors $\hat{\mathbf{x}}, \hat{\mathbf{y}}, \hat{\mathbf{z}}$ or the tangent vector $\hat{\mathbf{t}}(s)$ to a curve at position s. The symbol $\mathrm{d}^3\mathbf{r}$ is not a vector, but a volume element of integration.

A matrix (linear function of a vector) is denoted by sans serif type: $\mathsf{M} = \begin{bmatrix} M_{11} & M_{12} \\ M_{21} & M_{22} \end{bmatrix}$. For more details, see Section 9.3.1 on page 354.

Often the dimensionless form of a quantity will be given the same name as that quantity but with a bar on top.

The symbol $\equiv$ is a special kind of equals sign indicating that this equality serves as a *definition* of one of the symbols it contains. The symbol $\stackrel{?}{=}$ signals a provisional formula, or guess. The symbol $\approx$ means "approximately equal to;" $\sim$ means "has the same dimensions as." The symbol $\propto$ means "is proportional to."

The symbol $|x|$ refers to the absolute value of a quantity. The notation $\langle f \rangle$ refers to the average value of some function f, with respect to some probability distribution.

The symbol $\frac{\mathrm{d}S}{\mathrm{d}E}\big|_N$ refers to the derivative of S with respect to E, holding N fixed. But the symbol $\frac{\mathrm{d}}{\mathrm{d}\beta}\big|_{\beta=1}F$, or equivalently $\frac{\mathrm{d}F}{\mathrm{d}\beta}\big|_{\beta=1}$, refers to the derivative of F with respect to β, evaluated at the point $\beta = 1$.

The notation [X] denotes the concentration of some chemical species X, divided by the reference concentration of one mole per $10^{-3}\,\mathrm{m}^3$ (also written $1\,\mathrm{M}$). Square brackets with a quantity inside, $[x]$, refer to the dimensions of that quantity.

A dot over a quantity generally denotes that quantity's time derivative.

Electrical circuits

battery, —|⊢—. The wide end is maintained at a potential greater than that of the narrow end (if the battery voltage V is greater than zero).

resistor, —⋀⋀—.

capacitor, —|⊢—.

Named quantities

Roman alphabet

A or a area of some surface; A, generic constant

A bend persistence length of a polymer (bend modulus divided by $k_{\mathrm{B}}T$) [Equation 9.2 on page 346]

$\mathcal{A}$ autocorrelation function [Equation 9.30 on page 387]

a radius of an axon [Figure 12.4 on page 515]

B stretch modulus of a polymer divided by $k_{\mathrm{B}}T$ [Equation 9.2 on page 346]

B partition coefficient [Section 4.6.1 on page 135]

C generic constant

C twist persistence length of a polymer (twist modulus divided by $k_{\mathrm{B}}T$) [Equation 9.2 on page 346]

C capacitance [Equation 12.4 on page 513]

$\mathcal{C}$ capacitance per unit area [Section 12.1.2 on page 509]

c number density (for example, molecules per unit volume), also called concentration [Section 1.4.4 on page 22]; c_0, reference concentration [see Example 8A on page 296]; c_*, critical micelle concentration [Section 8.4.2 on page 317]

D twist-stretch coupling of a polymer [Equation 9.2 on page 346]

D diffusion constant [Equation 4.5 on page 115]; D_{r}, rotational diffusion constant [Problem 4.9 on page 156]

D separation between two objects

d generic distance, especially thickness of a layer

E energy (kinetic and/or potential); $\Delta E^{\ddagger}$, activation energy [Section 6.6.2 on page 220]

$\mathcal{E}$ electric field, units of $\mathrm{N\,coul}^{-1}$, or $\mathrm{volt\,m}^{-1}$ [Equation 7.20 on page 264]

e electric charge on a proton

$\mathbf{e}_{\pm}$ eigenvectors of a 2×2 matrix [Section 9.4.1 on page 358]

F Helmholtz free energy [Section 1.1.3 on page 8]

$\mathcal{F}$ force per unit volume [Section 7.3.1 on page 255]

f force

G conductance of a single object; G_i, conductance of an ion channel of type i [Section 12.3.2 on page 536]

G Gibbs free energy [Equation 6.37 on page 237]; $\Delta G^{\ddagger}$, activation (or transition state) free energy [Section 10.3.2 on page 423]; ΔG free energy change (net chemical force) [Equation 8.14 on page 303]; ΔG^0, standard free energy change [Equation 8.16 on page 304]; $\Delta G'^0$, standard transformed free energy change [Section 8.2.2 on page 301]

$\mathcal{G}$ shear modulus, same units as pressure [Equation 5.14 on page 172]

g acceleration of gravity

g_i conductance per area of a membrane to ions of type i [Section 11.2.2 on page 478]; g_{tot}, sum of all g_i [Equation 12.3 on page 512]; g_i^0, conductance at resting conditions

H enthalpy [Section 6.5.1 on page 210]

$\hbar$ Planck constant [Section 6.2.2 on page 200]

I disorder [Section 6.1 on page 196]

I electric current (charge per time) [Equation 12.5 on page 513]; I_x and I_r, axial and radial currents in a nerve axon [Figure 12.4 on page 515]

j number flux [Section 1.4.4 on page 22]; j_s number flux of solute molecules [Equation 4.21 on page 135]; $j^{(1d)}$ one-dimensional number flux [Equation 10.3 on page 419]

j_q charge flux (charge per time per area) [Equation 11.8 on page 480]; $j_{q,i}$, that part of the flux carried by ions of type i; $j_{q,r}(x)$, total charge flux across an axon's membrane (radial direction) at location x [Section 12.1.3 on page 514], considered to be positive when positive ions move outward.

j_Q flux of thermal energy [Section 4.6.4 on page 142]

j_v volume flux [Equation 7.15 on page 260]

K_M Michaelis constant for an enzyme [Equation 10.20 on page 435]

K the constant $1/\ln 2$ [Equation 6.1 on page 197]

K_{eq} dimensionless equilibrium constant of some chemical reaction [Section 8.2.2 on page 301]; $\hat{K}_{eq}$, dimensional form [Section 8.4.2 on page 317]

K_w ion product of water [Equation 8.25 on page 309]

k_B Boltzmann constant; $k_B T$, thermal energy at temperature T; $k_B T_r$, thermal energy at room temperature [Equation 1.12 on page 27]

k spring constant [Equation 9.11 on page 354]; k_t, torsional spring constant [Problem 9.9 on page 399]

k rate constant (probability per time) for a chemical reaction [Section 6.6.2 on page 220]

L, ℓ generic variables for lengths; $L_{seg}^{(1d)}$ effective (Kuhn) segment length of a polymer modeled as a one-dimensional, freely jointed chain [Equation 9.8 on page 353]; L_{seg}, segment length for three-dimensional freely jointed chain model [Equation 9.32 on page 388]

L_p filtration coefficient [Section 7.3.2 on page 259]

ℓ_B Bjerrum length in water [Equation 7.21 on page 266]

m mass of an object

N, n number of things

N_{mole} the dimensionless number $6.0 \cdot 10^{23}$ (Avogadro's number) [Section 1.5.1 on page 23]

P probability; $P_{2 \to 1}(t)dt$, probability of waiting in state S_2 until time t, then hopping to state S_1 before $t + dt$ [Equation 6.31 on page 223]

$\mathcal{P}$ permeability of a membrane [Equation 4.21 on page 135]; $\mathcal{P}_w$, to water; $\mathcal{P}_s$, to some solute

P membrane permeability matrix [Section 7.3.1' on page 283]

p pressure

p scaling exponent for a random walk [Problem 5.8 on page 192]

p momentum

Q volume flow rate [Equation 5.18 on page 181]

Q heat (transfer of thermal energy) [Section 6.5.4 on page 216]; Q_{vap}, heat of vaporization of water [Problem 1.6 on page 33]

q electric charge [Equation 1.9 on page 21]

R radius of a particle or pipe; radius of curvature of a bent rod

R electrical resistance; R_r, resistance in the radial direction (through an element of axon membrane); R_x resistance in the axial direction (through a neuron's axoplasm) [Figure 12.4 on page 515]

R_G radius of gyration of a polymer [Section 4.3.1 on page 122]

$\mathcal{R}$ Reynolds number [Equation 5.11 on page 168]

r position vector of an object, with components (x, y, z)

S entropy [Section 6.2.2 on page 200]

s arc length (also called contour length) [Section 9.1.2 on page 344]

s sedimentation coefficient [Equation 5.3 on page 160]

T absolute (Kelvin) temperature (unless otherwise specified). In illustrative calculations, we often use the value $T_r \equiv 295$ K ("room temperature"). T_m, midpoint temperature of a helix–coil transition [Equation 9.24 on page 368]

T transfer matrix [Section 9.4.1 on page 358]

t time

$\hat{\mathbf{t}}$ unit tangent vector to a curve [Section 9.1.2 on page 344]

U potential energy, for example, gravitational

u speed of a molecule, also written $|\mathbf{v}|$

u stretch (extensional deformation) of a rod [Section 9.1.2 on page 344]

V, v volume

$V(x)$ electrostatic potential at x [Equation 1.9 on page 21]; V_1, potential outside a cell; V_2, potential inside; $\Delta V = V_2 - V_1$, membrane potential difference [Equation 11.1 on page 474, abbreviated as V in Chapter 12]; $\tilde{V}(t)$, time course of potential at fixed location [Section 12.2.4 on page 524]; $\overline{V}$, dimensionless rescaled potential [Equation 7.22 on page 267]

$\mathcal{V}_i^{\text{Nernst}}$ Nernst potential of species i [Section 4.6.3 on page 139]; $\mathcal{V}^0$, quasi-steady resting potential [Equation 12.3 on page 512]

v_{max} maximum velocity of an enzyme-catalyzed reaction [Equation 10.20 on page 435]

$v(x, t)$ potential across a membrane, minus quasi-steady potential [Section 12.1.3 on page 514]; $\tilde{v}(t)$, time course of v at fixed location [Section 12.2.5 on page 527]; $\bar{v}$, dimensionless rescaled form [Section 12.2.5 on page 527]

v velocity vector, with components (v_x, v_y, v_z); v_{drift}, drift velocity [Section 4.1.4 on page 118]

W work (transfer of mechanical energy)

w weight (a force)

x generic variable

x some distance (for example, along the $\hat{\mathbf{x}}$ axis); x_0, the Gouy–Chapman length [Equation 7.25 on page 268]

Y degree of oxygen saturation [Your Turn 9M on page 376]

Z hydrodynamic resistance of a pipe [Section 5.3.4 on page 179]

Z partition function [Equation 6.33 on page 224]

$\mathcal{Z}$ grand partition function [Section 8.1.2 on page 298]

z generic distance, especially distance in the vertical direction; end-to-end length of a polymer [Section 9.2.1 on page 350]; z_*, scale height of a suspension [Section 5.1.1 on page 158]

z_i valence of an ion of type i, that is, its charge as a multiple of the proton charge, $z_i \equiv q_i/e$

Greek alphabet

α bias in a two-state chain, for example minus the free energy change to extend an alpha-helix by one unit [Equation 9.18 on page 359 and Equation 9.24 on page 368]

β parameter entering in the trial solution of the Poisson–Boltzmann equation [Section 7.4.4 on page 269]

$\boldsymbol{\beta}$ bending deformation of a rod [Section 9.1.2 on page 344]

Γ electrical resistance per unit length of a column of electrolyte [Problem 12.10 on page 555]

γ cooperativity parameter [Equation 9.17 on page 359 and Section 9.5.2 on page 366]

γ various constants of proportionality appearing in Equation 1.7 on page 15, Section 7.4.4' on page 286, and Your Turn 12B on page 520

Δ prefix indicating a small, but finite, change in the quantity following it. Thus for example Δt is a time step.

δ a small distance

ε permittivity of a medium; ε_0, permittivity of air or empty space [Section 7.4.1 on page 261]. (The dielectric constant of a medium is defined as the ratio $\varepsilon/\varepsilon_0$.)

ϵ internal stored energy; ϵ_α, internal energy of molecules of type α [Section 8.1.1 on page 295]

ζ coefficient of friction at low Reynolds number [Equation 4.13 on page 119]; ζ_r, rotational coefficient of friction [Problem 4.9 on page 156]

η viscosity [Section 5.1.2 on page 160]; η_w, viscosity of water; $[\eta]$, intrinsic viscosity of a polymer [Problem 5.8 on page 192]

Θ bending angle of one link relative to the next [Equation 9.36 on page 391]

θ optical rotation of a solution [Section 9.5.1 on page 364]

θ polar angle in spherical coordinates fixed in the lab [Problem 6.9 on page 243]

ϑ polar angle in spherical coordinates relative to some specified direction not fixed in the lab [Section 9.1.3' on page 386]

ϑ velocity of propagation of a traveling wave [Section 12.2.4 on page 524]

κ electrical conductivity [Section 4.6.4 on page 142]

κ bending stiffness of a membrane [Section 8.6.1 on page 322]

$\lambda_\pm$ eigenvalues of a 2×2 matrix [Section 9.4.1 on page 358]

λ_D Debye screening length in solution [Equation 7.35 on page 285]

λ_{axon} space constant of an axon [Equation 12.8 on page 517]

μ_α chemical potential of molecules of type α [Equation 8.1 on page 295]; μ_α^0, at standard concentration [Equation 8.3 on page 296]; μ_S, μ_P, chemical potential of enzyme substrate and of product [Section 10.3.4 on page 431]

ν_k stoichiometric coefficients [Equation 8.14 on page 303]

ν kinematic viscosity [Equation 5.21 on page 187]

ρ_m mass density (mass per unit volume) [Section 1.4.4 on page 22]; $\rho_{m,w}$, mass density of water; $\rho_m^{(1d)}$ linear mass density (mass per length) [Your Turn 12B on page 520]

ρ_q bulk charge density (charge per unit volume) [Equation 7.20 on page 264]; $\rho_{q,macro}$, charge density of impermeant macromolecules in a cell [Equation 11.3 on page 475]

Σ surface tension [Section 7.2.1 on page 248]

σ surface density (things per unit area) [Section 1.4.4 on page 22]; σ_q, surface density of electric charge [Section 7.4.2 on page 263]

σ_{chan} area density of ion channels in a membrane [Problem 12.8 on page 555]

σ width of a Gaussian distribution; standard deviation of any probability distribution; σ^2, variance of a distribution [Section 3.1.3 on page 73]

σ two-state variable describing the conformation of a monomer in a polymer chain [Section 9.2.2 on page 352; Section 9.5.3 on page 369]

τ torque

τ time constant for some relaxation process [see Example 4C on page 136]; time constant for electrotonus [Equation 12.8 on page 517]

ϕ volume fraction, dimensionless [Section 7.2.1 on page 248]

ϕ azimuthal angle in spherical coordinates, relative to some specified direction not fixed in the lab [Section 9.1.3′ on page 386]

φ azimuthal angle in spherical coordinates fixed in the lab [Problem 6.9 on page 243]

Ψ grand potential [Problem 8.8 on page 338]

ψ generic angle

Ω number of available states [Section 6.1 on page 196]

ω twist density (torsional deformation) of a polymer [Section 9.1.2 on page 344]

ω rotational angular velocity [Section 5.3.5 on page 182]

Dimensions

Most physical quantities carry dimensions. This book refers to abstract dimensions by the symbols $\mathbb{L}$ (length), $\mathbb{T}$ (time), $\mathbb{M}$ (mass), and $\mathbb{Q}$ (charge). (The abstract dimension for temperature has no symbol in this book.)

Some quantities are dimensionless, for example, geometrical angles: The angle of a pie wedge equals the circumference divided by the radius, so the dimensions cancel.

Units

> *There shall be standard measures of wine, beer, and corn throughout the whole of our kingdom ... and there shall be standard weights also.*
> — Magna Carta, 1215

See Section 1.4 on page 18. This book primarily uses the Système Internationale of units; but, when appropriate, convenient, or traditional, some outside units are also used.

SI base units Corresponding to the abstract dimensions previously listed, this book uses five of the seven SI base units:

length: The meter (m) has dimensions $\mathbb{L}$. It is defined as the length of the path traveled by light in vacuum during a time interval $(1/299\,792\,458)$s.

time: The second (s) has dimensions $\mathbb{T}$. It is defined as the duration of $9\,192\,631\,770$ periods of the radiation corresponding to the transition between the two hyperfine levels of the ground state of the cesium-133 atom.

mass: The kilogram (kg) has dimensions $\mathbb{M}$. It is defined as the mass of a particular object, called the international prototype of the kilogram.

electric current: The ampere (A) has dimensions $\mathbb{Q}\mathbb{T}^{-1}$. It is defined as the constant current which, if maintained in two straight parallel wires of infinite length placed 1 m apart in vacuum, would produce a magnetic force between these conductors equal to $2 \cdot 10^{-7}$ N per meter of length.

thermodynamic temperature: The kelvin (K) is defined as the fraction 1/273.16 of the thermodynamic temperature of the triple point of water counting up from absolute zero.

Prefixes The following prefixes modify the base units (and other units):

giga (G) $= 10^9$
mega (M) $= 10^6$
kilo (k) $= 10^3$
deci (d) $= 10^{-1}$
centi (c) $= 10^{-2}$
milli (m) $= 10^{-3}$
micro (μ) $= 10^{-6}$
nano (n) $= 10^{-9}$
pico (p) $= 10^{-12}$
femto (f) $= 10^{-15}$

SI derived units

volume: A liter (L) equals 10^{-3} m^3.
force: A newton (N) equals $1\,\text{kg}\,\text{m}\,\text{s}^{-2}$.
energy: A joule (J) equals $1\,\text{N}\,\text{m} = 1\,\text{kg}\,\text{m}^2\text{s}^{-2}$.
power: A watt (W) equals $1\,\text{J}\,\text{s}^{-1} = 1\,\text{kg}\,\text{m}^2\text{s}^{-3}$.
pressure: A pascal (Pa) equals $1\,\text{N/m}^2 = 1\,\text{kg}\,\text{m}^{-1}\text{s}^{-2}$.
charge: A coulomb (coul) equals $1\,\text{A}\,\text{s}$.
electrostatic potential: A volt (volt) equals $1\,\text{J}\,\text{s}^{-1}\text{A}^{-1} = 1\,\text{m}^2\,\text{kg}\,\text{s}^{-3}\text{A}^{-1}$. Its derived forms are abbreviated mV and so on.
capacitance: A farad (F) equals 1 coul/volt.
resistance: An ohm (Ω) equals $1\,\text{J}\,\text{s}\,\text{coul}^{-2} = 1$ volt A^{-1}.
conductance: A siemens (S) equals $1\,\Omega^{-1} = 1$ A/volt.

Traditional but non-SI units

length: An Ångstrom unit (Å) equals 0.1 nm.
time: A svedberg equals 10^{-13} s. (Some texts use the abbreviation S for svedberg, but we reserve this notation for the siemens.)
energy: A calorie (cal) equals 4.184 J. Thus $1\,\text{kcal}\,\text{mole}^{-1} = 0.043\,\text{eV} = 7 \cdot 10^{-21}\,\text{J} = 4.2\,\text{kJ}\,\text{mole}^{-1}$. An electron volt (eV) equals $e \times (1\text{ volt}) = 1.60 \cdot 10^{-19}$ J $= 96.5$ kJ/mole. An erg (erg) equals 10^{-7} J.

pressure: An atmosphere (atm) equals $1.01 \cdot 10^5$ Pa. 752 mm of mercury equals 10^5 Pa. (We also abbreviate this unit as "mm of Hg.")

viscosity: A poise (P) equals $1 \, \mathrm{erg \, s \, cm^{-3}} = 0.1 \, \mathrm{Pa \, s}$.

number density: A 1 M solution has a number density of $1 \, \mathrm{mole \, L^{-1}} = 1000 \, \mathrm{mole \, m^{-3}}$.

Dimensionless units

A degree of angle corresponds to $1/360$ of a revolution; a radian is $1/2\pi$ of a revolution.

In this book, the symbols mole and N_{mole} both refer to the dimensionless number $6.0 \cdot 10^{23}$.

APPENDIX B

Numerical Values

A single number has more genuine and permanent value than
an expansive library full of hypotheses.
— Robert Mayer, 1814–1878

Not all the values mentioned in this appendix are actually used in the text.

Fundamental constants

Boltzmann constant, $k_B = 1.38 \cdot 10^{-23}\,\mathrm{J\,K^{-1}}$. Thermal energy at room temperature ($T_r \equiv 295\,\mathrm{K}$): $k_B T_r = 4.1\,\mathrm{pN\,nm} = 4.1 \cdot 10^{-21}\,\mathrm{J} = 4.1 \cdot 10^{-14}\,\mathrm{erg} = 2.5\,\mathrm{kJ\,mole^{-1}} = 0.59\,\mathrm{kcal\,mole^{-1}} = 0.025\,\mathrm{eV}$.

Charge on a proton, $e = 1.6 \cdot 10^{-19}\,\mathrm{coul}$. (The charge on an electron is $-e$.) A useful restatement is $e = 40 k_B T_r/\mathrm{volt}$.

Permittivity of vacuum, $\varepsilon_0 = 8.9 \cdot 10^{-12}\,\mathrm{F\,m^{-1}}$ (or $\mathrm{coul^2 N^{-1} m^{-2}}$). The combination $e^2/(4\pi\varepsilon_0)$ equals $2.3 \cdot 10^{-28}\,\mathrm{J\,m}$. We treat water as a continuum dielectric with $\varepsilon \approx 80\varepsilon_0$.

Stefan–Boltzmann constant, $\sigma = 5.7 \cdot 10^{-8}\,\mathrm{W\,m^{-2}K^{-4}}$.

Magnitudes

Sizes (smallest to largest)

hydrogen atom (radius), 0.05 nm.

water molecule (radius), 0.135 nm.

covalent bond length, ≈ 0.1 nm.

H-bond (distance between centers of atoms flanking H), 0.27 nm.

sugar, amino acid, nucleotide (diameter), 0.5–1 nm.

electron microscope resolution, 0.7 nm.

Debye screening length (of physiological Ringer's solution), $\lambda_D \approx 0.7$ nm.

Bjerrum length of water at room temperature, $\ell_B \equiv e^2/(4\pi\varepsilon k_B T_r) = 0.71$ nm.

DNA (diameter), 2 nm.

globular protein (diameter), 2–10 nm (lysozyme, 4 nm; RNA polymerase, 10 nm).

bilayer membrane (thickness), ≈ 3 nm.

F-actin (diameter), 5 nm.

nucleosome (diameter), 10 nm.

E. coli flagellum (radius), 10 nm.

synaptic cleft in chemical synapse (width), 20–40 nm (neuromuscular junction, 50–100 nm).

poliovirus (diameter), 25 nm (smallest virus, 20 nm).

microtubule (diameter), 25 nm.

smallest feature that can be drawn with electron-beam lithography (width), 30 nm.

ribosome (diameter), 30 nm.

casein micelle (diameter), 100 nm.

thinnest wire in Pentium processor chip (width), $\approx$ 100 nm.

eukaryotic flagellum (diameter), 100–500 nm.

transistor in consumer electronics (diameter), $\approx$ 180 nm.

optical microscope resolution, $\approx$ 200 nm.

vertebrate axon (diameter), 0.2–20 μm.

wavelength of visible light, 400–650 nm.

smallest feature that can be created by photolithography, 0.5 μm.

typical bacterium (diameter), 1 μm (smallest, 0.5 μm).

myofibril (diameter), 1–2μm.

capillary (diameter), as small as 3 μm.

E. coli flagellum (length), 10 μm (20 000 subunits).

typical human cell (diameter), $\approx$ 10 μm (red blood cell, 7.5 μm).

lambda phage virus DNA (contour length) $\approx$ 16.5 μm.

T4 phage DNA (contour length), 54 μm (160 kbp); T4 capsid (length), $\approx$ 100 nm.

human hair (diameter), 100 μm.

naked eye resolution, 200 μm.

squid "giant" axon (diameter), 1 mm.

E. coli genome (length if extended), 1.4 mm.

human genome (total length), $\approx$ 1 m.

Earth (radius), $6.4 \cdot 10^6$ m.

Energies Most of the following values are expressed as multiples of $k_B T_r$, the thermal energy at room temperature.

complete oxidation of one glucose, $1159 k_B T_r$.

triple covalent bond (for example, C≡N), 9eV $\doteq$ $325 k_B T_r$; double bond (for example, C=C), $240 k_B T_r$; single bond (for example C–C), $140 k_B T_r$.

visible photon (green), $120 k_B T_r$.

streptavidin/biotin bond, $40 k_B T_r$.

ATP hydrolysis under normal cell conditions, $\Delta G = -11$ to -13 kcal mole^{-1} $\approx$ $-20 k_B T_r$/molecule. (The standard free energy change is $\Delta G'^0 = -12.4 k_B T_r$;

but cells are far from standard conditions.) ATP production in humans, $\approx$ 40 kg of ATP each day.

generic (van der Waals, or dispersion) attraction energy between atoms, 0.6–1.6$k_B T_r$.

human resting heat output, 100 W.

energy content: of glucose, $1.7 \cdot 10^7$ J/kg; of beer, $0.18 \cdot 10^7$ J/kg; of gasoline, $4.8 \cdot 10^7$ J/kg.

peak mechanical power of human athlete, 200 W; of bumblebee, 0.02 W.

solar energy output, $3.9 \cdot 10^{26}$ W; power density striking Earth, $1.4 \cdot 10^3$ W/m^2.

Specialized values

Viscosity

of water at 20°C, $1.0 \cdot 10^{-3}$ Pa s; of air, $1.7 \cdot 10^{-5}$ Pa s; of honey, 0.1 Pa s; of glycerol, 1.4 Pa s.

The effective viscosity of cell cytoplasm depends on the size of the object considered: For molecules smaller than 1 nm, it's similar to that of water; for particles of diameter 6 nm (such as a protein of mass 10^5 g mole^{-1}), it's about 3 times that of water. For 50–500 nm particles, it's 30–300 times that of water; the entire cell behaves as though its viscosity were a million times that of water.

viscous critical force: for water, 10^{-9} N, for air, $2 \cdot 10^{-10}$ N; for glycerine, 10^{-3} N.

More about water

energy to break an intramolecular hydrogen bond in water, 1–2$k_B T_r$ (hydrogen bond when two water molecules condense in vacuum, 8 $k_B T_r$).

electrostatic attraction energy of two 0.3 nm ions in water, $\approx k_B T_r$.

heat of vaporization of water, $Q_{vap} = 2.3 \cdot 10^6$ J kg^{-1}.

oil–water surface tension, $\Sigma = 0.04$ J m^{-2}; air–water surface tension, 0.072 J m^{-2}.

number density of water molecules in pure water, 55 M; mass density of water at 20°C, 998 kg m^{-3}.

diffusion constant for generic small molecules in water, $D \approx 1 \, \mu$m^2 ms^{-1}. Specifically, for O_2, it's 2 μm^{2}ms^{-1}; for water molecules themselves, 2.2 μm^2 ms^{-1}; for glucose, 0.67 μm^2 ms^{-1}; for globular protein in water, $D \approx 10^{-2} \, \mu$m^2 ms^{-1}.

heat capacity of water at room temperature, 4180 J kg^{-1}K^{-1} or 0.996 cal cm^{-3}K^{-1}.

thermal conductivity of water at 0°C, 0.56 J s^{-1} m^{-1} K^{-1}; at 100°C 6.8 J s^{-1} m^{-1} K^{-1}.

Rates

The turnover number for an enzyme can vary from about $5 \cdot 10^{-2}$ s^{-1} (chymotrypsin on N-acetylglycine ethyl ester) to $1 \cdot 10^7$ s^{-1} (catalase). For acetylcholinesterase, it's 25 000 s^{-1}.

Membranes (artificial)

bilayer bend stiffness (dimyristoyl phosphatidylcholine, or DMPC), $\kappa = 0.6 \cdot 10^{-19}$ J $= 14 k_B T_r$.

bilayer stretch modulus (DMPC), 144 mN m^{-1}.

rupture tension (DMPC), ≈ 5 mN/m.

Permeability to water, P_w: DMPC, $70\,\mu$m s^{-1}; dialysis tubing, $11\,\mu$m s^{-1}. (Filtration coefficient L_p for dialysis tubing, $3.4 \cdot 10^{-5}$ cm s^{-1}atm^{-1}.)

Permeability of a bilayer membrane to solutes, P_s: small inorganic cations, like sodium or potassium, $10^{-8}\,\mu$m s^{-1}; Cl$^-$, 10^{-6}; for glucose, $10^{-3}\,\mu$m s^{-1}. (For sucrose through 2 mil cellophane, $1.0\,\mu$m s^{-1}. For glucose through dialysis tubing, $1.8\,\mu$m s^{-1}.)

Membranes (cell)

permeability to water of human red blood cell membrane, $53\,\mu$m s^{-1}.

filtration coefficient L_p: human red blood cell membrane, $91 \cdot 10^{-7}$ cm s^{-1}atm^{-1}; capillary blood vessel walls, $69 \cdot 10^{-7}$ cm s^{-1}atm^{-1}.

Polymers

B-form DNA: bend persistence length, ≈ 50 nm $= 150$ basepairs (in 10 mM NaCl) (intrinsic, or high-salt limit, 40 nm); twist persistence length, 75–100 nm; stretch modulus ≈ 1300 pN; basepair rise, 0.34 nm/bp; helical pitch in solution, 10.3–10.6 bp.

Others:
microtubule diameter, 25 nm; persistence length, 1 mm.
intermediate filament persistence length, $0.1\,\mu$m; diameter 10 nm.
actin diameter, 7 nm; persistence length, 3–10 μm.
neurofilament persistence length, $0.5\,\mu$m.

Motors

Myosin:
myosin-II (fast skeletal muscles) speed in vitro, $8\,\mu$m s^{-1}; force, 2–5 pN.
myosin-V (vesicle transport) speed, $0.35\,\mu$m s^{-1}.
myosin-VIII and XI (cytoplasmic streaming in plants) speed, $60\,\mu$m s^{-1}.

Conventional (2-headed) kinesin: step size, 8 nm; fuel consumption, 44 ATP s^{-1} per head; stall force, 6–7 pN; speed in vitro, 100 steps/s $= 800$ nm s^{-1}; processivity, 100 steps/release.

E. coli flagellar motor: rotation rate, 100 revolutions/s (1200 protons/revolution); torque, 4000 pN nm.

F1 ATPase motor: stall torque, 100 pN nm; torque generated against frictional load, ≈ 40 pN nm.

RNA polymerase: stall force, 25 pN.

speed: *E. coli* Pol I, 16 basepair/s; Pol II, 0.05 times as great; Pol III, 15 times as great; eukaryotic polymerase, 50 bp/s; T7 virus polymerase, 250 bp/s.

DNA polymerase: stall force 34 pN.

speed: bacteria, 1000 nucleotides/s; eukaryotic cells, 100 nucleotides/s.

HIV reverse transcriptase: 20–40 nucleotides/s.

Ribosome: 2 amino acids/s (eukaryotic cells) or $20 \, s^{-1}$ (bacteria).

Neurons

pumps: $\approx 10^2$ ions/s per pump.

carriers: $\approx 10^4$ ions/s per carrier.

channels: $\approx 10^6$ ions/s per channel; density of sodium channels in squid axon, $\approx 300 \, \mu m^{-2}$; unit conductance of open channel, $G_{Na^+} = 25 \, pS = 25 \cdot 10^{-12} \, \Omega^{-1}$.

resting conductance per area, squid giant axon, $g_{tot}^0 \approx 5 \, \Omega^{-1} m^{-2}$; individual resting conductances follow $g_{K^+} \approx 25 g_{Na^+} \approx 2 g_{Cl^-}$. During an action potential, g_{Na^+} momentarily increases by about a factor of 500.

capacitance per area, $\approx 1 \cdot 10^{-2} \, F \, m^{-2}$.

conductivity of squid axoplasm, $\kappa \approx 3 \, \Omega^{-1} m^{-1}$.

human brain: power consumption, 10 W (about 10% of whole-body resting total). There are $\approx 10^{13}$ cells in the human body, of which $\approx 10^{11}$ are nerve cells, making $\approx 6 \cdot 10^{13}$ synapses.

Miscellaneous

acceleration of gravity at Earth's surface, $g = 9.8 \, m \, s^{-2}$.

typical acceleration in an ultracentrifuge, $3 \cdot 10^6 \, m \, s^{-2}$.

pH: human blood, 7.35–7.45; human stomach contents, 1.0–3.0; lemons, 2.2–2.4; drinking water, 6.5–8.0. Ion product of water at room temperature, 10^{-14}.

pK: dissociation of acetic acid, 4.76; of phosphoric acid, 2.15.

deprotonation of aspartic acid, 4.4; of glutamic acid, 4.3; of histidine, 6.5; of cysteine, 8.3; of tyrosine, 10.0; of lysine, 11.0; of arginine, 12; of serine, > 13.0.

Credits

COPYRIGHT PERMISSIONS

Cover: ©1996. Used by permission of the Optical Society of America.
Frontispiece: Used by permission of the Estate of Ruth Kavenoff.
Part I opener: ©1962. Used by permission of Mrs. Eric Sloane. Fig. 2.2: ©1993.
Used by permission of Springer-Verlag. Fig. 2.3: ©1993. Used by permission of
Springer-Verlag. Fig. 2.4: ©1993. Used by permission of Springer-Verlag. Fig. 2.6:
(b) ©1980. Used by permission of Elsevier Science. Fig. 2.7: Courtesy of Dr. John
Heuser. Fig. 2.8: G. F. Bahr/Biological Photo Service. Fig. 2.9: Courtesy of Dr. Ju-
lian Heath. Fig. 2.10: ©1982. Used by permission of Jones and Bartlett Publishers,
Sudbury MA. Fig. 2.11: ©1991 Larry Gonick. Fig. 2.14: ©1993. Used by permission
of Springer-Verlag. Fig. 2.15: ©1982, American Association for the Advancement
of Science. Used by permission. Fig. 2.16: Protein Data Base accession code 1EHZ
(H. Shi and P. B. Moore, *RNA* **6** 1091 (2000)). Fig. 2.17: Protein Database accession
code 1VII (C. J. McKnight, D. S. Doering, P. T. Matsudaira, P. S. Kim, *J. Mol. Biol.* **260**
126 (1996)). Fig. 2.18: (c) ©1996. Used by permission of Springer-Verlag. Fig. 2.20:
©1993. Used by permission of Springer-Verlag. Fig. 2.23: ©1993. Used by permission of Springer-Verlag. Fig. 2.24: ©1993. Used by permission of Springer-Verlag.
Part II opener: Omikron/Photo Researchers, Inc. Fig. 3.1: ©1991 by Larry Gonick.
Fig. 3.11: ©1961. Used by permission of Dover Publications. Fig. 3.13: ©1961. Used
by permission of Dover Publications. Fig. 3.14: (b) Andrew Syred/Science Photo
Library/Photo Researchers, Inc. Fig. 4.1: ©1961. Used by permission of Dover Publications. Fig. 4.8: ©1999. Used by permission of the American Physical Society.
Fig. 5.1: ©1972, MIT Press. Used by permission. Fig. 5.3: ©1968. Used by permis-
sion of Elsevier Science. Fig. 5.9: ©1998. Used by permission of Elsevier Science.
Fig. 6.12: ©2000. Used by permission of Lippincott Williams and Wilkins. Fig. 7.2:
©1993. Used by permission of Springer-Verlag. Fig. 7.4: ©1998, American Physical
Society. Used by permission. Fig. 7.12: (b) ©1999, Philip Ball. Reprinted by permission of Farrar Straus and Giroux LLC. Epigraph to Chapter 8: ©1936, Ogden Nash,
renewed. Reprinted by permission of Curtis Brown, Ltd. Fig. 8.2: ©1949, American
Association for the Advancement of Science. Used by permission. Fig. 8.5: ©1992.
Reprinted with permission from Elsevier Science. Fig. 8.10: ©2003, Miloslav Kalab;
used by permission. Part III opening: © HM Queen Elizabeth II; reproduced by
permission from the Royal Collection Picture Library. Fig. 9.2: ©2003 QuantomiX
Ltd. Used by permission. Fig. 9.11: ©1995, Mahlon Hoagland and Bert Dodson.
Used by permission of Times Books, a division of Random House, Inc. Fig. 9.12:
Reprinted with permission from *Nature.* ©1999, Macmillan Magazines Ltd. Epi-

graph to Chapter 10 appears translated in G. Gamow, *Thirty years that shook physics* (New York, Dover Publications, 1961). Used by permission of Dover Publications. Fig. 10.1: ©1984, Princeton University Press. Used by permission. Fig. 10.3: ©1993. Used by permission of the Biophysical Society. Fig. 10.4: ©1997, American Physical Society. Used by permission. Fig. 10.5: ©1989. Used by permission of The Rockefeller University Press. Fig. 10.19: ©1993. Used by permission of Springer-Verlag. Fig. 10.22: Adapted with permission from *Nature.* ©1999 by MacMillan Magazines Ltd. Table 10.2: ©1999 Elsevier Science. Used by permission. Fig. 11.9: ©1993. Used by permission of Springer-Verlag. Fig. 11.10: ©2000. Used by permission of Elsevier Science. Fig. 11.11: Reprinted with permission from *Nature.* ©1995, Macmillan Magazines Ltd. Fig. 12.6: Reprinted by permission from *Nature.* ©1939, Macmillan Magazines Ltd. Fig. 12.11: ©1991, Sinauer Associates. Used by permission. Fig. 12.13: ©1962. Used by permission of The Physiological Society. Fig. 12.17: Reprinted by permission from *Nature.* ©1980, Macmillan Magazines Ltd. Fig. 12.20: ©1985, Oxford University Press, Inc. Fig. 12.21: ©1993. Used by permission of Springer-Verlag. Color Figure 1: Conly S. Rieder/Biological Photo Service. Color Figure 2: Courtesy of Dr. Steve Nielsen. Color Figure 3: Courtesy of Dr. Scott Brady. Color Figure 4: ©1999, American Association for the Advancement of Science. Used by permission. Color Figure 5: ©1993. Used by permission of Springer-Verlag. Color Figure 6: ©1993. Used by permission of Springer-Verlag.

GRANT SUPPORT

This book is partially based on work supported by the United States National Science Foundation under Grants DUE–00-86511 and DMR–98-07156. Any opinions, findings, conclusions, levity, or recommendations expressed in this book are those of the author and do not necessarily reflect the views of the National Science Foundation. The Albert Einstein Minerva Center for Theoretical Physics and the Minerva Center for Nonlinear Physics of Complex Systems provided additional support for this project.

Bibliography

AIDLEY, D. J. 1998. *The physiology of excitable cells.* 4th ed. Cambridge, UK: Cambridge University Press.

ALBERTS, B. 1998. Preparing the next generation of molecular biologists. *Cell,* **92,** 291–294.

ALBERTS, B., BRAY, D., HOPKIN, K., JOHNSON, A., LEWIS, J., RAFF, M., ROBERTS, K., & WALTER, P. 2004. *Essential cell biology.* 2nd ed. New York: Garland Publishing.

ALBERTS, B., JOHNSON, A., LEWIS, J., RAFF, M., ROBERTS, K., & WALTER, P. 2002. *Molecular biology of the cell.* 4th ed. New York: Garland Publishing.

AMBEGAOKAR, V. 1996. *Reasoning about luck: Probability and its uses in physics.* Cambridge, UK: Cambridge University Press.

ARMSTRONG, C. M., & HILLE, B. 1998. Voltage-gated ion channels and electrical excitability. *Neuron,* **20,** 371–380.

ASTUMIAN, R. D. 1997. Thermodynamics and kinetics of a Brownian motor. *Science,* **276,** 917–922.

ATKINS, P. W. 1994. *The second law.* New York: W. H. Freeman.

ATKINS, P. W. 2001. *The elements of physical chemistry, with applications to biology.* New York: W. H. Freeman.

AUSTIN, R. H. 2002. Biological physics in silico. *In:* FLYVBJERG, H., JÜLICHER, F., ORMOS, P., & DAVID, F. (Eds.), *Physics of bio-molecules and cells.* New York: Springer.

AUSTIN, R. H., BEESON, K. W., EISENSTEIN, L., FRAUENFELDER, H., GUNSALUS, I. C., & MARSHALL, V. P. 1974. Activation energy spectrum of a biomolecule: Photodissociation of carbonmonoxy myoglobin at low temperatures. *Phys. Rev. Lett.,* **32,** 403–405.

AUSTIN, R. H., BRODY, J. P., COX, E. C., DUKE, T., & VOLKMUTH, W. 1997. Stretch genes. *Physics Today,* February, 32–38.

BAKER, P. F., HODGKIN, A. L., & SHAW, T. I. 1962. Replacement of the axoplasm of giant axon fibres with artificial solutions. *J. Physiol. (London),* **164,** 330–354.

BALL, P. 2000. *Life's matrix: A biography of water.* New York: Farrar Straus and Giroux.

BATCHELOR, G. K. 1967. *An introduction to fluid dynamics.* Cambridge, UK: Cambridge University Press.

BENEDEK, G. B., & VILLARS, F. M. H. 2000a. *Physics with illustrative examples from medicine and biology.* 2nd ed. Vol. 1. New York: Springer.

577

BENEDEK, G. B., & VILLARS, F. M. H. 2000b. *Physics with illustrative examples from medicine and biology*. 2nd ed. Vol. 2. New York: Springer.

BENEDEK, G. B., & VILLARS, F. M. H. 2000c. *Physics with illustrative examples from medicine and biology*. 2nd ed. Vol. 3. New York: Springer.

BERG, H. 1993. *Random walks in biology*. 2nd ed. Princeton, NJ: Princeton University Press.

BERG, H. C. 2000. Motile behavior of bacteria. *Physics Today*, January, 24–29.

BERG, H. C., & ANDERSON, R. 1973. Bacteria swim by rotating their flagellar filaments. *Nature*, **245**, 380–382.

BERG, H. C., & PURCELL, E. M. 1977. Physics of chemoreception. *Biophys. J.*, **20**, 193–219.

BERG, J. M., TYMOCZKO, J. L., & STRYER, L. 2002. *Biochemistry*. New York: W. H. Freeman.

BLAIR, D. F., & BERG, H. C. 1988. Restoration of torque in defective flagellar motors. *Science*, **242**, 1678–1681.

BLOOMFIELD, V. A., CROTHERS, D. M., & TINOCO, JR., I. 2000. *Nucleic acids: Structures, properties, and functions*. Sausalito, CA: University Science Books.

BOAL, D. 2001. *Mechanics of the cell*. Cambridge, UK: Cambridge University Press.

BORISY, G. G., & SVITKINA, T. M. 2000. Actin machinery: Pushing the envelope. *Curr. Opin. Cell Biol.*, **12**, 104–112.

BOUCHIAT, C., WANG, M. D., ALLEMAND, J.-F., STRICK, T., BLOCK, S. M., & CROQUETTE, V. 1999. Estimating the persistence length of a wormlike chain molecule from force-extension measurements. *Biophys. J.*, **76**, 409–413.

BOYER, P. D. 1997. The ATP synthase: A splendid molecular machine. *Annu. Rev. Biochem.*, **66**, 717–749.

BRADY, S. T., & PFISTER, K. K. 1991. Kinesin interactions with membrane bounded organelles in vivo and in vitro. *J. Cell Sci. Suppl.*, **14**, 103–108.

BRANDEN, C., & TOOZE, J. 1999. *Introduction to protein structure*. 2nd ed. New York: Garland Publishing.

BRAY, D. 2001. *Cell movements: From molecules to motility*. 2d ed. New York: Garland Publishing.

BRUINSMA, R. F. 2002. Physics of protein-DNA interaction. *In:* FLYVBJERG, H., JÜLICHER, F., ORMOS, P., & DAVID, F. (Eds.), *Physics of bio-molecules and cells*. New York: Springer.

BUSTAMANTE, C., SMITH, S. B., LIPHARDT, J., & SMITH, D. 2000. Single-molecule studies of DNA mechanics. *Curr. Opin. Struct. Biol.*, **10**, 279–285.

CALLADINE, C., & DREW, H. 1997. *Understanding DNA: The molecule and how it works*. 2nd ed. San Diego: Academic Press.

CALLEN, H. B. 1985. *Thermodynamics and introduction to thermostatistics*. 2nd ed. New York: John Wiley and Sons.

CHAIKIN, P., & LUBENSKY, T. C. 1995. *Principles of condensed matter physics.* Cambridge, UK: Cambridge University Press.

CHANDLER, D. 1987. *Introduction to modern statistical mechanics.* New York: Oxford University Press.

CIZEAU, P., & VIOVY, J.-L. 1997. Modeling extreme extension of DNA. *Biopolymers,* **42**, 383–385.

CLUZEL, P., LEBRUN, A., HELLER, C., LAVERY, R., VIOVY, J.-L., CHATENAY, D., & CARON, F. 1996. DNA: An extensible molecule. *Science,* **271**, 792–794.

COLQUHOUN, D., & HAWKES, A. G. 1983. Principles of the stochastic interpretation of ion-channel mechanisms. *In:* SAKMANN, B., & NEHER, E. (Eds.), *Single-channel recording.* New York: Plenum Publishing.

COOPER, G. M. 2000. *The cell: A molecular approach.* 2nd ed. Sunderland, MA: Sinauer Associates.

COTTERILL, R. 2002. *Biophysics: An introduction.* New York: John Wiley and Sons.

COUTANCEAU, M. 1968. Mouvement uniforme d'une sphère dans l'axe d'un cylindre contenant un liquide visqueux. *J. Mécanique,* **7**, 49–67.

COWAN, W. M., SÜDHOF, T. C., & STEVENS, C. F. (Eds.). 2001. *Synapses.* Baltimore, MD: Johns Hopkins University Press.

COWLEY, A. C., FULLER, N. L., RAND, R. P., & PARSEGIAN, V. A. 1978. Measurements of repulsive forces between charged phospholipid bilayers. *Biochemistry,* **17**, 3163–3168.

DEAMER, D. W., & FLEISCHAKER, G. R. 1994. *Origins of life: The central concepts.* Boston, MA: Jones and Bartlett.

DEGENNES, P.-G., & BADOZ, J. 1996. *Fragile objects.* New York: Springer.

DEPABLO, P. J., SCHAAP, I. A. T., & SCHMIDT, C. F. 2003. Observation of microtubules with scanning force microscopy in liquid. *Nanotechnology,* **14**, 143–146.

DEROSIER, D. 1998. The turn of the screw: The bacterial flagellar motor. *Cell,* **93**, 17–20.

DICKERSON, R. E., DREW, H. R., CONNER, B. N., WING, R. M., FRATINI, A. V., & KOPKA, M. L. 1982. The anatomy of A-, B-, and Z-DNA. *Science,* **216**, 476–485.

DILL, K. A. 1990. Dominant forces in protein folding. *Biochemistry,* **29**, 7133–7155.

DILL, K. A., & BROMBERG, S. 2002. *Molecular driving forces: Principles of statistical thermodynamics in chemistry and biology.* New York: Garland Publishing.

DINSMORE, A. D., WONG, D. T., & YODH, A. G. 1998. Hard spheres in vesicles: Curvature-induced forces and particle-induced curvature. *Phys. Rev. Lett.,* **80**, 409–412.

DISCHER, D. E. 2000. New insights into erythrocyte membrane organization and microelasticity. *Curr. Opin. Hematol.,* **7**, 117–122.

DIXON, M., & WEBB, E. C. 1979. *Enzymes.* 3rd ed. New York, NY: Academic Press.

DODGE, J. D. 1968. *An atlas of biological ultrastructure.* London, UK: Edward Arnold Ltd.

DOI, M. 1996. *Introduction to polymer physics.* Oxford, UK: Oxford University Press.

DOWLING, J. E. 2001. *Neurons and networks: An introduction to behavioral neuroscience.* 2nd ed. Cambridge, MA: Harvard University Press.

DRESSLER, D., & POTTER, H. 1991. *Discovering enzymes.* New York: Scientific American Press.

DUKE, T. 2002. Modelling motor protein systems. *In:* FLYVBJERG, H., JÜLICHER, F., ORMOS, P., & DAVID, F. (Eds.), *Physics of bio-molecules and cells.* New York: Springer.

EATON, W. A., HENRY, E. R., HOFRICHTER, J., & MOZZARELLI, A. 1999. Is cooperative oxygen binding by hemoglobin really understood? *Nature Struct. Biol.,* **6**, 351–358.

EINSTEIN, A. 1956. *Investigations on the theory of the Brownian movement.* Mineola, NY: Dover Publications. Contains reprints of "On the movement of small particles suspended in a stationary liquid demanded by the molecular theory of heat," *Ann. Phys.* **17** (1905), 549; "On the theory of the Brownian movement," *Ibid.* **19** (1906), 371–381; "A new determination of molecular dimensions," *Ibid.* 289–306 (Erratum *Ibid.* **34** (1911), 591–592); and two other papers.

EISENBERG, D., & CROTHERS, D. 1979. *Physical chemistry.* Menlo Park, CA: Benjamin/Cummings.

ELLIS, R. J. 2001. Macromolecular crowding: An important but neglected aspect of the intracellular environment. *COSB,* **11**, 114–119.

EVANS, D. F., & WENNERSTRÖM, H. 1999. *The colloidal domain: Where physics, chemistry, and biology meet.* 2d ed. New York: Wiley-VCH.

FEYNMAN, R. P. 1965. *The character of physical law.* Cambridge, MA: MIT Press.

FEYNMAN, R. P., LEIGHTON, R., & SANDS, M. 1963a. *The Feynman lectures on physics.* Vol. 1. San Francisco: Addison-Wesley.

FEYNMAN, R. P., LEIGHTON, R., & SANDS, M. 1963b. *The Feynman lectures on physics.* Vol. 2. San Francisco: Addison-Wesley.

FEYNMAN, R. P., HEY, J. G., & ALLEN, R. W. 1996. *Feynman lectures on computation.* San Francisco: Addison-Wesley.

FINER, J. T., SIMMONS, R. M., & SPUDICH, J. A. 1994. Single myosin molecule mechanics—piconewton forces and nanometer steps. *Nature,* **368**, 113–119.

FINKELSTEIN, A. 1987. *Water movement through lipid bilayers, pores, and plasma membranes: Theory and reality.* New York: John Wiley and Sons.

FINZI, L., & GELLES, J. 1995. Measurement of lactose repressor-mediated loop formation and breakdown in single DNA molecules. *Science,* **267**, 378–380.

FISHER, M. E., & KOLOMEISKY, A. B. 2001. Simple mechanochemistry describes the dynamics of kinesin molecules. *Proc. Natl. Acad. Sci. USA,* **98**, 7748–7753.

FLORY, PAUL J. 1953. *Principles of polymer chemistry.* Ithaca, NY: Cornell University Press.

FRANK-KAMENETSKII, M. D. 1997. *Unraveling DNA: The most important molecule of life.* 2nd ed. Reading, MA: Addison-Wesley.

FRANKS, F. 2000. *Water: A matrix of life.* 2nd ed. Cambridge, UK: Royal Society of Chemistry.

FRAUENFELDER, H., & WOLYNES, P. G. 1985. Rate theories and puzzles of hemeprotein kinetics. *Science,* **229,** 337–345.

FUNG, D. C., & BERG, H. C. 1995. Powering the flagellar motor of *Escherichia coli* with an external voltage source. *Nature,* **375,** 809–812.

FUNG, Y. C. 1997. *Biomechanics: Circulation.* 2nd ed. New York: Springer.

FYGENSON, D. K., MARKO, J. F., & LIBCHABER, A. 1997. Mechanics of microtubule-based membrane extension. *Phys. Rev. Lett.,* **79,** 4497–4500.

GAMOW, G. 1961. *One, two, three, infinity.* Mineola, NY: Dover Publications.

GELBART, W. M., BRUINSMA, R. F., PINCUS, P. A., & PARSEGIAN, V. A. 2000. DNA-inspired electrostatics. *Physics Today,* **53,** 38–44.

GILBERT, S. P., MOYER, M. L., & JOHNSON, K. A. 1998. Alternating site mechanism of the kinesin ATPase. *Biochemistry,* **37,** 792–799.

GONICK, L., & SMITH, W. 1993. *Cartoon guide to statistics.* New York: HarperCollins.

GONICK, L., & WHEELIS, M. 1991. *Cartoon guide to genetics.* New York: HarperCollins.

GOODSELL, D. S. 1993. *The machinery of life.* New York: Springer.

GOODSELL, D. S. 1996. *Our molecular nature.* New York: Springer.

GRABERT, H. 1982. *Projection operator techniques in nonequilibrium statistical mechanics.* New York: Springer.

GROSBERG, A. YU., & KHOKHLOV, A. R. 1994. *Statistical physics of macromolecules.* New York: AIP Press.

GROSBERG, A. YU., & KHOKHLOV, A. R. 1997. *Giant molecules: Here, and there, and everywhere.* San Diego, CA: Academic Press.

HÄNGGI, P., TALKNER, P., & BORKOVEC, M. 1990. Reaction rate theory: Fifty years after Kramers. *Rev. Mod. Phys.,* **62,** 251–342.

HAPPEL, J., & BRENNER, H. 1983. *Low Reynolds-number hydrodynamics: With special applications to particulate media.* Norwell, MA: Kluwer Academic Publishers.

HARTSHORNE, R. P., KELLER, B. U., TALVENHEIMO, J. A., CATTERALL, W. A., & MONTAL, M. 1985. Functional reconstitution of the purified brain sodium channel in planar lipid bilayers. *Proc. Natl. Acad. Sci. USA,* **82,** 240–244.

HILLE, B. 2001. *Ionic channels of excitable membranes.* 3d ed. Sunderland, MA: Sinauer Associates.

HIROKAWA, N., PFISTER, K. K., YORIFUJI, H., WAGNER, M. C., BRADY, S. T., & BLOOM, G. S. 1989. Submolecular domains of bovine brain kinesin identified by electron microscopy and monoclonal antibody decoration. *Cell,* **56,** 867–878.

HOAGLAND, M., & DODSON, B. 1995. *The way life works.* New York: Random House.

HOBBIE, R. K. 1997. *Intermediate physics for medicine and biology*. 3rd ed. New York: AIP Press.

HODGKIN, A. 1992. *Chance and design: Reminiscences of science in peace and war*. Cambridge, UK: Cambridge University Press.

HODGKIN, A. L., & HUXLEY, A. F. 1939. Action potentials recorded from inside a nerve fibre. *Nature*, **144**, 710–711.

HODGKIN, A. L., & HUXLEY, A. F. 1952a. Currents carried by sodium and potassium ions through the membrane of the giant axon of *Loligo*. *J. Physiol. (London)*, **116**, 449–472. *Reprinted in:* I. COOKE & M. LIPKIN, JR., (Eds.). 1972. *Cellular neurophysiology: A source book*. (New York: Holt, Rinehart, and Winston).

HODGKIN, A. L., & HUXLEY, A. F. 1952b. A quantitative description of membrane current and its applicaiton to conduction and excitation in nerve. *J. Physiol. (London)*, **117**, 500–544. *Reprinted in:* I. COOKE & M. LIPKIN, JR., (Eds.). 1972. *Cellular neurophysiology: A source book*. (New York: Holt, Rinehart, and Winston).

HODGKIN, A. L., & KATZ, B. 1949. Effect of sodium ions on the electrical activity of the giant axon of the squid. *J. Physiol. (London)*, **108**, 37–77. *Reprinted in:* I. COOKE & M. LIPKIN, JR., (Eds.). 1972. *Cellular neurophysiology: A source book*. (New York: Holt, Rinehart, and Winston).

HODGKIN, A. L., & KEYNES, R. D. 1955. Active transport of cations in giant axons from *Sepia* and *Loligo*. *J. Physiol*, **128**, 28–60. *Reprinted in:* I. COOKE & M. LIPKIN, JR., (Eds.). 1972. *Cellular neurophysiology: A source book*. (New York: Holt, Rinehart, and Winston).

HODGKIN, A. L., & RUSHTON, W. A. H. 1946. The electrical constants of a crustacean nerve fibre. *Proc. R. Soc. London, Ser. B*, **133**, 444–479. *Reprinted in:* I. COOKE & M. LIPKIN, JR., (Eds.). 1972. *Cellular neurophysiology: A source book*. (New York: Holt, Rinehart, and Winston).

HOPFIELD, J. J. 2002. Form follows function. *Physics Today*, November, 10–11.

HOPPENSTEADT, F. C., & PESKIN, C. S. 2002. *Modeling and simulation in medicine and the life sciences*. 2nd ed. New York: Springer.

HOWARD, J. 2001. *Mechanics of motor proteins and the cytoskeleton*. Sunderland, MA: Sinauer Associates.

HOWARD, J., HUDSPETH, A. J., & VALE, R. D. 1989. Movement of microtubules by single kinesin molecules. *Nature*, **342**, 154–158.

HOWELL, M. L., SCHROTH, G. P., & HO, P. S. 1996. Sequence-dependent effects of spermine on the thermodynamics of the B-DNA to Z-DNA transition. *Biochemistry*, **35**, 15373–15382.

HUA, W., CHUNG, J., & GELLES, J. 2002. Distinguishing inchworm and hand-over-hand processive kinesin movement by neck rotation measurements. *Science*, **295**, 844–848.

ISRAELACHVILI, J.N. 1991. *Intermolecular and surface forces*. 2nd ed. London: Academic Press.

JUDSON, H. F. 1995. *The eighth day of creation: The makers of the revolution in biology.* 2nd ed. Cold Spring Harbor, NY: Cold Spring Harbor Laboratory Press.

JÜLICHER, F., AJDARI, A., & PROST, J. 1997. Modeling molecular motors. *Rev. Mod. Phys.*, **69**, 1269–1282.

KANDEL, E. R., SCHWARTZ, J. H., & JESSELL, T. M. (Eds.). 2000. *Principles of neural science.* 4th ed. New York: McGraw-Hill.

KARP, G. 2002. *Cell and molecular biology: Concepts and experiments.* 3rd ed. New York: John Wiley and Sons.

KATZ, B. 1966. *Nerve, muscle, and synapse.* New York: McGraw-Hill.

KEENER, J., & SNEYD, J. 1998. *Mathematical physiology.* New York: Springer.

KOCH, C. 1999. *Biophysics of computation: Information processing in single neurons.* New York: Oxford University Press.

KORNBERG, A. 1989. *For the love of enzymes.* Cambridge, MA: Harvard University Press.

KRAMERS, H. 1940. Brownian motion in a field of force and the diffusion model of chemical reactions. *Physica (Utrecht)*, **7**, 284–304.

LACKIE, J. M., & DOW, J. A. T. (Eds.). 1999. *The dictionary of cell and molecular biology.* 3rd ed. San Diego, CA: Academic Press.

LAIDLER, K. 1972. Unconventional applications of the Arrhenius law. *J. Chem. Educ.*, **49**, 343–344.

LAMB, T. D., MATTHEWS, H. R., & TORRE, V. 1986. Incorporation of calcium buffers into salamander retinal rods: A rejection of the calcium hypothesis of phototransduction. *J. Physiol. (London)*, **372**, 315–349.

LANDAU, L. D., & LIFSHITZ, E. M. 1980. *Statistical Physics, Part 1.* 3rd ed. Oxford, UK: Pergamon Press.

LANDAU, L. D., & LIFSHITZ, E. M. 1986. *Theory of elasticity.* 3rd ed. Oxford, UK: Butterworth and Heinemann.

LANDAU, L. D., & LIFSHITZ, E. M. 1987. *Fluid mechanics.* 2nd ed. Oxford, UK: Pergamon Press.

LÄUGER, P. 1991. *Electrogenic ion pumps.* Sunderland, MA: Sinauer Associates.

LEFF, H. S., & REX, A. F. 1990. *Maxwell's demon: Entropy, information, computing.* Princeton, NJ: Princeton University Press.

LEUBA, S. H., & ZLATANOVA, J. (Eds.). 2001. *Biology at the single molecule level.* New York: Pergamon Press.

LI, H., DEROSIER, D. J., NICHOLSON, W. V., NOGALES, E., & DOWNING, K. H. 2002. Microtubule structure at 8Å resolution. *Structure*, **10**, 1317–1328.

LIDE, D. R. (Ed.). 2001. *CRC handbook of chemistry and physics.* 82d ed. Boca Raton, FL: CRC Press.

LIPHARDT, J., ONOA, B., SMITH, S. B., TINOCO, JR., I., & BUSTAMANTE, C. 2001. Reversible unfolding of single RNA molecules by mechanical force. *Science*, **292**, 733–737.

LIPOWSKY, R., & SACKMANN, E. (Eds.). 1995. *Handbook of biological physics,* Vols. 1A, B: *Structure and dynamics of membranes.* Amsterdam: Elsevier.

LODISH, H., BERK, A., ZIPURSKY, S. L., MATSUDAIRA, P., BALTIMORE, D., & DARNELL, J. 2000. *Molecular cell biology.* New York: W. H. Freeman.

LUMRY, R., SMITH, E. L., & GLANTZ, R. R. 1951. Kinetics of carboxypeptidase action. 1. Effect of various extrinsic factors on kinetic parameters. *J. Am. Chem. Soc.,* **73**, 4330–4340.

MAGNASCO, M. O. 1993. Forced thermal ratchets. *Phys. Rev. Lett.,* **71**, 1477–1481.

MAGNASCO, M. O. 1994. Molecular combustion motors. *Phys. Rev. Lett.,* **72**, 2656–2659.

MAGNASCO, M. O. 1996. Brownian combustion engines. *In:* MILLONAS, M. (Ed.), *Fluctuations and order.* New York: Springer.

MAHADEVAN, L., & MATSUDAIRA, P. 2000. Motility powered by supramolecular springs and ratchets. *Science,* **288**, 95–97.

MAIER, B., & RÄDLER, J. O. 1999. Conformation and self-diffusion of single DNA molecules confined to two dimensions. *Phys. Rev. Lett.,* **82**, 1911–1914.

MALKIEL, B. G. 1996. *A random walk down Wall Street.* 6th ed. New York: W. W. Norton.

MANNING, G. S. 1968. Binary diffusion and bulk flow through a potential-energy profile: A kinetic basis for the thermodynamic equations of flow through membranes. *J. Chem. Phys.,* **49**, 2668–2675.

MARKO, J. F., & SIGGIA, E. D. 1994. Bending and twisting elasticity of DNA. *Macromolecules,* **27**, 981–988. Erratum *Ibid.,* **29** (1996), 4820.

MARKO, J. F., & SIGGIA, E. D. 1995. Stretching DNA. *Macromolecules,* **28**, 8759–8770.

MARTIN, P., MEHTA, A. D., & HUDSPETH, A. J. 2000. Negative hair-bundle stiffness betrays a mechanism for mechanical amplification by the hair cell. *Proc. Natl. Acad. Sci. USA,* **97**, 12026–12031.

MCBAIN, J. W. 1944. Solutions of soaps and detergents as colloidal electrolytes. *In:* ALEXANDER, J. (Ed.), *Colloid chemistry: Pure and applied,* Vol. 5. New York: Reinhold.

MCGEE, H. 1984. *On food and cooking: The science and lore of the kitchen.* New York: Collier Books.

MCMAHON, T. A. 1984. *Muscles, reflexes, and locomotion.* Princeton, NJ: Princeton University Press.

MEHTA, A. D., RIEF, M., SPUDICH, J. A., SMITH, D. A., & SIMMONS, R. M. 1999. Single-molecule biomechanics with optical methods. *Science,* **283**, 1689–1695.

MEYERHOFF, G., & SCHULTZ, G. V. 1952. Molekulargewichtsbestimmungen an polymethacrylsaureestern mittels sedimentation in der ultrazentrifuge und diffusion. *Makromol. Chem.*, **7**, 294.

MILLER, R. C., & KUSCH, P. 1955. Velocity distributions in potassium and thallium atomic beams. *Phys. Rev.*, **99**, 1314–1321.

MILLS, F. C., JOHNSON, M. L., & ACKERS, G. K. 1976. Oxygenation-linked subunit interactions in human hemoglobin. *Biochemistry*, **15**, 5350–5362.

MOGILNER, A., FISHER, J. J., & BASKIN, R. J. 2001. Structural changes in the neck linker of kinesin explain the load dependence of the motor's mechanical cycle. *J. Theor. Biol.*, **211**, 143–157.

MOGILNER, A., ELSTON, T., WANG, H., & OSTER, G. 2002. Chapters 12–13. *In:* FALL, C. P., MARLAND, E., TYSON, J., & WAGNER, J. (Eds.), *Joel Keizer's computational cell biology*. New York: Springer.

MORTIMER, R. G. 2000. *Physical chemistry*. 2nd ed. San Diego, CA: Harcourt Academic.

MURRAY, J. D. 2002. *Mathematical Biology*. 3rd ed. New York: Springer.

NATIONAL RESEARCH COUNCIL. 2003. *Bio2010: Transforming undergraduate education for future research biologists*. Washington, DC: National Academies Press.

NEHER, E., & SAKMANN, B. 1992. The patch clamp technique. *Scientific American*, March, 44–51.

NELSON, D. L., & COX, M. M. 2000. *Lehninger principles of biochemistry*. 3d ed. New York: W. H. Freeman.

NICHOLLS, J. G., MARTIN, A. R., WALLACE, B. G., & FUCHS, P. A. 2001. *From neuron to brain*. 4th ed. Sunderland, MA: Sinauer Associates.

NIELSEN, S. O., & KLEIN, M. L. 2002. A coarse grain model for lipid monolayer and bilayer studies. *In:* NIELABA, P., MARESCHAL, M., & CICCOTTI, G. (Eds.), *Bridging the time scales: Molecular simulations for the next decade*. New York: Springer-Verlag Telos.

NOJI, H., YASUDA, R., YOSHIDA, M., & KINOSITA, JR., K. 1997. Direct observation of the rotation of F1–ATPase. *Nature*, **386**, 299–302.

OKADA, Y., & HIROKAWA, N. 1999. A processive single-headed motor: Kinesin superfamily protein KIF1A. *Science*, **283**, 1152–1157. See also the Supplemental Materials cited in footnote 12 of the paper.

OSTER, G., & WANG, H. 2000. Reverse-engineering a protein: The mechanochemistry of ATP synthase. *Biochim. Biophys. Acta (Bioenergetics)*, **1458**, 482–510.

PAIS, A. 1982. *Subtle is the lord: The science and the life of Albert Einstein*. Oxford, UK: Oxford University Press.

PARSEGIAN, V. A., RAND, R. P., & RAU, D. C. 2000. Osmotic stress, crowding, preferential hydration, and binding: Comparison of perspectives. *Proc. Natl. Acad. Sci. USA*, **97**, 3987–3992.

PAULING, L., ITANO, H. A., SINGER, S. J., & WELLS, I. C. 1949. Sickle cell anemia, a molecular disease. *Science*, **110**, 543–548.

PENNYCUICK, C. J. 1992. *Newton rules biology: A physical approach to biological problems.* Oxford, UK: Oxford University Press.

PERRIN, J. 1948. *Les atomes.* 3rd ed. Paris: Presses Universitaires de France.

PERUTZ, M. F. 1998. *I wish I'd made you angry earlier: Essays.* Plainview, NY: Cold Spring Harbor Laboratory Press.

PESKIN, C. S., ODELL, G. M., & OSTER, G. F. 1993. Cellular motions and thermal fluctuations: The Brownian ratchet. *Biophys. J.*, **65**, 316–324.

PESKIN, C. S., ERMENTROUT, G. B., & OSTER, G. F. 1994. The correlation ratchet: A novel mechanism for generating directed motion by ATP hydrolysis. *In:* Mow, V. C., GUILAK, F., TRAN-SON-TAY, R., & HOCHMUTH, R. (Eds.), *Cell mechanics and cellular engineering.* New York: Springer.

POLAND, D., & SCHERAGA, H. A. 1970. *Theory of helix–coil transition in biopolymers.* New York: Academic Press.

POLLARD, T. D., & EARNSHAW, W. C. 2002. *Cell biology.* Philadelphia, PA: W. B. Saunders.

POWERS, T. R., HUBER, G., & GOLDSTEIN, R. E. 2002. Fluid-membrane tethers: Minimal surfaces and elastic boundary layers. *Phys. Rev. E*, **65**. Article no. 041901.

PROST, J. 2002. The physics of *Listeria* propulsion. *In:* FLYVBJERG, H., JÜLICHER, F., ORMOS, P., & DAVID, F. (Eds.), *Physics of bio-molecules and cells.* New York: Springer.

PURCELL, E. 1977. Life at low Reynolds number. *Am. J. Physics*, **45**, 3–10.

RAMÓN Y CAJAL, S. 1995. *Histology of the nervous system of man and vertebrates.* Vol. 1. New York: Oxford University Press. Two-volume set translated by N. Swanson and L. W. Swanson. Originally published in 1909.

REIF, F. 1965. *Fundamentals of statistical and thermal physics.* New York: McGraw-Hill.

RICE, S., LIN, A. W., SAFER, D., HART, C. L., NABER, N., CARRAGHER, B. O., CAIN, S. M., PECHATNIKOVA, E., WILSON-KUBALEK, E. M., WHITTAKER, M., PATE, E., COOKE, R., TAYLOR, E. W., MILLIGAN, R. A., & VALE, R. D. 1999. A structural change in the kinesin motor protein that drives motility. *Nature*, **402**, 778–784.

ROSSI-FANELLI, A., & ANTONINI, E. 1958. Studies on the oxygen and carbon monoxide equilibria of human myoglobin. *Arch. Biochem. Biophys.*, **77**, 478–492.

RUELLE, D. 1991. *Chance and chaos.* Princeton, NJ: Princeton University Press.

SACKMANN, E., BAUSCH, A. R., & VONNA, L. 2002. Physics of composite cell membrane and actin based cytoskeleton. *In:* FLYVBJERG, H., JÜLICHER, F., ORMOS, P., & DAVID, F. (Eds.), *Physics of bio-molecules and cells.* New York: Springer.

SAFRAN, S. A. 1994a. *Statistical thermodynamics of surfaces, interfaces, and membranes.* Boulder, CO: Westview Press.

SCHIEF, W. R., & HOWARD, J. 2001. Conformational changes during kinesin motility. *Curr. Opin. Cell Biol.*, **13**, 19–28.

SCHNITZER, M. J., VISSCHER, K., & BLOCK, S. M. 2000. Force production by single kinesin motors. *Nature Cell Biol.*, **2**, 718–723.

SCHRÖDINGER, E. 1967. *What is Life? The physical aspect of the living cell.* Cambridge, UK: Cambridge University Press.

SCHROEDER, D. V. 2000. *An introduction to thermal physics.* San Francisco: Addison-Wesley.

SEGRÈ, G. 2002. *A matter of degrees: What temperature reveals about the past and future of our species, planet, and universe.* New York: Viking.

SEIFERT, U. 1997. Configurations of fluid membranes and vesicles. *Adv. Physics*, **46**, 12–137.

SHANKAR, R. 1995. *Basic training in mathematics: A fitness program for science students.* New York: Plenum Publishing.

SHAPIRO, A. H. (Ed.). 1972. *Illustrated experiments in fluid mechanics: The National Committee for Fluid Mechanics Films book of film notes.* Cambridge, MA: MIT Press.

SHIH, C. Y., & KESSEL, R. 1982. *Living images.* Boston, MA: Science Books International.

SIGWORTH, F. J., & NEHER, E. 1980. Single sodium channel currents observed in cultured rat muscle cells. *Nature*, **287**, 447–449.

SILVERMAN, M., & SIMON, M. 1974. Flagellar rotation and mechanism of bacterial motility. *Nature*, **249**, 73–74.

SKLAR, L. 1993. *Physics and chance: Philosophical issues in the foundations of statistical mechanics.* Cambridge, UK: Cambridge University Press.

SKOU, J. 1957. The influence of some cations on an adenosine triphosphatase from peripheral nerves. *Biochim. Biophys. Acta*, **23**, 394–401.

SKOU, J. C. 1989. The identification of the sodium pump as the membrane-bound sodium–potassium ATPase. *Biochim. Biophys. Acta*, **774**, 91–95.

SMITH, A. D., DATTA, S. P., SMITH, G. H., & CAMPBELL, P. N. (Eds.). 2000. *Oxford dictionary of biochemistry and molecular biology.* Oxford, UK: Oxford University Press.

SMITH, S., CUI, Y., & BUSTAMANTE, C. 1996. Overstretching B-DNA: The elastic response of individual double-stranded and single-stranded DNA molecules. *Science*, **271**, 795–799.

SOUTHALL, N. T., DILL, K. A., & HAYMET, A. D. J. 2002. A view of the hydrophobic effect. *J. Phys. Chem. B*, **106**, 521–533.

STONG, C. L. 1956. Amateur scientist. *Scientific American*, **194**, 149–158.

SVOBODA, K., & BLOCK, S. M. 1994. Biological applications of optical forces. *Annu. Rev. Biophys. Biomol. Struct.*, **23**, 247–285.

Svoboda, K., Mitra, P. P., & Block, S. M. 1994. Fluctuation analysis of motor protein movement and single enzyme kinetics. *Proc. Natl. Acad. Sci. USA*, **91**, 11782–11786.

Svoboda, K., Denk, W., Knox, W. H., & Tsuda, S. 1996. Two-photon laser scanning fluorescence microscopy of living neurons using a diode-pumped Cr:LiSAF laser mode-locked with a saturable Bragg reflector. *Opt. Lett.*, **21**, 1411–1413.

Tanaka, K. 1980. Scanning electron microscopy of intracellular structures. *Int. Rev. Cytol.*, **68**, 97–125.

Tanford, C. 1961. *Physical chemistry of macromolecules.* New York: John Wiley and Sons.

Tanford, C. 1980. *The hydrophobic effect: Formation of micelles and biological membranes.* 2nd ed. New York: John Wiley and Sons.

Tanford, C. 1989. *Ben Franklin stilled the waters.* Durham, NC: Duke University Press.

Tilney, L., & Portnoy, D. 1989. Actin filaments and the growth, movement, and spread of the intracellular bacterial parasite, *Listeria monocytogenes. J. Cell Biol.*, **109**, 1597–1608.

Timoféeff-Ressovsky, N. W., Zimmer, K. G., & Delbrück, M. 1935. Über die Natur der Genmutation und der Genstruktur. *Nachrichten Gesselshaft der Wissenschaft Göttingen*, **6**(NF(13)), 189–245.

Tinoco, Jr., I., Sauer, K., Wang, J. C., & Puglisi, J. D. 2001. *Physical chemistry: Principles and applications in biological sciences.* 4th ed. Upper Valley, NJ: Prentice Hall.

Uemura, S., Kawaguchi, K., Yajima, J., Edamatsu, M., Toyoshima, Y. Y., & Ishiwata, S. 2002. Kinesin–microtubule binding depends on both nucleotide state and loading direction. *Proc. Natl. Acad. Sci. USA*, **99**, 5977–5981.

Vale, R. D. 1999. Millennial musings on molecular motors. *Trends Cell Biol.*, **9**, M38–M42.

Vale, R. D., & Milligan, R. A. 2000. The way things move: Looking under the hood of molecular motor proteins. *Science*, **288**, 88–95.

van Dyke, M. 1982. *An album of fluid motion.* Stanford, CA: Parabolic Press.

van Holde, K. E., Johnson, W. C., & Ho, P. S. 1998. *Principles of physical biochemistry.* Upper Saddle River, NJ: Prentice Hall.

Viovy, J. L. 2000. Electrophoresis of DNA and other polyelectrolytes: Physical mechanisms. *Rev. Mod. Phys.*, **72**, 813–872.

Visscher, K., Schnitzer, M. J., & Block, S. M. 1999. Single kinesin molecules studied with a molecular force clamp. *Nature*, **400**, 184–189.

Voet, D., & Voet, J. G. 2003. *Biochemistry.* 3rd ed. New York: John Wiley and Sons.

Vogel, S. 1994. *Life in moving fluids: The physical biology of flow.* 2nd ed. Princeton, NJ: Princeton University Press.

VOGEL, S. 2003. *Comparative biomechanics.* Princeton, NJ: Princeton University Press.

VON BAEYER, H. C. 1999. *Warmth disperses and time passes: The history of heat.* New York: Random House.

WADA, Y., SAMBONGI, Y., & FUTAI, M. 2000. Biological nano motor, ATP synthase F_0F_1: From catalysis to $\gamma\epsilon c_{10-12}$ subunit assembly rotation. *Biochim. Biophys. Acta,* **1459**, 499–505.

WANG, M. D., YIN, H., LANDICK, R., GELLES, J., & BLOCK, S. M. 1997. Stretching DNA with optical tweezers. *Biophys. J.,* **72**, 1335–1346.

WANG, M. D., SCHNITZER, M. J., YIN, H., LANDICK, R., GELLES, J., & BLOCK, S. M. 1998. Force and velocity measured for single molecules of RNA polymerase. *Science,* **282**, 902–907.

WEISS, T. F. 1996. *Cellular Biophysics.* Cambridge MA: MIT Press. 2 volumes.

WIDOM, B. 2002. *Statistical mechanics: A concise introduction for chemists.* Cambridge, UK: Cambridge University Press.

WOLFE, S. L. 1985. *Cell ultrastructure.* Belmont, CA: Wadsworth Publishing.

ZIMM, B. H., DOTY, P., & ISO, K. 1959. Determination of the parameters for helix formation in poly-[γ-benzyl-L-glutamate]. *Proc. Natl. Acad. Sci. USA,* **45**, 1601–1607.

Index

Bold references are the defining instance of a key term.